Plastics Processing Data Handbook

Plastics Processing Data Handbook

Second edition

Dominick V. Rosato, P.E.

Rhode Island School of Design
Chatham, MA, 02633
USA

CHAPMAN & HALL

London · Weinheim · New York · Tokyo · Melbourne · Madras

**Published by Chapman & Hall, 2–6 Boundary Row,
London SE1 8HN, UK**

Chapman & Hall, 2–6 Boundary Row, London SE1 8HN, UK

Chapman & Hall GmbH, Pappelallee 3, 69469 Weinheim, Germany

Chapman & Hall USA, 115 Fifth Avenue, New York, NY 10003, USA

Chapman & Hall Japan, ITP-Japan, Kyowa Building, 3F, 2-2-1
Hirakawacho, Chiyoda-ku, Tokyo 102, Japan

Chapman & Hall Australia, 102 Dodds Street, South Melbourne, Victoria
3205, Australia

Chapman & Hall India, R. Seshadri, 32 Second Main Road, CIT East,
Madras 600 035, India

First edition published by Van Nostrand Reinhold 1990
Reprinted by Chapman & Hall 1995

Second edition 1997

© 1997 Dominick V. Rosato

Typeset in 10/12 Palatino by Best-set Typesetter Ltd., Hong Kong
Printed in Great Britain by The University Press, Cambridge

ISBN   0 412 80190 6

A catalogue record for this book is available from the British Library

∞ Printed on permanent acid-free text paper, manufactured in accord-
ance with ANSI/NISO Z39.48-1992 and ANSI/NISO Z39.48-1984
(Permanence of Paper).

Contents

Preface

This second edition modifies and updates the previous book with more concise and useful information which interrelates processes with plastic materials' product requirements, and costs. It includes more details on the different processes and continues to simplify the understanding of how plastics behave during processing. The past few years have seen the continuing development of many advances in plastic materials and corresponding advances in processing techniques.

This book has been prepared with an awareness that its usefulness will depend greatly upon its simplicity. The overall guiding premise has therefore been to provide all essential information. Each chapter is organized to best present a methodology for processing with plastics.

This comprehensive book is useful to those in any industry to expand their knowledge in one or more processes. It has been prepared to be useful to those using plastics as well as those contemplating their use. To this end, the presentations are comprehensive yet simplified, so that even the specialist will obtain useful information.

The book may be used and understood by people in fabrication and production, design, engineering, marketing, quality control, R&D, and management. Sufficient information is presented to ensure the reader has a sound understanding of the principles involved, and thus recognizes what problems could exist, or more importantly, how to eliminate or compensate for potential problems. Knowledge of all these processing methods, including their capabilities and limitations, helps one to decide whether a given product can be fabricated and by which process.

Many guidelines are provided for maximizing plastics processing efficiency in the manufacture of all types of products, using all types of plastics. A practical approach is employed to present fundamental yet comprehensive coverage of processing concepts. The information and data presented by 290 figures and 137 tables relate the different variables that affect the processes listed in the table of contents.

The text presents a great number of problems pertaining to different phases of processing. Solutions are provided that will meet product performance requirements at the lowest cost. Many of the processing variables and their behaviors in the different processes are the same, as they all involve the basic conditions of temperature, time, and pressure. The book begins with information applicable to all processes, on topics such as melting, melt flow, and controls; all processes fit into an overall scheme that requires the interaction and proper control of systems. Individual processes are reviewed to show the effects of changing different variables to meet the goal of zero defects. The content is arranged to provide a natural progression from simple to complex situations, which range from control of a single manually operating machine to simulation of sophisticated computerized processes which inteface with many different processing functions.

Many of the procedures used in an individual process are identical or similar to those of other processes. A reader interested in a process can gain a different perspective by reviewing other processes. When an obvious correlation exists, a cross-reference is made to the other processes. Other pertinent information includes processing guidelines for trouble-shooting, etc.

Both practical and theoretical viewpoints are presented. Persons dealing with practicalities will find the theoretical explanations enlightening and understandable. And theorists will gain insight into the practical limitations of equipment and plastics (which are not perfect). The various process limitations are easily understood.

Processing is interdisciplinary. It calls for the ability to recognize situations in which certain techniques may be used and to develop problem-solving methods to fit specific processing situations.

It is important to reemphasize that all the data on plastic properties are to be used as guides. Obtain the latest, most complete data from material suppliers and data banks.

Information contained in this book may be covered by US and world-wide patents. No authorization to utilize these patents is given or implied; they are discussed for information only. Any disclosures are neither a license to operate nor a recommendation to infringe any patent. No attempt has been made to refer to patents by number, title, or ownership.

As trade and competition increase throughout the world, maintaining a competitive advantage depends on the ability to use the technology contained in this book most efficiently and bring the advanced technologies rapidly to the marketplace.

D. V. Rosato
Waban, MA, USA
October, 1996

1

Fundamentals

INTRODUCTION

This chapter reviews information pertinent to the processing of a diverse group of plastics that are summarized in Figs 1.1 to 1.3. There are many different types of plastics processed by different methods to produce products meeting many different performance requirements, including costs. This chapter provides guidelines and information that can be followed when processing plastics and understanding their behavior during processing. The basics in processing relate to temperature, time, and pressure. In turn they interrelate with product requirements, including plastics type and the process to be used (Fig. 1.4). Worldwide plastics consumption is at least 200 billion tons; the estimated use by process is shown in Table 1.1. In the United States it is estimated that about 70 000 injection molding machines (IMMs), 14 000 extruders and 6000 blow molders (BMs) are in about 22 000 plants with annual sales of about $20 billion [1–93].

The various processes reviewed in this book are used to fabricate all types and shapes of plastic products; household convenience packages, electronic devices and many others, including the strongest products in the world, used in space vehicles, aircraft, building structures, and so on. Proper process selection depends upon the nature and requirements of the plastic, the properties desired in the final product, the cost of the process, its speed, and product volume. (Note that a plastic also may be called a polymer or a resin.) Some materials can be used with many kinds of processes; others require a specific or specialized machine. Numerous fabrication process variables play an important role and can markedly influence a product's esthetics, performance, and cost.

This book will provide information on the effects on performance and cost of changing individual variables during processing, including upstream and downstream auxiliary equipment. Many of these variables and their behaviors are the same in the different processes, as they all

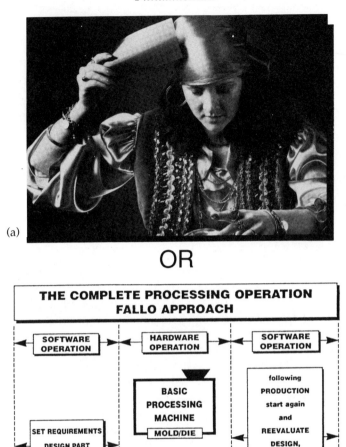

Figure 1.1 Developing a procedure: (a) the mythical appraoch, (b) the factual approach.

relate to temperature, time, and pressure. This chapter contains information applicable to all processes characterized by certain common variables or behaviors: plastic melt flow, heat controls, and so forth. It is essential to recognize that, for any change in a processing operation, there can be advantages and/or disadvantages. The old rule still holds: for every action there is a reaction. A gain in one area must not be allowed to cause

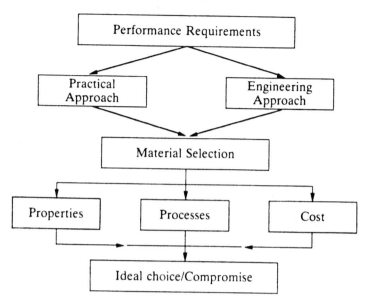

Figure 1.2 Product manufacture: a simplified flow diagram.

Table 1.1 Plastics consumption by processes

Extrusion	36%
Injection	32%
Blow	10%
Calendering	6%
Coating	5%
Compression	3%
Powder	2%
Others	6%

a loss in another; changes must be made that will not be damaging in any respect.

All processes fit into an overall scheme that requires interaction and proper control of different operations. An example is shown in Fig. 1.1(b), where a complete block diagram pertains to a process. This FALLO (Follow All Opportunities) approach can be used in any process by including those 'blocks' that pertain to the fabricated product's requirements.

The FALLO concept has been used by many manufacturers to produce acceptable products at the lowest cost. Computer programs featuring this type of layout are available. The FALLO approach makes one aware that many steps are involved in processing, and all must be coordinated. The

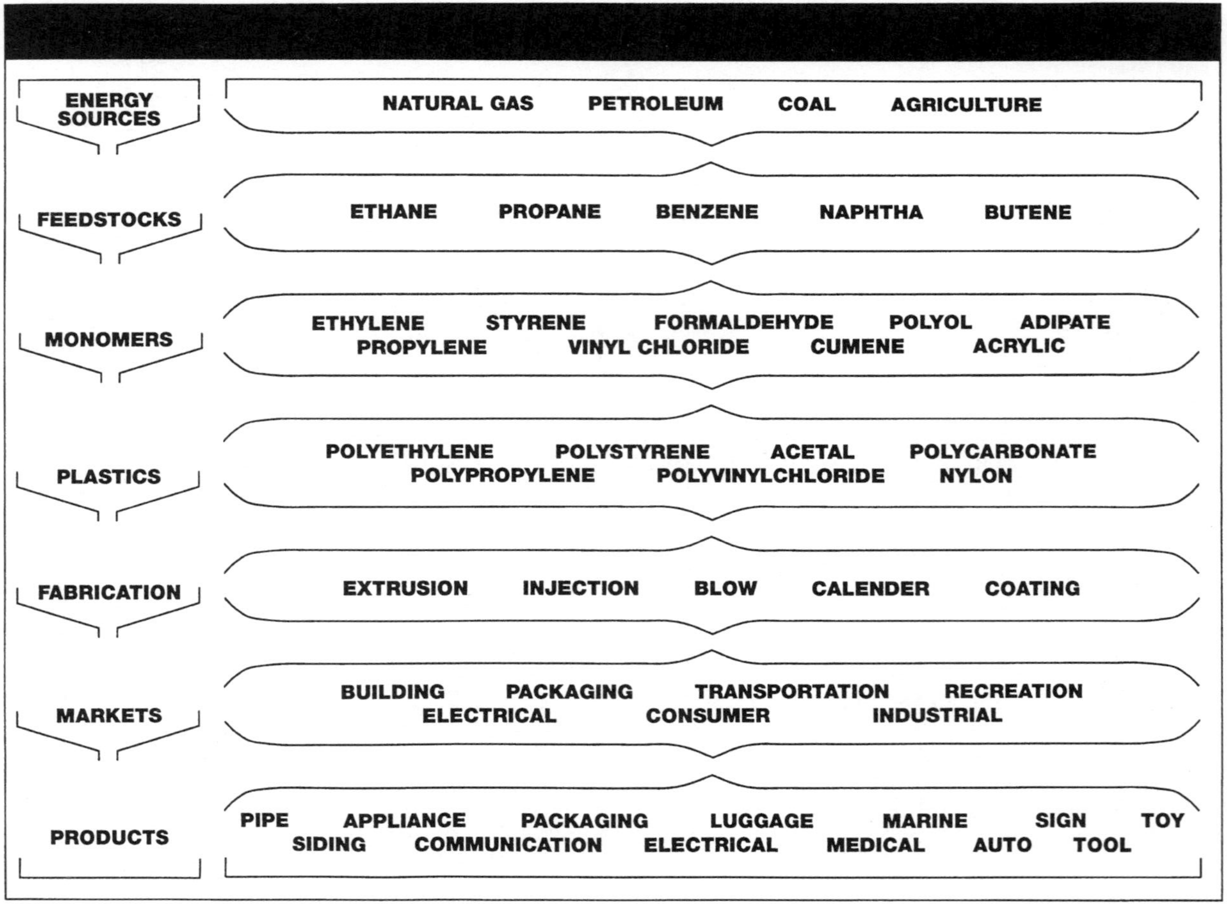

Figure 1.3 Simplified flowchart of plastics from raw materials to products.

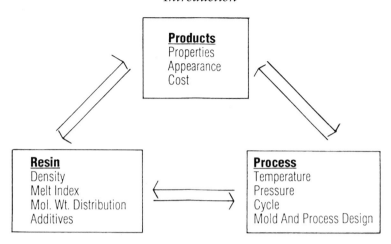

Figure 1.4 Interrelation of products, resin, and process.

specific process (injection, etc.) is an important part of the overall scheme and should not be problematic. The process depends on several interrelated factors: (1) designing a part to meet performance and manufacturing requirements at the lowest cost; (2) specifying the plastic; (3) specifying the manufacturing process, which requires (a) designing a tool 'around' the part, (b) putting the 'proper performance' fabricating process around the tool, (c) setting up necessary auxiliary equipment to interface with the main processing machine, and (d) setting up 'completely integrated' controls to meet the goal of zero defects; and (4) 'properly' purchasing equipment and materials, and warehousing the materials [1–9].

Major advantages of using plastics include formability, consolidation of parts, and providing a low cost-to-performance ratio. For the majority of applications that require only minimum mechanical performance, the product shape can help to overcome the limitations of commodity resins such as low stiffness; here improved performance is easily incorporated in a process. However, where extremely high performance is required, reinforced plastics or composites are used (Chapter 12).

Knowledge of all processing methods, including their capabilities and limitations, is useful to a processor in deciding whether a given part can be fabricated and by which process. Certain processes require placing high operating pressures on plastics, such as those used in injection molding, where pressures may be 2000–30 000 psi (13.8–206.9 MPa). High pressures coupled with the creation of three-dimensional parts make injection molding the most complex process; nevertheless, it can be easily controlled. Lower pressures are used in extrusion and compression, ranging from 200 to 10 000 psi (1.4 to 69 MPa); and some processes, such as thermoforming and casting, operate at relatively low pressure. Higher pressures allow the development of tighter dimensional tolerances with higher mechanical performance; but there is also a tendency to develop undesirable stresses (orientations) if the processes are not properly understood or controlled. A major exception is reinforced plastics processing at low or contact pressures (Chapter 12). Regardless of the process used, its proper control will maximize performance and minimize undesirable process characteristics.

PRODUCT REQUIREMENTS AND MACHINE PERFORMANCE

Almost all processing machines can provide useful products with relative ease, and certain machines have the capability of manufacturing products to very tight dimensions and performances. The coordination of plastic and machine facilitates these processes. This interfacing of product and process requires continual updating because of continuing new developments in manufacturing operations. The information presented in this book should make past, present and future developments understandable in a wide range of applications.

Most products are designed to fit processes of proven reliability and consistent production. Various options may exist for processing different shapes, sizes and weights (Table 1.2). Parameters that help one to select the right options are (1) setting up specific performance requirements; (2) evaluating material requirements and their processing capabilities; (3) designing parts on the basis of material and processing characteristics, considering part complexity and size (Fig. 1.5) as well as a product and process cost comparison; (4) designing and manufacturing tools (molds, dies, etc.) to permit ease of processing; (5) setting up the complete line,

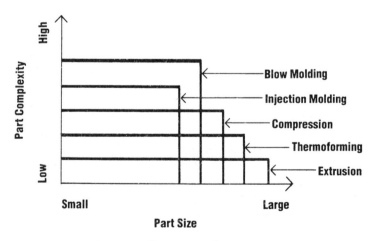

Figure 1.5 Processing characteristics.

Table 1.2 Competitive processes

	Injection molding	Extrusion	Blow molding	Thermo-forming	Reaction injection molding	Rotational molding	Compression and transfer molding	Matched mold, sprayup
Bottles, necked containers, etc.	2, A		1	2, A		2		2
Cups, trays, open containers, etc.	1			1	1		1	2
Tanks, drums, large hollow shapes, etc.			1	2, A		1		2
Caps, covers, closures, etc.	1			2	2		1	
Hoods, housings, auto parts, etc.	1		2	2	2		1	1
Complex shapes, thickness changes, etc.	1						1	2
Linear shapes, pipe, profiles, etc.	2, B	1					2, B	
Sheets, panels, laminates, etc.		1, C					2	2

1 = prime process.
2 = secondary process.
A = combine two or more parts with ultrasonics, adhesives, etc. (Chapter 17).
B = short sections can be molded.
C = also calendering process.

including auxiliary equipment (Fig. 1.1 and Chapter 16); (6) testing and providing quality control, from delivery of the plastics, through production, to the product (Chapter 16); and (7) interfacing all these parameters by using logic and experience and/or obtaining a required update on technology.

PROCESSING FUNDAMENTALS

Polymers are usually obtained in the form of granules, powder, pellets, and liquids. Processing mostly involves their physical change (thermoplastics), though chemical reactions sometimes occur (thermosets). A variety of processes are used. One group consists of the extrusion processes (pipe, sheet, profiles, etc.). A second group takes extrusion and sometimes injection molding through an additional processing stage (blow molding, blown film, quenched film, etc.). A third group consists of injection and compression molding (different shapes and sizes), and a fourth group includes various other processes (thermoforming, calendering, rotational molding, etc.).

The common features of these groups are (1) mixing, melting, and plasticizing; (2) melt transporting and shaping; (3) drawing and blowing; and (4) finishing. Mixing melting, and plasticizing produce a plasticized melt, usually made in a screw (extruder or injection). Melt transport and shaping apply pressure to the hot melt to move it through a die or into a mold. The drawing and blowing technique stretches the melt to produce orientation of the different shapes (blow molding, forming, etc.). Finishing usually means solidification of the melt.

The most common feature of all processes is deformation of the melt with its flow, which depends on its rheology. Another feature is heat exchange, which involves the study of thermodynamics. Changes in a plastic's molecular structure are chemical. These properties are reviewed briefly in the following paragraphs, and will be discussed in detail throughout the book, with a focus on how they influence processes.

PROCESSABILITY

Processability does not mean the same thing to all processors. It describes quite generally the ease or difficulty with which a plastic can be handled during its fabrication into film, molded products, pipe, etc. A plastic with good processability possesses the properties necessary to make it easy to process the plastics into desired shapes. The main characteristics or properties which determine a plastic's processability are molecular weight, uniformity, additive type and content, and plastic feed rates.

However, processability is usually considered in less tangible terms, using properties derived from the basic characteristics above. In extrusion, for example, these characteristics include drawdown (hot melt extensibility), pressure and temperature sensitivity, smoke and odor, product stability during hauloff, and flow rate (which is an operating condition). And there are other factors, too (Chapter 3). Often, however, it is not the plastic but unfavorable operating conditions that lead to inadequate plastic performance.

PROCESSES

Overview

The type of process to be used depends on a variety of factors, including product shape and size, plastic type, quantity to be produced, quality and accuracy (tolerances) required, design load performance, cost limitation, and time schedule. Each of the processes reviewed provides different methods to produce different products. As an example, extrusion with its many methods produces films, sheet, pipe, profile, wire coating, etc. Some of the process overlap since different segments of the industry use them. Also terms such as molding, embedding, casting, potting, impregnation and encapsulation are sometimes used interchangeably and/or allowed to overlap. However, they each have their specific definitions [9].

Almost all processing machines can provide useful products with relative ease, and certain machines have the capability of manufacturing products to very tight dimensions and performances. The coordination of plastic and machine facilitates these processes; this interfacing requires

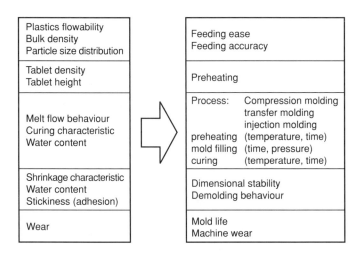

Figure 1.6 Processing behavior: some examples.

continual updating because of continuing new developments in processing operations.

During the conversion of plastics into products, various melt-processing procedures are involved. The inherent viscoelastic properties of each plastic type can lead to certain undesirable processing defects, so additives are used to ease these processing-related problems. Examples are heat and light stabilizers, antioxidants, and lubricants. There are usually many different additives already in the plastics [3, 9]. Some plastics have limited applications due to their undesirable physical properties and/or their poor processing (Fig. 1.6). These limitations have greatly been reduced through the use of plastic blends and alloys. By mixing two or more plastics, the overall balance of physical and processing properties can be optimized. These processing aids are used in such small quantities that their effect on the final physical properties of the major plastic or plastic mixture is minimized.

Most processes fit into an overall scheme that requires interaction and proper control of its different operations. There are machines that perform multifunctions. Figure 1.7 shows the world's first machine to combine more that seven processes; it includes injection molding, flow molding, stamping, coinjection, gas injection, compression molding, flow forming and proprietary processes. This one-of-a-kind machine is 75 ft (23 m) long, 50 ft (15 m) wide, and 55 ft (17 m) tall; it can produce parts up to 6 ft (1.8 m) square.

Some products are limited by the economics of the process that must be used to make them. For example, hollow parts, particularly very large ones, may be produced more economically by the rotational process than by blow molding. Thermosets (TSs) cannot be blow molded (to date) and they have limited extrusion possibilities. The need for a low quantity may allow certain processes to be eliminated by going to casting. Thermoplastics (TPs) present fewer problems for extrusion than for injection molding, but extrusion suffers greater problems over dimensional control and shape. However, as the plastic leaves the extrusion die, it can be relatively stress-free. It is drawn in size and passed through downstream equipment from the extruder to form its shape as it is cooled, usually by air and/or water. The dimensional control and the die shape required to achieve the desired part shape are usually solved by trial-and-error settings; the more experience one has in the specific plastic and equipment being used, the less trial time will be needed.

Other analyses can be made. Compression and injection molds, which are expensive and relatively limited in size, are employed when the production volume required is great enough to justify the mold costs and the sizes are sufficient to fit available equipment limitations. Extrusion produces relatively uniform profiles at unlimited lengths. Casting is not limited by pressure requirements, so large sheets can be produced.

Figure 1.7 Alpha 1 is a multipurpose machine in the GE Plastics Polymer Process-ing Development Center, Pittsfield MA.

Calendered sheets are limited in their width by the width of the material rolls, but are unlimited in length. Vacuum forming is not greatly limited by pressure, although even a small vacuum distributed over a large area can build up an appreciable load. Blow molding is limited by equipment that is feasible for the mold sizes. Rotational molding can produce relatively large parts.

Injection molding and extrusion tend to align long-chain molecules in the direction of flow. This produces markedly greater strength in the direction of flow than at right angles to the flow. An extruded pressure pipe could have its major strength in the axial/machine direction when the major stresses in the pipe wall are circumferential. Proper controls on processing conditions allow the required directional properties to be obtained. If in an injection mold the plastic flows in from several gates, the melts must unite or weld where they meet. But this process may not be complete, especially with filled plastics, so the welds may be points of weakness. Careful gating with proper process control can allow welds to occur where stresses will be minimal.

Certain products are most economically produced by fabricating them with conventional machining out of compression-molded blocks, laminates or extruded sheets, rods or tubes. It may be advantageous to design a product for the postmolding assembly of inserts, to gain the benefit of fully automatic molding and automatic insert installation.

The choice of molder and fabricator places no limits on a design. There is a way to make a part if the projected values justify the price; any job can be done at a price. The real limiting factors are tool-design considerations, material shrinkage, subsequent assembly or finishing operations, dimensional tolerances, allowances, undercuts, insert inclusions, parting lines, fragile sections, the production rate or cycle time, and the selling price.

Applying the following principles, applicable to virtually all manufacturing processes, will aid in specifying parts that can be produced at minimum cost: (1) maintaining simplicity; (2) using standard materials and components; (3) specifying liberal tolerances; (4) employing the most processable plastics; (5) collaborating with manufacturing people; (6) avoiding secondary operations; (7) designing what is appropriate to the expected level of production; (8) utilizing special process characteristics; and (9) avoiding processing restrictions.

Processing and properties

In order to understand potential problems and their solutions, it is helpful to consider the relationships of machine capabilities, plastics processing variables, and part performance. A distinction should be made between machine conditions and processing variables as summarized in Fig. 1.8.

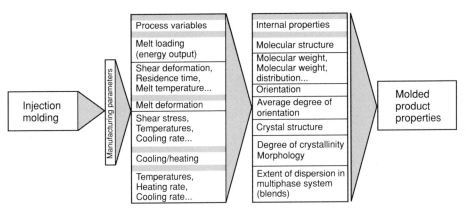

Figure 1.8 Relationship between manufacturing process and properties of injection-molded product.

Machine conditions are temperature, pressure, and processing time (such as screw rotation in RPM for a screw plasticator), die and mold temperature and pressure, machine output rate (lb h^{-1}), etc. Processing variables are more specific than machine conditions: the melt in the die or the mold temperature, the flow rate, and the pressure used.

The distinction between machine conditions and fabricating variables is important when considering cause-and-effect relationships. If the processing variables are properly defined and measured, not necessarily the machine settings, they can be correlated with the part properties. For example, if one increases cylinder temperature, melt temperatures do not necessarily increase, too. Melt temperature is also influenced by screw design, screw rotation rate, back pressure, and dwell times (Chapter 2). It is much more accurate to measure melt temperature and correlate it with properties than to correlate cylinder settings with properties.

The problem-solving approach that ties the processing variables to part properties includes considering melt orientation, polymer degradation, free volume/molecular packing and relaxation and cooling stresses. The most influential of these four conditions is melt orientation, which can be related to molded-in stress or strain.

Polymer degradation can occur from excessive melt temperatures or abnormally long times at temperature, called the heat history from plasticator to cooling of the part. Excessive shear can result from poor screw design, too much screw flight-to-barrel clearance, cracked or worn-out flights, etc.

Orientation in plastics refers simply to the alignment of the melt-processing variables that definitely affect the intensity and performance of orientation.

Processing methods

Table 1.3 is a very brief review and guide on the different processes. The relevant chapters give more details, the many exceptions, and any changes that can occur, particularly limitations.

Machine operation terminology

Terminology in the plastics industry regarding the operation of machinery is as follows:

Manual operation

Each function and the timing of each function is controlled manually by an operator.

Semiautomatic operation

A machine operating semiautomatically will stop after performing a complete cycle of programmed molding functions automatically. It will then require an operator to start another complete cycle manually.

Automatic operation

A machine operating automatically will perform a complete cycle of programmed molding functions repetitively; it will stop only for a malfunction on the part of the machine or mold, or when it is manually interrupted.

Plastics memory and processing

Thermoplastics can be bent, pulled, or squeezed into various useful shapes. But eventually, especially if heat is added, they return to their original form. This behavior, known as plastic memory, can be annoying. If properly applied, however, plastic memory offers interesting design possibilities for all types of fabricated parts.

When most materials are bent, stretched, or compressed, they somehow alter their molecular structure or grain orientation to accomodate the deformation permanently, but this is not so with polymers. Polymers temporarily assume the deformed shape, but they always maintain the internal stresses and keep wanting to force the material back to its original shape. This desire to change shape is what is usually called plastic memory.

This memory is often unwelcome. Sometimes we prefer that thermoplastic parts forget their original shape and stay put, especially when

Table 1.3 Plastic processing methods preliminary guide

Process	Description	Limitations
Blow	An extruded parison tube of heated thermoplastic is positioned between two halves of an open split mold and expanded against the sides of the closed mold via air pressure. The mold is opened and the part ejected. Low tool and die costs, rapid production rates, and ability to mold fairly complex hollow shapes in one piece.	Generally limited to hollow or tubular parts; some versatile mold shapes, other than bottles and containers.
Calendering	Dough-consistent thermoplastic mass is formed into a sheet of uniform thickness by passing it through and over a series of heated or cooled rolls. Calenders are also utilized to apply plastic covering to the backs of other materials. Low cost, and sheet materials are virtually free of molded-in stresses.	Limited to sheet materials and very thin films are not possible.
Casting	Liquid plastic which is generally thermoset except for acrylics is poured into a mold without pressure, cured, and taken from the mold. Cast thermoplastic films are produced via building up the material (either in solution or hot-melt form) against a highly polished supporting surface. Low mold cost, capability to form large parts with thick cross sections, good surface finish, and convenient for low-volume production.	Limited to relatively simple shapes. Most thermoplastics are not suitable for this method. Except for cast films, method becomes uneconomical at high volume production rates.
Centrifugal casting	Reinforcement is placed in mold and is rotated. Resin distributed through pipe; impregnates reinforcement through centrifugal action. Utilized for round objects, particularly pipe.	Limited to simple curvatures in single-axis rotation. Low production rates.

Table 1.3 *Continued*

Process	Description	Limitations
Coating	Process methods vary. Both thermoplastics and thermosets widely used in coating of numerous materials. Roller coating similar to calendering process. Spread coating employs blade in front of roller to position resin on material. Coatings also applied via brushing, spraying, and dipping.	Economics generally depends on close tolerance control.
Cold pressure molding	Similar to compression molding in that material is charged into a split mold; it differs in that it employs no heat, only pressure. Part cure takes place in an oven in a separate operation. Some thermoplastic billets and sheet material are cold formed in a process similar to drop-hammer die forming or fast cold-form stamping of metals. Low-cost matched-tool moldings exist which utilize a rapid exotherm to cure moldings on a relatively rapid cycle. Plastic or concrete tooling can be used. With process comes ability to form heavy or tough-to-mold materials; simple, inexpensive, and often has rapid production rate.	When compared to process such as injection molding it is limited to relatively simple shapes, and few materials can be processed in this manner.
Compression molding	Principally polymerized thermoset compound, usually preformed, is positioned in a heated mold cavity; the mold is closed (heat and pressure are applied) and the material flows and fills the mold cavity. Heat completes polymerization and the part is ejected. The process is sometimes used for thermoplastics, e.g. vinyl phonograph records. Little material waste is attainable; large, bulky parts can be molded; process is adaptable to rapid automation (racetrack techniques, etc.).	Extremely intricate parts containing undercuts, side draws, small holes, delicate inserts, etc.; very close tolerances are difficult to produce.

Process	Description	Limitations
Encapsulation	Mixed compound is poured into open molds to surround and envelope components; cure may be at room temperature with heated postcure. Encapsulation generally includes several processes such as potting, embedding and conformal coating.	Low-volume process subject to inherent limitations on materials which can lead to product defect caused by exotherm, curing or molding conditions, low thermal conductivity, high thermal expansion, and internal stresses.
Extrusion molding	Widely used for continuous production of film, sheet, tube, and other profiles; also used in conjunction with blow molding. Thermoplastic or thermoset molding compound is fed from a hopper to a screw and barrel where it is heated to plasticity then forwarded, usually via a rotating screw, through a nozzle possessing the desired cross section. Production lines require input and takeoff equipment that can be complex. Low tool cost, numerous complex profile shapes possible, very rapid production rates, can apply coatings or jacketing to core materials (such as wire).	Usually limited to sections of uniform cross section.
Filament winding	Excellent strength-to-weight. Continuous, reinforced filaments, usually glass, in the form of roving are saturated with resin and machine-wound onto mandrels having shape of desired finished part. Once winding is completed, part and mandrel are cured; mandrel can then be removed through porthole at end of wound part. High-strength reinforcements can be oriented precisely in direction where strength is required. Good uniformity of resin distribution in finished part; mainly circular objects such as pressure bottles, pipes, and rocket cases.	Limited to shapes of positive curvature; openings and holes can reduce strength if not properly designed into molding operations.

Table 1.3 *Continued*

Process	Description	Limitations
Injection molding	Very widely used. High automation of manufacturing is standard practice. Thermoplastic or thermoset is heated to plasticity in cylinder at controlled temperature, then forced under pressure through a nozzle into sprues, runners, gates, and cavities of mold. The resin undergoes solidification rapidly, the mold is opened, and the part ejected. Injection molding is growing in the making of glass-reinforced parts. High production runs, low labor costs, high reproducibility of complex details, and excellent surface finish.	High initial tool and die costs; not economically practical for small runs.
Laminating	Material, usually in form of reinforcing cloth, paper, foil, metal, wood, glass fiber, plastic, etc., preimpregnated or coated with thermoset resin (sometimes a thermoplastic) is molded under pressure greater than 1000 psi (7 MPa) into sheet, rod, tube, or other simple shapes. Excellent dimensional stability of finished product; very economical in large production of parts.	High tool and die costs. Limited to simple shapes and cross sections.
Matched-die molding	A variation of the conventional compression molding this process employs two metal molds possessing a close-fitting, telescoping area to seal in the plastic compound being molded and to allow trim of the reinforcement. The mat or preform reinforcement is positioned in the mold and the mold is closed and heated under pressures of 150–400 psi (1–3 MPa). The mold is then opened and the part is removed.	Prevalent high mold and equipment costs. Parts often require extensive surface finishing.

Process	Description	Limitations
Pultrusion	This process is similar to profile extrusion, but it does not provide flexibility and uniformity of product control, and automation. Used for continuous production of simple shapes (rods, tubes, and angles) principally incorporating fiberglass or other reinforcement. High output possible.	Close tolerance control requires diligence. Unidirectional strength usually the rule.
Rotational molding	A predetermined amount of powdered or liquid thermoplastic or thermoset material is poured into mold; mold is closed, heated, and rotated in the axis of two planes until contents have fused to inner walls of mold; mold is then opened and part is removed. Low mold cost, large hollow parts in one piece can be produced, and molded parts are essentially isotropic in nature.	Limited to hollow parts; production rates are usually slow.
Slush molding	Powdered or liquid thermoplastic material is poured into a mold to capacity; mold is closed and heated for a predetermined time in order to achieve a specified buildup of partially cured material on mold walls; mold is opened and unpolymerized material is poured out; and semifused part is removed from mold and fully polymerized in oven. Low mold costs and economical for small production runs.	Limited to hollow parts; production rates are very slow; and limited choice of materials that can be processed.
Thermoforming	Heat-softened thermoplastic sheet is positioned over male or female mold; air is evacuated from between sheet and mold, forcing sheet to conform to contour	Limited to parts of simple configuration, high scrap, and limited number of materials from which to choose.

Table 1.3 *Continued*

Process	Description	Limitations
	of mold. Variations are vacuum snapback, plug assist, drape forming, etc. Tooling costs are generally low, large part production with thin sections possible, and often comes out economical for limited part production.	
Transfer molding	Related to compression and injection molding processes. Thermoset molding compound is fed from hopper into a transfer chamber where it is then heated to plasticity; it is then fed by a plunger through sprues, runners, and gates into a closed mold where it cures; mold is opened and part ejected. Good dimensional accuracy, rapid production rate, and very intricate parts can be produced.	High mold cost; high material loss in sprues and runners; size of parts is somewhat limited.
Wet-layup or contact molding	Several layers, consisting of a mixture of reinforcement (generally glass cloth) and thermosetting resin are positioned in mold and roller contoured to mold's shape; assembly is usually oven-cured without the application of pressure. In spray molding, a modification, resin systems and chopped fiber are sprayed simultaneously from a spray gun against the mold surface. Wet-layup parts are sometimes cured under pressure, using vacuum bag, pressure bag, or autoclave, and depending on the method employed, wet-layup can be called open molding, hand layup, sprayup, vacuum bag, pressure bag, or autoclave molding. Little equipment required, efficient, low cost, and suitable for low-volume production of parts.	Not economical for large-volume production; uniformity of resin distribution difficult to control; only one good surface; limited to simple shapes.

the parts must be coined, formed, machined, or rapidly cooled. Occasionally, however, this memory or instability can be used advantageously.

Most plastic parts can be produced with a built-in memory. That is, their tendency to move into a new shape is included as an integral part of the design. So after the parts are assembled in place, a small amount of heat can coax them to change shape. Plastic parts can be deformed during assembly then allowed to return to their original shape. In this case the parts can be stretched around obstacles or made to conform to unavoidable irregularities without their suffering permanent damage.

The time- and temperature-dependent change in mechanical properties results from stress relaxation and other viscoelastic phenomena that are typical of polymers. When the change is an unwanted limitation, it is called creep. When the change is skillfully adapted to use in the overall design it is called plastic memory.

Potential memory exists in all thermoplastics. Polyolefins, neoprenes, silicones and other cross-linkable polymers can be given memory either by radiation or by chemical curing. Fluorocarbons, however, need no such curing. When this phenomenon of memory is applied to fluorocarbons such as TFE, FEP, ETFE, ECTFE, CTFE, and PVF_2, interesting high-temperature or wear-resistant applications become possible.

Drying plastics

There is much more to drying resins for processes such as injection molding, extrusion, and blow molding than blowing 'hot air' into a hopper. Effects of excessive moisture cause degradation of the melt viscosity and in turn can lead to surface defects, production rejects, and even failure of parts in service (Table 1.4). First one determines from a supplier, or experience, the resin's moisture content limit. Next one must determine which procedure will be used in determining water content, such as 'weighing, drying, and reweighing.' This procedure has definite limitations. Fast automatic analyzers, suitable for use with a wide variety of resin systems, are available that provide quick and accurate data for achieving in-plant control of this important parameter.

Drying or keeping moisture content at designated low levels is important, particularly for hygroscopic resins (nylon, PC, PUR, PMMA, ABS, etc.). Most engineering resins are **hygroscopic** and must be 'dry' prior to processing. Usually the moisture content is <0.02 wt%. In practice, a drying heat 30 °C below the softening heat has proved successful in preventing caking of the resin in the dryer. Drying time varies in the range 2–4 h, depending on moisture content. As a rule of thumb, the drying air should have a dew point of −30 °F (−34 °C) and be capable of being heated up to 250 °F (121 °C). It takes about $1\,ft^3\,min^{-1}$ of air for every $lb\,h^{-1}$ of resin

Table 1.4 Moisture troubleshooting guide

Symptom	Possible causes	Solutions
Silver streaks, splay	Wet material due to improper drying, high percent regrind, over wet virgin resin	Follow resin manufacturer's drying instructions and dryer manufacturer's operating and maintenance instructions Use desiccant dryer
Brown streaks/burning	Contamination	Purge barrel/screw and clean dryer/auxiliary equipment Check resin Check molding equipment settings and controls
	Overheating of material	Check resin manufacturer's instructions about processing temperatures
Bubbles	High moisture content	Check each step of drying process Check dried resins exposure to air
	Trapped air	Force air out of feed vent Increase screw speed and/or back pressure
Brittle parts	Wet resin or overdried resin	Check drying instructions and conditions
	Molded-in stresses	Increase melt temperature/reduce injection pressure Review part design
	Poor part design	Review design for notches and other stress concentrators
Flash	Wet material	Check drying procedures
	Insufficient clamp tonnage	Use larger machine
	Excessive vent depth	Change mold design
	High injection pressure	Decrease injection pressure
	Damaged mold	Repair damage
	Misaligned platen	Realign platen

Problem	Cause	Remedy
Material in drying hopper caking or meltdown occurring	Material temperature too high	Decrease material temperature by lowering cylinder temperature, decrease screw speed/lower back pressure (screw machine)
	Cycle time too long	Decrease overall cycle time
	Plunger pushing forward too long	Decrease plunger forward time
	Process temperature set too high	Check resin data sheet for meltdown temperature
		Make sure operators know correct process temperature set point
Dew point reading too high	Dirty process/auxiliary filter(s)	Clean or replace filters[a]
	Desiccant saturated	Dry cycle machine for several complete cycles. (This is common with equipment which is not operated on a continual basis)
	Material residence time in hopper too short	Replace with larger hopper[b]
	Return air temperature too high	Add 'after-cooler' to return air line
	Heaters burned out	Replace
	Bad heater thermostat or thermocouple	Replace
	Cycle timer malfunctioning	Adjust or replace
	Air control valves not seating properly	Adjust
	Contaminated or wornout desiccant	Replace[c]

Table 1.4 *Continued*

Symptom	Possible causes	Solutions
	Incorrect blower rotation	Check and correct rotation
	Regeneration heating elements inoperative	Check electrical connections; replace elements if needed
	Desiccant assembly not transferring	If valve system, check and repair valve/drive assembly. If rotational system, adjust drive-assembly. Check electrical connections on motor and replace motor if needed
	Moist room air leading into dry process air	Check hopper lid, all hose connections, hoses and filters. Tighten, replace, repair as needed
	Dew point meter incorrect	Check meter and recalibrate
Dew point cycling from high to low	Electrical malfunctions	Check electrical connections on heaters/controller. Repair/replace[c]
	Desiccant bed(s) contaminated	replace[c]
Process air temperature too high	Incorrect temperature setting	Reset for correct temperature
	Thermocouple not properly located	Secure thermocouple probe into coupling at inlet of hopper
	Electrical malfunctions	Check electrical connections and replace if necessary Insulate hopper and hopper inlet air line
Excessive changeover temperature	Insufficient reactivation airflow	See 'Insufficient airflow'
	Malfunctioning cycle time	Adjust or replace
	Blades of blower wheel dirty	Clean
Process air temperature too low	Incorrect temperature setting on controller	Reset for correct temperature
	Controller malfunctioning	Check electrical connections. Replace/repair if needed
	Process heating elements	Check electrical connections Replace/repair if needed

Problem	Possible cause	Remedy
	Thermostat malfunction	Replace or repair
	Voltage differentials	Check supply voltage
	Inadequate airflow	Check/clean filters, check blower rotation and correct, check and repair airflow meter
	Hose connections incorrect	Check connections. Delivery hose should enter hopper at bottom
	Inadequate insulation	Insulate hopper and hopper inlet air line
	Dryer inadequate for required temperatures	Replace with high temperature dryer
Insufficient airflow (Dew point reading could be good but resin is still wet)	Process or auxiliary filter(s) blocked	Clean or replace[a]
	Blower rotation incorrect	Check manufacturer's electrical instructions, and change blower rotation
	Air ducts blocked	Remove obstruction
	Airflow meter incorrect	Disengage line exiting dryer repair if needed
	Desiccant bed contaminated	Replace desiccant[c]
	Tightly packed material in hopper	Increase hopper size or drain hopper and refill
Heater burn out	Excessive vibration	Relocate dryer, reduce vibration
	High voltage condition	Reduce voltage, relocate dryer, or use heaters rated for actual voltage
	Malfunction in heater thermostat	Adjust or replace
	Blades on blower wheel dirty	Clean

[a] An inexpensive pressure-differential switch, common option for almost every brand of dehumidifying dryer, will signal when a filter is restricting airflow.

[b] Since drying systems tend to be designed for a specific material, different materials may need longer residence times or higher drying temperatures.

[c] Plastic dust contaminants, because of their flash point, can ignite during regeneration of the desiccant bed causing a fire inside the dehumidifier.

processed when using a desiccant dryer. The pressure drop through the bed should be less than 1 mm H_2O per millimeter of bed height.

Simple tray dryers or mechanical convection, hot-air dryers, while adequate for some resins, are simply incapable of removing enough water for proper processing of hygroscopic resins, particularly during periods of high ambient humidity. The most effective and efficient drying system for these resins incorporates an air-dehumidifying system in the material storage/handling network. Consistently and adequately it has to provide moisture-free air in order to dry the 'wet' resin.

Initially expensive, it does give improved production rates and helps to achieve zero defects. There are several manufacturers and systems from which to choose. All the systems are designed to accomplish the same end results, but the approaches to regeneration of the desiccant beds vary.

Hygroscopic resins are commonly passed through dehumidifying hopper dryers before they enter a screw plasticator. However, except where extremely expensive protective measures are taken, the drying may be inadequate, or the moisture regain may be too rapid to avoid product defects unless barrel venting is provided (Chapters 2 and 3 review venting). To ensure proper drying for 'delicate' parts such as lenses, some plants use drying prior to entry into the barrel as well as venting. Although it is much less hygroscopic than the usual resin (ABS, PC, etc.), PS too is usually vented during processing to protect against surface defects.

The effect of having excess moisture manifests itself in various ways, depending on the process being employed. The common result is a loss in mechanical properties (Fig. 1.9) and physical properties, with splays, nozzle drool between shot-size control, sinks, and other losses that may occur during processing. The effects during extrusion can also include gels,

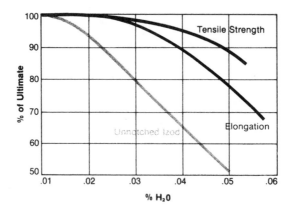

Figure 1.9 Moisture affects the mechanical properties of hygroscopic PET plastic during molding.

trails of gas bubbles in the extrudate, arrowheads, waveforms, surging, lack of size control, and poor appearance.

Air entrapment

Air entrapment is a common problem in most processes. With screw plasticators (injection, extrusion, blow molding) it is caused by air being taken in with plastics from the feedhopper. Compression of the solid material in the screw feed section (Chapter 2) will normally force air out of the 'solid' melt bed. However, there are times that the air cannot exit back to the hopper. Thus, it moves forward with the melt until it exits from the mold or die. These air pockets can cause problems such as inclusions or surface imperfections.

There are solutions to this situation. The more popular and simple approach is to change the temperature in the solids-conveying zone to achieve a more positive compacting of the solid bed. Often a temperature increase of the first barrel section reduces the air entrapment; however, a lower temperature sometimes gives an improvement. In any case, the temperatures in the solids-conveying zone are important parameters for air entrapment. The barrel and screw temperatures are both important. The next step is an increase in the mold or diehead pressure to alter the pressure profile and hopefully to achieve a more rapid compacting of the solid bed. Another possible solution is to starve-feed the extruder, but this will probably reduce the output and requires additional hardware such as, an accurate feeding device.

If the problem still exists, a change in plastic particle size or shape could help. Other options include a vacuum hopper system (rather complex and expensive), a grooved barrel section (rapid compaction occurs), reduced friction on the screw (applying a coating), and increased compression ratio of the screw (Chapter 2).

Note that the problem may not be air entrapment. It could be volatiles from the plastic, plastic degrading, additives or surface agents, and/or moisture. By determining which problem exists, it may be easier to resolve the problem of air. Vented barrels may be the solution.

Release agents

Release agents (RAs) are substances that control or eliminate the adhesion between two surfaces. They are known by a variety of terms indicative of their function: adherents, antiblocking agents, antistick agents, external or surface lubricants, parting agents, and slip aids. Processors use RA in a wide variety of general handling operations, such as calendering, casting, molding, packaging and labeling, and protection of equipment. Processors can be confronted with parts adhering to mold or die surfaces.

Plastics most frequently needing RAs include polyurethanes, polyesters (TPs and TSs), polyolefins, polycarbonates and epoxies.

RAs come in a wide variety of forms and modes of application. Both neat liquids and solvent solutions, as well as solids such as powders and flakes, compounded mixtures and pastes, emulsions, dispersions, pre-formed films, *in situ* film formers, and integral migratory additives are available. Many formulated products serve more than one processing function and contain other additives, such as antioxidants. Combined lubricant–stabilizer packages are often used with polyvinyl chloride. The external treatment types can be applied by any of the standard coating methods, including brushing, dipping, dusting, spraying, electrostatic coating, and plasma arc coating.

Despite the diversity of products, RAs have some features in common, notably inertness to at least one of the surfaces in question at the tempera-ture of the release process and low surface tension. Low surface tension is important in obtaining good wetting of a mold. It is also a reflection of the low intermolecular forces desirable in effective release compositions. Not surprisingly, inertness and weak cohesion are the opposite of good adhe-sive properties.

The release agent should be chosen with subsequent events in mind, as well as for ease of release and stability. In addition to ease of release, other selection criteria include prevention of buildup; mold cleanability; compatibility with secondary operations such as painting, plating, and ultrasonic welding; mold compatibility; type and time available for appli-cation; health and safety requirements and cost. As an example, when using silicone as an RA, it becomes difficult or impossible to paint on it or to attach it adhesively.

Heat history, residence time, and recycling

The process of heating and cooling TPs can be repeated indefinitely by granulating scrap, defective parts, and so on. During the heating and cooling cycles of injection, extrusion, and so on, the material develops a 'time-to-heat' history, or residence time. With only limited repeating of the recycling, the properties of certain plastics are not significantly af-fected by residence time. However, some TPs can significantly lose certain properties. Heat can also develop significantly if improper granulating/ recycling systems are used, producing quick degradation of plastics. Chapter 16 gives more details on recycling.

The residence time and its distribution only partially determine the chance of degradation in screw plasticizing systems. The other factors that play an important role are the actual stock temperatures and the strain rates occuring during melting. These two characteristics are closely re-lated. Residence time will be reviewed throughout this book.

Heat profile

To obtain the best processing melts for any plastics, one starts with the plastic manufacturer's recommended heat profile and/or one's own experience. These are starting points for various types of plastics, as shown in Fig. 1.10 and Table 1.5. The time and effort spent on startup make it possible to achieve maximum efficiency of performance versus cost for the processed plastics. By the application of logic, the information gained can be stored and applied to future setups. In all probability, similar machines (even from the same manufacturer) will not permit duplication of a process, but knowledge thus gained will guide the processor in future setups.

An amorphous material usually requires a fairly low initial heat in a screw plasticator; its purpose is to preheat material but not melt it in the feed section before it enters the compression zone of the screw. On the other hand, crystalline material requires higher initial heating to ensure that it melts before reaching the compression zone; otherwise satisfactory melting will not occur. Careful implementation of these procedures produces the best melt, which in turn produces the best part. (Filled plastics, particularly those with thermally conductive fillers, usually require different heat profiles, i.e., a reverse profile where the area of the feed throat is better than the front zone.)

Example of a Thermoset Processing Heat-Time Profile Cycle

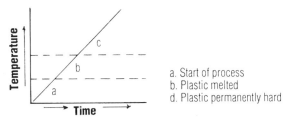

a. Start of process
b. Plastic melted
d. Plastic permanently hard

Example of a Thermoplastic Processing Heat-Time Profile Cycle

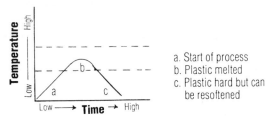

a. Start of process
b. Plastic melted
c. Plastic hard but can be resoftened

Figure 1.10 Melting characteristics of TPs and TSs.

Table 1.5 Melt processing temperatures for TPs[a]

Material	Processing temperature rate	
	°C	°F
ABS	180–240	356–464
Acetal	185–225	365–437
Acrylic	180–250	356–482
Nylon	260–290	500–554
Polycarbonate	280–310	536–590
LDPE	160–240	320–464
HDPE	200–280	392–536
Polypropylene	200–300	392–572
Polystyrene	180–260	356–500
PVC, rigid	160–180	320–365

[a] Values are typical for injection molding and most extrusion operations. Extrusion coating is done at higher temperatures (i.e., about 600 °F for LDPE). See the appendix for metric conversion charts (page 642).

Processing and tolerances

Processing is extremely important for tolerance control, sometimes it is the most influential factor. The dimensional accuracy of the finished part relates to the process, the accuracy of mold or die production, and the process controls, as well as the shrinkage behavior of the plastic. Changing the dimensions of a mold or a die can cause wear to arise during production; this should always be taken into account.

The mold or die is one of the most important pieces of production equipment in the plant. This controllable, complex device must be an efficient heat exchanger and provide the part's shape. The mold or die designer needs to understand how to produce the tooling required for the part and to meet the appropriate tolerances. To maximize control in setting tolerances there is usually a minimum-to-maximum limit on thickness based on the process used. Examples are shown in Tables 1.6 to 1.8.

Adequate process control and its associated instrumentation are essential for product quality control. Sometimes, the goal is precise adherence to a control point. Sometimes it is simply to maintain the temperature within a comparatively narrow range.

A knowledge of processing methods can help to determine the tolerances obtainable. With such high-pressure methods as injection and compression molding that use 13.8–206.9 MPa (2000–30 000 psi) it is possible to develop tighter tolerances, but there is also a tendency to develop undesirable stresses (i.e., orientations) in different directions. The low-pressure

Table 1.6 Guidelines for wall thicknesses of TS molding materials

	Minimum thickness, in. (mm)	Average thickness, in. (mm)	Maximum thickness, in. (mm)
Alkyd, glass filled	0.040 (1.0)	0.125 (3.2)	0.500 (13)
Alkyd, mineral filled	0.040 (1.0)	0.187 (4.7)	0.375 (9.5)
Dially phthalate	0.040 (1.0)	0.187 (4.7)	0.375 (9.5)
Epoxy, glass filled	0.030 (0.76)	0.125 (3.2)	1.000 (25.4)
Melamine, cellulose filled	0.035 (0.89)	0.100 (2.5)	0.187 (4.7)
Urea, cellulose filled	0.035 (0.89)	0.100 (2.5)	0.187 (4.7)
Phenolic, general purpose	0.050 (1.3)	0.125 (3.2)	1.000 (25.4)
Phenolic, flock filled	0.050 (1.3)	0.125 (3.2)	1.000 (25.4)
Phenolic, glass filled	0.030 (0.76)	0.093 (2.4)	0.750 (19)
Phenolic, fabric filled	0.062 (1.6)	0.187 (4.7)	0.375 (9.5)
Phenolic, mineral filled	0.125 (3.2)	0.187 (4.7)	1.000 (25.4)
Silicone glass	0.050 (1.3)	0.125 (3.2)	0.250 (6.4)
Polyester premix	0.040 (1.0)	0.070 (1.8)	1.000 (25.4)

Table 1.7 Guidelines for wall thicknesses of TP molding materials

	Minimum, in. (mm)	Average, in. (mm)	Maximum, in. (mm)
Acetal	0.015 (0.38)	0.062 (1.6)	0.125 (3.2)
ABS	0.030 (0.76)	0.090 (2.3)	0.125 (3.2)
Acrylic	0.025 (0.63)	0.093 (2.4)	0.250 (6.4)
Cellulosics	0.025 (0.63)	0.075 (1.9)	0.187 (4.7)
FEP fluoroplastic	0.010 (0.25)	0.035 (0.89)	0.500
Nylon	0.015 (0.38)	0.062 (1.6)	0.125 (3.2)
Polycarbonate	0.040 (1.0)	0.093 (2.4)	0.375 (9.5)
Polyester TP	0.025 (0.63)	0.062 (1.6)	0.500 (12.7)
Polyethylene (LD)	0.020 (5.1)	0.062 (1.6)	0.250 (6.4)
Polyethylene (HD)	0.035 (0.89)	0.062 (1.6)	0.250 (6.4)
Ethylene vinyl acetate	0.020 (0.51)	0.062 (1.6)	0.125 (3.2)
Polypropylene	0.025 (0.63)	0.080 (2.0)	0.300
Polysulfone	0.040 (1.0)	0.100 (2.5)	0.375 (9.5)
Noryl (modified PPO)	0.030 (0.76)	0.080 (2.0)	0.375 (9.5)
Polystyrene	0.030 (0.76)	0.062 (1.6)	0.250 (7.6)
SAN	0.030 (0.76)	0.062 (1.6)	0.250 (7.6)
PVC, Rigid	0.040 (1.0)	0.093 (2.4)	0.375 (9.5)
Polyurethane	0.025 (0.63)	0.500 (12.7)	1.500 (38.1)
Surlyn (ionomer)	0.025 (0.63)	0.062 (1.6)	0.750 (19.1)

Table 1.8 Guide to tolerances of TP extrusion profiles

	HIPS	PC, ABS	PP	PVC Rigid	PVC Flexible	LDPE
Wall thickness (%)	8	8	8	8	10	10
Angle (°)	2	3	3	2	5	5
Profile dimensions (in., ±)						
<0.125	0.007	0.010	0.010	0.007	0.010	0.012
0.125–0.500	0.012	0.020	0.015	0.010	0.015	0.025
0.500–1.00	0.017	0.025	0.020	0.015	0.020	0.030
1.0–1.5	0.025	0.027	0.027	0.020	0.030	0.035
1.5–2.0	0.030	0.035	0.035	0.025	0.035	0.040
2–3	0.035	0.037	0.037	0.030	0.040	0.045
3–4	0.050	0.050	0.050	0.045	0.065	0.065
4–5	0.065	0.065	0.065	0.060	0.093	0.093
5–7	0.093	0.093	0.093	0.075	0.125	0.125
7–10	0.125	0.125	0.125	0.093	0.150	0.150

processes, including contact and casting with no pressure, rarely lend themselves to tight tolerances. There are exceptions, such as certain reinforced plastics (RPs) that are processed at quite low pressures. Regardless of the process, exercising appropriate control will maximize the achievable tolerances and increase their repeatability.

For example, certain injection-molded parts can be molded to extremely close tolerances of less than a thousandth of an inch, or down to 0.0%, particularly when TPs are filled with additives or when TS compounds are used. To eliminate shrinkage and provide a smooth surface, one should use a small amount of a chemical blowing agent (<0.5wt%) and a regular packing procedure. For conventional molding, tolerances can be met of ±5% for a part 0.020in. (0.5mm) thick, ±1% for 0.050in. (1.27mm), ±0.5% for 1.000in. (25.4mm), ±0.25% for 5.000in. (127mm), and so on. Thermosets are generally more suitable than TPs for meeting the tightest tolerances.

Economical production requires that tolerances are not specified to be tighter than necessary. However, after a production target is met, one should mold 'tighter' if possible, for greater profit by using less material. Table 1.9 reviews factors affecting tolerances. Many plastics change dimensions after molding, principally because their molecular orientations or molecules are not relaxed. To ease or eliminate the problem, one can change the processing cycle so that the plastic is 'stress relieved,' even

Table 1.9 Parameters that influence part tolerance

Part design	Part configuration (size/shape). Relate shape to flow of melt in mold to meet performance requirements that should at least include tolerances.
Material	Chemical structure, molecular weight, amount and type of fillers/additives, heat history, storage, handling.
Mold design	Number of cavities, layout and size of cavities/runners/gates/cooling lines/side actions/knockout pins/etc. Relate layout to maximize proper performance of melt and cooling flow patterns to meet part performance requirements; pre-engineer design to minimize wear and deformation of mold (use proper steels); lay out cooling lines to meet temperature-to-time cooling rate of plastics (particularly crystalline types).
Machine capability	Accuracy and repeatability of temperature, time, velocity, and pressure controls of injection unit, accuracy and repeatability of clamping force, flatness and parallelism of platens, even distribution of clamping on all tie-rods, repeatability of controlling pressure and temperature of oil, oil temperature variation minimized, no oil contamination (by the time you see oil contamination, damage to the hydraulic system could have already occurred), machine properly leveled.
Molding cycle	Set up the complete molding cycle to repeatedly meet performance requirements at the lowest cost by interrelating material, machine and mold controls.

though that may extend the cycle time, or heat-treat according to the resin supplier's suggestions.

Shrinkage

One factor in tolerances is shrinkage. Generally, shrinkage is the difference between the dimensions of a fabricated part at room temperature and the cooled part, usually checked 12–24 h after fabrication. Having an elapsed time is necessary for many plastics, particularly the commodity TPs, to allow parts to complete their shrinkage behavior. The extent of this postshrinkage can be almost zero for certain plastics or may vary considerably. Shrinkage can also depend on climatic conditions such as temperature and humidity, under which the part will exist in service, as well as its conditions of storage.

Plastics suppliers can provide the initial information on shrinkage that has to be added to the design shape and will influence its processing. The shrinkage and postshrinkage will depend on the types of plastics and fillers. Compared to the TPs, the TSs generally have more filler. The type and amount of filler, such as its reinforcement, can significantly reduce shrinkage and tolerances.

Inspection

Inspection variations are often the most critical and most overlooked aspect of the tolerance of a fabricated part. Designers and processors base their development decisions on inspection readings, but they rarely determine the tolerances associated with these readings. The inspection variations may themselves be greater than the tolerances for the characteristics being measured, but this can go unnoticed without a study of the inspection method capability.

Inspection tolerance can be divided into two major components: the accuracy variability of the instruction and the repeatability of the measuring method. The calibration and accuracy of the instrument are documented and certified by its manufacturer, and the instrument is periodically checked. Understanding the overall inspection process is extremely useful in selecting the proper method for measuring a specific dimension. When all the inspection methods available provide an acceptable level of accuracy, the most economical method should be used.

As the overall fabricating tolerance is analyzed into the sources of its variation components, the potential advantage of analytical programs comes into play with their ability to process all these factors efficiently. All the empirical tolerance ranges for each tooling method and inspection method are stored in data files for easy retrieval. For each critical dimension the program sums all the component tolerances and computes a ± overall tolerance for each critical dimension. The program then provides a tabulated estimate of the achievable processing tolerances and pinpoints the areas that contribute most to the required overall tolerance. This information is useful in identifying the needed tolerances, which usually exceed the initial design tolerances.

Screw plasticating

The screw plasticator is an important device that plasticates or melts plastic. Many methods are used, but by far the most common is the single plasticating screw/barrel system. The twin-screw is primarily used in compounding plastics. The single-screw is used for injection molding, extrusion and blow molding. A helically flighted, hard steel shaft rotates within a barrel to process the melt mechanically and to advance it through a die or into a mold. Screw plasticating systems are reviewed in Chapters 2 and 3. They include the use of two-stage vented or devolatilizing screw systems.

Intelligent processing

To remain competitive on a worldwide basis, processors must continue to improve productivity and product quality. What is needed is a way to cut ineffciency and the costs associated with it. One approach that promises to overcome these difficulties is called intelligent processing of materials. This technology utilizes new sensors, expert systems, and process models that control processing conditions as materials are produced, without the need for human control or monitoring.

Sensors and expert systems are not new in themselves. What is novel is the manner in which they are tied together. In intelligent processing, new nondestructive evaluation sensors are used to monitor the development of a material's microstructure as it evolves during production in real time. These sensors can indicate whether the microstructure is developing properly. Poor microstructure will lead to defects in materials. In essence, the sensors are inspecting the material online, before the end product is produced.

Next, the information these sensors gather is communicated, along with data from conventional sensors that monitor temperature, pressure, and other variables, to a computerized decision-making system. This decision maker includes an expert system and a mathematical model of the process. The system then makes any changes necessary in the production process to ensure the material's structure is forming properly. These might include changing temperature or pressure, or altering other variables that will lead to a defect-free end product.

There are several significant benefits that can be derived from intelligent processing. There is a marked improvement in overall product quality and a reduction in the number of rejected products. And the automation concept that is behind intelligent processing is consistent with the broad, systematic approaches to planning and implementation being undertaken by industries to improve quality.

Intelligent processing involves building in quality rather than attempting to obtain it by inspecting a product after it's made. Thus, industry can expect to reduce post-manufacturing inspection costs and time.

Being able to change manufacturing processes or the types of material being produced is another potential benefit of the technique. The technology will also help to shorten the long lead time needed to bring new materials from R&D to mass production. Although much effort has gone into applying this technology to advanced materials, it also holds promise in making such conventional materials as steel and cement.

Processing diagrams

The processor setting up a machine, regardless of the type of controls available, uses a systematic approach that should be outlined in the

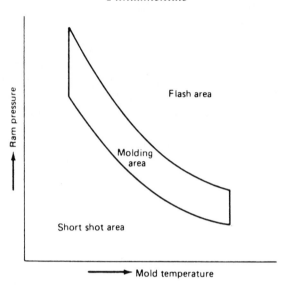

Figure 1.11 Two-dimensional molding area diagram (MAD) that plots injection molding ram pressure versus mold temperature.

machine and/or control manuals. Once the machine is operating, the processor methodically makes one change at a time to determine the result. Two basic examples are presented to show a logical approach to evaluating changes made with any processing machine. As the injection molding machine is very complex with the all controls required to set it up, these examples refer to the injection molding process.

Figure 1.11 shows what happens when changing mold temperature and injection (ram) pressure. This molding area diagram (MAD) provides information on the best combination. Any setting outside the diagram produces defective molded parts that will not meet the required performances. The size of the diagram shows the latitude available to produce good parts. However, to mold at the fastest cycle and/or lowest cost, the machine would be set at the lowest temperature and highest pressure, or the upper left corner of the diagram. One must be aware that there could be variations in the material (thermoplastic), basic machine functions, and/or controls; so even with operation in that corner of the diagram, defective parts can occur. Unless the original cycles were run with all these variables at their worst, all parts should be okay. It is statistically probable that rejects will occur; so one must determine how far away from optimum and in what directions the molding conditions can deviate and still provide good parts. If a thermoset were used and a diagram were developed, its best operation would be at high pressure and high temperature, or the upper right corner of the diagram. In spite of the

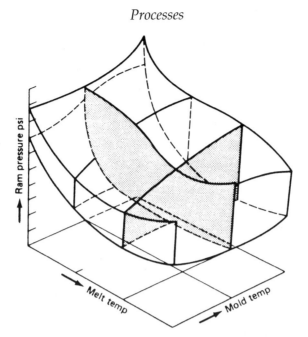

Figure 1.12 Three-dimensional molding volume diagram (MVD) that plots melt temperature, mold temperature, and ram pressure.

different shapes developed for a TP and a TS, the approach is similar for both.

The second example is the use of a three-dimensional diagram (Fig. 1.12). This molding volume diagram (MVD) compares the behavior of a thermoplastic in the mold based on varying melt temperature, mold temperature, and ram pressure. Thus all parts molded within the volume produce good parts. To operate with maximum efficiency, one should plan to operate the machine at the highest possible pressure, the lowest possible melt temperature, and the lowest possible mold temperature. This approach can be continued with any controls available in the machine, easier said than done. A costly amount of time will initially be required, but eventually this becomes second nature so that many shortcuts can be taken. The fact remains that some logical approach has to be used to determine the best setting for the machine. Proper record keeping is vital. With the computer control systems that are available, a systematic approach can be readily conducted, process parameters can be stored, and an 'expert system' can be developed.

Selecting a process

The processing information presented throughout this book provides a variety of useful selection guides. Summary selection guides are provided in Figs 1.13 and 1.14 as well as Tables 1.10 to 1.14. The most important

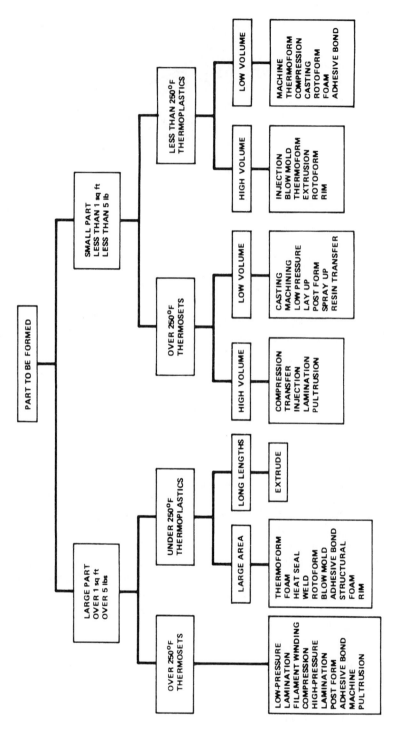

Figure 1.13 Guide to process selection.

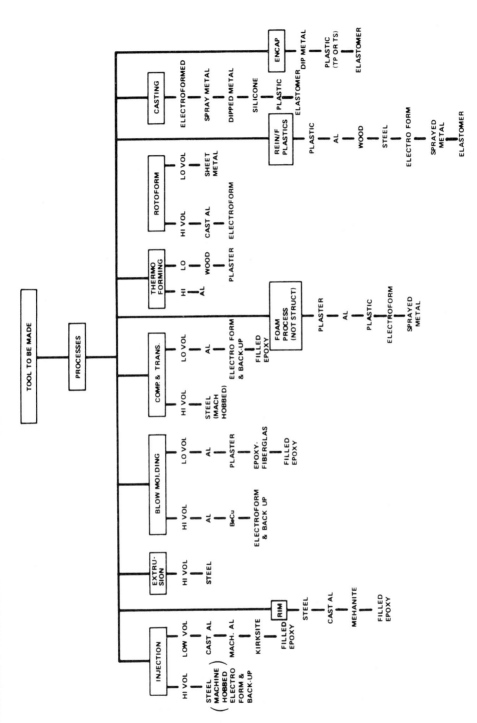

Figure 1.14 Guide to tool selection.

Table 1.10 Molding process guide to plastic materials[a]

Material family	Injection	Compression	Transfer	Casting	Cold molding	Coating	Structural foam	Extrusion	Laminating	Sheet forming	RP molding FRP	Filament	Dip and slush	Blow	Rotational
ABS	X						X	X		X				X	X
Acetal	X						X	X		X				X	X
Acrylic	X	X		X		X		X	X	X				X	
Allyl			X	X	X	X			X			X			
ASA	X						X	X		X					
Cellulosic	X					X	X	X	X	X				X	X
Epoxy	X	X	X	X	X	X			X			X			
Fluoroplastic	X	X	X		X	X		X		X					X
Melamine-formaldehyde	X	X	X	X		X			X						
Nylon	X		X	X		X	X	X						X	X
Phenol-formaldehyde	X	X	X	X	X	X	X		X						
Poly (amide-imide)	X	X	X					X							
Polyarylether	X							X							
Polybutadiene	X		X					X						X	
Polycarbonate	X	X					X	X		X				X	X
Polyester (TP)	X					X		X						X	X

Polyester-fiberglass (TS)	X		X			X		X		X
Polyethylene	X	X		X	X	X	X	X(TP)		X
Polyimide	X	X			X		X	X(TP)		X
Polyphenylene oxide	X			X	X		X	X		X
Polyphenylene sulfide	X			X		X		X		
Polypropylene	X				X	X	X	X	X	X
Polystyrene	X				X	X	X	X	X	X
Polysulfone	X	X		X	X		X	X		X
Polyurethane (TS) (TP)	X	X	X	X		X	X	X(TP)	X	X
SAN	X				X		X	X		X
Silicone	X	X	X		X			X		
Styrene-butadiene	X			X	X			X		
Urea-formaldehyde	X				X		X	X		
Vinyl	X	X	X	X	X	X	X	X	X	X

[a] Compounding permits using other processes.

Table 1.11 Guide to compatibility of processes and materials

	Thermosets					Thermoplastics										
	Polyester	Polyester SMC	Polyester BMC	Epoxy	Polyurethane	Acetal	Nylon-6	Nylon-6,6	Polycarbonate	Polpropylene	Polyphenylene sulfide	ABS	Polyphenylene oxide	Polystyrene	Polyester PBT	Polyester PET
Injection molding	X		X	X	X	X	X	X	X	X	X	X	X	X	X	
Hand layup	X				X											
Sprayup	X				X											
Compression molding	X	X	X	X							X					
Preform molding	X				X											
Filament winding	X				X											X
Pultrusion	X				X											
Resin transfer molding	X														X	X
Reinforced reaction injection molding	X				X	X	X									

processing requirements should be based on the plastic to be processed, the quantity of the part, its dimensions, and the tolerances. Process selection is a critical step in product design. Failure to select a viable process during the initial design stages can dramatically increase development costs and timing. The process can have a significant effect on the performance of the finished part. The following examples are based on material considered throughout this book.

1. The nature of the process may have a profound influence on a product's mechanical strength.
2. Excessive heat during processing can consume sacrificial heat stabilizers for certain plastics, rendering stabilization levels insufficient to ensure long life at elevated temperatures. Thermal degradation usually

Table 1.12 General information relating processes and materials to properties of plastics

Thermosets	Properties	Processes
Polyesters Properties shown also apply to some polyesters formulated for thermoplastic processing by injection molding	Simplest, most versatile, economical and most widely used family of resins, having good electrical properties, good chemical resistance, especially to acids	Compression molding Filament winding Hand layup Mat molding Pressure bag molding Continuous pultrusion Injection molding Sprayup Centrifugal casting Cold molding Comoform[a] Encapsulation
Epoxies	Excellent mechanical properties, dimensional stability, chemical resistance (especially alkalis), low water absorption, self-extinguishing (when halogenated), low shrinkage, good abrasion resistance, very good adhesion properties	Compression molding Filament winding Hand layup Continuous pultrusion Encapsulation Centrifugal casting
Phenolics	Good acid resistance, good electrical properties (except arc resistance), high heat resistance	Compression molding Continuous laminating
Silicones	Highest heat resistance, low water absorption, excellent dielectric properties, high arc resistance	Compression molding Injection molding Encapsulation
Melamines	Good heat resistance, high impact strength	Compression molding
Diallyl phthalate	Good electrical insulation, low water absorption	Compression molding
Thermoplastics		
Polystyrene	Low cost, moderate heat distortion, good dimensional stability, good stiffness, impact strength	Injection molding Continuous laminating
Nylon	High heat distortion, low water absorption, low elongation, good impact strength, good tensile and flexural strength	Injection molding Blow molding. Rotational molding
Polycarbonate	Self-extinguishing, high dielectric strength, high mechanical properties	Injection molding

Table 1.12 *Continued*

Thermosets	Properties	Processes
Styrene-acrylonitrile	Good solvent resistance, good long-term strength, good appearance	Injection molding
Acrylics	Good gloss, weather resistance, optical clarity, and color; excellent electrical properties	Injection molding Vacuum forming Compression molding Continuous laminating
Vinyls	Excellent weatherability, superior electrical properties, excellent moisture and chemical resistance, self-extinguishing	Injection molding Continuous laminating Rotational molding
Acetals	Very high tensile strength and stiffness, exceptional dimensional stability, high chemical and abrasion resistance, no known room temperature solvent	Injection molding
Polyethylene	Good toughness, light weight, how cost, good flexibility, good chemical resistance; can be 'welded'	Injection molding Rotational molding Blow molding
Fluorocarbons	Very high heat and chemical resistance, nonburning, lowest coefficient of friction, high dimensional stability	Injection molding Encapsulation Continuous pultrusion
Polyphenylene oxide, modified	Very tough engineering plastic, superior dimensional stability, low moisture absorption, excellent chemical resistance	Injection molding
Polypropylene	Excellent resistance to stress or flex cracking, very light weight, hard, scratch-resistant surface, can be electroplated; good chemical and heat resistance; exceptional impact strength; good optical qualities	Injection molding Continuous laminating Rotational molding
Polysulfone	Good transparency, high mechanical properties, heat resistance, electrical properties at high temperatures; can be electroplated	Injection molding

[a]Comoform is an extension of the cold molding process which utilizes a thermoformed plastic skin to impart excellent surface to a cold-molded laminate.

produces embrittlement (tests can be conducted to determine the re-maining levels).

3. The slow cooling of crystalline polymers, such as HDPE and PP, can allow large crystal formations to develop. Such crystals embrittle the resin and make it prone to stress cracking.

4. The rapid cooling of certain plastic parts can produce 'frozen-in' stresses and strains (particularly with injection molding). The stresses may decay with time, in a viscoelastic manner. However, they will act like any other sustained stress to aggravate cracking or crazing in the presence of aggressive media and hostile environments like UV radiation.

5. Annealing at temperatures below the T_g (glass transition temperature) where material becomes leathery is not necessarily beneficial. For example, annealing a PC greatly accelerates both its crazing and rup-ture under sustained loading. In general, the annealing of a plastic lowers its properties; however, its dimensional stability may be improved. Heating a material to above its T_g relieves the internal stresses.

6. Knit or weld lines form where the melt flow during processing meets after flowing through separate gates in an injection mold or after being parted by either 'spiders' in an extruder die or bosses in an injection mold. Because the material is not well mixed in the zone of the knit or weld line, the seam thus formed can be weak or brittle under long-term or impact loads. This problem can easily arise with fiber-reinforced plastics, where 40–60% of their strength can be lost, since fibers fail to knit together at their seams.

7. In reinforced plastics, insufficient compaction and consolidation of a composite before resin cure will produce air pockets, incomplete wet-out and encapsulation of the fibers, and/or insufficient fiber or uniform fiber content. These deficiencies lead to loss of strength and stiffness and susceptibility to deterioration by water and aggressive agents.

In summary, when considering alternative processes for producing plastic and composite products, the major concerns usually involve (1) limitations that may be imposed by the material, because not all mate-rials can be processed by all methods; (2) limitations imposed by the design, such as the size, single-piece versus multiple-piece construction, a closed or open shape, and the level of dimensional and tolerance accuracy required; (3) the number of products required; and (4) the avail-able capital equipment. Certain equipment may already be available and in use, although it may not necessarily give the lowest production cost.

Table 1.13 Specific processing methods as a function of part design[a]

Process	Ribs	Bones	Vertical walls	Spherical shape	Box sections	Slides/cores	Weldable	Good finish, both sides	Varying cross section
Thermoplastics									
Injection	Y	Y	Y	N	N	Y	Y	Y	Y
Injection compression	Y	Y	N	N	N	Y	Y	Y	Y
Hollow injection	Y	Y	Y	N	Y	Y	Y	Y	Y
Foam injection	Y	Y	Y	N	Y	Y	Y	Y	Y
Sandwich molding	Y	Y	Y	N	N	Y	Y	Y	Y
Compression	Y	Y	Y	N	N	Y	Y	Y	Y
Stamping	N	N	N	N	N	N	Y	Y	N
Extrusion	Y	N	N/A	N	Y	Y	Y	Y	Y
Blow molding	N	N	Y	Y	Y	N	Y	N	N
Twin-sheet forming	N	N	Y	Y	Y	N	Y	N	N
Twin-sheet stamping	N	N	N	N	Y	N	Y	Y	N
Thermoforming	N	N	Y	N	N	Y	Y	N	N
Filament winding	Y	N	Y	Y	Y	N	Y	N	Y
Rotational casting	N	N	Y	Y	N	N	Y	N	N
Thermosets									
Compression									
Powder	Y	Y	Y	N	N	Y	N	Y	Y
Sheet molding compound	Y	Y	Y	N	N	Y	N	Y	Y
Cold-press molding	N	Y	Y	N	N	N	N	Y	Y
Hot-press molding	N	Y	Y	N	N	N	N	Y	Y

Process							
High-strength sheet molding compound	Y	Y	Y	N	N	N	Y
Prepreg	N	N	Y	N	Y	N	Y
Vacuum bag	N	Y	Y	N	Y	N	Y
Hand layup	N	Y	Y	N	Y	N	Y
Injection							
Powder	Y	Y	Y	N	N	Y	Y
Bulk molding compound	Y	Y	Y	N	N	Y	Y
ZMC	Y	Y	Y	N	N	Y	Y
Stamping	N	N	Y	N	N	N	N
Reaction injection molding	Y	Y	N	N	Y	N	Y
Resin transfer molding, or resinject	Y	N	Y	N	Y	N	Y
High-speed resin transfer molding, or fast resinject	Y	N	Y	N	Y	N	Y
Foam polyurethane	Y	Y	Y	Y	Y	N	Y
Reinforced foam	Y	Y	Y	N	Y	N	Y
Filament winding	Y	N	Y	Y	Y	N	Y
Pultrusion	Y	N	N/A	N	Y	N	Y

[a] Y = yes; N = no; N/A = not applicable.

Table 1.14 Basic processing methods as a function of part design

Part design	Blow molding	Casting	Compression	Extrusion	Filament winding	Injection	Matched die molding	Rotational	Thermoforming	Transfer compression	Wet layup (contact molding)
Major shape characteristics	Hollow bodies	Simple configurations	Moldable in one plane	Constant cross section profile	Structure with surfaces of revolution	Few limitations	Moldable in one plane	Hollow bodies	Moldable in one plane	Simple configurations	Moldable in one plane
Limiting size factor	Material	Material	Equipment	Material	Equipment	Equipment	Equipment	Material	Material	Equipment	Mold size
Max. thickness. in. (mm)	>0.25 (6.4)	–	0.5 (12.7)	6 (150)	3 (76)	6 (150)	2 (51)	0.5 (12.7)	3 (76)	6 (150)	0.5 (12.7)
Min. inside radius. in. (mm)	0.125 (3.18)	0.01–0.125 (0.25–3.18)	0.125 (3.18)	0.01–0.125 (0.25–3.18)	0.125 (3.18)	0.01–0.125 (0.25–3.18)	0.06 (1.5)	0.01–0.125 (0.25–3.18)	0.125 (3.18)	0.01–0.125 (0.25–3.18)	0.25 (6.4)
Min. draft (deg.)	0	0–1	>1	NR[b]	2–3	<1	1	1	1	1	0
Min. thickness. in. (mm)	0.01 (0.25)	0.01–0.125 (0.25–3.18)	0.01–0.125 (0.25–3.18)	0.001 (0.02)	0.015 (0.38)	0.005 (0.1)	0.03 (0.8)	0.02 (0.5)	0.002 (0.05)	0.01–0.125 (0.25–3.18)	0.06 (1.5)
Threads	Yes	Yes	Yes	No	No	Yes	No	Yes	No	Yes	No
Undercuts	Yes	Yes[a]	NR[b]	Yes	NR[b]	Yes[a]	NR[b]	Yes[c]	Yes[a]	NR[b]	Yes
Inserts	Yes	Yes	Yes	Yes	Yes	Yes	Yes	Yes	NR[b]	Yes	Yes
Built-in cores	Yes	Yes	No	Yes[d]	Yes	Yes	Yes	Yes	Yes	Yes	Yes
Molded-in holes	Yes	Yes	Yes	Yes	Yes	Yes	Yes	Yes	No	Yes	Yes
Bosses	Yes	Yes	Yes	Yes	No[e]	Yes	No[f]	Yes	Yes	Yes	Yes
Fins or ribs	Yes	Yes	Yes	No	No[e]	Yes	Yes	Yes	Yes	Yes	Yes
Molded in designs and numbers	Yes	Yes	Yes		No	Yes		Yes	Yes	Yes	Yes
Surface finish[g]	1–2	2	1–2	1–2	5	1	4–5	2–3	1–3	1–2	4–5
Overall dimensional tolerance (±)	0.01	0.001	0.001	0.005	0.005	0.001	0.005	0.01	0.01	0.001	0.02

[a] Special mold required.
[b] Not recommended.
[c] Only flexible material.
[d] Only direction of extrusion.
[e] Possible with special techniques.
[f] Fusing premix/yes.
[g] Rated 1 to 5 (1 = very smooth. 5 = rough).

PLASTIC MATERIALS

Introduction

Plastics (polymers, resins, etc.) are organic materials with high molecular weight, produced by combining highly purified simple molecules under controlled heat and pressure, frequently in the presence of catalysts, accelerators, or promoters [9]. The profound impact of plastics in all industries is due to the intelligent application of modern chemistry and engineering principles that utilized the versatility and vast array of inherent plastic properties as well as high-speed/low-energy processing techniques. The result has been the development of cost-effective products that compete well with conventional and new products of metal, glass, etc.

There are literally hundreds of classes of polymers with about 20 principal classes used in production. New subclasses exist and are always being created by ingenious combinations of polymers, additives, fillers, alloying, etc. Development of desired properties for a specific application depends upon a good understanding of comparative molecular achitectures of candidate polymers; modifications made via additives, grafting, etc., have created about 17 000 processable plastics worldwide. The following overview points out some plastic features that influence processing properties and product performance. As shown in Fig. 1.3 (page 4), feedstocks are used to produce monomers. Monomers are chemical compounds consisting of simple molecules having different molecular structures which can be joined together by polymerization. The result is to produce a plastic composed of much more complex molecules. A monomer is the basic material from which plastics are made; it is a simple molecule capable of reacting with like or unlike molecules to form the polymer (plastic); it is the smallest repeating structure of a polymer (a mer). Addition polymers are produced from a monomer that is the original unpolymerized compound. Styrene is a monomer for polystyrene plastic, vinyl chloride is a monomer for polyvinyl chloride plastic, etc.

Polymerization is a chemical reaction in which the molecules of a monomer are linked together to form large molecules with a molecular weight that is a multiple of the molecular weight of the original substance. When two or more monomers are involved, the process is called copolymerization. The molecular structure of a polymer is determined during its formation by polymerization; the conditions (temperature, time, monomer concentration, catalyst or initiator concentration, etc.) must therefore be chosen so that the polymer with the desired structure is obtained. The repeat unit structure is determined by the choice of monomer, but the degree of polymerization (DP) depends on polymerization conditions. The polymer may be contaminated by unreacted monomer or other materials, especially solvent required for polymerization.

Fundamentals

Strength

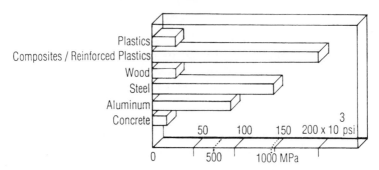

Modulus of Elasticity

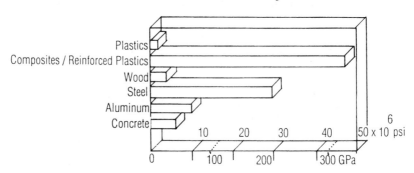

Specific Gravity

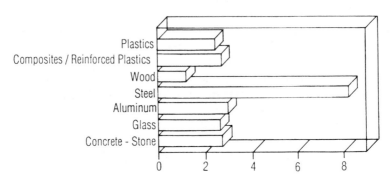

Figure 1.15 A general comparison of different materials.

Plastics commercially provide more types than all the other materials put together. Many variations are available, providing a wide range of properties (Figs 1.15 and 1.16). Like other materials, plastics are variously identified, such as plastics, resins, polymers, elastomers, foams, reinforced plastics, and composites. The terms *polymers*, *plastics* and *resins* are usually taken as synonymous. This book treats them as synonymous but there are technical distinctions. A polymer is a pure unadulterated material that is usually taken as the family name for a group of materials; a polymer is a NEAT material (nothing else added to). Pure polymers are seldom used on their own. The terms *plastic* or *resin* are used when additives are included. Resin tends to be used with thermoset 'plastic' materials.

Elastomers are plastics (or polymers) that are flexible. Reinforced plastics (RPs) or composites are plastics (or resins) with reinforcing additives such as fibers and whiskers to increase mechanical properties (Chapter 12). Throughout this book these terms are used according to their respective areas of interest.

The term *plastics* is not a definitive one. Metals, for instance, are also deformable and are therefore plastic. How else could roll aluminum be made into foil, or tungsten wire be drawn into filament for an incandescent lightbulb, or a 100 ton ingot of steel be forged into a rotor for an electric generator?

The term *plastics* became attached to polymeric materials because they are capable of being molded or formed, as are clay or plaster. Potters use wet clay to create their art, although objects are not called plastics. Despite this seeming contradiction in the use of the term, *plastics* definitely identifies the materials described in this book and those produced by the world-wide plastics industry.

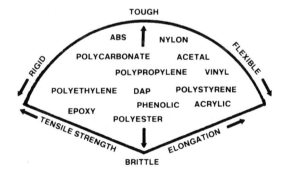

Note: With formulation changes (via additives, fillers, reinforcements, alloying, etc.) position of plastic can move practically any place in the "pie."

Figure 1.16 A range of mechanical properties for several plastics.

Table 1.15 Types of plastics

Acetal (POM)
Acrylics
  Polyacrylonitrile (PAN)
  Polymethylmethacrylate (PMMA)
Acrylonitrile butadiene styrene (ABS)
Acrylonitrile/methylmethacrylate
  (AMMA)
Acrylonitrile/styrene/acrylate (ASA)
Alkyd
Allyl diglycol carbonate (CR-39)
Allyls
  Diallyl isophthalate (DAIP)
  Diallyl phthalate (DAP)
Aminos
  Melamine formaldehyde (MF)
  Urea formaldehyde (UF)
Carboxymethyl cellulose (CMC)
Casein (CS)
Cellulosics
  Cellulose acetate (CA)
  Cellulose acetate butyrate (CAB)
  Cellulose acetate propionate (CAP)
  Cellulose nitrate (CN)
  Cellulose plastics, general (CE)
  Cellulose propionate (CP)
  Cellulose triacetate (CTA)
  Ethyl cellulose (EC)
Chlorinated polyether
Cresol formaldehyde (CF)
Epoxy (EP)
Ethylene vinyl acetate (EVA)
Ethylene vinyl alcohol (EVOH)
Fluorocarbons
  Ethylene-tetrafluoroethylene
    copolymer (ETFE)
  Fluorinated ethylene propylene
    (FEP)
  Polymonochlorotrifluoroethylene
    (CTFE)
  Polytetrafluoroethylene (PTFE)
  Polyvinyl fluoride (PVF)
  Polyvinylidene fluoride (PVDF)
Furan
Furan formaldehyde (FF)
Ionomer

Ketone
Liquid crystal polymer (LCP)
  Aromatic copolyester (TP polyester)
Melamine formaldehyde (MF)
Nylon (Polyamide) (PA)
Parylene
Phenolic
  Phenol formaldehyde (PF)
Phenoxy
Polyallomer
Polyamide (nylon) (PA)
Polyamide-imide (PAI)
Polyarylethers
  Polyaryletherketone (PAEK)
  Polyaryl sulfone (PAS)
Polyarylate (PAR)
Polybenzimidazole (PBI)
Polycarbonate (PC)
Polyesters
  Aromatic polyester (TS polyester)
  Thermoplastic polyesters
    Crystallized PET (CPET)
    Polybutylene terephthalate (PBT)
    Polyethylene terephthalate (PET)
  Unsaturated polyester (TS
    polyester)
Polyetherketone (PEK)
Polyetheretherketone (PEEK)
Polyetherimide (PEI)
Polyimide (PI)
  Thermoplastic PI
  Thermoset PI
Polymethylmethacrylate (acrylic)
  (PMMA)
Polymethylpentene
Polyolefins (PO)
  Chlorinated PE (CPE)
  Cross-liked PE (XLPE)
  High-density PE (HDPE)
  Ionomer
  Linear LDPE (LLDPE)
  Low-density PE (LDPE)
  Polyallomer
  Polybutylene (PB)
  Polyethylene (PE)

Table 1.15 *Continued*

Polypropylene (PP)	Styrene acrylonitrile (SAN)
Ultrahigh molecular weight PE (UHMWPE)	Styrene butadiene (SB)
	Sulfones
Polyoxymethylene (POM)	Polyether sulfone (PES)
Polyphenylene ether (PPE)	Polyphenyl sulfone (PPS)
Polyphenylene oxide (PPO)	Polysulfone (PSU)
Polyphenylene sulfide (PPS)	Urea formaldehyde (UF)
Polyurethane (PUR)	Vinyls
Silicone (SI)	Chlorinated PVC (CPVC)
Styrenes	Polyvinyl acetate (PVAc)
Acrylic styrene acrylonitrile (ASA)	Polyvinyl alcohol (PVA)
Acrylonitrile butadiene styrene (ABS)	Polyvinyl butyrate (PVB)
	Polyvinyl chloride (PVC)
General-purpose PS (GPPS)	Polyvinylidene chloride (PVDC)
High-impact PS (HIPS)	Polyvinylidene fluoride (PVF)
Polystyrene (PS)	Etc. . . .

Table 1.16 Properties of some plastics

Property	Thermoplastics	Thermosets
Low temperature	TFE	DAP
Low cost	PP, PE, PVC, PS	Phenolic
Low gravity	Polypropylene methylpentene	Phenolic/nylon
Thermal expansion	Phenoxy glass	Epoxy-glass fiber
Volume resistivity	TFE	DAP
Dielectric strength	PVC	DAP, polyester
Elasticity	EVA, PVC, TPR	Silicone
Moisture absorption	Chlorotrifluorethylene	Alkyd-glass fiber
Steam resistance	Polysulfone	DAP
Flame resistance	TFE, P1	Melamine
Water immersion	Chlorinated polyether	DAP
Stress craze resistance	Polypropylene	All
High temperature	TFE, PPS, P1, PAS	Silicones
Gasoline resistance	Acetal	Phenolic
Impact	UHMWPE	Epoxy-glass fiber
Cold flow	Polysulfone	Melamine-glass fiberglass
Chemical resistance	TFE, FEP, PE, PP	Epoxy
Scratch resistance	Acrylic	Allyl diglycol carbonate (C-39)
Abrasive wear	Polyurethane	Phenolic-canvas
Colors	Acetate, PS	Urea, melamine

Each plastic has its own distinct or special properties and advantages. See Tables 1.15 and 1.16 for names and properties typical of plastics. They fall into two groups: thermoplastics (TPs) and thermosets (TSs). The dividing line between a TP and a TS is not always distinct. For instance, cross-linked TSs are TPs during their initial heat cycle and before chemical cross-linking. Others, such as a cross-linked polyethylene (XLPE), are normally TPs that have been cross-linked either by high-energy radiation or chemically, during processing.

In addition to the broad categories of TPs and TSs, TPs can be further classified in terms of their structure, as either crystalline, amorphous, or liquid crystalline. Other classes include elastomers, copolymers, compounds, commodity resins, and engineering resins. Additives, fillers, and reinforcements are other classifications that relate directly to plastic properties and performance.

Molecular properties

The size and flexibility of a polymer molecule explain how an individual molecule would behave if it were completely isolated from its neighbors. Such isolated molecules are encountered only in theoretical studies on dilute solutions. In practice, polymer molecules always occur in a mass, and the behavior of each individual molecule is very greatly affected by its relationship to adjacent polymer molecules in the mass. These intermolecular relationships between adjacent molecules may be divided into two groups: intermolecular order which describes the geometrical arrangement of adjacent molecules in space, and intermolecular bonding which describes the attractive forces between adjacent molecules in the polymeric mass. Together these two relationships modify the simple effects of molecular size and flexibility, and determine the overall behavior of homogeneous plastic materials.

Polymer science recognizes three distinctly different states of order, namely amorphous, crystalline, and oriented for TPs. When molecules are arranged in completely random, intertwined coils, this completely unordered structure is known as an **amorphous** state. When they are **neatly** arranged so that each of their atoms falls into a precise position in a tightly packed, repeating, regular structure, the molecules are in a **crystalline** state. When molecules are stretched into a rather linear conformation and lie fairly parallel to each other in the mass, this partially ordered structure is described as an **oriented** state.

Three basic molecular properties of density, average molecular weight, and molecular weight distribution affect most of the mechanical and thermal properties essential for processing plastics and obtaining the required performance of fabricated parts. Small variations in these basics may improve or impair some of these properties considerably.

Increasing the MW will generally increase a plastic's tensile and compressive strength. Similarly, copolymerization and alloying can improve a plastic's mechanical properties. Alloys or composite polymers sometimes contain a weaker polymer, which has an adverse effect on mechanical properties, but it may be used to increase lubricity or frictional characteristics.

Put simply, processability is best at low molecular weight (MW), whereas properties of the finished product are best at high MW. Even though there are exceptions to this simple rule, it is sufficiently important to play a major role in the design for practical use, such as (1) the largest volume of plastic usage is based upon thermoplastic at a constant MW and this MW is chosen to represent the best compromise between processability and final part properties; (2) on the other hand, the greatest variety of plastics and processes, particularly in thermosets, elastomers, and coatings, start with a low MW for easy processing and convert it into a high MW for best ultimate properties.

In particular, processability requires an optimum MW for each processing technique that is used. Stretching, orientation, and thermoforming involve the ability of a plastic film or fiber to withstand high mechanical stress and permit considerable elongation without failure; they are therefore best at quite high MW. Extrusion in the form of continuous profile requires sufficient strength so that the molten product coming from the hot die in the unsupported form will be strong enough to retain its shape until it is cooled to the point of substantial strength.

When the melt is injection molded, the final product is well supported on all sides by the cold mold and does not require such a high melt strength; all that is required is the lowest MW to provide optimum melt flow in the mold. Coating applications require all of the fluidity of adhesives (using a fairly low MW) but place fewer demands upon quick strength; they are generally best carried out with a still lower average MW, most often obtained by the addition of large amounts of monomeric solvents. Each type of process involves its own optimum MW range for optimum processability.

Thermoplastics

Thermoplastics (TPs) are resins that repeatedly soften when heated and harden when cooled (Fig. 1.17 and Table 1.17). Many are soluble in specific solvents and burn to some degree. Their softening temperatures vary with the polymer type and grade. Care must be taken to avoid degrading, decomposing, or igniting them. Chemical changes rarely take place during processing. An analogy would be a block of ice that can be softened (turned back to a liquid), poured into any shape of mold or die, then cooled to become a solid again. Compared with TSs, TPs generally offer

Thermoplastic:

These plastics become soft when exposed to sufficient heat and harden when cooled, no matter how often the process is repeated.

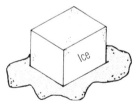

Thermosetting:

The plastics materials belonging to this group are set into permanent shape when heat and pressure are applied to them during forming. Reheating will not soften these materials.

Figure 1.17 Characteristics of thermoplastics (TPs) and thermosets (TSs).

Table 1.17 Melt-processing temperatures for thermoplastics[a]

Material	Processing temperature rate	
	°C	°F
ABS	180–240	356–464
Acetal	185–225	365–437
Acrylic	180–250	356–482
Nylon	260–290	500–554
Polycarbonate	280–310	536–590
LDPE	160–240	320–464
HDPE	200–280	392–536
Polypropylene	200–300	392–572
Polystyrene	180–260	356–500
PVC, rigid	160–180	320–365

[a] Values are typical for injection molding and most extrusion operations. Extrusion coating is done at higher temperatures (i.e., about 600 °F for LDPE).

higher impact strength, easier processing, and better adaptability to complex designs.

Most TP molecular chains can be thought of as independent, intertwined strings resembling spaghetti. When heated, the individual chains slip, causing a plastic flow. Upon cooling, the chains of atoms and molecules are once again held firmly. With subsequent heating the slippage again takes place. There are practical limitations to the number of heating and cooling cycles before appearance or mechanical properties are affected.

Thermosets

Thermosets (TSs) are resins that undergo chemical change during processing to become permanently insoluble and infusible (Figs 1.18 and 1.19) Figure 1.18 shows viscosity change during the processing of TSs. The B-stage represents the start of the heating cycle that is followed by a chemical reaction (cross-linking) and solidification of the plastics. Also used in the TS family are such natural and synthetic rubbers (elastomers) as latex, nitrile, millable polyurethanes, silicone butyl, and neoprene, which attain their properties through the process of vulcanization. The best analogy with TSs is that of a hard-boiled egg whose yolk has turned from a liquid to a solid and cannot be converted back to a liquid. In general, with their tightly cross-linked structure, TSs resist higher temperatures and provide greater dimensional stability than most TPs.

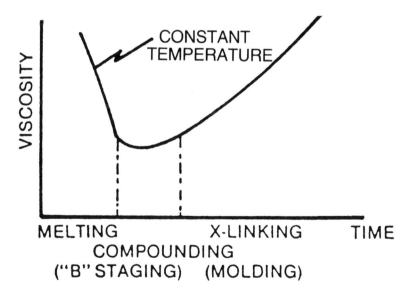

Figure 1.18 Thermoset viscosity during temperature rise.

Similar to TPs, TSs have a chainlike structure. Before molding, TSs resemble TPs. Cross-linking is the principal difference between TSs and TPs. During curing or hardening of TSs the cross-links are formed between adjacent molecules, producing a complex, interconnected network that can be related to its viscosity and performance (Figs 1.18 and 1.19). These cross-bonds prevent the slippage of individual chains, thus preventing plastic flow under the addition of heat. If excessive heat is added after cross-linking has been completed, degradation rather than melting will occur.

TSs on their own can seldom be used as structural components; they must, be filled or reinforced with materials such as calcium carbonate, talc, or glass fiber. The most common reinforcement is glass fiber, but others are also used.

Curing may be defined as changing the properties of resin or rubber formulations by chemical reaction. This is usually accomplished by the action of heat under varying conditions of pressure. Although the term *cure* is used in connection with thermosets and TS rubbers, it is also applied to thermoplastics; this usage is incorrect since no chemical reaction occurs in thermoplastics.

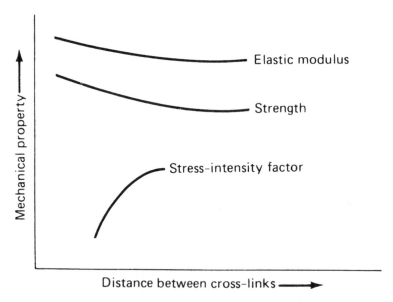

Figure 1.19 The effect of distance between TS cross-linked sites on compressive properties.

A-, B- and C-stages

The various stages of a catalyzed thermoset plastic are known as **A-stage**, **B-stage**, and **C-stage**. A-stage material is still soluble in certain liquids and is fusible. B-stage material softens when heated and swells when in contact with certain liquids, but does not fully fuse or dissolve. Thermoset molding compounds and prepregs are examples of B-stage materials. C-stage plastic is fully cured; it is relatively insoluble and infusible. A-stage is uncured, B-stage is partially cured, and C-stage is fully cured.

Properties influencing processes

Structure and morphology

In addition to the size of the molecules and their distribution, the shapes or structures of individual polymer molecules also play an important role in determining the properties and processability of plastics. Some are formed by aligning themselves into long chains of molecules, others with branches or lateral connections to form complex structures. All these forms exist in either two or three dimensions. Because of the geometry, or morphology, of these molecules some can come closer together than others. These are identified as crystalline, all others as amorphous. Morphology influences mechanical properties, thermal properties, swelling and solubility, specific gravity, chemical properties, and electrical properties.

The morphology of polymers is concerned with the shape, arrangement, and function of crystals alone or embedded in the solid. Polymers are either truly homogeneous, amorphous solids or heterogeneous, semicrystalline solids. Polymer morphology or structure dramatically affects processability and product performance. The structure of an amorphous material is characterized by the absence of a regular three-dimensional arrangement of molecules that are large compared to atomic dimensions, i.e., there is no long-range order. However, due to the close packing of the particles in the condensed state, certain regularity of the structure exists on a local scale.

The structure is not static, it is subject to thermally driven fluctuations. The local structure changes continuously as a function of time due to orientational and translational molecular motions. The timescale of these motions may range from nanoseconds up to several hundred years.

This behavior of morphology occurs with TP but not TS plastics. When TSs are processed, their individual chain segments are strongly bonded together during a chemical reaction that is irreversible.

Crystalline and amorphous plastics

Plastic molecules that can be packed closer together can more easily form crystalline structures in which the molecules align themselves in some orderly pattern. During processing they tend to develop higher strength in the direction of the molecules. Since commercially perfect crystalline polymers are not produced, they are technically identified as semi-crystalline TPs, but in this book they are called crystalline (as used by the plastics industry).

Amorphous TPs, which have their molecules going in all different directions, are normally transparent. Compared to crystalline types, they undergo only small volumetric changes when melting or solidifying during processing. Tables 1.18 to 1.22 compare the basic performance of crystalline and amorphous plastics. Exceptions exist, particularly with respect to the plastic compounds that include additives and reinforcements.

As symmetrical molecules come within a critical distance of each other, crystals begin to form in the areas where they are most densely packed. A crystallized area is stiffer and stronger, a noncrystallized (amorphous)

Table 1.18 General morphology of thermoplastics

Crystalline		Amorphous
No	Transparent	Yes
Excel	Chemical resistance	Poor
No	Stress-craze	Yes
High	Shrinkage	Low
High	Strength	Low[a]
Low	Viscosity	High
Yes	Melt temperature	No
Yes	Critical T/T[b]	No

[a] Major exception is PC.
[b] T/T = Temperature/time.

Table 1.19 Distinctive characteristics of polymers

Crystalline	Amorphous
Sharp melting point	Broad softening range
Usually opaque	Usually transparent
High shrinkage	Low shrinkage
Solvent resistant	Solvent sensitive
Fatigue/wear resistant	Poor fatigue/wear

Table 1.20 Examples of crystalline (semicrystalline) and amorphous TPs

Crystalline	Amorphous
Acetal (POM)	Acrylonitrile-butadiene-styrene (ABS)
Polyester (PET, PBT)	Acrylic (PMMA)
Polyamide (nylon) (PA)	Polycarbonate (PC)
Fluorocarbons (PTFE, etc.)	Modified polyphenylene oxide (PPO)
Polyethylene (PE)	Polystyrene (PS)
Polypropylene (PP)	Polyvinyl chloride (PVC)

Table 1.21 Examples of key properties for engineering TPs

Crystalline	Amorphous
Acetal	*Polycarbonate*
Best property balance	Good impact resistance
Stiffest unreinforced thermoplastic	Transparent
Low friction	Good electrical properties
Nylon	*Modified PPO*
High melting point	Hydrolytic stability
High elongation	Good impact resistance
Toughest thermoplastic	Good electrical properties
Absorbs moisture	
Glass reinforced	
High strength	
Stiffness at elevated temperatures	
Mineral reinforced	
Most economical	
Low warpage	
Polyester (glass reinforced)	
High stiffness	
Lowest creep	
Excellent electrical properties	

area tougher and more flexible. Increased crystallinity causes other effects. Polyethylene exhibits increased resistance to creep, heat, and stress cracking as well as increased mold shrinkage.

A polymeric solid can be partially crystalline, such as polyethylene and polyethylene terephthalate, or noncrystalline, such as commercial polymethyl methacrylate (PMMA) and polystyrene. Partially crystalline polymers, called crystalline polymers, are constructed by a complicated

Table 1.22 General properties of TPs during and after processing

Property	Crystalline[a]	Amorphous[b]
Melting or softening	Fairly sharp melting point	Softens over a range of temperature
Density (for the same material)	Increases as crystallinity increases	Lower than for crystalline material
Heat content	Greater	Lower
Volume change on heating	Greater	Lower
After-molding shrinkage	Greater	Lower
Effect of orientation	Greater	Lower
Compressibility	Often greater	Sometimes lower

[a] Typical crystalline plastics are polyethylene, polypropylene, nylon, acetals, and thermoplastic polyesters.
[b] Typical amorphous plastics are polystyrene, acrylics, PVC, SAN, and ABS.

aggregation of crystalline and amorphous regions. In the amorphous region the molecular chains are in a conformationally random state.

Crystalline plastics are generally more difficult to process. They require more precise control during fabrication, they have higher melting temperatures and melt viscosities, and they tend to shrink and warp more than amorphous plastics. They have a relatively sharp melting point, i.e., they do not soften gradually with increasing temperature but remain hard until a given quantity of heat has been absorbed, then change rapidly into a low-viscosity liquid. If the amount of heat is not applied properly during processing, product performance can be drastically reduced or there may be an increase in processing costs. This is not necessarily a problem, because the qualified processor will know what to do. Amorphous plastics soften gradually as they are heated, but during molding they do not flow as easily as crystalline materials.

Processing conditions influence the performance of plastics. For example, heating a crystalline material above its melting point then quenching it can produce a polymer that has a far more amorphous structure. Its properties can be significantly different than if it is cooled properly (slowly) and allowed to recrystallize, so it becomes amorphous. The effects of time are similar to those of temperature in the sense that any given plastic has a preferred or equilibrium structure in which it would prefer to arrange itself. However, it is prevented from doing so instantaneously or at short notice. If given enough time, the molecules will rearrange themselves into their preferred pattern. Heating causes this to happen sooner, perhaps accompanied by severe shrinkage and property changes in all directions in the processed plastics. This characteristic morphology of plastics can be identified by tests. It provides excellent control as soon as material is received in the plant, during processing and after fabrication [3].

Liquid crystalline polymers

Liquid crystalline polymers (LCPs) are best thought of as being a separate, unique class of TPs. Their molecules are stiff, rodlike structures organized in large parallel arrays or domains in both the melted and solid states. These large, ordered domains provide LCPs with characteristics that are unique compared to those of the basic crystalline or amorphous plastics (Table 1.23).

Unlike many high-temperature plastics, LCPs have a low melt viscosity and are thus more easily processed, and in faster cycle times, than those with a high melt viscosity. They have the lowest warpage and shrinkage of all the TPs. When they are injection molded or extruded, their molecules align into long, rigid chains that in turn align in the direction of flow, acting like reinforcing fibers and giving LCPs their high strength and stiffness. As the melt solidifies during cooling, the molecular orientation freezes into place. The volume changes only minutely, with virtually no frozen-in stresses.

Molded parts experience very little shrinkage or warpage in service. They have high resistance to creep. Their fiberlike molecular chains tend to concentrate near the surface, producing parts that are anisotropic, meaning they have greater strength and modulus in the flow direction, typically on the order of 3–6 times those of the transverse direction. However, adding fillers or reinforcing fibers to LCPs significantly reduces their anisotropy, more evenly distributing strength and modulus, even

Table 1.23 General properties of crystalline, amorphous, and liquid crystalline polymers

Property	Crystalline	Amorphous	Liquid crystalline
Specific gravity	Higher	Lower	Higher
Tensile strength	Higher	Lower	Highest
Tensile modulus	Higher	Lower	Highest
Ductility, elongation	Lower	Higher	Lowest
Resistance to creep	Higher	Lower	High
Max. usage temperature	Higher	Lower	High
Shrinkage and warpage	Higher	Lower	Lowest
Flow	Higher	Lower	Highest
Chemical resistance	Higher	Lower	Highest

boosting them. Most fillers and reinforcements also reduce overall cost and place mold shrinkage to almost zero. Consequently, parts can be molded to tight tolerances. These low melt viscosity LCPs thus permit the design of parts with long or complex flow paths and thin sections.

Elastomers

An elastomer may be defined as a natural or synthetic material that exhibits the rubberlike properties of high extensibility and flexibility. Although the term *rubber* originally meant the TS elastomeric material obtained from the rubber tree (*Hevea brasiliensis*), it now identifies any thermoset elastomer (TSE) or thermoplastic elastomeric (TPE) material. Synthetics such as neoprene, nitrile, styrene butadiene, and butadiene are grouped with TS natural rubber.

They may be formulated to produce rubbers with certain characteristic properties, including high resilience; high tensile strength and elongation; resistance to tear, flexing, freezing, and abrasion; and low permanent set. Elastomer composition can range from relatively uncomplicated homopolymer elastomers such as TS natural rubber to very complex multimonomer copolymers, EPDM elastomers.

Elastomers are generally lower-modulus flexible materials that can be stretched repeatedly to at least twice their original length at room temperature, but will return to their approximate original length when the stress is released. TS elastomeric or rubber materials will always be required to meet certain desired properties, but TPEs are replacing traditional TS natural and synthetic rubbers. TPEs are also widely used to modify the properties of rigid TPs, usually by improving their impact strength. TPEs offer a combination of strength and elasticity as well as exceptional processing versatility. They present creative designers with endless new and unusual product opportunities. More than 100 major groups of TPEs are produced worldwide, with new grades continually being introduced to meet different electrical, chemical radiation, wear, swell, and other requirements.

Previous work concerning polymerization reactions produced polymers that in turn were heated via processing to produce products. With natural rubber (TS) vulcanization is used to provide their high strengths and other properties. These materials forcibly retract approximately to their original shape after a large mechanical deformation. Vulcanization can be defined as a process that increases the retractive force and reduces the amount of permanent deformation remaining after removal of the deforming force. Thus, vulcanization increases elasticity and decreases plasticity. It is generally accomplished by the formation of a cross-linked molecular network.

Important characteristics related to the vulcanization process are the time elapsed before cross-linking starts, the rate of cross-link formation, and the extent of cross-linking at the end of the process to complete the cure. There must be sufficient delay or scorch resistance (resistance to premature vulcanization) to permit mixing, shaping, forming, and flowing in the mold. Then the formation of cross-links should be rapid and the extent of cross-linking must be controlled.

Quite large elastic strains are possible with minimal stress in TPEs; these TPEs are the synthetic rubbers. TPEs have two specific characteristics: their glass transition temperature (T_g) tends to be below their usage temperature and their molecules are highly kinked, as in natural TS rubber (isoprene). When a stress is applied, the molecular chain uncoils and the end-to-end length can be extended several hundred percent, with minimum stresses. Some TPEs have an initial modulus of elasticity of less than 10 MPa (1500 psi); once the molecules are extended, the modulus increases.

The modulus of metals decreases with an increase in temperature. The opposite is true of stretched TPEs because at higher temperatures there is increasingly vigorous thermal agitation in their molecules. Therefore, the molecules more strongly resist the tension forces attempting to uncoil them. To resist requires greater stress per unit of strain, so the modulus

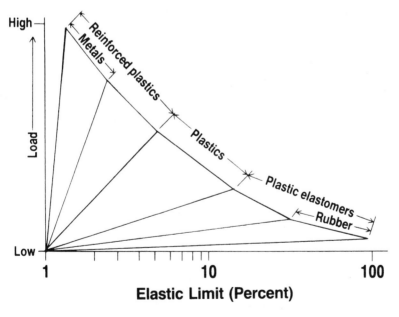

Figure 1.20 Strength and elasticity of different materials.

increases with temperature. When stretched into molecular alignment many rubbers can form crystals, an impossibility when they are relaxed and kinked.

To date, with the exception of vehicle tires, TPEs have been replacing TS rubbers in virtually all applications. Unlike natural TS rubbers, most TPEs can be reground and reused, thereby reducing overall cost. The need to cure or vulcanize them is eliminated, reducing cycle times, and parts can be molded to tighter tolerances. Most TPEs can be colored, whereas natural rubber is available only in black. TPEs also weigh 10–40% less than rubbers.

TPEs range in hardness from as low as 25 Shore A up to 82 Shore D (ASTM test). They span a temperature range of −34 to 177 °C (−29 to 350 °F), dampen vibration, reduce noise, and absorb shock (Fig. 1.20). However, designing with TPEs requires care; unlike TS rubber, which is isotropic, TPEs tend to be anisotropic during processing and injection molding. Tensile strengths in TPEs can vary as much as 30–40% with direction.

Copolymers

Polymer properties can be varied during polymerization. The basic chemical process is carried out at the resin company, during which the polymer is formed under the influence of heat, pressure, a catalyst, or a combination thereof, inside vessels or tubular systems called reactors. One special form of property variation involves the use of two or more different monomers as comonomers, copolymerizing them to produce copolymers (two comonomers) or terpolymers (three monomers). Their properties are usually intermediate between those of homopolymers, which may be made from the individual monomers, and sometimes superior or inferior to them. (A polymer such as polyethylene is formed from its monomer ethylene, polyvinyl chloride polymer from its vinyl chloride monomer, and so on.)

Compounds and additives

Since the first cellulosic was produced in 1868, there has been as ever growing demand for specially compounded plastics. Using a postreactor technique, resins can be compounded by alloying or blending polymers, using colorants, flame retardants, heat or light stabilizers, lubricants, and so on, and adding fillers and reinforcements, or a combination thereof. The resulting reinforced compounds are usually called reinforced plastics (RPs) or composites (Fig. 1.21).

In compounding, a polymer is modified physically and chemically by additives that change its properties. For example, the incorporation of glass fibers improves impact strength, and the incorporation of a pigment

Plastic Composition

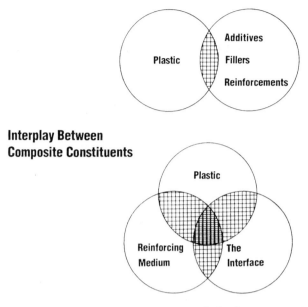

**Interplay Between
Composite Constituents**

Figure 1.21 Composition of plastics.

imparts color. The ways in which a polymer can be modified are manifold; consequently, a polymer is usually available from the compounder in a wide range of grades. These grades provide a plethora of material choices from which to make products.

Compounding also embraces the physical mixing of two or more polymers to form blends. Blends may be miscible, e.g., polyphenylene oxide in polystyrene, or immiscible, e.g., polyurethane in polyoxymethylene.

The importance of compounding to the plastics industry cannot be overstated. A particular goal can be reached by employing various polymers, additives, compounding processes, and a variety of routes. The compounding line may be arranged in several ways; it involves many operations, each of which can be executed on more than one specific piece of equipment. This offers numerous choices, the best of which is frequently the simplest.

The precise quantity of additive is very important to ensure the required properties. To this end, the ingredients must be precisely weighed before compounding, using scales in a batch operation, a weight-belt feeder or continuous weighing (or metering) for direct additions to a continuous process. The choice is not solely a technical matter, because the economics of the process, available equipment, and workloads must also be considered. Processing equipment includes nonintensive mixers, medium-intensity mixers, and intensive-fluxing mixers.

Additives are a diverse group of speciality materials. They may be added to plastic formulations or to the surfaces of finished products (Chapter 17). Sometimes they are added by the material supplier before the formulation is processed, sometimes they are added by the processor. Among the additives used as processing aids are blowing (or foaming) agents, organic peroxides, mold-release agents, and lubricants. Additives intended to improve properties of fabricated goods include antimicrobial, antioxidants, antistatic agents, colorants, flame retardants, impact modifiers, and ultraviolet light stabilizers. Some additives, such as plasticizers and heat stabilizers, enhance both processability of plastic resins and the properties of finished products made from them.

Others include antiblocking agents, antifoaming agents, antifogging agents, antistatic agents, biocides, coupling agents, curing agents, low-profile additives, mold-release agents, odorants, slip agents, urethane catalysts, viscosity-control agents.

Each additive can provide special characteristics. The primary role of plasticizers is to increase the flexibility, softness, and extensibility of inherently rigid thermoplastic resins, thermosetting resins, rubbers, coatings, and other compositions. Secondary benefits include improved melt processability, greater impact resistance, and depressed brittle point. The performance benefits obtained from plasticizer additives can be offset by unwanted effects, including reduced tensile strength, lowered heat distortion temperature, increased flammability, and the occurence of plasticizer volatilization and permanence problems.

Plasticizers can also function as vehicles for plastisols (liquid dispersions of resins which solidify upon heating) and as carriers for pigments and other plastics additives. Some offer synergistic heat and light stabilization or flame retardance.

Colorants

Plastics can be colored by dyes and inorganic and organic pigments. Dyes soluble under the conditions of use must be completely dissolved, leaving no color streaks and little or no haze. Pigments are insoluble and consist of particles that must be dispersed by physical means.

Dispersion of a dry pigment into a liquid vehicle or molten plastic takes place in two steps: the pigment agglomerates are first broken up into much smaller particles; the air is then displaced from the particle surfaces to obtain a complete pigment–vehicle interface.

TPs are completely polymerized, and are processed, i.e., colored, at or near their melting points. Scrap may be reground and remolded. TS resins are only partially polymerized when the colorants are incorporated. Polymerization is completed when the resin is processed. The result is a nonmeltable cross-linked plastic that cannot be reworked. For these plastics, it is customary to incorporate colorants into the less/reactive ingredient.

A reaction may occur between a plastic and a colorant at processing temperatures. More often allowance must be made for thermal stability and additives, such as antioxidants, flame retardants, ultraviolet light absorbers, and fillers. The final use of the resin often indicates the selection of colorant. Colorants need to be evaluated for later processing conditions in view of the subsequent fate of the fabricated parts.

Alloys and blends

Alloys are combinations of polymers that are mechanically blended. They do not depend on chemical bonds, but they often require special compatibilizers. Plastic alloys are usually designed to retain the best characteristics of each constituent. Most often the property improvements are in areas such as impact strength, weather resistance, improved low-

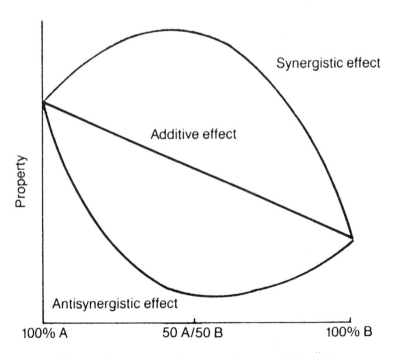

Figure 1.22 Compounding aims for synergistic effects.

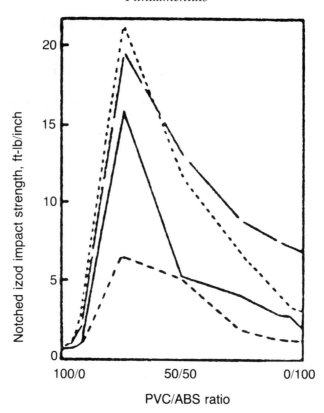

Figure 1.23 How alloying may produce synergistic gains. The curves in this graph reflect four different poly blends.

temperature performance, and flame retardation (Figs 1.22 to 1.25 and Tables 1.24 to 1.26)

The classic objective of alloying and blending is to find two or more polymers whose mixture will have synergistic property improvements beyond those that are purely additive in effect (Figs 1.22 and 1.23). Among the techniques used to combine dissimilar polymers are cross-linking, to form what are called interpenetrating networks (IPNs), and grafting, to improve the compatibility of the resins. Grafting two dissimilar plastics often involves a third plastic whose function is to improve the compatibility of the principal components. This compatibilizer material is a grafted copolymer that consists of one of the principal components and is similar to the other component. The mechanism is similar to having soap improve the solubility of a greasy substance in water. The soap contains components that are compatible with both the grease and the water.

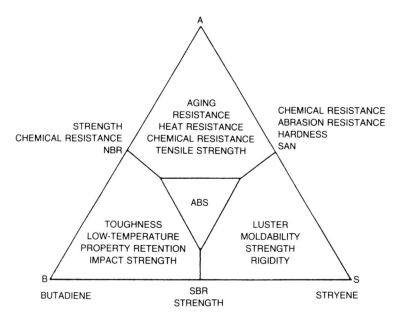

ACRYLONITRILE

A

AGING
RESISTANCE
HEAT RESISTANCE
CHEMICAL RESISTANCE
TENSILE STRENGTH

STRENGTH
CHEMICAL RESISTANCE
NBR

CHEMICAL RESISTANCE
ABRASION RESISTANCE
HARDNESS
SAN

ABS

TOUGHNESS
LOW-TEMPERATURE
PROPERTY RETENTION
IMPACT STRENGTH

LUSTER
MOLDABILITY
STRENGTH
RIGIDITY

B

S

BUTADIENE

SBR
STRENGTH

STRYENE

Figure 1.24 ABS terpolymer properties may be influenced by individual constituent polymer properties.

	☐ Unmodified resin	▨ Alloy	
Plastic	Cost index	Yield strength index	Impact strength index
Polypropylene			
Polystyrene Impact styrene (alloy)			
ABS ABS/PVC (alloy) ABS/Polycarbonate (alloy)			
Rigid PVC PVC/acrylic (alloy)			
Polyphenyleneoxide (Noryl)			
Polycarbonate			
Polysulfone Polysulfone/ABS (alloy)			

100   500   100   200   100   450 1250   3000

Figure 1.25 Different plastics can be combined to provide cost-to-performance improvements.

Table 1.24 Upgrading PVC by alloying and blending

Upgraded property	Blending polymer
Impact resistance	ABS, methacyrylate-butadiene-styrene, acrylics, polycaprolactone, polyimide, polyurethanes, PVC-ethyl acrylate
Tensile strength	ABS, methacyrylate-butadiene-styrene, polyurethanes, ethylene-vinyl acetate
Low-temperature toughness	Styrene-acrylonitrile, polyurethanes, polyethylene, chlorinated polyethylene, copolyester
Dimensional stability	Styrene-acrylonitrile, methacrylate-butadiene-styrene
Heat-distortion temperature	ABS, methacyrylate-butadiene-styrene, polyimide, polydimethyl siloxane
Processability	Styrene-acrylonitrile, methacrylate-butadiene-styrene, chlorinated polyethylene, PVC-ethyl acrylate, ethylene-vinyl acetate, chlorinated polyoxymethylenes (acetals)
Moldability	Acrylics, polycaprolactone
Plasticization	Polycaprolactone, polyurethanes, nitrile rubber, ethylene-vinyl acetate, copolyester, chlorinated polyoxymethylenes (acetals)
Transparency	Acrylics, polymide
Chemical/oil resistance	Acrylics
Toughness	Nitrile rubber, ethylene-vinyl acetate
Adhesion	Ethylene-vinyl acetate

Table 1.25 Outstanding properties of some commercial plastic alloys

Alloy	Properties
PVC/acrylic	Flame, impact, and chemical resistance
PVC/ABS	Flame resistance, impact resistance, processability
Polycarbonate/ABS	Notched impact resistance, hardness, heat-distortion temperature
ABS/polysulfone	Lower cost
Polypropylene/ethylene-propylene-diene	Low-temperature impact resistance and flexibility
Polyphenylene oxide/polystyrene	Processability, lower cost
Styrene acrylonitrile/olefin	Weatherability
Nylon/elastomer	Notched Izod impact resistance
Polybutylene terephthalate/polyethylene terephthalate	Lower cost
Polyphenylene sulfide/nylon	Lubricity
Acrylic/polybutylene rubber	Clarity, impact resistance

Table 1.26 Plastic blends with advantages and applications

System	Property advantages	Applications
PVC-ABS	Better processibility and toughness than PVC, better fire retardance than ABS	Mass-transit interiors, appliance housings
PVC-acrylic	Impact-modified, similar to PVC-ABS	Mass-transit interiors, appliance housings
PVC-chlorinated PE	Better impact than PVC	Pipe and siding
PC-ABS	Better toughness and HDT than ABS, better processibility and lower cost than PC	Appliance and business-machine housings, automotive components
PSF-ABS	Similar to PC-ABS, composition can be electroplated, lower cost than polysulfone (PSF)	Plumbing fixtures, food-service trays, appliances
PC-PE	Better flow and energy absorption than PC	Automotive applications
PC-PET	Better chemical resistance and processibility and lower cost than PC	Tubing, auto bumpers, business-machine housings
PC-PBT	Better solvent resistance and processibility than PC	Tubing, auto bumpers, business-machine housings
PET-PMMA	Lower cost than PMMA, lower warp and shrink than PET	Electrical and electronic applications
PC-SMA	Impact-modified, better toughness and ductility than SMA, better retention of properties upon aging at high temperature and lower cost than PC	Automotive applications, appliances, cookware
PP-EPDM	Better impact and toughness than PP	Wire and cable insulation, auto bumpers, hose and gaskets
PE-ethylene copolymers	Better chemical resistance, impact, and toughness than PE	Film
Nylon-ethylene copolymers	Better toughness and impact	Transport containers, sports equipment

Reactive polymers

A reactive polymer is simply a device to alloy different materials by changing their molecular structure inside a compounding machine. True reactive alloying induces an interaction between different phases of an incompatible mixture and assures the stability of the mixture's morphology. The concept is not new; this technology is now capable of producing thousands of new compounds to meet specific design requirements. The relatively low capital investment associated with compounding machinery (usually less than $1 million for a line, compared with many millions for a conventional reactor), coupled with a processing need for small amounts of tailored materials, now allows small and mid-sized compounding companies to take advantage of it.

A variety of reactive alloying techniques are now available to the compounder. They typically involve the use of a reactive agent or compatibilizer to bring about a molecular change in one or more of the blend's components, thereby facilitating bonding. They include the grafting process mentioned earlier and copolymerization interactions, whereby a functional material is built into the polymer chain of a blend component as a comonomer, with the resultant copolymer then used as a compatibilizer in ternary bonds, such as a PP–AA copolymer that bonds polypropylene and acrylic acid. Another technique is to use solvent-based interactions of materials such as polycaprolactone, which is miscible in many materials and exhibits strong polarity, as well as hydrogen bonding, using the simple polarity of alloy components.

The manufacture of products made from plastics traditionally involved two operations: reaction and processing. Polymerization reactors made monomer molecules into polymer (plastic) molecules. Processing equipment fabricated the plastic molecules into shaped products.

Reactive processing combines these two operations by conducting polymerization and polymer (plastic) modification reactions in processing equipment. This type of processing can be done by reactive extrusion (REX) and injection molding (Chapter 3).

Reinforced plastics and composites

Reinforced plastics (RPs) or composites hold a special place in the design and manufacturing industry because they are quite simply unique materials (Fig. 1.26). During the 1940s reinforced plastics (or low-pressure laminates, as they were then commonly known) were easy to identify. The basic definition then, as now, is simply that of a plastic reinforced with either a fibrous or nonfibrous material.

What essentially characterizes RPs is their ability to be molded into extremely small but also large shapes well beyond the basic capabilities of

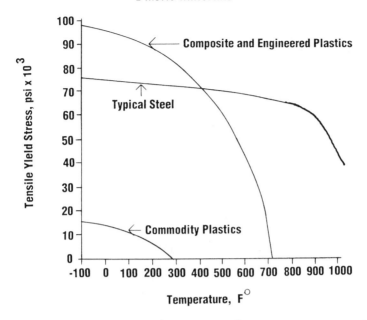

Figure 1.26 Guide to maximum short-term tensile stress versus temperature.

other processes, at little or no pressure. Also, there are instances in which less heat is required (Chapter 12). Consequently, RPs went by the name **low-pressure laminates**.

In the past, the term **high-pressure laminates** was reserved for TS melamine- and phenolic-impregnated papers or fabrics compressed under high pressures (about 13.8–34.5 MPa, 2000–5000 psi) and heated to form either decorative laminates (e.g., Formica and Micarta) or industrial laminates for electrical and other industries.

By the early 1960s, the processing of RPs had begun to involve higher pressures, and 'low-pressure laminates' was dropped in favor of RPs. But even then, the name referred primarily to reinforced TSs and encompassed specialized RP molding processes.

By 1970 major changes had occurred. Reinforcements other than glass fiber were in use and TPs as well as TSs were being reinforced in volume. The application of RTS and RTP methods of processing began to increase, using conventional processing techniques like injection molding and rotational molding. By this time the industry required a more inclusive term to describe RPs, so **composite** was added. (For some of the different composites that exist see Table 1.27). The fiber reinforcements included higher-modulus glasses, carbon, graphite, boron, and aramid (Du Pont's Kevlar aramid is the strongest synthetic fiber in the world, five times as

Table 1.27 Comparison of theoretically possible and actual values of fibers (1944 data)

Type of material	Modulus of elasticity			Tensile strength		
	Theoretical, $N\,mm^{-2}$ (kpsi)	Fiber, $N\,mm^{-2}$ (kpsi)	Nonfiber, $N\,mm^{-2}$ (kpsi)	Theoretical, $N\,mm^{-2}$ (kpsi)	Fiber, $N\,mm^{-2}$ (kpsi)	Nonfiber, $N\,mm^{-2}$ (kpsi)
Polyethylene	300 000 (43 500)	100 000 (33%)[a] (14 500)	1000 (0.33%) (145)	27 000 (3900)	1500 (5.5%) (218)	30 (0.1%) (4.4)
Polypropylene	50 000 (7250)	20 000 (40%) (2900)	1600 (3.2%) (232)	16 000 (2300)	1300 (8.1%) (189)	38 (0.24%) (5.5)
Polyamide-66	160 000 (23 200)	5000 (3%) (725)	2000 (13%) (290)	27 000 (3900)	1700 (6.3%) (246)	50 (0.18%) (7.2)
Glass	80 000 (11 600)	80 000 (100%) (11 600)	70 000 (87.5%) (10 100)	11 000 (1600)	4000 (36%) (580)	55 (0.5%) (8.0)
Steel	210 000 (30 400)	210 000 (100%) (30 400)	210 000 (100%) (30 400)	21 000 (3050)	4000 (19%) (580)	1400 (6.67%) (203)
Aluminum	76 000 (11 000)	76 000 (100%) (11 000)	76 000 (100%) (11 000)	7600 (1100)	800 (10.5%) (116)	600 (7.89%) (87)

[a] For an experimental value the percentage of the theoretically calculated value is given in parentheses, e.g. (33%).

strong as steel on an equal-weight basis). Plastics include use of the heat-resistant TPs: the polimides, polyamide-imide, and so on [3, 4].

Commodity and engineering plastics

About 90% of plastics (by weight) can be classified as commodity resins; the others are engineering plastics. The five commodities of LDPE, HDPE, PP, PVC, and PS account for about two-thirds of all the resins consumed. The engineering resins – nylon, PC, acetal, and so on – are characterized by improved performance in higher mechanical properties, better heat resistance, higher impact strength, and so forth. Thus, they demand a higher price. There are commodity resins with certain reinforcements and/or alloys with other resins that put them into the engineering category. Many TSs are engineering resins.

Multilayer materials

All materials (plastics, metals, glass, wood, paper, etc.) have certain strengths and weaknesses, advantages and disadvantages. Two or more materials can often be layered and combined to overcome weaknesses economically. Multilayer materials, made via coinjection, coextrusion, blow molding and other processes (Chapters 2, 3, 4), provide better strength and improve on other properties of any single plastic. Improvements may be gas barriers, clarity, use of recycled plastics, product compatibility, and so on; costs may also be reduced. Polyethylene (PE) or polypropylene (PP) coextrusions produce lower-cost products for food contact, excellent barriers to water vapor, barriers to oxygen, etc. Polyethylene vinyl alcohol is a relatively high-cost material that provides an excellent oxygen barrier but is very sensitive to water, which can deteriorate its properties. The water problem is eliminated by a thin layer between the layers of PE or the layers of PP; this construction has been used for many decades in different applications.

A common problem with certain multilayers is that different layers will not adhere (stick) to each other. A thin layer of 'adhesive' is therefore used to create the bond. Multilayers offer a wide latitude for material selection and also allow the use of recycled materials. An example of a coextruded product is a seven-layer gas tank containing PE as the structural material and other layers to prevent the emission of gasoline fumes, etc.

Melt flow and rheology

Rheology is the science that deals with the deformation and flow of matter under various conditions (such as plastic melt flow). The rheology of plastics, particularly TPs, is complex but manageable. These materials combine the properties of an ideal viscous liquid (pure shear deformations) with those of an ideal elastic solid (pure elastic deformation). Plastics are therefore said to be viscoelastic. The mechanical behavior of plastics is dominated by viscoelastic phenomena such as tensile strength, elongation at breaks, and rupture energy, which are often the controlling factors. The viscous attributes of polymer melt flow are also important considerations in plastics processing and fabrication.

Viscoelasticity

The flow of plastics is compared to that of water in Fig. 1.27 to show their different behaviors. With plastics there are two types of deformation or flow: in viscous flow the energy causing the deformation is dissipated, in elastic flow that energy is stored. The combination produces viscoelastic plastics. Viscosity is a material's resistance to viscous deformation (flow).

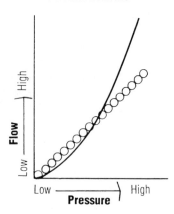

Figure 1.27 Rheology and flow properties of plastics (solid line).

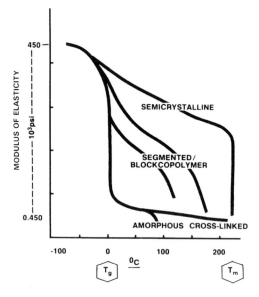

Figure 1.28 Modulus of elasticity is a dynamic mechanical property.

The resistance to elastic deformation is the modulus of elasticity (E); its range for a plastic melt is 1000–7000 kPa (145–1015 psi), called the rubbery range (Figs 1.28 and 1.29).

Not only are there two classes of deformation; there are also two modes in which deformation can be produced: simple shear and simple tension. The actual action during melting, as in a screw plasticator, is extremely complex, with all types of shear–tension combinations. Together with engineering design, deformation determines the pumping efficiency of a

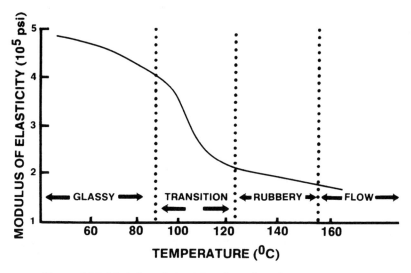

Figure 1.29 Modulus of elasticity plotted against temperature.

screw plasticator and controls the relationship between output rate and pressure drop through a die system or into a mold.

Shear rate

When a melt moves in a direction parallel to a fixed surface, such as a screw barrel, mold runner, or die wall, it is subject to a shearing force. As the screw speed increases, so does the shear rate, with potential advantages and disadvantages. Advantages of an increased shear rate are a less viscous melt and easier flow. This shear-thinning action is required to 'move' plastic. When water (a Newtonian liquid) is in an open-ended pipe, pressure can be applied to move it and by doubling the water pressure, the flow rate of the water is doubled. Water does not have a shear-thinning action. However, in a similar situation but using a plastic melt (a non-Newtonian liquid), if the pressure is doubled, the melt flow may increase from 2 to 15 times, depending on the plastic used. As an example, linear low-density polyethylene (LLDPE), with a low shear-thinning action, experiences a very low rate increase, which explains why it can cause more processing problems compared to other PEs in certain equipment. The higher-flow melts include polyvinyl chloride (PVC) and polystyrene (PS).

A disadvantage observed with the higher shear rates is that too high a heat increase may occur, potentially causing problems in cooling, as well as degradation and discoloration. A high shear rate can lead to a rough product surface (melt fracture, etc.). For each plastic and every processing condition, there is a maximum shear rate beyond which such problems can develop.

Shear in the channel of the screw is equal to $\pi DN/60h$ (where D = average barrel inside diameter, N = screw RPM, and h = average screw channel depth). This formula does not include the melt slippage between the barrel wall and screw surfaces, but the shear rate obtained is still useful for purposes of comparison. A $2\frac{1}{2}$ in. (63.5 mm) screw with a 0.140 in. (3.6 mm) channel rotating at 100 RPM produces a shear rate of $93.5 s^{-1}$. This value is approximately the desired value in most extrusion processes, with $100 s^{-1}$ generally the target.

The same formula can be used to determine the shear rate of the slippage between the barrel and screw. With a new barrel, which usually has a small clearance of 0.005 in. (0.127 mm), a very high shear rate of about $2618 s^{-1}$ can exist. With this small clearance only a small amount of melt is subject to the higher heat, so that any overheating is overcome by the mass of melt it encounters (mixes with). As the screw wears, more melt flows through enlarged clearances, but the shear rate is lower. The effect of wear on overheating is usually very small and is not the main reason why the complete melt overheats.

Shear rates can also be determined in melt flow through mold cavities and particularly in extrusion dies. The formulas applicable to the different die shapes seldom account for slippage of melt on the die surfaces, but they can be used to compare the processability of melts and to control melt flow. The formula for a die extruding a rod is $4Q/\pi R^3$, for a long slit it is $6Q/wh^2$, and for an annulus die it is $6Q/\pi Rh^2$ (where Q = volumetric flow rate, R = radius, w = width, and h = die gap).

Molecular weight distribution

Plastics consist of molecules arranged in long flexible chains which become entangled with each other. These entanglements are largely responsible for high viscosity in melts. Shear can be envisioned as sliding molecules in rotation; this rotation causes the chains to disentangle. At low shear, molecular chains become entangled; as the shear rate increases, they gradually disentangle, and the viscosity is reduced. The result, expressed as a so-called flow curve (Fig. 1.27), is related to the processability of the plastic material.

One method of defining plastics uses their molecular weight (MW), a reference to the plastic molecules' weight or size. Here MW refers to the average weight of a plastic, which is always composed of molecules having different weights. These differences are important to the processor, who uses the molecular weight distribution (MWD) to evaluate materials. A narrow MWD enhances the performance of plastic products (see later). Melt flow rates depend on the MWD, as illustrated in Fig. 1.30.

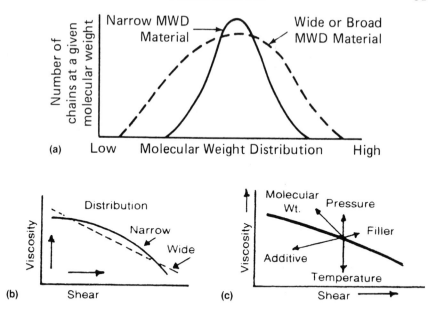

Figure 1.30 (a) Molecular weight distribution (MWD) curves; (b) viscosity versus shear rate as related to MWD; (c) factors influencing viscosity.

Elasticity

As a melt is subjected to a fixed stress (or strain), the deformation – time curve will show an initial rapid deformation followed by a continuous flow (Fig. 1.31). The relative importance of elasticity (deformation) and viscosity (flow) depends on the timescale of the deformation. For a short time, elasticity dominates; over a long time, the flow becomes purely viscous. This behavior influences processes. When a part is annealed, it will change its shape; or swelling occurs with postextrusion (Chapter 10). Deformation contributes significantly to process flow defects. Melts with small deformation have proportional stress–strain behavior. As the stress on a melt is increased, the recoverable strain tends to reach a limiting value. It is in the high-stress range, near the elastic limit, that processes operate.

Molecular weight, temperature, and pressure have little effect on elasticity; the main controlling factor is MWD. Practical elasticity phenomena often exhibit little concern for the actual value of the modulus and the viscosity. Although the modulus is influenced only slightly by MW and temperature, these parameters have a great effect on viscosity and can therefore alter the balance of a process.

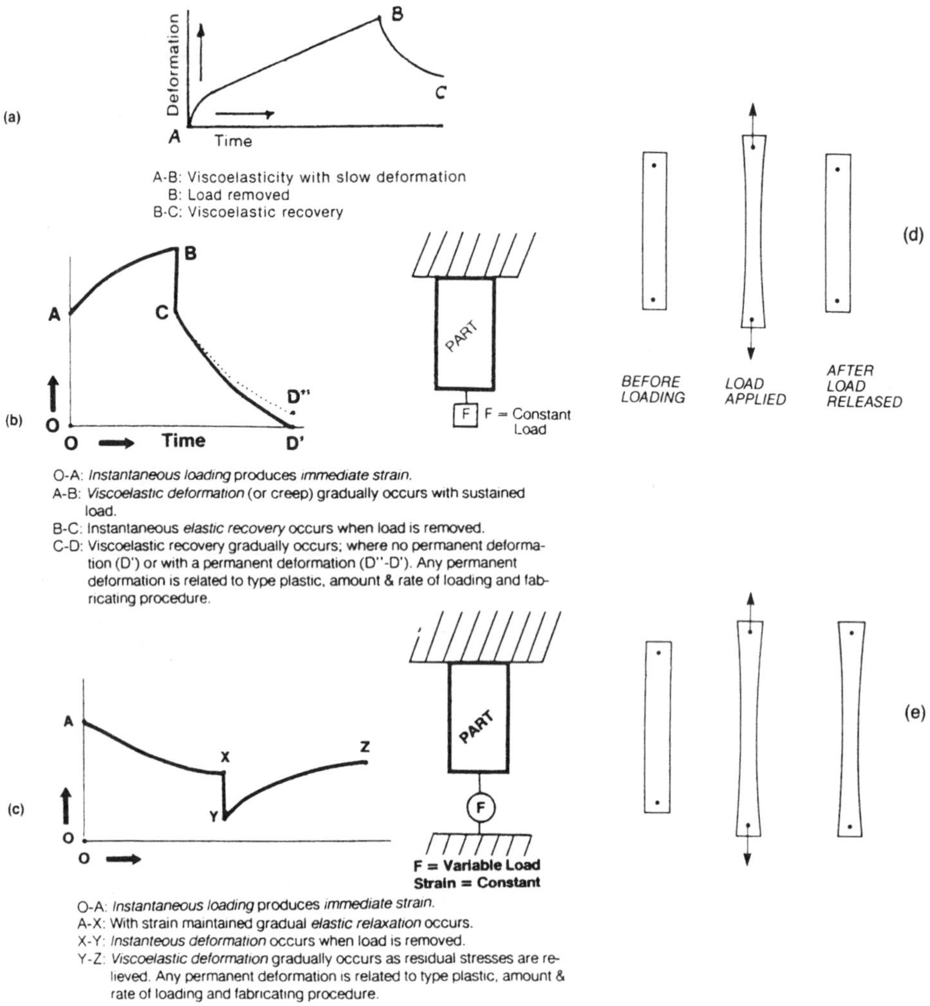
(a)
Deformation
B
C
A
Time
A-B: Viscoelasticity with slow deformation
B: Load removed
B-C: Viscoelastic recovery
(b)
B
A
C
D''
O
O
Time
D'
PART
F
F = Constant Load
O-A: Instantaneous loading produces immediate strain.
A-B: Viscoelastic deformation (or creep) gradually occurs with sustained load.
B-C: Instantaneous elastic recovery occurs when load is removed.
C-D: Viscoelastic recovery gradually occurs; where no permanent deformation (D') or with a permanent deformation (D''-D'). Any permanent deformation is related to type plastic, amount & rate of loading and fabricating procedure.
(c)
A
X
Z
Y
O
O
PART
F
F = Variable Load
Strain = Constant
O-A: Instantaneous loading produces immediate strain.
A-X: With strain maintained gradual elastic relaxation occurs.
X-Y: Instanteous deformation occurs when load is removed.
Y-Z: Viscoelastic deformation gradually occurs as residual stresses are relieved. Any permanent deformation is related to type plastic, amount & rate of loading and fabricating procedure.
(d)
BEFORE LOADING
LOAD APPLIED
AFTER LOAD RELEASED
(e)

Figure 1.31 Elasticity and strain: (a) basic deformation versus time curve; (b) stress–strain deformation versus time (the creep effect); (c) stress–strain deformation versus time (the stress-relaxation effect); (d) material exhibiting elasticity; and (e) material exhibiting plasticity.

Flow performance

In any practical deformation there are local stress concentrations. Should the viscosity increase with stress, the deformation at the stress concentration will be less rapid than in the surrounding material; the stress concentration will be smooth, and the deformation will be stable. However, when the viscosity decreases with increased stress, any stress concentration will cause catastrophic failure.

Flow defects

Flow defects, especially as they affect the appearance of a product, play an important role in many processes. Flow defects are not always undesirable, as, for example, in producing a matt finish. Six important types of defects can be identified, and are applied here to extrusion because of its relative simplicity. These flow analyses can be related to other processes; and even to the complex flow of injection molding.

Nonlaminar flow Ideally, a melt flows in a steady, streamlined pattern in and out or a die. In practice the extrudate is distorted, causing defects called melt fracture or elastic turbulence. To reduce or eliminate this problem, the entry to the die is tapered or streamlined.

Sharkskin During flow through a die, the melt next to the die tends not to move, whereas that in the center flows rapidly. When the melt leaves the die, its flow profile is abruptly changed to a uniform velocity. This change requires a rapid acceleration of the surface layer, resulting in a high local stress. If this stress exceeds some critical value, the surface breaks, giving a rough appearance (sharkskin). With the rapid acceleration, the deformation is primarily elastic. Thus the highest surface stress, and worst sharkskin, will occur in plastics with a high modulus and high viscosity, or in high molecular weight plastics of narrow MWD at low temperatures and high extrusion rates. The addition of die-lip heating, locally reducing the viscosity, is effective in reducing sharkskin.

Nonplastication This condition produces uneven stress distribution, with consequent lumpiness. The product could appear ugly or have a fine matt finish. With a wide MWD there could be a lack of gloss.

Volatiles Many plastics contain small quantities of material that boil at processing temperatures; or they may be contaminated by water absorbed from the atmosphere. These volatiles may cause bubbles, a scarred surface, and other defects. Chapters 2 and 3 describe methods for removing volatiles (vented barrels and dryers).

Shrinkages The transition from room temperature to a high processing temperature may decrease a plastic's density up to 25%. Cooling causes

possible shrinkage (up to 3%) and may cause surface distortions or voiding with internal frozen strains. As reviewed in other chapters, this situation can be reduced or eliminated by special techniques, such as cooling
under pressure in the injection molding process.

Melt structures High shear at a temperature not far above the melting point
may cause a melt to take on too much molecular order. In turn, distortion
could result. This subject is discussed further in this and other chapters.

Thermodynamics

With the heat exchange that occurs during processing, thermodynamics
becomes important. It is the high heat content of melts (about $400\,J\,g^{-1}$)
combined with the low rate of thermal diffusion ($10^{-3}\,cm^2\,s^{-1}$) that
limits the cycle time of many processes. Also important are density
changes, which, for crystalline plastics, may exceed 25% as melts
cool. Melts are highly compressible; a 10% volume change for 10 000 psi
(70 MPa) is typical. Surface tension of about $20\,g\,cm^{-1}$ may be typical
for film and fiber processing when there is a large surface-to-volume
ratio.

Chemical changes

The chemical changes which can occur during processing include
(1) polymerization and cross-linking, which increases viscosity;
(2) depolymerization or damaging of molecules, which reduces viscosity; and (3) complete changes in the chemical structure, which may
cause color changes. Already degraded plastics may catalyze further
degradation.

Trends

Because melts have many different properties and there are many ways to
control processes, detailed and factual predictions of final output are
difficult. Research and hands-on operation have been directed mainly at
explaining the behavior of melts or plastics. Modern equipment and controls are overcoming some of the unpredictability. Processes and equipment should ideally be designed to take advantage of the novel properties
of plastics rather than to overcome them.

Thermal properties

In order to select materials that will maintain acceptable mechanical characteristics and dimensional stability, one must be aware of both the normal and extreme operating environments to which a product will be subjected, including the environment (conditions) during processing to manufacture the product. Processes are influenced by the thermal characteristics of plastics, such as melt temperature (T_m), glass transition temperature (T_g), thermal conductivity, thermal diffusivity, heat capacity, coefficient of linear thermal expansion and decomposition temperature (T_d) (Tables 1.28 and 1.29). All these properties relate to the selection of the optimum processing conditions. There is a maximum processing temperature – or to be more precise, a maximum time-to-temperature cycle – for all materials prior to their decomposition or destruction. Thermal properties change according to the types of additives in plastics (metal powders, glass fibers, calcium carbonate, etc.).

Most plastics are processed and fabricated at elevated temperatures, often as a melt, necessitating the transfer of substantial quantities of heat first to the material to reach the desired temperatures and then to return the formed or cured product to a solidified state at room temperature. Because of the low thermal conductivities and thermal diffusivities of plastics in general, this process can be both slow and expensive, especially when large masses are involved. The heaters and heat-exchange equipment must therefore be carefully designed to ensure efficient and effective heat transfer, particularly if phase changes take place.

After formation or curing processes, most plastic products are cooled from relatively high temperatures, accompanied by shrinkage and dimensional changes. These effects must be carefully considered during the design stage of processing equipment and can become ever more critical with multiphase products, such as filled plastics or composite systems. And accurate thermal property data are required for modeling processing steps such as molding or extrusion.

Other more subtle heating and cooling effects may also play an important role. For example, when a mass of crystallizable plastic is cooled from its melt, the inner regions of the mass tend to cool more slowly than the other surfaces because of a slow overall heat-transfer process. This leads to a more highly crystalline, and hence brittle, inner core. This cooling-induced morphology may be beneficial or detrimental, depending on the use of the product. Thus, an intimate relationship exists between thermal properties and energy or dimensional considerations on the one hand and equipment design and fabrication techniques on the other. Ultimately, process optimization rests on these relationships.

Table 1.28 Examples of thermal properties of TPs (properties of some common materials included for comparison)

Plastic and (morphology)[a]		Density, $g\,cm^{-3}$ ($lb\,ft^{-3}$)	Melt temperature (T_m), °C (°F)	Glass transition temperature (T_g), °C (°F)	Thermal conductivity, $10^{-4}\,cal\,s^{-1}\,cm^{-1}\,°C^{-1}$ ($Btu\,lb^{-1}\,°F^{-1}$)	Heat capacity, $cal\,g^{-1}\,°C^{-1}$ ($Btu\,lb^{-1}\,°F^{-1}$)	Thermal diffusivity, $10^{-4}\,cm^2\,s^{-1}$ ($10^{-3}\,ft^2\,h^{-1}$)	Thermal expansion, $10^{-6}\,°C^{-1}$ ($10^{-6}\,°F^{-1}$)
PP	(C)	0.9 (56)	168 (334)	5 (41)	2.8 (0.068)	0.9 (0.004)	3.5 (1.36)	81 (45)
HDPE	(C)	0.96 (60)	134 (273)	−110 (−166)	12 (0.290)	0.9 (0.004)	13.9 (5.4)	59 (33)
PTFE	(C)	2.2 (137)	330 (626)	−115 (−175)	6 (0.145)	0.3 (0.001)	9.1 (3.53)	70 (39)
PA	(C)	1.13 (71)	260 (500)	50 (122)	5.8 (0.140)	0.075 (0.003)	6.8 (2.64)	80 (44)
PET	(C)	1.35 (84)	250 (490)	70 (158)	3.6 (0.087)	0.45 (0.002)	5.9 (2.29)	65 (36)
ABS	(A)	1.05 (66)	105 (221)	102 (215)	3 (0.073)	0.5 (0.002)	3.8 (1.47)	60 (33)
PS	(A)	1.05 (66)	100 (212)	90 (194)	3 (0.073)	0.5 (0.002)	5.7 (2.2)	50 (28)
PMMA	(A)	1.20 (75)	95 (203)	100 (212)	6 (0.145)	0.56 (0.002)	8.9 (3.45)	50 (28)
PC	(A)	1.20 (75)	266 (510)	150 (300)	4.7 (0.114)	0.5 (0.002)	7.8 (3.0)	68 (38)
PVC	(A)	1.35 (84)	199 (390)	90 (194)	5 (0.121)	0.6 (0.002)	6.2 (2.4)	50 (128)
Aluminum		2.68 (167)	1000		3000 (72.5)	0.23	4900 (1900)	19 (10.6)
Copper/bronze		8.8 (549)	1800		4500 (109)	0.09	5700 (2200)	18 (10)
Steel		7.9 (493)	2750		800 (21.3)	0.11	1000 (338)	11 (6.1)
Maple wood		0.45 (28.1)	400 (burns)		3 (0.073)	0.25	27 (10.5)	60 (33)
Zinc alloy		6.7 (418)	800		2500 (60.4)	0.10	3700 (1430)	27 (15)

[a] C = crystalline resin, A = amorphous resin.

Table 1.29 General properties of a few plastics

Polymer	Shear modulus G' (GPa)	Young's modulus E' (GPa)	Tensile stress σ_{UT} (MPa)	Tensile strain ε_{UT} (%)	Poisson's ratio μ'	Specific heat C ($J\,kg^{-1}\,K^{-1}$)	Thermal conductivity λ at 2 K ($mW\,m^{-1}\,K^{-1}$)	Integral thermal expansion $\Delta L/L$ (%)
Epoxy resin	2.6	8.1	179	2.1	0.37	1.6	70	−1.2
Thermoplastic polymers								
HDPE	3.5	9.7	175	4.0	0.30	1.0	26	−1.8 to −2.2
PTFE	2.8	8.5	81	1.2				−1.6
PS	2.1	5.0	68	2.1	0.35	4–6	25	−1.5
PSU	1.9	6.0	150	3.0	0.37			−1.1
PVC	2.6	7.9	110				20	
PC	2.2	6.0	170	3.5	0.39		25	−1.4
POM	5.0	13.0	170	1.5	0.32	1.1	30	−1.5
Polyamide	2.55	7.5	190	2.2	0.30	1.6		−0.8
Polyimide	2.35	6.6			0.39		40	−0.8

Melt properties

The melt temperature (T_m) occurs at a relatively sharp point for crystalline materials. The amorphous materials do not have a T_m; they start melting as soon as the heat cycle begins. In reality there is no single T_m point, but a range. It is often taken as the peak of a DSC curve [3].

T_m depends on the processing pressure and time at heat, particularly during a slow temperature change for relatively thick melts. Also if T_m is too low, the melt's viscosity is high, and more power is required to process it. Degradation will occur if the viscosity is too high.

Glass-transition temperature

The glass-transition temperature (T_g) is the point below which plastic behaves like glass – it is brittle but very strong and rigid. Above this temperature it is neither as strong or rigid as glass, but neither is it brittle. At T_g the plastic's volume or length increases (Figs 1.32 and 1.33). The amorphous TPs have a more definite T_g.

A plastic's thermal properties, particularly its T_g, influence its processability in many different ways. The selection of a plastic should take these properties into account. A more expensive plastic could cost

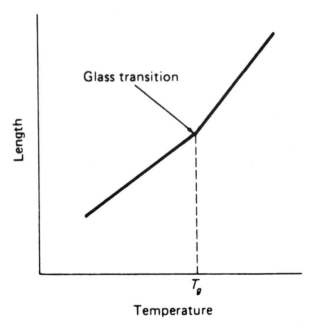

Figure 1.32 Effect of T_g on volume or length of plastics.

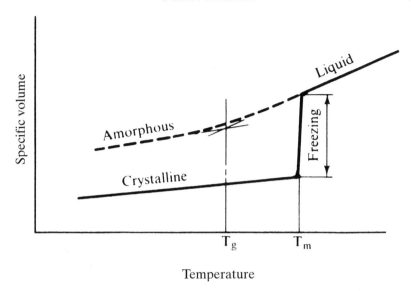

Figure 1.33 Solidification during processing of glassy/amorphous and crystalline TPs.

less to process because of its shorter processing time, requiring less energy for a particular weight.

The T_g is unique to amorphous TPs. It occurs at a specific temperature that depends on pressure and specific volume and is lower than the melting point. Designers should know that, above T_g, the mechanical properties are reduced. Most noticeable is a reduction in stiffness by a factor that may be as high as 1000. Therefore, the operating temperature of an amorphous TP is usually limited to below its T_g. Amorphous TPs generally have several transitions.

The glass transition generally occurs over a relatively narrow temperature span and is similar to the solidification of a liquid to a glassy state; it is not a phased transition. Not only do hardness and brittleness undergo rapid changes in this temperature region, but other properties, such as the coefficient of thermal expansion and specific heat, also change rapidly. This phenomenon has been called second-order transition, rubber transition, and rubbery transition. The world **transformation** has also been used instead of **transition**. When more than one amorphous transition occurs in a polymer, the transition associated with segmental motions of the polymer backbone chain, or accompanied by the largest change in properties, is usually considered to be the glass transition.

The glass-transition temperature can be determined readily only by observing the temperature at which a significant change takes place in a specific electrical, mechanical, or other physical property. Moreover, the observed temperature can vary significantly, depending on the specific property chosen for observation and on details of the experimental technique (e.g., the rate of heating, or frequency). Therefore, the observed T_g should be considered only as an estimate. The most reliable estimates are normally obtained from the loss peak observed in dynamic mechanical tests or from dilatometric data (ASTM D-20).

Mechanical properties and T_g

As can be seen from Table 1.28 (page 86), the value of T_g for a particular plastic is not necessarily a low temperature, which immediately helps explain some of the differences we observe in plastics. For example, because at room temperature polystyrene and acrylic are below their respective T_g values, we observe them in their glassy state. But at room temperature, natural rubber is above its T_g [$T_g = -75\,°C$ ($-103\,°F$); $T_m = 30\,°C$ ($86\,°F$)], with the result that it is very flexible. When it is cooled below its T_g, natural rubber becomes hard and brittle.

Dimensional stability

Dimensional stability is an important thermal property for the majority of plastics. It is the temperature above which plastics lose their dimensional stability. For most plastics the main determinant of dimensional stability is their T_g. Only with highly crystalline plastics is T_g not a limitation.

Substantially crystalline plastics in the range between T_g and T_m are described as leathery; this is because they are made up of a combination of rubbery noncrystalline regions and stiff crystalline areas. Plastics such as PE and PP are still useful at room temperature and nylon is useful in moderately elevated temperatures, even though they may be above their respective glass-transition temperatures.

Thermal conductivity and thermal insulation

Thermal conductivity is the rate at which a material will conduct heat energy along its length or through its thickness. ASTM tests give an indication of how much heat must be added to a unit mass of plastic in order to raise its temperature by $1\,°C$. This is an important factor, since plastics are often used as effective heat insulation in heat-generating applications and in structures where heat dissipation is important. The high degree of molecular order for crystalline TPs means their values tend to be twice those of the amorphous types.

The thermal conductivity of plastics depends on several variables and cannot be reported as a single factor. But it is possible to ascertain the two principal dependences, temperature and molecular orientation. In fact, molecular orientation may vary within a product, producing a variation in thermal conductivity. Thus, it is important for the designer to recognize such a situation. Certain products require skill to estimate a part's performance under steady-state heat flow, especially products made of composites. The method and repeatability of the processing technique can have a significant effect.

In general, thermal conductivity is low for plastics and their structures do not alter their values significantly. To increase it, the usual approach is to add metallic fillers, glass fibers, or electrically insulating fillers such as alumina. Thermal conductivity can be decreased by foaming.

Heat capacity

The heat capacity or specific heat of a unit mass of material is the amount of energy required to raise its temperature by 1 °C. It can be measured either at constant pressure or constant volume. If at constant pressure it is larger than at constant volume, because additional energy is required to bring about a volume change against external pressure. The specific heat of amorphous plastics increases with temperature in an approximately linear fashion below and above T_g, but a steplike change occurs near T_g. No such stepping occurs with crystalline types.

The heat capacity of plastics is usually reported during constant pressure heating. Plastics differ from traditional engineering materials because their specific heat is temperature sensitive.

Thermal diffusivity

Whereas heat capacity is a measure of energy, thermal diffusivity is a measure of the rate at which energy is transmitted through a given plastic. It relates directly to processability. In contrast, metals have values hundreds of times larger than those of plastics. Thermal diffusivity determines a plastic's rate of energy change with time. This function depends on thermal conductivity, specific heat at constant pressure, and density, all of which vary with temperature; nevertheless, thermal diffusivity is relatively constant.

Coefficient of linear thermal expansion

Like metals, plastics generally expand when heated and contract when cooled. For a given temperature change, TPs usually have a greater change than metals. The coefficient of linear thermal expansion (CLTE) is the ratio between the change of a linear dimension to the original dimension of the material per unit change in temperature (per ASTM standards). It is generally given as $°C^{-1}$ ($°F^{-1}$).

The CLTE is an important consideration if dissimilar materials like one plastic to another or a plastic to metal and so forth are to be assembled. The CLTE is influenced by the type of plastic (e.g., liquid crystal) and composite, particularly the glass-fiber content and its orientation. It is especially important if the temperature range includes a thermal transition such as T_g. All this activity with dimensional changes is normally available from material suppliers, allowing the designer to apply a logical approach and to understand what could happen.

The design of products has to take into account the dimensional changes that can occur during fabrication and during its useful service life. With a mismatched CLTE there could be destruction of plastics from factors such as cracking or buckling.

Expansion and contraction can be controlled in plastic by its orientation, cross-linking, adding fillers or reinforcements, and so on. With certain additives the CLTE value could be zero or almost zero. For example, plastic with a graphite filler contracts rather than expands during a temperature rise. Composites with only fiberglass reinforcement can be used to match those of metal and other materials. In fact, TSs are especially compounded to have little or no change.

In a TS the ease or difficulty of thermal expansion is dictated for the most part by the degree of cross-linking as well as the overall stiffness of the units between the cross-links. The less flexible units are also more resistant to thermal expansion. Influences such as secondary bonds have much less effect on the thermal expansion of TSs.

Any cross-linking has a substantial effect on TPs. It reduces expansion in amorphous TPs, but in a crystalline TP the decreased expansion as a result of cross-linking may be partially offset by a loss of crystallinity.

Thermal or residual stresses

Processing operations are usually carried out under highly nonisothermal conditions in which the melt passes through a complicated temperature history with solidification. Depending on the process, cooling rates can be as high as several thousand degrees per minute. During the initial stage of cooling, the surface layer cools so fast that the temperature difference between the surface layer and the core attains a maximal value. Then the core cools faster than the surface layer until uniform temperature is attained. During the initial stage of cooling, the contraction of the surface layer is greater than the contraction of the core. This introduces tensile stresses at the surface and compressive stresses in the core. Later the core contracts more than the surface, leading to tensile stresses in the core and compressive stresses in the surface.

Stresses that arise during rapid inhomogeneous cooling of the melt through the glass-transition temperature or melting point are called thermal or cooling stresses (stresses due to a change in temperature). In this case the thermal stresses are introduced into a material that was originally stress-free and externally unconstrained.

Thermal stresses also arise in externally constrained polymeric products subjected to a change in temperature (cooling or heating) even without the occurrence of a solidification or phase transition. Thermal stresses that arise in products are developed from a melt which is not originally stress-free. They occur in the presence of unrelaxed flow stresses. Accordingly, development of residual stresses in products during processing is due to the coupled effect of flow stresses and of thermal stresses arising from a temperature gradient and an external constraint at the boundary. Thus, the term **residual stresses** identifies the system of stresses that are in effect locked into a product, even without external forces acting on it.

It is well known that the presence of residual stresses greatly affects the mechanical properties and performance characteristics of products. Thus, understanding the factors governing the development of residual stress in products is of great importance to the design of plastic products. It is also important to be able to measure and predict the development of residual stress during polymer processing.

Degradation

Plastics degradation is a deleterious change in characteristics such as the chemical structure, physical and mechanical properties, and/or appearance of plastics. A degraded appearance usually means discoloration. Degradation can occur during processing. Factors that are important in determining the rate of degradation are (a) residence time, (b) stock (melt) temperature and distribution of stock temperature, (c) deformation rate and deformation rate distribution, (d) presence of oxygen or other degradation-promoting additive, and (e) presence of antioxidants and other stabilizers.

Orientation: defect or benefit?

Plastic molecular orientation can be accidental or deliberate. (Here, *accidental* refers to orientations that occur in processing plastics that may be acceptable. However, excessive frozen-in stresses can be extremely damaging if parts are subject to environmental stress cracking or crazing in the presence of chemicals, heat, etc.) The molecules are initially relaxed; molecules in amorphous regions are in random coils, and those in crystalline regions relatively straight and folded. During processing, the moelcules tend to be more oriented than relaxed, particularly when sheared, just as during injection molding and extrusion. After temperature–time–pressure is applied, and the melt goes through restrictions (molds, dies, etc.), molecules tend to be stretched and aligned in a parallel form. The result is a change in directional properties and dimensions. The amount of change depends on the type of TP, the amount of restriction, and most important, the rate of cooling. The faster the cooling rate, the greater the retention of frozen orientation. After processing, parts could be subject to stress relaxation, with changes in performance and dimensions. With certain plastics and processes there is an insignificant change. If changes are significant, one must take action to change the processing conditions, particularly increasing the cooling rate.

By deliberate stretching, the molecular chains of a plastic are drawn in the direction of the stretching, and inherent strengths of the chains are more nearly realized than they are in their naturally relaxed configuration. Stretching can take place with heat during or after processing (blow molding, extruding film, thermoforming, etc.). Products can be drawn in one direction (uniaxial) or in two opposite directions (biaxial), in which case many properties significantly increase uniaxially or biaxially (Table 1.30 and Fig. 1.34) [3].

Products with designed-in orientation are increasingly important in the polymer industry. They include filaments with uniaxial orientation and films and bottles with biaxial orientation. A variety of molding operations

Table 1.30 Effects of orientation on polypropylene films

Property	*Stretch (%)*				
	None	*200*	*400*	*600*	*900*
Tensile strength, psi	5600	8400	14000	22000	23000
(MPa)	(38.6)	(58.0)	(96.6)	(152.0)	(159.0)
Elongation at break (%)	500	250	115	40	40

Property[a]		*As cast*	*Uniaxial orientation*	*Balanced orientation*
Tensile strength, psi (MPa)				
MD		5700 (39.3)	8000 (55.2)	26000 (180)
TD		3200 (22.1)	40000 (276)	22000 (152)
Modulus of elasticity, psi (MPa)				
MD		96000 (660)	150000 (1030)	340000 (2350)
TD		98000 (680)	400000 (2760)	330000 (2280)
Elongation at break (%)				
MD		425	300	80
TD		300	40	65

[a] MD = machine direction, TD = transverse direction and direction of uniaxial orientation.

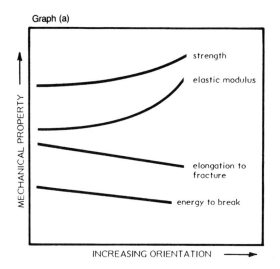

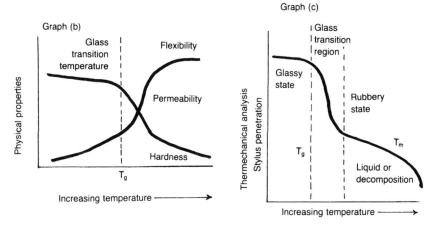

Figure 1.34 Effect of orientation on the properties of plastics.

have been developed in which multiaxial orientation is built in during the process. Uniaxial orientation produces high tensile modulus and high tensile strength in one direction, whereas equal biaxial orientation produces equal properties in all directions in the film plane. Biaxial orientation in bottles balances the mechanical properties between the bottle-axis direction and the hoop direction.

Efforts toward orientation processing has led to new molding technology. A process for extruding, reheating, and inflating tubes has produced orientation. In an injection molding process the core of an annular mold

Table 1.31 Examples of shrink film properties

Film type	Tensile strength, psi (MPa)	Elongation (%)	Tear strength, gf mil⁻¹ (mN m⁻¹)	Maximum shrink (%)	Shrink tension, psi (MPa)	Film shrink temperature range, °F (°C)
Polyethylene (low density)	9000 (62)	130	8 (3.1)	85	250–400 (1.7–2.8)	150–250 (65–120)
Polyethylene, (low density, irradiated)	8000–13000 (55–90)	115	5–10 (1.9–3.9)	80	400 (2.8)	170–250 (75–120)
Polyethylene (copolymer)	19000 (131)	130	7 (2.7)	60	450 (3.1)	180–260 (85–125)
Polypropylene	26000 (179)	50–100	5 (1.9)	80	600 (4.1)	250–330 (120–165)
Polyester	30000 (207)	130	10–60 (3.9–23.2)	55	700–1500 (4.8–10.3)	170–300 (75–150)
Poly(vinyl chloride)	9000–14000 (62–97)	140	variable	60	150–300 (1–2.1)	150–300 (65–150)

rotates to produce a helically oriented molded part [3]. The Du Pont work on biaxially oriented polyethylene terephthalate (PET) involved the molding of an isotropic glassy parison that was heated and inflated with air in a mold to form a biaxially oriented bottle. This led to a new technology called stretch blow molding [2].

Table 1.32 Specific gravity and density comparisons of different materials

Materials	Specific gravity[a]	Density (lb in.$^{-3}$)
Thermoplastics		
ABS	1.06	0.0383
Acetal	1.43	0.0516
Acrylic	1.19	0.0430
Cellulose acetate	1.27	0.0458
Cellulose acetate butyrate	1.19	0.0430
Cellulose propionate	1.21	0.0437
Ethyl cellulose	1.10	0.0397
Methyl methacrylate	1.20	0.0433
Nylon, glass-filled	1.40	0.0505
Nylon	1.12	0.0404
Polycarbonate	1.20	0.0433
Polyethylene	0.94	0.0339
Polypropylene	0.90	0.0325
Polybutylene	0.91	0.0329
Polystyrene	1.07	0.0386
Polyimides	1.43	0.0516
PVC, rigid	1.20	0.0433
Polyester	1.31	0.0473
Thermosets		
Alkyds, glass-filled	2.10	0.0758
Phenolic, GP	1.40	0.0505
Polyester, glass-filled	2.00	0.0722
Rubber	1.25	0.0451
Metals		
Aluminum SAE-309 (360)	2.64	0.0953
Brass, Yellow (#403)	8.50	0.3070
Steel, CR Alloy (Strip & Bar)	7.85	0.2830
Steel, Stainless 304	7.92	0.2860
Magnesium AZ-91B	1.81	0.0653
Iron, Pig, Basic	7.10	0.2560
Zinc, SAE-903	6.60	0.2380

[a]The number of grams per cubic centimeter is the same as the specific gravity. For example, if the specific gravity is 1.47, that substance has a density of 1.47 g cm^{-3}.

New polymers have been biaxially oriented by older technologies. Tentering-frame and double-bubble technologies have been used to make a biaxially oriented polypropylene film.

An example of the many oriented products is the heat-shrinkable material found in flat or tubular film or sheets. The orientation in this case is terminated downstream of an extrusion–stretching operation when a cold enough temperature is achieved. Reversing the operation, or shrinkage, occurs when a sufficiently high temperature is introduced. The reheating and subsequent shrinking of these oriented plastics can produce a useful property (Table 1.31). It is used in heat-shrinkable tubular or flat communication cable wrap, heat-shrinkable furniture webbing, pipe fittings, medical devices, and many other products.

Density and specific gravity

The density of any material is a measure of its mass, per unit volume, usually expressed in grams per cubic centimeter ($g\,cm^{-3}$) or pounds per cubic inch ($lb\,in.^{-3}$) (Tables 1.32 and 1.33 and Fig. 1.35). Figure 1.36 shows how to determine the specific gravity of filled compounds. It is necessary to know the density of a particular plastic in order to calculate the relationship between the weight and volume of the material in a specific product.

Specific gravity is the ratio of the mass in air of a given volume (of plastic) compared to the mass of the same volume of water, both being measured at room temperature ($23\,°C$, $73.4\,°F$); in other words, it is the density of the plastic divided by the density of the water. Since this is a dimensionless quantity, it is convenient for comparing different materials. Like density, specific gravity is used extensively in determining part costs, weight, and quality control.

The ASTM D792 standard provides the relationship of density to specific gravity at $23\,°C$:

$$\text{Density}\left(g\,cm^{-3}\right) = \text{specific gravity} \times 0.9975$$

Also

$$\text{Specific gravity} \times 0.0361 = lb\,in.^{3}$$

Selecting plastics

The information and data presented in this chapter and other chapters provide a variety of useful selection guides. Figures 1.37 to 1.40 present general overall data, ratings and guides; Fig. 1.40 provides general heat-resistance properties of plastics retaining 50% of properties at room temperature with resin exposure and testing at elevated temperature.

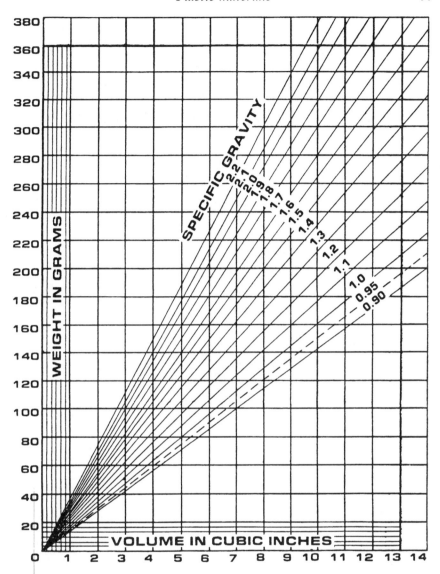

Figure 1.35 Weight in grams and volume in cubic inches versus specific gravity.

Zone 1: Acrylics, cellulose esters, LDPE, PS, PVC, SAN, SBR, UF, etc. **Zone 2**: Acetals, ABS, chlorinated polyether, ethyl cellulose, EVA, ionomer, PA, PC, HDPE, PET, PP, PVC, PUR, etc. **Zone 3**: PCTFE, PVDF, etc. **Zone 4**: Alkyds, fluorinated ethylene–propylene, MF, polysulfone, etc. **Zone 5**: TS acrylic, DAP, epoxy, PF, TS polyester, PTFE, etc. **Zone 6**: Parylene, polybenzimidazole, silicone, etc. **Zone 7**: PAI, PI, etc. **Zone 8**: Plastics in

Table 1.33 Specific gravity as a function of mass per volume[a]

Specific gravity	Density in oz in.$^{-3}$	Density in g in.$^{-3}$	Specific gravity	Density in oz in.$^{-3}$	Density in g in.$^{-3}$	Specific gravity	Density in oz in.$^{-3}$	Density in g in.$^{-3}$
0.90	0.5220	14.748	1.31	0.7569	21.467	1.71	0.9880	28.022
0.91	0.5258	14.912	1.32	0.7627	21.631	1.72	0.9938	28.186
0.92	0.5316	15.076	1.33	0.7685	21.795	1.73	0.9996	28.350
0.93	0.5374	15.240	1.34	0.7743	21.959	1.74	1.0054	28.513
0.94	0.5431	15.404	1.35	0.7800	22.122	1.75	1.0112	28.677
0.95	0.5489	15.568	1.36	0.7858	22.286	1.76	1.0169	28.841
0.96	0.5447	15.732	1.37	0.7916	22.450	1.77	1.0227	29.005
0.97	0.5605	15.895	1.38	0.7974	22.614	1.78	1.0285	29.169
0.98	0.5662	16.059	1.39	0.8031	22.778	1.79	1.0343	29.333
0.99	0.5720	16.223	1.40	0.8089	22.942	1.80	1.0400	29.497
1.00	0.5778	16.387	1.41	0.8147	23.106	1.81	1.0458	29.660
1.01	0.5836	16.551	1.42	0.8205	23.269	1.82	1.0516	29.824
1.02	0.5894	16.715	1.43	0.8263	23.433	1.83	1.0574	29.988
1.03	0.5951	16.879	1.44	0.8320	23.597	1.84	1.0632	30.152
1.04	0.6009	17.042	1.45	0.8378	23.761	1.85	1.0689	30.316
1.05	0.6067	17.206	1.46	0.8436	23.925	1.86	1.0747	30.480
1.06	0.6125	17.370	1.47	0.8494	24.089	1.87	1.0805	30.644
1.07	0.6182	17.534	1.48	0.8551	24.253	1.88	1.0862	30.808
1.08	0.6240	17.698	1.49	0.8609	24.417	1.89	1.0920	30.971
1.09	0.6298	17.862	1.50	0.8667	24.581	1.90	1.0978	31.135
1.10	0.6356	18.026	1.51	0.8725	24.745	1.91	1.1036	31.299
1.11	0.6414	18.189	1.52	0.8783	24.908	1.92	1.1094	31.463
1.12	0.6471	18.353	1.53	0.8840	25.072	1.93	1.1152	31.627
1.13	0.6529	18.517	1.54	0.8898	25.236	1.94	1.1209	31.791

1.14	0.6587	18.681	1.55	0.8956	25.400	1.95	1.1267	31.955
1.15	0.6645	18.845	1.56	0.9014	25.564	1.96	1.1325	32.119
1.16	0.6702	19.009	1.57	0.9071	25.726	1.97	1.1383	32.282
1.17	0.6760	19.173	1.58	0.9129	25.891	1.98	1.1440	32.446
1.18	0.6818	19.337	1.59	0.9187	26.055	1.99	1.1498	32.610
1.19	0.6876	19.501	1.60	0.9245	26.219	2.00	1.1556	32.774
1.20	0.6934	19.664	1.61	0.9303	26.383	2.01	1.1614	32.938
1.21	0.6991	19.828	1.62	0.9360	26.547	2.02	1.1672	33.102
1.22	0.7049	19.992	1.63	0.9418	26.711	2.03	1.1729	33.266
1.23	0.7107	20.156	1.64	0.9476	26.875	2.04	1.1787	33.429
1.24	0.7165	20.320	1.65	0.9534	27.039	2.05	1.1845	33.593
1.25	0.7222	20.484	1.66	0.9591	27.202	2.06	1.1903	33.757
1.26	0.7280	20.648	1.67	0.9649	27.366	2.07	1.1960	33.921
1.27	0.7338	20.811	1.68	0.9707	27.530	2.08	1.2018	34.085
1.28	0.7396	20.975	1.69	0.9765	27.694	2.09	1.2076	34.249
1.29	0.7454	21.139	1.70	0.9823	27.858	2.10	1.2134	34.413
1.30	0.7511	21.303						

[a] The number of grams per cubic centimeter is the same as the specific gravity. For example, if the specific gravity is 1.47, that substance has a density of $1.47\,\mathrm{g\,cm^{-3}}$.

Factor used in converting to ounces per cubic inch = specific gravity multiplied by 0.5778.

Factor used in converting to grams per cubic inch = specific gravity multiplied by 16.387.

To compute:

Specific gravity: multiply pounds per cubic foot by 0.01604.
Pounds per cubic foot: multiply specific gravity by 62.4.
Pounds per cubic inch: multiply specific gravity by 0.0361.

1 oz = 28.3495 g
1 g = 0.0352793 oz

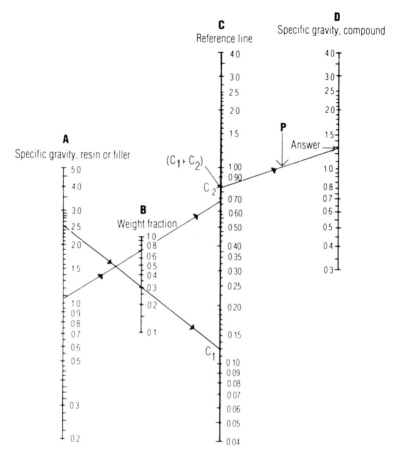

Figure 1.36 Nomograph for determining the specific gravity of filled compounds by using various fillers and reinforcements.

R&D, etc. Since plastics compounding is rather extensive, certain basic resins can be modified to meet different heat-resistance properties.

Remember that the values given here and elsewhere in this book are representative rather than precise. Values vary depending on the specific type of plastics, the manufacturing settings, and the methods of evaluation. The procedure to follow when a specific material and process is to be used requires proper identification and documentation of the 'facts' such as machine settings, operator, etc. The data in this book are not intended to be used as a substitute for the more accurate information obtained from developing and/or past performances.

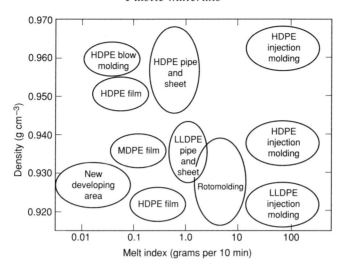

Figure 1.37 Example of PE markets based on melt index versus density [3].

PRODUCT DESIGN

There is a practical and easy approach to designing with plastics. Figures 1.41 and 1.42 illustrate its similarity to designing with other materials: steel, aluminum, wood, glass, etc.

Many different products can be designed using plastics. They will take low to extremely high loads and operate in widely differing environments, from the highly corrosive to those involving electrical insulation. They challenge the designer with a combination of often unfamiliar advantages and limitations. The designer can meet this challenge by understanding the many different structures and properties as well as the design and fabrication capabilities, as demonstrated by the existence of the many different products made from plastics.

Although plastics and composites may appear to some observers to be new, because the industry has an unlimited capacity to produce new plastics to meet new performance and processing requirements, plastics have been used in no-load to extremely high-load situations for over a century. The ever evolving technology does not mean that plastics will automatically replace other materials. Each material (plastics, metals, wood, aluminum, etc.) will be used in favorable cost-to-performance situations. Since the early 1980s, more plastics were used worldwide on a volumetric basis than any other materials except wood and concrete. And by the end of this century, it will also hold on a weight basis.

With plastics and composites, to a greater extent than with other materials, an opportunity exists to optimize design by focusing on a material's composition and orientation as well as the geometry of its structural

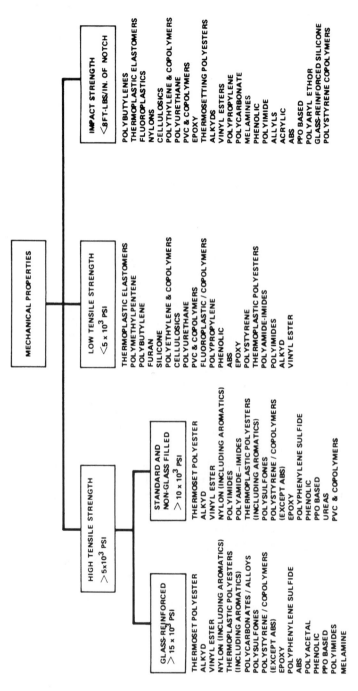

Figure 1.38 Overview of mechanical properties.

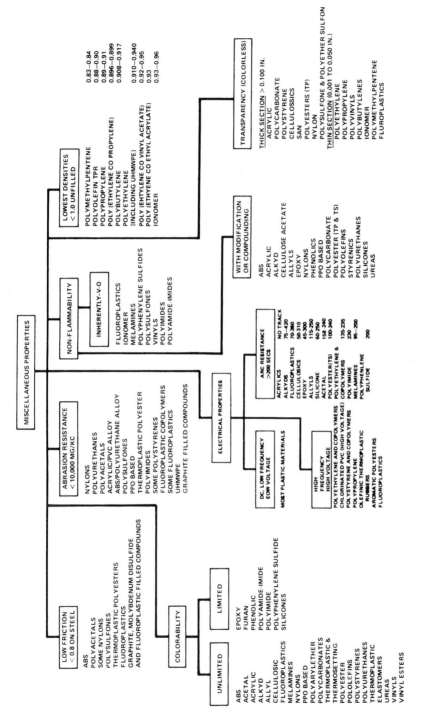

Figure 1.39 Overview of other properties.

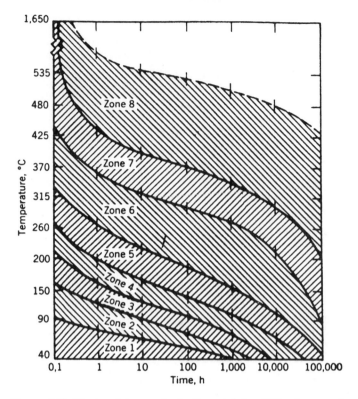

Figure 1.40 Heat resistance of plastics retaining 50% of properties.

members. There are also important interrelationships among shape, material selection (reinforced plastics, elastomers, foams, etc.), the consolidation of parts, manufacturing selection, and other factors that provide low cost-to-performance products. For the many applications that require only minimal mechanical performance, shaping through processing techniques can help to overcome limitations such as low stiffness with commodity (lower cost-to-performance) plastics. And when extremely high performance is required, reinforced plastic (RP), composites, and other engineering plastics are available. In this book the term *plastics* also refers to composites. Examples of these plastic products are shown in Figs 1.42 to 1.44.

Design features that influence processing and performance

The successful design and fabrication of good plastic products requires a combination of sound judgment and experience. Designing good products requires a knowledge of plastics that includes their advantages and disadvantages and some familiarity with processing methods. Until the designer becomes familiar with processing, a fabricator must be taken into the designer's confidence early in development and consulted frequently during those early days. The fabricator and mold or die designer should advise the product designer on materials behavior and how to simplify processing. The designer should not become restricted by understanding only one process, method, particularly just a certain narrow aspect of it.

One of the earliest steps in product design is to establish the configuration of the parts that will form the basis on which strength calculations will be made and a suitable material selected to meet the anticipated requirements. During the sketching and drawing phase of working with shapes and cross sections, there are certain design features with plastics that have to be kept in mind to avoid degradation of the properties. Such features may be called property detractors or constraints. Most of them are responsible for the unwanted internal stresses that can reduce the available stress level for load-bearing purposes. Other features, which are covered in this chapter, may be classified as precautionary measures that may influence the favorable performance of a part if they are properly incorporated.

Although there is no theoretical limit to the shapes that can be created, there are practical considerations which need to be met. These relate not only to part design, but also to mold or die design, since they must be considered as one entity in the total creation of a usable, economically feasible part. In the sections that follow, various phases considered important in the creation of such parts are examined for their contribution to and effect on design and function.

Before designing a part, the designer should understand such basic factors as those summarized in Fig. 1.45. Success with plastics, or any other material, for that matter, is directly related to observing design details (Figs 1.46 and 1.47). For example, something as simple as a stiffening rib is different for an injection molded or structural foam part, even though both parts may be molded from the same plastic (Fig. 1.45). However, a stiffening rib that is to be molded in a low mold shrink, amorphous TP will differ from a high mold shrink, crystalline TP rib, even though both plastics are just injection molded. Ribs molded in RP or composite plastics have their own distinct requirements:

The important factors to consider in designing can be categorized as follows: part thickness, tolerance, ribs, bosses and studs, radii and fillets, draft or taper, holes, threads, color, surface finish and gloss level, decorating operations, the parting line, gate locations, molded part shrinkage, assembly techniques, mold or die design, production volume, the tooling and other equipment amortization period, as well as the plastic and machine selected. The order that these factors follow can vary, depending on

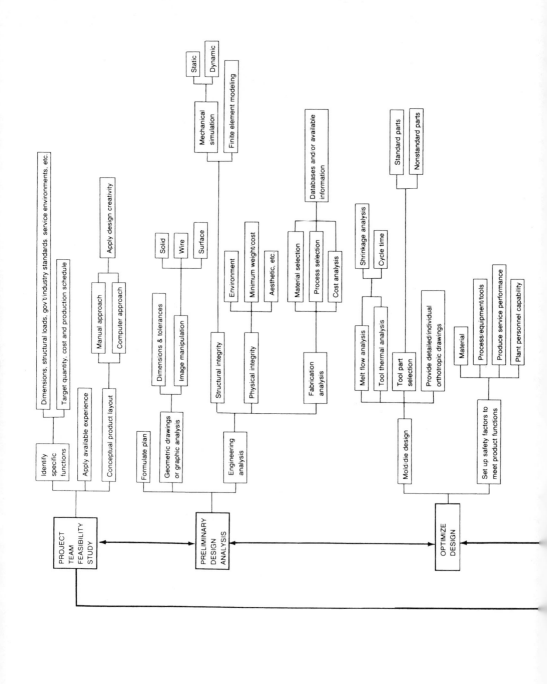

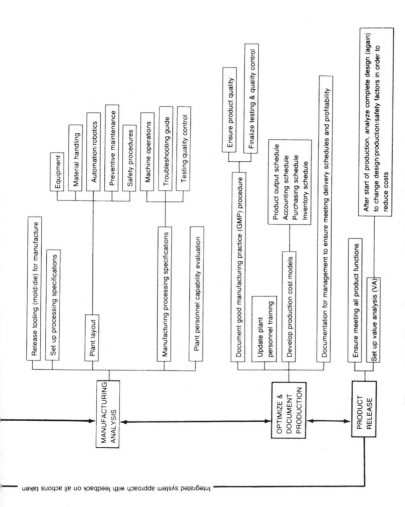

Figure 1.41 Flow diagram with feedback loops; design through processing.

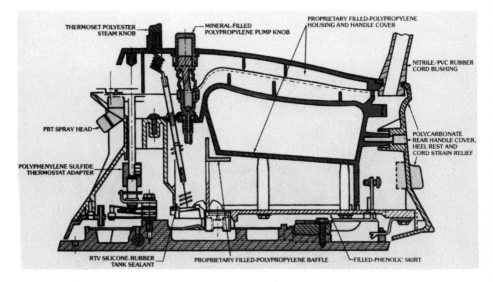

Figure 1.42 An electric iron using different plastics and processes.

Figure 1.43 Injection-molded plastic and RP open layup produces the leisure boat.

the product to be designed and the designer's familiarity with particular materials and processes.

Preparing a complete list of design constraints is a crucial first step in plastic part design; failure to take this step can lead to costly errors. For example, a designer might have an expensive injection mold prepared, designed for a specific material's shrink value, only to discover belatedly that the initial material chosen did not meet some overlooked design

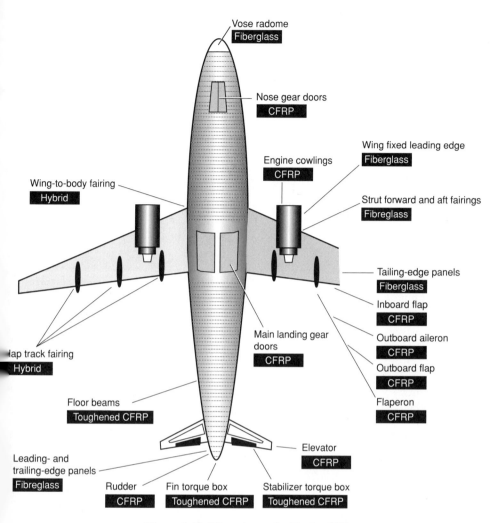

Figure 1.44 RP parts on the Boeing 777.

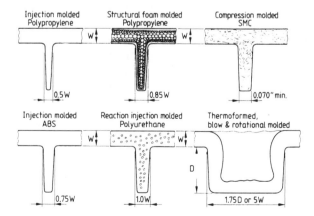

Figure 1.45 How different plastics and molding processes can affect the design details of a stiffening rib.

Fundamentals

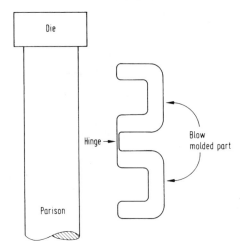

Figure 1.46 An extrusion blow-molded container with a living hinge in the as-molded position.

constraint. Flammability, fiberglass fillers to provide a higher modulus, and other requirements are best considered before a tool is made-otherwise, the designer may have the difficult if not impossible task of finding a plastic that does meet all the design constraints, including the important appropriate shrink value for the existing mold. Such desperation in the last stages of a design project can and should be avoided. As emphasized from one end of this book to the other, it is vital to set up a complete checklist of product requirements, to preclude the possibility that a critical requirement may be overlooked initially. Apparently impossible approaches and 'approaches' to be avoided can often produce excellent products, but it is easier to take the direction with the fewest problems.

Establishing initial design tolerances is often done on an arbitrary, uninformed basis. If the initial estimated tolerance proves too great, a lower-shrink plastic could be used to reduce the shrinkage range. But if the key dimension is across a main parting line, the tooling could be redesigned to eliminate the condition and consequently reduce variation from tool construction. Even with all these data processed by a computer, estimating tolerances is difficult if they are not properly interrelated with the highly dependent factors of the part and processing capability.

PROCESS CONTROLS

Adequate process control (PC) and its associated instrumentation are essential for product quality control (Figs 1.48 and 1.49). Sometimes the

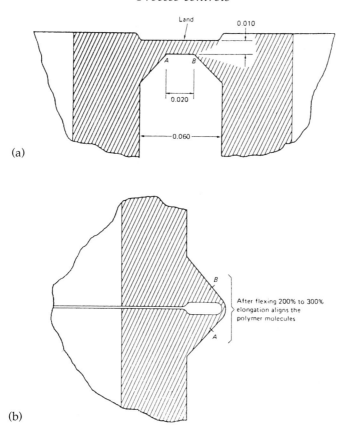

Figure 1.47 Molded living hinge in (a) the as-molded position and (b) a flexed position (the dimensions are in inches).

goal is precise adherence to a control point, other times it is sufficient to maintain the temperature within a comparatively narrow range. For effortless controller tuning and lowest initial cost, the processor should select the simplest controller (of temperature, time, pressure, melt flow rate, etc.) that will produce the desired results.

PCs for the individual machines and the complete fabricating line can range from unsophisticated equipment to extremely sophisticated devices. They can (1) provide closed-loop control of temperature and/or pressure, (2) maintain preset parameters for a process, (3) monitor and/or correct machine operation, (4) constantly fine tune equipment, and (5) provide consistency and repeatability in the operations. PC is not a toy or a panacea; it demands a high level of expertise from the processor.

Based on the process control settings, different behaviors of the plastic will occur. Regardless of the type of controls available, the processor

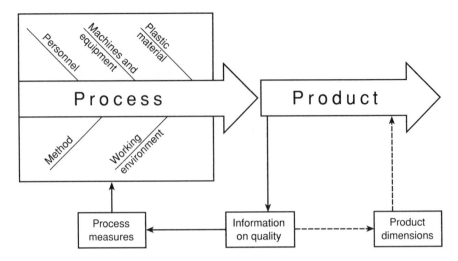

Figure 1.48 Summarizing process contol in a simplified pattern.

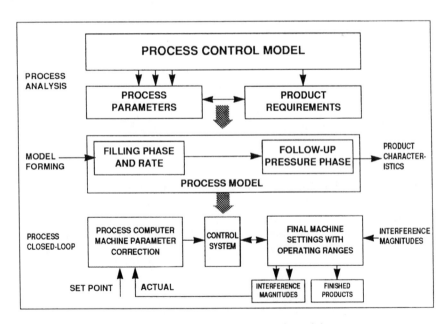

Figure 1.49 A process control model.

setting up a machine uses a systematic approach that should be outlined in the machine or control manuals. Once the machine is operating, the processor methodically makes one change at a time, to determine the result.

Temperature controls

Heat is usually applied in various amounts and in different locations, whether in a metal plasticating barrel (extrusion, injection molding, etc.) or in a metal mold/die (compression, injection, thermoforming, extrusion, etc.). With barrels a thermocouple is usually embedded in the metal to send a signal to a temperature controller. In turn it controls the electric power output device regulating the power to the heater bands in different zones of the barrel. The placement of the thermocouple temperature sensor is extremely important. The heat flow in any medium sets up a temperature gradient in that medium, just as the flow of water in a pipe sets up a pressure drop, and the flow of electricity in a wire causes a voltage drop.

The introduction of automatic reset into controllers for the plastic processor has made it possible to hold the temperature constant even in the presence of extremely long lags. Automatic reset is a characteristic added to a proportional controller that functions as an integrating or averaging system, looking at the droop, or temperature error, over a period of time and adjusting the output so the droop goes to zero. As a result, the actual temperature goes to the set point. Automatic reset is almost always used with an additional 'rate' term, which adds an anticipatory characteristic that does not affect steady-state performance but does speed up the response to changes in operating conditions. A modern proportional plus automatic reset plus rate, a three-mode controller, is capable of controlling within 1 °F (0.6 °C) of the set point all the way from full heating to full cooling, even when controlling from a deep-well sensor.

Timing and sequencing

Most processes operate more efficiently when functions must occur in a desired time sequence or at prescribed intervals of time. In the past, mechanical timers and logic relays were used. Now electronic logic and timing devices predominate, based on the so-called programmable logic controller. These devices provide sophisticated sequencing and timing, and lend themselves easily to reprogramming if it becomes necessary. Different suppliers provide special consoles that can be plugged in, and logic sequences can be added using 'ladder diagrams', which represent the desired functions and/or timing.

Screw speed control

Many processes require speed controls. Performance and reliability of these controls are very similar to those of the temperature controls – you get what you purchase. Like temperature controllers, the early speed controllers were mechanical. Speeds were held within 5%, producing poor plastic melt control. Where better speed control is desired, the solution is the same as in temperature control; only the equipment names are changed. A device is added to the motor, and an integral characteristic is provided, corresponding to the automatic reset with heat. It brings the speed closer to the set point. A derivative characteristic, corresponding to the rate in heat, ensures a prompt response to any upsets.

The arguments for the use of integral or derivative control of speed are the same as for temperature. Different systems are available, the latest being the all-digital speed control on plastic processing machines that require speed control. These speed controls permit accuracies of 0.5% or less. An all-digital phase-locked loop system permits all motors in a machine and/or a processing line to be synchronized with each other exactly or in a desired speed ratio, just as if they were mechanically geared together.

Weight, thickness, and size controls

By controlling important factors such as final product weight, thickness, and/or size, enhanced product performance and a lower product cost can definitely be achieved. Many different systems are used to ensure product performance (from different suppliers), so product controls will directly influence process control [1–4]. In controlling the feeding of many processes, it is very important to meter material accurately by volume and/or weight.

Microprocessor control systems

Many microprocessor systems are available to control all of the different sequences required, including those described in this book for all the different processes. Some are dedicated and have been designed for specific processes. All these systems consist of electronic devices that can receive and store all types of information (upstream activities like machine operation, melt process, and materials handling, as well as downstream operations like takeoff, weighing, quality control, decorating, and packaging, plus other activities besides). Some microprocessors can be programmed to make decisions on inputs and/or machine operation responses. These devices have memories that can store and retrieve all the pertinent data.

Today's microprocessors react and scan inputs quickly; a time of 10 ms includes all this as well as mathematical analysis. Tomorrow's systems will do even more, so it is important to consider using microprocessors that can be easily updated and integrated with new developments to improve performance and cost of products.

Problems in process control

Purchasing a sophisticated process control system is not a foolproof solution that will guarantee perfect parts. Solving problems requires a full understanding of their causes, which may not be as obvious as they first appear. Failure to identify contributing factors when problems arise can easily prevent the microprocessor from doing its job. The conventional place to start troubleshooting a problem is with the basics: temperature, time, and pressure limits. A problem may often be very subtle, such as a faulty control device or an operator making random control adjustments. Process controls cannot usually compensate for such extraneous conditions; however, if desired, they may be included in a program that provides the capability to add functions as needed.

Most controls, particularly the older ones, are open-loop controls. They merely set mechanical or electrical devices to some operating temperature, time, and pressure. If this is all that is required, the control may remain in operation. However, this setup is subject to a variety of hard-to-observe disturbances that are not compensated by open-loop controls. Process control must therefore close the loop to eliminate the effect of process disturbances.

There are two basic approaches to problem solving: (1) find and correct the problem, applying only the controls needed; or (2) overcome the problem with an appropriate process control strategy. The approach one selects depends on the nature of the processing problem, and whether enough time and money are available to correct it. Process controls usually provide the most economical solution. To make the right decision, one must systematically measure the magnitude of the disturbances, relate them to product quality, and identify their cause so that proper control action can be taken.

Before investing in a more expensive system, the processor should methodically determine the exact nature of the problem, to decide whether or not a better control system is available and will solve it. For example, the temperature differential across a mold (for injection molding, etc.) can cause uneven thermal mold growth. The mold growth can also be influenced by uneven heat on tie-bars (uppers can be hotter, causing platens to bend, and this, change may be reflected on the mold). Once the cause is determined, appropriate corrective action can be taken. (For example, if the mold heat has varied, perhaps all that is needed is to close a large garage door nearby, to eliminate the flow of air over the problem mold; or perhaps it is sufficient to change the direction of flow from an air-conditioning duct.)

PROCESSORS

This is the part of the plastics industry where advances in materials and machinery are combined to create the finished products that the consuming public buys, products that range from the triviality of swizzle sticks to the lifesaving glory of an artificial heart. There are three types of processor: custom, captive, and proprietary.

Custom

These are operations that in the metalworking field might be known as job shops. They process plastics into components for other manufacturers to use in their products. For example, a manufacturer of refrigerators may retain a custom thermoformer to make inner door liners. Or a typewriter manufacturer may have keys made by a custom molder. Custom processors typically have a close relationship with the companies for whom they work. They may be involved (to varying degrees) in the design of the part and the mold, they may have a voice in material selection, and in general they assume a reasonable level of responsibility for the work they turn out. There is a subgroup in custom processing known as 'contract' molders, They have little involvement in the business of their customers. In effect, they just sell machine time.

Captive

These are operations of manufacturers who have acquired plastics processing equipment to make parts they need for the product they manufacture. For example, a car maker may install equipment to mold accelerator pedals, or instrument cluster housings (rather than have a custom molder produce them). A refrigerator manufacturer may acquire a thermoforming machine to produce inner door liners. Generally speaking, manufacturers will install a captive plastics processing operation only when their component requirements are large enough to make it economical. Some manufacturers who are big enough to run their own plastics shops will nevertheless place a portion of their requirements with outside vendors to keep their own capital investment down, to avoid internal single-source supply, and to maintain contact with the market and the pricing intelligence it provides. Automobile makers are a good example of this type of operation.

Proprietary

These are operations where the molder makes a product for sale directly to the public under his or her own name. Boonton Molding Co. (no longer in existence) made melamine dinnerware sold to the public under the Boontonware label.

ADVANTAGES AND DISADVANTAGES

Like any process (metals, etc.), materials (steel, etc.), products, people, governments, etc., nothing reaches 'perfection', so it is best to recognize that limitations exist. Disadvantages exist when limitations are not recognized.

Achievable program plans begin with the recognition that *smooth* does not mean *perfect*. Perfection is an unrealistic ideal. It is a fact of life that the further someone is removed from a task, the more they are apt to expect perfection from those performing it. The expectation of perfection blocks genuine communication between workers, departments, management, customers, and vendors. Therefore, one can define a smoothly run program as one that creates a product which meets the specifications, is delivered on time, falls within the price guidelines, and stays close to budget. Perfection is never reached; there is always room for more development. To live is to change, and to approach perfection is to have changed often (in the right direction).

Troubleshooting

With all the types of plastics processes, troubleshooting guides are set up to take fast, corrective action when products do not meet their performance requirements. Guides are provided in the following chapters. A simplified approach to troubleshooting is to develop a checklist that incorporates the rules of problem solving such as (a) have a plan and keep updating it based on the experience gained; (b) watch the processing conditions; (c) change only one condition or control at a time; (d) allow sufficient time for each change, keeping an accurate log of each; (e) check housekeeping, storage areas, granulators, etc.; (f) narrow the range of areas in which the problem belongs, e.g. machine, molds or dies, operating controls, material, part design, environment (humidity, ventilation, etc.), people and management.

When a startup for a process is conducted, the operator will set the operation in all its different modes of operation going from not enough (or too little) to more than what is required to fabricate acceptable products. This approach follows the FALLO approach (Fig. 1.1, page 2). By going through all these setups, one becomes exposed to many, perhaps all, problems that will develop when the process is not producing acceptable products. One could record all the problems encountered and how they were corrected. This would set up a troubleshooting guide, targeting for what is known as total control.

2

Injection molding

BASIC PROCESS

The injection molding machine (IMM) basically goes through the following stages:

Plasticizing: heating and melting of the plastics and venting of the melt.
Injection: injecting under pressure of the melt into the closed mold, solidification of the plastics begins on the cavity walls.
After-filling: maintaining the injected material under pressure for a specific time period to prevent backflow of melt and to compensate for the decrease in volume of melt during solidification.
Cooling: cooling the molded part until it is sufficiently rigid to be ejected.
Mold release: opening the mold, ejection of the molding and closing the mold so it is ready for the next cycle.

Thus, the IMM features two basic components: an injection unit to melt and transfer the plastic into the mold and a clamping unit to close and open the mold.

The IM process is extremely useful since it permits the manufacture of very complex shapes and their three dimensions can be more accurately controlled and predicted than with other processes. Its operation is more complex than other processes, so IM needs to be thoroughly understood [1]. Figures 2.1 to 2.4 are schematics of the load profile and the molding cycle, schematics which highlight the way in which the melt is plasticized (softened) and forced into the mold.

The injection unit (also called plasticator or extruder) melts the plastic then injects it into the mold with controlled pressure, temperature, and rate (time). It uses two basic injection units: a two-stage unit (screw preplasticator) and a reciprocating screw. A two-stage unit uses a fixed plasticating screw (first stage) to feed melted plastic into a chamber (second stage). A plunger then forces the melt into the mold. Its advantages

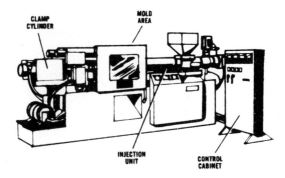

Figure 2.1 General layout for an injection molding machine.

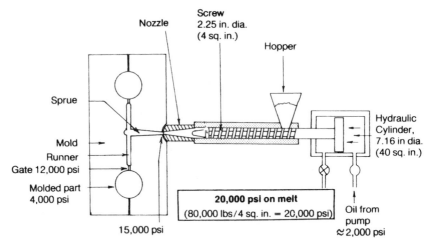

Figure 2.2 Pressure loading on a plastic melt during injection molding.

are consistent melt quality, high pressures, fast rates, accurate shot size control, product clarity, molding very thin-walled parts, etc. Disadvantages include uneven residence time, higher equipment costs, and increased maintenance.

The reciprocating screw injection unit is most commonly used; it melts and injects the plastic without a plunger (no separate second-stage unit). Plastic is melted in the machine's barrel and transferred to the nozzle end of the machine by a rotating screw. The accumulation of melt at the screw tip forces the screw towards the rear of the machine until enough material is collected for a shot (shot represents the volume of melt required in the mold cavity or cavities). The screw is then driven forward, forcing the melt into the mold (the screw is normally not turning; it acts like a

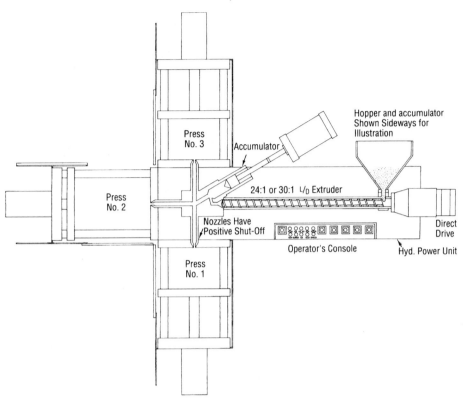

Figure 2.3 A two-stage machine with three clamping presses.

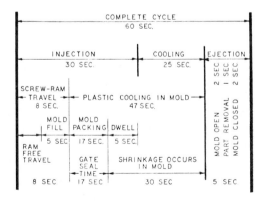

Figure 2.4 An injection molding cycle for processing thermoplastics (TPs).

(a)

(b)

(c)

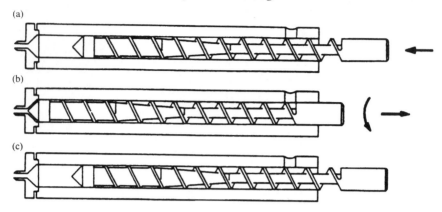

Figure 2.5 Reciprocating screw sequence of operations: (a) injection, screw moves axially forward; (b) shot preparation, screw rotates and retracts; (c) soak or idle, no screw movement.

plunger, as shown in Fig. 2.5). Different designs of screw-tip nonreturn valves are used at the end of the barrel to prevent melt from flowing back along the screw when it acts as a plunger (Fig. 2.6).

The advantages of reciprocating screw units include reduced residence time, self-cleaning screw action, accurate and responsive injection control. These advantages are key to processing heat-sensitive materials or when making color or resin changes. They offer repeatable part-to-part consistency and the capability to produce increasingly complex parts with faster cycle times and meeting tight tolerances.

Even though most of the literature on processing (even in this book) specifically identifies or refers to TPs (representing at least 90 wt%) some TSs are processed. During injection molding the TPs reach maximum heat before entering a cool mold, whereas TSs reach maximum temperature in hot molds (Fig. 1.10, page 29).

Machine characteristics

IMMs are characterized by their shot capacity, plasticizing capacity, rate of injection, injection pressure and clamp pressure (or mold-locking force). Shot capacity may be given in terms of the maximum weight that can be injected per shot, usually quoted (as a standard) in grams or ounces of general-purpose polystyrene. This depends on the swept volume of the cylinder during one stroke of the plunger and on the volumetric capacity of the feed mechanism, so it is necessary to correct for bulk density and specific gravity when dealing with other materials. A better way of expressing shot capacity is in terms of the volume of material which can be injected into a mold at a specific pressure.

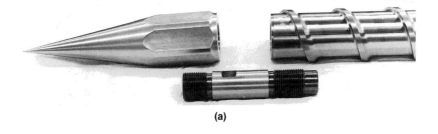

(a)

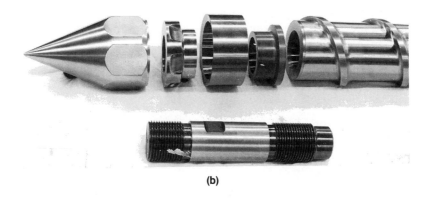

(b)

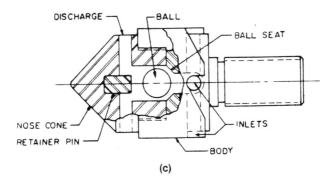

(c)

Figure 2.6 Basic screw-tip design: (a) plain or smearhead (disassmebled); (b) sliding ring (disassembled); (c) ball check (assembled).

One of the factors affecting the output of an IMM is the plasticizing capacity, usually defined as the plastic weight per hour of a particular material that can be brought to temperature required for molding. Injection rate is also an important factor in determining the output of the machine; it is usually expressed as the volume of material discharged per second through the nozzle during a normal injection stroke. This obviously depends on pressure, temperature, material used, and on the size of the smallest aperture in the flow line. With low-viscosity materials and fairly large gates, etc., the plunger speed may be a limiting factor.

The injection pressure is the pressure exerted by the face of the plunger and can vary from about 2000 psi (14 MPa) to 30 000 psi (200 MPa). Pressure losses in the system mean that pressure in the mold cavity or cavities is much less than the plunger pressure. Nevertheless, the actual force exerted on the inside surface of the mold is large if the projected area (in line with the pressure) of the mold is large. The clamp pressure is therefore an important factor in determining the maximum projected area that can be molded on a particular machine.

Machine types

The range of injection molding machines is extremely wide. Apart from their different shot capacities – from a few grams to many kilograms (Fig. 2.7) – there are IMMs that run with or without preplasticizing, IMMs that run automatically or semiautomatically and IMMs using shuttle or rotary molds. And the preplasticizer can be inline, piggyback, or in a variety of other designs. Giant machines with kilogram shot capacities use at least two or three plasticizers operating in a parallel flow pattern.

Development of the larger IMMs has not been due solely to the demand for larger single moldings; economies of scale also apply. One machine with a 4 oz (113 g) shot size is slightly cheaper than two machines with a 2 oz (57 g) shot size, so there is a tendency to use larger machines with multicavity molds. However, the flexibility of operation is a point in favor of single-mold faster-cycle machines. Another factor to be considered is where several different-colored moldings are required, and it may be better to run several smaller machines rather than one large machine using multicavity molds. The larger machines with multicavity molds have a cost advantage because they produce parts at a faster rate. The approach to be used on deciding size and number of machines depends on one's knowledge and capability of IM; the mulicavity molds (128 cavities and upwards) have continued to be very popular. Another important factor is quantity and future orders.

A different approach is the use of shuttle and rotary mold machines [1], where a single IMM feeds several molds. Another approach is multicolor

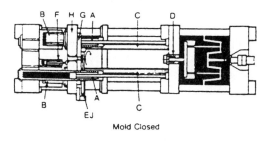

Mold Closed

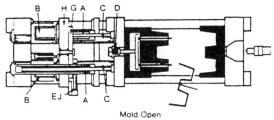

Mold Open

Figure 2.7 Injection shot capacity for this Billion machine is 350lb (177 kg); its clamping capacity is 10000 tons. A, approach cylinders; B, clamp cylinders; C, locking cylinders; D, ejector; E, pivoting cylinder (closing); F, returning cylinder; G, pivoting plate; H, support plate; and J, pivoting cylinder (opening).

molding using a rotary horizontal table or rotary vertical platform on a vertical IMM platen. Two or more colors are used (with two or more plasticators for each color or resin). Take three colors as an example. The first color is injected and the mold opens, then the mold with plastic automatically rotates 120°, matching a mold half that provides a cavity only for the second color. It rotates a further 120° to the third position, matching a cavity only for the third color. There is great controversy over the economics of these machines, but they continue to be used where they appear to be feasible.

Machine variables

For any particular material, successful molding is determined by part and mold design, and the correct setting of the following variables.

Injection pressure

Injection pressures can vary quite widely; they depend on mold design and the machine size. In general, the aim is to use minimum pressure to produce full-shot moldings free from such defects as surface sink marks or voids. Excess pressure should be avoided as this could lead to flashing (escape of material from the mold parting lines).

Cylinder temperature

The object of heating the molding material is to bring it to a suitably plasticized condition by the time it is ready for injection to the mold cavities. The temperature of the material depends not only on the temperature of the cylinder but also on the rate at which the material passes through it. Uniform heating of the material therefore depends on accurate temperature control of the heating cylinder and on strict control of the time cycle. Heating is usually carried out by band heaters on the outside walls of the heated cylinder. The band heaters are arranged in two or three zones, each controlled separately. Thin-walled moldings require higher cylinder temperatures than thick sections. This is because they require less time to freeze (solidify) in the mold, reducing the overall time cycle and, consequently, the residence time of the molding material in the heating cylinder (Chapter 3).

Cycle time

From the standpoint of economics, the aim should be the minimum time for each part of the molding cycle consistent with good-quality molding. The injection stroke speed was mainly governed by the viscosity of the material (and therefore the material temperature), the ram pressure, and the minimum aperture size in the flow line. The minimum ram-forward time is governed by the required duration of the pressure. If pressure is removed too early from the molding, surface sinking or internal voiding can occur; it is therefore essential that the gate should freeze off completely before the ram is retracted.

Mold temperature

The aim is constant mold temperature, below the softening point of the material; this is normally carried out by circulation of a constant-temperature fluid through channels in the mold. With TSs it is above the melt temperature, usually using electric heating elements in the mold.

Productivity and cycle time

Productivity is directly related to cycle time. There is usually considerable common knowledge about part geometry and process conditions that will provide a minimum cycle time. Solidification time will be decreased by using thinner wall sections, cold versus hot runners for TPs, or hot versus cold for TSs, narrow sprues and runners, optimal size and location of coolant (or heat) channels, and lower melt/mold heat when possible.

Numerous factors affect the elapsed time required to eject a part, as different plastics can have dramatically different melt behaviors. Many of these influences are poorly understood. Some critical factors are the coefficient of expansion, melt rheology, thermal diffusivity, and the thermomechanical spectrum (Chapter 1). Although the usual and important ways to optimize time are based on part design and process conditions, it can be shown that additional and significant decreases occur by using modified molding compounds via additives, alloying ratios, molecular weight distribution, and so on (Chapter 1).

Mathematical models of part geometry and melt flow within the mold cavity, which are available for mold design, are useful tools for optimizing cycle times. They allow a wide variety of plastics materials and process parameters to be evaluated in a convenient and cost-effective manner. Computer-aided design (CAD) and computer-aided engineering (CAE) algorithms offer a continually higher level of sophistication in determining the best heat distribution throughout the part. In practice the parts are not ejected according to a measurement of the internal or wall temperature. Ejection times are set through secondary thermal characteristics of a part, such as its ability to withstand the forces of ejection, the occurrence of sink marks or other thermal warpage, and the overall gloss or appearance. The prediction of sink marks appears amenable to CAD/CAE. Part appearance still plays a larger role in the art of IM.

Material movement and setup

Plastic moves from the hopper onto the feeding portion of the reciprocating extruder screw. The flights of the rotating screw cause the material to move through a heated 'extruder' barrel, where it softens (is made fluid) so it can be fed into the shot chamber (front of screw). This motion generates pressure, usually 50–300 psi (0.3–2 MPa), which causes the screw to retract. When the preset limit is reached, the shot size is met and the screw stops rotating; at a preset time the screw acts as a ram to push the melt into the mold. Injection takes place at high pressure, up to 30 000 psi (200 MPa) melt pressure in the nozzle. Adequate clamping pressure must be used to eliminate mold opening (flashing). The melt pressure within the mold cavity ranges from 1–15 tons in.^{2} (14–210 MPa), and depends on the plastic's rheology or flow behavior (Chapter 1).

Time, temperature, and pressure controls indicate whether performance requirements of a molded part are met. Time factors include rate of injection, duration of ram pressure, time of cooling, time of plastication, and screw RPM (Fig. 2.4, page 123). Pressure factors are injection high and low pressure, back pressure on the extruder screw, and pressure loss before the plastic enters the cavity, which can be caused by a variety of restrictions in the mold (Fig. 2.8). Temperature factors are mold (cavity and core), barrel and nozzle temperatures, as well as the melt temperature due to back pressure, screw speed, frictional heat, and so on (Fig. 2.9).

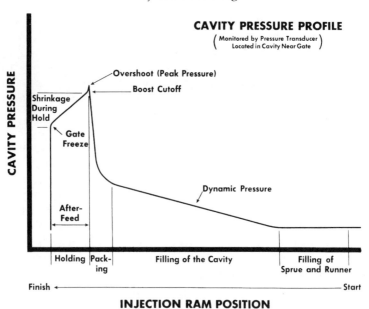

Figure 2.8 Cavity pressure profile.

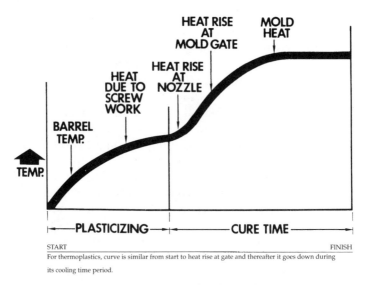

Figure 2.9 Thermal load profile during injection molding of thermosets (TSs). (For TPs the curve is similar from the start to the heat rise at the gate; thereafter it descends during its cooling time period.)

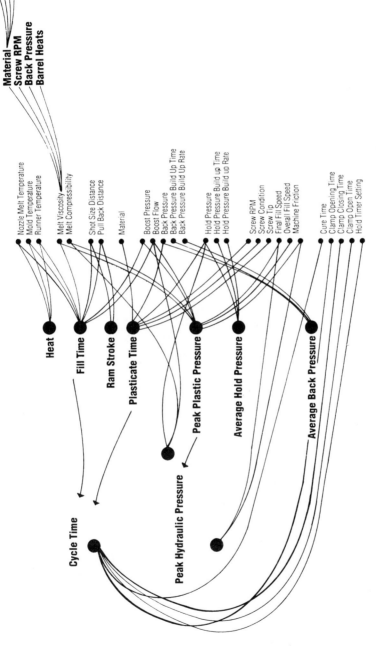

Figure 2.10 Injection molding machine controls.

Table 2.1 Guide for injection molding and extrusion machine settings[a]

Resin data[b]	Specific gravity (g cm⁻³)	Density (lb ft⁻³)	Specific volume (in.³ lb⁻¹)	Specific volume (cm³ g⁻¹)	Extrusion temperature (°F)	Injection temperature (°F)	Linear mold shrinkage	Specific heat (Btu lb⁻¹ °F⁻¹)	Water absorption in 24 h (%)	Maximum water content allowable for molding
ABS, extrusion	1.02	64.0	27.0	0.980	435		0.005	0.34	0.25	
ABS, injection	1.05	65.0	26.0	0.952		500	0.005	0.40	0.40	0.20
Acetal, injection	1.41	88.0	19.7	0.709		390	0.020	0.35	0.25	
Acrylic, extrusion	1.19	74.3	23.3	0.839	375		0.004	0.35	0.30	
Acrylic, injection	1.16	72.0	24.1	0.868		450	0.005	0.35	0.20	0.08
CAB	1.20	74.6	23.1	0.833		440	0.004	0.35	1.50	0.15
Cellulose acetate, extrusion	1.28	80.2	21.6	0.781	380		0.005	0.40	2.50	
Cellulose acetate, injection	1.26	79.0	21.9	0.794	380		0.005	0.36	2.40	0.20
Cellulose proprionate, extrusion	1.22	76.1	22.7	0.821	380		0.004	0.40	1.70	
Cellulose proprionate, injection	1.22	75.5	22.9	0.828		425	0.004	0.40	2.00	0.25
CTFE	2.11	134.0	13.1	0.473		550	0.008	0.22	0.01	
FEP	2.11	134.0	12.9	0.465	600	600	0.010	0.28	<0.01	
Ionomer, extrusion	0.95	59.6	29.0	1.050	500		0.007	0.54	0.07	
Ionomer, injection	0.95	59.1	29.2	1.060		420	0.007	0.54	0.20	
Nylon-6	1.13	70.5	24.5	0.886	520	550	0.013	0.40	1.60	0.15

Nylon-6,6	1.14	71.2	24.3	0.878	510	510	0.015	0.40	1.50	0.15
Nylon-6,10	1.08	67.4	25.6	0.927		450	0.011	0.40	0.40	0.15
Nylon-6,12	1.07	66.8	25.9	0.935	475	500	0.011	0.40	0.40	0.20
Nylon-11	1.04	64.9	26.6	0.962	460	450	0.005	0.47	0.30	0.10
Nylon-12	1.02	63.7	27.1	0.980	450	445	0.003		0.25	0.10
Phenylene oxide based	1.08	67.5	25.6	0.926	480	525	0.006	0.32	0.07	
Polyallomer	0.90	56.2	30.7	1.110	405	405	0.015	0.50	0.01	
Polyarylene ether	1.06	66.2	30.7	0.940	460	535	0.006		0.10	
Polycarbonate	1.20	74.9	23.1	0.832	550	575	0.006	0.30	0.20	0.02
Polyester PBT	1.34	83.6	20.7	0.746		460	0.020		0.08	0.04
Polyester PET	1.31	8.18	21.1	0.746	480	490	0.002	0.40	0.10	0.005
HD polyethylene, extrusion	0.96	59.9	28.8	1.040	410		0.025	0.55	<0.01	
HD polyethylene, injection	0.95	59.3	29.1	1.050		480	0.025	0.55	<0.01	
HD polyethylene, blow molding	0.95	56.9	28.8	1.040	410		0.025	0.55	<0.01	
LD polyethylene, film	0.92	57.44	30.1	1.090	350		0.032	0.55	<0.01	
LD polyethylene, injection	0.92	57.4	30.1	1.090		400	0.032	0.55	<0.01	
LD polyethylene, wire	0.92	57.4	30.1	1.090	400		0.025	0.55	<0.01	
LD polyethylene, ext. coating	0.92	57.1	30.0	1.090	600		0.025	0.55	<0.01	
LLD polyethylene, extrusion	0.92	57.4	30.1	1.087	500					
LLD polyethylene, injection	0.93	58.0	29.8	1.075		425				
Polypropylene, extrusion	0.91	56.8	30.4	1.100	450		0.005	0.50	0.03	
Polypropylene, injection	0.90	56.2	30.7	1.110		490	0.018	0.50	<0.01	
Polystyrene, impact sheet	1.04	64.9	26.6	0.963	450		0.005	0.34	0.10	
Polystyrene, gp crystal	1.05	65.5	26.2	0.943	410	425	0.004	0.32	0.03	
Polystyrene, injection impact	1.04	64.9	26.6	0.968		440	0.006	0.34	0.10	
Polysulfone	1.25	77.4	22.3	0.807	650	680	0.007	0.28	0.30	0.05
Polyurethane	1.20	74.9	23.1	0.834	400	400	0.020	0.40	0.10	0.03
PVC, rigid profiles	1.39	86.6	19.9	0.720	365		0.025	0.25	0.02	
PVC, pipe	1.44	87.5	19.7	0.714	380		0.025	0.25	0.10	

Table 2.1 *Continued*

Resin data[b]	Specific gravity (g cm⁻³)	Density (lb ft⁻³)	Specific volume (in.³ lb⁻¹)	Specific volume (cm³ g⁻¹)	Extrusion temperature (°F)	Injection temperature (°F)	Linear mold shrinkage	Specific heat (Btu lb⁻¹ °F⁻¹)	Water absorption in 24 h (%)	Maximum water content allowable for molding
PVC, rigid injection	1.29	83.6	21.0	0.756		380	0.025	0.25	0.10	0.07
PVC, flexible wire	1.37	85.5	20.2	0.731	365		0.025			
PVC, flexible extruded shapes	1.23	76.8	22.5	0.814	350		0.025			
PVC, flexible injection	1.29	80.5	21.4	0.776		300	0.025			
PTFE	2.16	134.8	12.9	0.464				0.25	<0.01	
SAN	1.08	67.4	25.6	0.927	420	470	0.005	0.31	0.03	0.02
TFE	1.70	106.1	16.3	0.589		610	0.040	0.46	0.01	
Urethane elastomers	0.83	51.6	33.5	1.210	390	400	0.001	0.46	0.07	0.03

[a] Specific information on all machine settings and plastic properties is initially acquired by using the resin supplier's data sheet on the particular compound or resin to be used.
[b] These are strictly typical average values for a resin class; consult your resin supplier for values and more accurate information.

The large number of variables summarized in Fig. 2.10 (page 131) will cause part changes if not controlled properly. Basic settings for these variables are provided by the plastic producer: the injection barrel temperature (Table 2.1), the cavity melt pressure, and so on. However, the final setting are determined by the processor on a specific machine and mold.

SCREW/BARREL PERFORMANCE

The screw is a helically flighted, hard steel shaft which rotates within a plasticizing barrel to mechanically process and advance a plastic being prepared for processes such as IM, extrusion, or blow molding (Chapters 3 and 4). Regarding the conventional reciprocating single screw, Figs 2.11 and 2.12 explain the melt action of the plastics as it travels around the screw inside the barrel.

This review concerns processing TPs. TS plastics are also processed through IMMs, but the screw is usually limited in design having an L/D of 1 (L/D to be explained later) and the barrel provides heat usually via hot water. The TS cannot be permitted to overheat in the barrel as it will solidify (Fig. 1.10, page 29). If it does, the screw must be removed from the barrel and the hard TS removed from the screw. With TPs the material cannot be permitted to cool in the cylinder as it will solidify. If it solidifies, the screw is generally removed from the barrel, in turn removing the solid TP.

With nearly all machines, only the cylinder temperature is directly controlled. The actual heat of the melt, within the screw and as it is ejected from the nozzle, can vary considerably, depending on the efficiency of the screw design and the method of operation. Factors affecting melt heat include the time plastic remains in the cylinder (the residence time); the internal surface heating area of the cylinder and the screw per volume of material being heated; the thermal conductivity of the cylinder, screw, and plastic (Table 1.28, page 86); the heat differential between the cylinder and the melt; and the amount of melt turbulence in the cylinder. In designing the screw, a balance must be maintained between the need to provide adequate time for heat exposure and the need to maximize output most economically.

In general, heat transfer problems have led screw designers to concentrate on turning screws into more efficient heat transfer devices. As a result, the internal design and performance of screws vary considerably to accommodate the different plastics that are used. Most machines are single, constant-pitch, metering-type screws to handle the majority of plastics (Figs 2.11 and 2.12).

Plastic in the screw channel is subject to changing experiences as the screw operation changes during the cycle. Each operation of the screw,

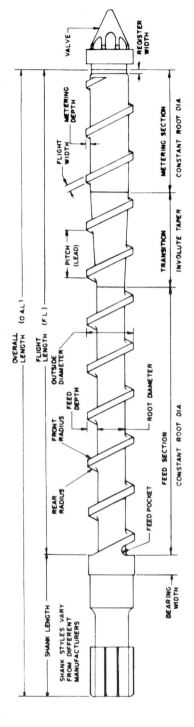

Figure 2.11 Screw nomenclature.

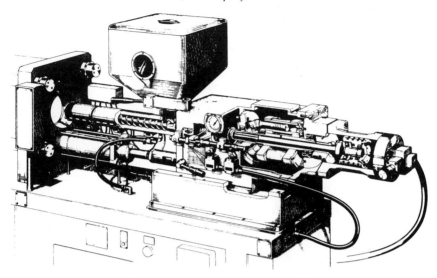

Figure 2.12 Cutout of a Negri Bossi injection unit (plasticizing unit) using a reciprocating screw with a squash-plate hydraulic drive.

(a)   (b)   (c)   (d)   (e)   (f)

Figure 2.13 Melt model for a standard screw.

whether it is moving forward, rotating and retracting during shot preparation or static during an idle period, subjects the plastic to different thermal and shear situations. Consequently, the IM plasticating process becomes rather complex, but it is controllable and repeatable within the limits of equipment capability. At a fixed screw speed, the screw pitch, diameter, and channel depth determine output. A deep-channel screw is much more sensitive to pressure changes than a shallow-channel screw. In the lower pressure range, a deep channel will provide more output; however, the reverse is true at high pressures. Shallower channels tend to give better mixing and flow patterns. Although the screw is usually a simple-looking device, its three sections accomplish many different operations at the same time: (1) solids conveying or feeding; (2) compressing, melting, and pressurizing the melt; and (3) mixing, melt refinement, and pressure–temperature stabilization (Fig. 2.13).

Hypothetical data on screws are given in Table 2.2, which provides some examples of variations on the same length-to-diameter screw processing different plastics. In reality, the **L/D ratios** (flight length/out-

Table 2.2 Hypothetical screw designs for general types of plastic

Dimensions (in.)	Rigid PVC	Impact polystyrene	Low-density polyethylene	High-density polyethylene	Nylon	Cellulose acetate/butyrate
Diameter	$4\frac{1}{2}$	$4\frac{1}{2}$	$4\frac{1}{2}$	$4\frac{1}{2}$	$4\frac{1}{2}$	$4\frac{1}{2}$
Total length	90	90	90	90	90	90
Feed zone (F)	$13\frac{1}{2}$	27	$22\frac{1}{2}$	36	$67\frac{1}{2}$	0
Compression zone	$76\frac{1}{2}$	18	45	18	$4\frac{1}{2}$	90
Metering zone (M)	0	45	$22\frac{1}{2}$	36	18	0
Depth (M)	0.200	0.140	0.125	0.155	0.125	0.125
Depth (F)	0.600	0.600	0.600	0.650	0.650	0.600

side diameter) vary according to the rheology of the resin. Advantages of a short L/D are lower residence time in the barrel, so heat-sensitive resins need not be kept at melt heat for so long, lessening the chance of degradation; the design occupies less space; it requires less torque, making the screw strength and amount of horsepower less important; and it requires a smaller investment cost, both initially and for replacement parts. Advantages of a long L/D are the screw can be designed for a greater output or recovery rate, provided that sufficient torque is available; it can be designed for more uniform output and greater mixing; it will pump at higher pressures and give greater melting with less shear, as well as providing more conductive heat from the barrel.

The **compression ratio (CR)** relates to compression that occurs on the resin in the transition section; it is the ratio of the volume at the start of the feed section divided by the volume in the metering section (determined by dividing the feed depth by the metering depth). The CR should be high enough to compress the low-bulk unmelted resin into a solid melt without air pockets. A low ratio will tend to entrap air bubbles. High percentages of regrind, powders, and other low-bulk materials will be achieved by a high compression ratio. A high CR can overpump the metering section. A common misconception is that engineering and heat-sensitive resins should use a low CR. This is true only if it is decreased by deepening the metering section, not by making the feed section shallower. The problem of overheating is related more to channel depths and shear rates than to CR. As an example, a high CR in polyolefins can cause melt blocks in the transition section, leading to rapid wear of the screw and/or barrel. CR is usually zero for processing TS material; this helps to prevent overheating, which could cause the melt to solidify in the barrel. But should there be any melt solidification, the zero CR makes it easy to remove – remove the nozzle, and the solid can be 'unscrewed'. Zero ratios are also used for TPs when the rheology so requires.

The output of a metering screw is fairly predictable, provided the melt is under control. With a square-pitch screw (a conventional screw where the distance from flight to flight is equal to the diameter), a simplified formula for **rate** or **output** is

$$R = 2.3 D^2 h g N$$

where R = rate or output (lb h^{-1}), D = screw diameter (in.), h = depth of the metering section (in.) (for a two-stage screw use the depth of the first metering section), g = gravity of the melt, and N = screw RPM.

This formula does not take into account backflow and leakage flow over the flights. These flows are not usually a significant factor unless the resin has a very low viscosity during processing, or the screw is worn out. The formula assumes pumping against low pressure, giving no consideration to melt quality and leakage flow of screws with a severely worn outside diameter (OD). Despite these and other limitations, the formula can still provide a general guide to output. If the output is significantly greater, caused by a high CR that overpumps the metering section, it may sometimes be desirable but it can lead to surging and rapid screw wear if it is excessive. If the output is a lot less than estimated, it usually indicates a feed problem or a worn screw and/or barrel. A feeding problem can sometimes be corrected by changes in the barrel heat. More often the probem is caused by factors such as screw design, shape and bulk density of the feedstock, surface conditions of the screw root, and the screw heat. The problem of screw/barrel wear can be assessed by measuring the screw/barrel.

An accurate method used to determine **output loss** due to screw wear is to compare the worn screw's current output with the initial production benchmark, originally determined by shooting into a bucket to check the shot weight. Another approach is to measure the worn screw's clearance to the barrel wall (W), which is used along with the original measured screw clearance (O) and the metering depth (M) from the screw root to the barrel wall. Here the approximate percentage output loss (OL) with constant RPM is calculated from

$$OL = (W - O)/M \times 100$$

Mixing devices

Processors have developed an almost universally accepted model for melting in a single screw (Fig. 2.11, page 136). This model, which is the basis for most computer simulations, has been demonstrated to be correct by many freeze tests. The model shows how the plastic goes from the solid state to a melt as it moves through the action of the controlled screw and barrel. The results describe such relationships as high output via deep screws, low melt temperatures via deep screws, and melt quality via shallow screws. For these different situations, solutions have been developed to provide good mixing and product uniformity at high production rates without excessive melt heat.

A variety of different mixing and barrier screws (principally used in extruders, Chapter 3), designed to improve melt processability with high output, can be used to meet the different flow requirements of the various plastics. The Dulmage Mixer (Dow Chemical), one of the first, is usually located at the end of the screw. It has a series of semicircular grooves cut on a long helix in the direction of the screw flights. These grooves interrupt laminar flow by dividing and recombining the melt many times, like a static mixer. Mixing pins placed radially in the screw root, using different patterns and shapes, improves performance. The Maddock Mixer (Union Carbide) is a series of opposed, semicircular grooves along the screw exit. Alternate grooves are open to the upstream entry, and the other grooves are open to the downstream discharge. Its ribs and flutes, which divide the alternating entry and discharge grooves, also alternate. This mixer does an effective job of mixing and screening unmelted plastics. Unmelted material gets trapped in the grooves, mixes with the melt, and itself becomes molten. In the Pulsar Mixer (Spirex Corp.) the metering section is divided into constantly changing sections, which are either deeper or shallower than the average metering depth. The resin alternates many times between the shallower depth with a somewhat higher shear and deeper channels with a lower shear. This tumbling and massaging action produces excellent mixing and melt uniformity without high shear [1].

Barrier screws

The original barrier screw was a very important design. The first patent, in 1959 (Maillefer and Geyer of Uniroyal), had broad claims; so all subsequent barrier screw designs pay a royalty under the original patent. Barrier screws have two channels in the barrier section, which are located mainly in the transition section (Fig. 2.14). A secondary flight is started, usually at the beginning of the transition, to create two distinct channels, a solids channel and a melt channel. The barrier flight is undercut below the primary flight, allowing melt to pass over it.

There are many different barrier screw designs, each offering unique advantages. They include the Uniroyal (the original barrier screw); the MC-3 of Hartig; the Efficient of New Castle Industries; the Barr 2 of R. Barr Inc.; the VPB of Davis Standard; the Willert II of W. H. Willert Inc.; and the Double Wave of HPM Corp. [1].

Custom-designed screws

The specially designed barrier screws meet special requirements. Plastics tailored for specific applications usually have a high or very high molecu-

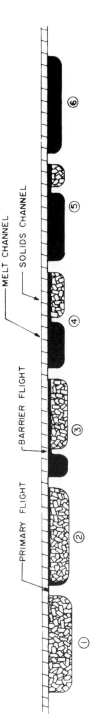

MELT CHANNEL
SOLIDS CHANNEL

PRIMARY FLIGHT   BARRIER FLIGHT

(1) The feed section establishes the solids conveying in the same way as a conventional screw.

(2) At the beginning of the transition (compression), a second flight is started. This flight is called the barrier or intermediate flight, and it is undercut below the primary flight OD. This barrier flight separates the solids channel from the melt channel.

(3) As melt progress down the transition, melting continues as the solids are pressed and sheared against the barrel, forming a melt film. The barrier flight moves under the melt film and the melt is collected in the melt channel. In this manner, the solid pellets and melted polymer are separated and different functions are performed on each.

(4) The melt channel is deep, giving low shear and reducing the possibility of overheating the already melted polymer. The solids channel becomes narrower and/or shallower forcing the unmelted pellets against the barrel for efficient frictional melting. Break up of the solids bed does not occur to stop this frictional melting.

(5) The solids bed continues to get smaller and finally disappears into the back side of the primary flight.

(6) All of the polymer has melted and gone over the barrier flight. Melt refinement can continue in the metering section. In some cases mixing ssections are also included downstream of the barrier section. In general, the melted plastic is already fairly uniform upon exit from the barrier section.

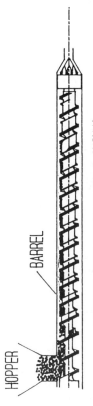

HOPPER

BARREL

BARRIER TYPE SCREW (BARR-2 SHOWN)

Figure 2.14 Melt model for the barrier screw.

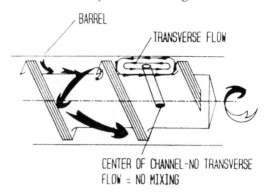

Figure 2.15 Typical flow profile for a conventional screw.

lar weight; they are often sensitive to heat and shear. Single-flighted screws traditionally provide melts via relatively high shear rates. As the solid plastic moves down the screw, a thin melt film forms between the barrel and the solid plastic bed. Due to screw rotation, the shear on the film is quite high, generating a great amount of shear-energy heat. This heat is transferred to the solid bed, promoting additional melting. The film becomes thicker as it moves to the rear of the channel, where it is scraped off the barrel by the screw flight. Then the only place it can go is down into the channel to create the usual flow profile in the screw (Fig. 2.15). The melting barrier type of screw is now a well-known means of improving melt processability. But in spite of its advantages, it has difficulty running at a low enough melt heat for high molecular weight plastics; this is because it relies on high-shear melting.

Mechanical requirements

Screws always run inside a stronger, more rigid barrel, so they are not subjected to high bending forces. The critical strength requirement is resistance to torque. This is particularly true of smaller screws with diameters of $2\frac{1}{2}$ in. (64 mm) or less. Unfortunately, the weakest area of a screw is the portion subject to the highest torque. This is the feed section, which has the smallest root diameter. A rule of thumb is that a screw's ability to resist twisting failure is proportional to the cube of the root diameter in the feed section. The maximum torque that can be applied to a screw can then be calculated [1].

Valves and screw tips

When the melt is forced into the mold, the plunger action could cause the melt to flow back into the screw flights. With heat-sensitive resins such as

Table 2.3 Comparison of sliding-ring valve with ball-check valve

Advantages	Disadvantages
Sliding-ring valve	
Greater streamlining for less degrading of materials; best for heat-sensitive materials	Less positive shutoff, especially in $4\frac{1}{2}$ in. dia. and larger sizes; less shot control
Less barrel wear	More expensive than front-discharge ball-check
Less pressure drop across valve	
Best for vented operation	
Easier to clean	
Less expensive than side-discharge ball-check	
Ball-check valve	
More positive shutoff; better shot control	Less streamlined; more degrading of heat-sensitive materials
Front-discharge ball-check less expensive than sliding-ring valve	More barrel wear
	Side-discharge type more expensive than sliding-ring type
	Greater pressure drop, creating more heat
	Poor for vented operation
	Harder to clean

PVC and thermosets, a plain or smearhead screw tip is generally used. But this is not adequate for other resins; several different check valves are used instead. They work in the same manner as a check valve in a hydraulic system, allowing fluid to pass only in one direction. These check valves, which are basically a sliding-ring or ball-check design (Fig. 2.6, page 125) are supplied by many manufacturers. Table 2.3 shows some comparisons and Table 2.4 shows the influence of check valves on plastics performance.

Vented barrels

Moisture retention in and on plastics has always been a problem for all processors. Surface moisture or moisture absorbed within the plastic can cause splay, an unsightly surface defect of the molded part, and reduce mechanical properties. The increased use of hygroscopic plastics (Chapter 1) also requires care and the assurance of proper drying of material via the usual technique, using dryers and/or vented barrels (Fig. 2.16). There are advantages of using vented barrels as opposed to the more familiar dryers [54].

Table 2.4 Effect of varying check valves and processing conditions on final physical properties of dry blended polystyrene copolymer

	Control	Check valve	Gate size	Back pressure	Screw speed	Fill-time
Check valve	ring	ball	ring	ring	ring	ring
Gate size (in.)	0.13 × 0.25	0.13 × 0.25	0.062 × 0.063	0.13 × 0.25	0.13 × 0.25	0.13 × 0.25
Back pressure (psi)	0	0	0	125	0	0
Screw speed (RPM)	73	73	73	73	53	73
Fill-time (s)	1	1	1	1	1	4
Notched Izod impact strength (ftlb^{-1}in.$^{-1}$)	3.2	2.6	1.9	1.5	4.0	3.9
Flexural strength (kpsi)	17	17	17	17	19	18
Flexural modulus (Mpsi)	0.98	0.98	0.98	0.98	1.00	0.98

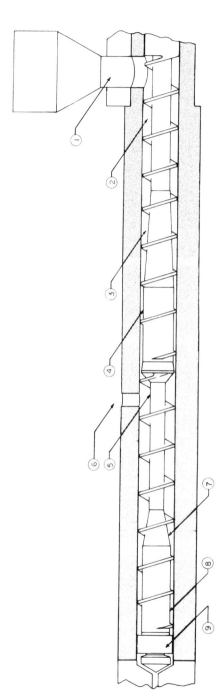

(1) Wet material enters from a conventional hopper. (2) The pellets are conveyed forward by the screw feed section, and are heated by the barrel and by some frictional heating. Some surface moisture is removed here. (3) The compression or transition section does most of the melting. (4) The 1st metering section accomplishes final melting and evens flow to the vent section. (5) Resin is pumped from the 1st metering section to a deep vent or devolitizing section. This vent section is capable of moving quantities well in excess of the material delivered to it by the 1st metering section. For this reason, the flights in the vent section run partially filled and at zero pressure. It is here that volatile materials such as water vapor, and other nondesirable materials, escape from the melted plastic. The vapor pressure of water at 500°F is 666 psi. These steam pockets escape the melt, and travel spirally around the partially filled channel until they escape out the vent hole in the barrel. (6) Water vapor and other volatiles escape from the vent. (7) The resin is again compressed and pressure is built in the 2nd transition section. (8) The 2nd metering section evens the flow and maintains pressure so that the screw will be retracted by the pressure in front of the non-return valve. (9) A low resistance, sliding ring, non-return valve works in the same manner as it does with a nonvented screw.

Figure 2.16 Basic operation of a vented barrel.

Elimination of predrying: a vented injection unit removes moisture more completely without a dryer. Often a dryer cannot do the job completely in a reasonable time period.

Rapid startup and color or material changes: you do not have to wait for hours when starting up or when changing colors or materials. This increases machine and personnel utilization.

Superior parts: the improved melt, free of volatiles, renders higher-quality parts with excellent appearance and better physical properties. Splay marks are eliminated from appearance parts and parts to be plated.

Energy efficiency: the vented machine uses less energy. BTUs are not lost while material stands in large hoppers at elevated temperatures for long periods. Dryers are large users of energy.

Removal of other volatiles: water vapor is not always the only volatile contaminant that should be removed. The vent removes other undesirable materials that come off at temperatures not possible in a dryer. Escape of volatiles is easier from a melted and agitated plastic. This has been very effective in solving mold and ejector pin plate-out problems.

Elimination of dryer maintenance: dryers are high-maintenance items with clogged filters, heater-element burnout, and contaminated desiccant beds. Even in shops with good routine maintenance programs, it is common to operate with ineffective dryers for long periods before they are noticed. When this happens, quality goes down, and scrap accumulates.

Contamination and material-handling cut: there is no need to clean out larger complicated hopper dryer systems on every material or color change. The simple, lightweight, standard hopper is easier to clean.

Less space required: the hopper dryer requires a large volume in order to get up to 5 h of drying time. This means a heavy, large, and high hopper that may not fit into the space available.

Elimination of dryer variability: the variation in part quality and appearance due to changes in dryer performance is eliminated. The vent operates the same all the time.

Greater use of regrind: the improved moisture removal ability of the vent allows the use of percentages of regrind. The vent also allows the storage of materials in open containers.

Reduced mold venting: the removal of volatiles from the vent reduces the mold-venting problem. It can also eliminate the problem of clogged mold vents.

Care is required when a plug is placed in the vent opening so that a vented barrel can be used in the same way as a solid nonvented barrel. Internal barrel pressure can develop that exceeds the strength of the bolt retaining the plug, causing the plug to be released with an explosive action. Operators may be hit by the plug and/or the hot melt. One safety precaution is to rotate the barrel so the plug is pointing downward or away from the operator. The most important safety measure is to use retaining bolts (or another method of plug attachment) that will provide more than enough strength to prevent plug release. The injection molding or extrusion barrel manufacturer should be able to supply the information needed to ensure safe operation. Another precaution is to install a pressure gauge at the head of the barrel and to establish a maximum pressure at which the barrel can be safely operated. To eliminate any runaway situation, shear bolts or a rupture disk can be installed near the plug.

During startup and operation of a vented barrel, make sure the barrel in front of the vent is at a temperature above the 'freezing' point of the plastic. To help retain heat in the metal barrel, thermal insulation should be used to cover exposed metal surfaces (even if the barrel is not vented).

Air entrapment

Air can be entrapped in the melt during processing. This can happen when plastic (pellets, flakes, etc.) is melted in a normal air environment (as in a plasticating extrusion process or in an injection barrel, compression mold, casting form, spray system, etc.), and the air cannot escape (Chapter 1). The melt is generally subject to a compression load, or even a vacuum, which causes release of air; but sometimes the air is trapped. If air entrapment is acceptable, no further action is required, although performance and esthetics usually mean that air entrapment is unacceptable.

Changing the initial melt temperature in either direction may solve the problem. With a barrel and screw, it is important to study the effects of temperature changes. Another approach is to increase the pressure in processes which use process controls. Particle size, melt shape, and the melt delivery system may have to be changed or better controlled. A vacuum hopper feed system may be useful. With screw plasticators, changes in screw design may be helpful. A vented barrel will usually solve the problem.

The presence of bubbles could be due to air alone or moisture, plastic surface agents or volatiles, degradation, or the use of contaminated regrind. With molds such as those used for injection, compression, casting, or reaction injection, air or moisture in the mold cavity will be the culprit. So the first step to resolving a bubble or air problem is to be sure what problem exists. A logical troubleshooting approach can be used.

Shot-to-shot variation

Shot-to-shot variations can occur during IM. Major causes of inconsistency are worn nonreturn valves, bad seating of a nonreturn valve, a broken valve ring, a worn barrel in the valve area, and/or a poor heat profile. To identify the cause, one follows a logical procedure [1]. Any problem created by the valve will cause the screw to rotate in the reverse direction during injection. To locate the trouble, one must pull and inspect the valve, and check the OD of the ring for wear. The inspector looks for a broken valve stud (caused by cold startup when the screw is full of plastic), bad seating of the ring or ball (angles of the ring ID and the seat must be different, in order to ensure proper shutoff action at the ID of the ring), and a broken ring. One checks the dimensions of the valve and compares them with those determined before using the machine.

A poor heat profile for crystalline resins can cause unmelted material to be caught between the ring and the seat, holding the valve open and allowing leakage. A change in the heat profile (Chapter 1) or the machine's plasticizing capacity is not sufficient to correct the problem. For any resin, if the problem does not occur with every shot, the cause may be improper adjustment or damaged barrel heat controls.

Nonuniform melt density could be caused by nonuniform feeding to the screw and/or the regrind blend, which could have a different bulk density. Increasing the back pressure may help [1]. This throughput condition, the residence time of the plastic in the barrel, and the barrel heat profile are all important in obtaining the best melt quality. The heat profile is the most important parameter and varies from resin to resin, as well as with different cycle times and shot sizes. As the following example shows, a screw operating under two different conditions will produce different results.

Consider a screw that has a 2 in. (51 mm) diameter, $L/D = 20$, and a 20 oz (0.57 kg) melt screw capacity. With a 15 s cycle and a shot size of 2 oz (57 g), it operates as follows:

20 oz (screw capacity) ÷ 2 oz = 10 cycles
15 s cycle = 4 cycles per minute
10 cycles ÷ 4 = 2.5 min of residence time, from the time plastic starts
   through the screw until it enters the mold

Another set of requirements uses a 6 oz (170 g) shot size with the same 15 s cycle:

20 oz ÷ 6 oz = 3.33 cycles
3.33 cycles ÷ 4 = 0.83 min of residence time

In the second case, a higher rate of melting will be required, with the probability that the screw will be inadequate for the melt, and problems will develop.

The inventory in a screw will run between $1\frac{1}{2}$ and 2 times the maximum shot size rating in polystyrene. With other resins, calculate the differences in density to arrive at the maximum shot size and the expected inventory.

Purging

Purging has always been a necessary evil, consuming substantial amounts of materials, labor, and machine time, all nonproductive. In IM (extrusion, blow molding, etc.), it is sometimes necessary to run hundreds of pounds of resin to clean out the last traces of a dark color before changing to a lighter one. Sometimes there is no choice but to pull the screw for a thorough cleaning. Although there are few generally accepted rules on how to purge, the following tips should be considered: (1) try to follow less viscous with more viscous resins; (2) try to follow a lighter color with a darker color resin; (3) maintain the equipment; (4) keep the materials-handling equipment clean; and (5) use an intermediate resin to bridge the

Table 2.5 Guidelines for purging agents

Material to be purged	Recommended purging agent
Polyolefins	HDPF
Polystyrene	Cast acrylic
PVC	Polystyrene, general-purpose, ABS, case acrylic
ABS	Cast acrylic, polystyrene
Nylon	Polystyrene, low melt index HDPE, cast acrylic
PBT polyester	Next material to be run
PET polyester	Polystyrene, low melt index HDPE, cast acrylic
Polycarbonate	Cast acrylic or polycarbonate regrind; follow with polycarbonate regrind; do not purge with ABS or nylon
Acetal	Polystyrene; avoid any contact with PVC
Engineering resins	Polystyrene, low melt index, HDPE, cast acrylic
Fluoropolymers	Cast acrylic, followed by polyethylene
Polyphenylene sulfide	Cast acrylic, followed by polyethylene
Polysulfone	Reground polycarbonate, extrusion-grade PP
Polysulfone/ABS	Reground polycarbonate, extrusion-grade PP
PPO	General-purpose polystyrene, cast acrylic
Thermoset polyester	Material of similar composition without catalyst
Filled and reinforced materials	Cast acrylic
Flame-retardant compounds	Immediate purging with natural, non-flame-retardant resin, mixed with 1% sodium stearate

temperature gap (such as that encountered in going from acetal to nylon), and use a PS as a purge.

Ground/cracked cast acrylic and PE-based materials are the main purging agents, but others are commercially available for certain machines and materials. Cast acrylic, which does not melt completely, is suitable for virtually any resin. About one pound (450 g) for each ounce (28 g) of

Table 2.6 Guidelines for resin changes

Material in machine	Material changing to	Mix with rapid purge and soak	Temperature-bridging material	Follow with
ABS	PP	ABS	–	PP
ABS	SAN	SAN	–	SAN
ABS	Polysulfone	ABS	PE	Polysulfone
ABS	PC	ABS	PE	PC
ABS	PBT	ABS	PE	PBT
Acetal	PC	Acetal	PE	PC
Acetal	Any material	PE	–	New material
Acrylic	PP	Acrylic	–	PP
Acrylic	Nylon	Acrylic	–	Nylon
TPE	Any material	PE	–	New material
Nylon	PC	PC	–	PC
Nylon	PVC	Nylon	PE	PVC
PBT	ABS	PBT	PE	ABS
PC	Acrylic	PC	–	Acrylic
PC	ABS	PC	PE	ABS
PC	PVC	PC	PE	PVC
PE	Ryton	PE	PE	Ryton
PE	PP	PP	–	PP
PE	PE	PE	–	PE
PE	PS	PS	–	PS
PETG	Polysulfone	PETG	–	Polysulfone
Polysulfone	ABS	Polysulfone	PE	ABS
Polysulfone	ABS	Cracked acrylic	–	ABS
PP	ABS	ABS	–	ABS
PP	Acrylic	Acrylic	–	Acrylic
PP	PE	PE	–	PE
PP	PP	PP	–	PP
PS	PP	PP	–	PP
PVC	Any material	LLDPE or HDPE	–	New material
PVC	PVC	LLDPE or HDPE	–	PVC
PPS	PE	PPS	PE	PE
SAN	Acrylic	Acrylic	–	Acrylic
SAN	PP	SAN	–	SAN

injection capacity will be needed (5–10 lb in.$^{-1}$ or 17.83 g mm^{-1} of screw diameter in an extruder). With extruders, special conditions and preparations are required, which suppliers of the purging compound can recommend (remove dies, screen packs, etc.).

PE-based compounds usually contain abrasive and release agents. They are used to purge the 'softer' TPs (polyolefins, styrenes, some PVCs, etc.). With extruders, many of the requirements and restrictions do not apply.

These purging agents function by mechanically pushing and scouring residue out of the machines. Others also apply chemical means. Tables 2.5 and 2.6 provide information on purging.

Wear

Wear in screw plasticators generally causes an increase in the clearance between screw flight and barrel. It often occurs towards the end of the compression section. This type of wear is more likely to occur when the screw has a high compression ratio. Wear in the compression section of the screw reduces the melting capacity, and will lead to temperature nonuniformities and pressure fluctuations. Wear in the metering section of the screw will reduce the pumping capacity; however, the reduction in pumping capacity is generally quite small as long as the wear does not exceed 2–3 times the design clearance. An increased flight clearance will also reduce the effectiveness of the heat transfer from the barrel to the plastic melt and vice versa; this may contribute to temperature nonuniformities in the plastic melt.

Wear can only be detected by disassembling the unit and by inspection of the screw and barrel. If the wear is serious enough to affect performance, it will often be noticeable with the naked eye. However, it is recommended to measure the ID of the barrel and the OD of the screw over the length of the machine. If this is done regularly, one can determine how fast wear is progressing with time. By extrapolating to the maximum allowable wear, one can determine when the screw and/or barrel should be replaced or rebuilt. If replacement as a result of wear is necessary after several years of operation, the easiest solution is simply to replace the worn parts. However, if replacement as a result of wear becomes necessary within a short period of time, perhaps several months, then simple replacement will not provide an acceptable solution. In short-term wear problems, the cost of downtime and replacement parts can easily become unacceptable and the solution has to be found in reducing the actual wear rate instead of simply replacing the worn parts. To reduce the wear rate, one has to understand the wear mechanism(s) in order to determine the most effective way to reduce wear. Mechanisms that cause wear include adhesive wear (metal-to-metal contact under high stresses), abrasive wear (galling), laminar wear (thin outer layers of the metal interface wear), surface-fatigue wear (micro- or macroscopic separation from the surfaces) and corrosive wear (chemical reaction and mechanical attack of the sliding surfaces) [63].

Output

The objective of a screw design is generally to deliver the largest amount of output at acceptable melt quality. Unfortunately, high output and mixing quality are, to some extent, conflicting requirements. As output goes up, residence time goes down and so does the mixing quality. However, the mixing quality can generally be restored by incorporating mixing sections; either a mixing section along the screw or a static mixing section.

It is also important to realize that all functional zones of the extruder are interdependent. It makes little sense to drastically increase the pumping capacity if the melting rate is the real bottleneck in the process. Thus, before designing a new screw to replace an existing screw, one should determine which part of the extruder is limiting the rate.

CLAMPING

Basic types

The IMM clamp is used (1) to close the mold, (2) to hold the mold closed during the melt injection and 'curing' (solidification), and (3) to open the mold for the removal of molded product [1, 64–72, 80, 86–88]. Types of design include toggle (Fig. 2.17), hydraulic (Fig. 2.18), and hydromechanical (Fig. 2.19). There are also hydraulic–electric and the increasingly popular all-electric IMMs that will be reviewed later. Toggle clamps are very popular on low-tonnage machines because they are less expensive to manufacture and provide operating (molding) advantages. Their features include high mechanical advantage at lockup, inherent built-in clamp slowdown, and rapid clamp operation. A hydraulic cylinder moves the toggle's crosshead forward, extending the toggle links and moving the platen forward. As the clamp closes, the mechanical advantage is low, producing rapid platen movement. As the platen approaches the mold-close position, the toggle links undergo a transition from high speed and low mechanical advantage to low speed and high mechanical advantage. Low speed is critical for mold protection and high mechanical advantage is needed to build clamp tonnage.

When the linkage is fully extended, locking the mold closed, hydraulic pressure is not needed to hold tonnage; with the hydraulic IMM the pressure has to be held. Since the toggle linkage must be at full stroke to achieve tonnage, adjusting the clamp to different mold heights (width of

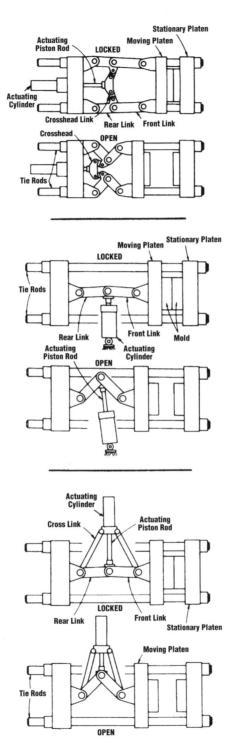

Figure 2.17 Mechanical clamping systems use a variety of toggles.

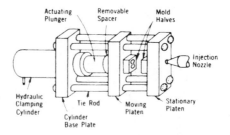

Figure 2.18 Hydraulic clamp, schematic.

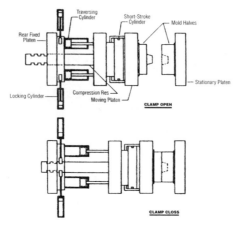

Figure 2.19 Hydromechanical clamp, schematics.

mold between platens) is accomplished by moving the entire toggle mechanism and moving the platen assembly to accommodate the mold.

Advantages of the toggle design include fast operation, reduced energy consumption and lower machine costs. But compared to hydraulic designs it has relatively complicated mold setup and startup, maintenance for the link pins and bushings, particularly with older machines, and there is a potential difficulty in retaining parallelism, especially with large molds. However, major advances in toggle designs have reduced toggle maintenance and setup problems. Among the improvements have been graphite-impregnated oilless bushings, automatic tonnage adjustment to ease mold setup and oversized platen-support skates to improve platen parallelism. Another advance has been the totally **electric machine**, which uses AC servo motors to replace hydraulic power on toggle clamp machines. Electric power greatly reduces the energy required to mold parts and also offers a high degree of machine accuracy.

Hydraulic clamps are used predominately in machines of 150–4000 tons. Rapid closing is accomplished by either a booster tube internal to the clamp cylinder or outboard hydraulic cylinders of relatively small diameter. Applying large amounts of oil to a small area produces faster clamp speeds. Using outboard cylinders for clamp transversing simplifies clamp design; it also eliminates internal booster tubes and the need for piston rings on the main ram. Prefill valve designs have also been simplified, allowing higher oil flows and faster clamp speeds.

By eliminating the booster tube and leakage over the ram piston rings, reliability has been improved and the hydraulic pressure for clamp tonnage is effectively trapped inside the ram cylinder. This eliminates the need for a hydraulic pump to maintain clamp tonnage, improves clamp reliability, and reduces energy requirements. Regardless of design, the hydraulic system offers ease and flexibility in machine setup and operation (compared to toggle). Since tonnage can be developed anywhere along the clamp stroke, mold setup is performed by simply setting a position that corresponds to where the mold halves touch. Other advantages include fewer moving parts, less lubrication, and more sensitive low-pressure mold protection.

Potential disadvantages include greater oil requirements, higher energy consumption, slower clamp operation, more difficult clamp slowdown control, and ram-seal oil leakage. However, advances in the relatively newer designs have minimized these disadvantages. The use of small-diameter, outboard, rapid-transverse cylinders and regenerative hydraulic circuit designs increases clamp speeds and reduces energy consumption, making them comparable to toggle systems. Accumulator-assisted tonnage is also used for short-cycle production.

Hydromechanical clamps combine mechanical and hydraulic functions to move the clamp and build up tonnage. They are used typically in clamp sizes of 1000 tons and larger. The hydromechanical design includes the following parts: (1) hydraulic cylinders to traverse the moving platen to a position where the mold halves almost touch, (2) a mechanical locking plate to prevent rearward movement during tonnage buildup, and (3) one or more short-stroke cylinders to move the mold the final distance to closing and tonnage buildup. Advantages include fast tonnage build time, lower oil requirements, and with certain designs, small machine footprints. A disadvantage is a high degree of complexity.

Variable-speed, brushless DC **electric motors** have been used in conventional toggle or hydraulic-clamping IMMs for injection screw rotation. They can provide significant energy savings. And variable-speed, brushless DC motors are used in place of hydraulic-drive AC motors. These variable-speed motors with variable-volume hydraulic pumps can provide up to 70% energy savings and improve power factors up to a factor of 0.98 [1, 60–62, 67, 69, 71, 87–90].

Different systems

The usual IMM has three platens (Figs 2.17 to 2.19): the stationary platen, the moving platen, and the cylinder base plate/platen. There are also two-platen IMMs, currently providing economical (energy-saving), space-saving (one-third floor space) hydraulic and hydromechanical clamp designs for machine over 500 tons; these machines use tie-bars as hydraulic pistons. They usually have four high-pressure cylinders mounted on the fixed platen for locking, and short-stroke clamping cylinders on the moving platen for quick buildup of clamping force and fast mold opening. During the 1960s this type of design was used by various machine builders (Fig. 2.20). They are now becoming more popular.

Figure 2.20 This 2500 ton hydraulic machine has a special retractable tie-rod system; the car is included for scale.

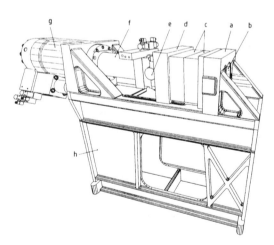

Figure 2.21 A tie-barless injection molding machine.

Another design is the tie-barless IMM. Figure 2.21 (Engel hydraulic machine) identifies (a) stationary platen, (b) opening for the injection unit, (c) mold, (d) movable platen, (e) rotary joint, (f) clamping piston, (g) clamping cylinder, and (h) IMM frame. The clamping is a fully hydraulic, low-friction plunger design with internal jack ram for fast closing. The clamping force is contained in a very rigid C-frame structure open at the top. Compensation is made for any slight rotation of the stationary and movable platens with respect to one another as a result of frame deformation under the action of the clamping force. A rotary joint with a precisely adjustable positive stop is used between the clamp ram and the movable platen. As the clamp force builds up, the movable platen adjusts itself to be exactly parallel with the stationary platen [1, 56].

Clamping forces

The clamping force required to keep the mold closed during injection must exceed the force given by the plastic melt pressure in the 'live' cavities and the total projected area of all impressions and runners. The projected area can be defined as the area of the shadow cast by the molded shot when it is held under a light source, with the shadow falling on a plane surface parallel to the parting line.

With cold-runner systems for TPs (or hot-runner systems for TSs), the projected areas include runners and sprues. For hot-runner TPs (cold-runner TSs), runners and sprue are not included. As an example, if the projected area is $132 \, in.^2$ ($852 \, cm^2$), and a pressure of 4000 psi (28 MPa) is required, the clamp force is

$$132 \, in.^2 \times 4000 \, psi = 528\,000 \, lb \qquad 0.0852 \, m^2 \times 28 \, MPa = 2.38 \, MPa$$
$$528\,000 \, lb \div 2000 \, lb = 264 \, tons$$

One should consider including a safety factor of 10–20% to ensure sufficient clamping force, particularly when not familiar with the operation and/or material. Then the clamping force would be 290–317 tons (2.62–2.86 MPa) in the cavity. However, because of partial hardening of plastic as it flows through the relatively restrictive sprue and runners in a cold-runner TP system (or a hot-runner TS system), the actual pressure in the cavity is less than the applied plunger pressure (Fig. 2.2, page 122).

The actual pressure developed within a mold cavity varies directly with the thickness of the molded section and inversely with the melt viscosity. Thick sections require greater clamping force than thin sections because the melt in the thick sections stays semifluid longer. Similarly, a higher stock heat, hotter mold, larger gates, or a faster rate of injection will require a higher clamping force [1].

Whereas the projected area determines the clamping force, the weight or volume of a shot determines the capacity of the IM machine required.

For the hot runners of TPs, the shot size includes the gate and runners. Capacities of machines are generally rated in ounces of general-purpose PS; with other resins, convert to the correct capacity by relating the resin densities to that of PE. If the shot size is based on volume, densities are not involved.

Using too much clamping force has drawbacks: (1) a slower cycle time; (2) possible damage to the mold; (3) reduced venting; (4) possible damage to platens if a small mold is used in a large-platen machine (the machine builder should provide information); and (5) extra energy consumption in hydraulics

To determine the proper clamping force, one must start at low force and begin molding at a reduced injection pressure, gradually building up melt pressure until the mold fills without flashing. If flash occurs prior to mold fill, it is necessary to increase the clamp tonnage gradually until no flash exists. Then the mold lockup is set correctly. If the melt temperature is lowered, the injection pressure and time will have to be changed, with a possible increase of clamp force.

If the mold still flashes, usually in the center, the mold-clamping area may be too large. To determine if this is the problem, one rubs machinist bluing on the face of one mold half, then clamps and opens the mold to see if bluing was transferred. With incomplete transfer, the processor puts the mold in a machine with higher clamp tonnage, and operating costs go up. Running the mold in the original machine would require removal of probably 0.010 in. (254 μm) of metal in the cavity areas that were not blued.

TIE-BAR GROWTH

One problem that most controls do not consider involves the effect of heat on tie-bars, which can directly influence mold performance, particularly at startup. If the heat differs from top and bottom bars, it is necessary to insulate the mold from the platens. The insulator pad used also confines heat more to the mold, producing savings in heat and/or better heat control. Tie-bar elongation calculations can be made and related to IM performance [1].

HYDRAULIC AND ELECTRIC IMMs

IMMs in the high clamping force range have for many years been following a trend towards individual electric drives. This development is not new; three-phase synchronous motors with intermediate gears have been used since the start of the screw drive [1, 63, 69, 87, 88, 90]. Advances in three-phase servoengineering, in particular frequency converters, expanded the use of electric drives in IMMs and other processes (extrusion, blow molding, etc.). These individual electric drives have advantages such as lower energy consumption, the possibility of precise control, accurate reproducibility, and high dynamics of startup and shutdown movements during the IM process. Their disadvantage to date has been their very high price (a factor of about 1.6) which tends to restrict their use. The main reason for the price differential is their expensive drive components.

Not only do all drive components offer a good cost-to-performance ratio, they also provide technical advantages. The logical result has been the combination of these two systems. The combination provides advantages in each separated system. The range today extends from all-electric IMMs (desired and used in certain product markets such as medical) to hybrid IMMs (hydraulic–electric). The all-electric drives have individual electric drives for all functions of the IMM, including a separate electric motor for screw plasticizing.

The concept of the hybrid machine represents a technically effective and economically reasonable compromise. Because of their low energy consumption and their rapid movements, hybrid machines are suitable for molding complicated and thin-walled parts. At the current rate of technical development, it is predicted that more economical and efficient electric drives will become available in the near future, making it possible to build

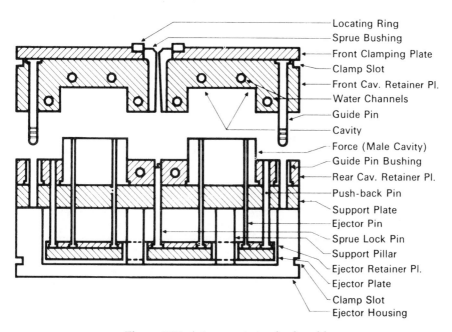

Locating Ring
Sprue Bushing
Front Clamping Plate
Clamp Slot
Front Cav. Retainer Pl.
Water Channels
Guide Pin
Cavity
Force (Male Cavity)
Guide Pin Bushing
Rear Cav. Retainer Pl.
Push-back Pin
Support Plate
Ejector Pin
Sprue Lock Pin
Support Pillar
Ejector Retainer Pl.
Ejector Plate
Clamp Slot
Ejector Housing

Figure 2.22 A two-part standard mold.

an all-electric IMM having a reasonable and competitive cost-to-performance capability [1, 69, 88].

MOLDS

A mold must be considered as one of the most important pieces of production equipment in the plant. It is a controllable complex device (Fig. 2.22) that must be an efficient heat exchanger. If not properly handled and maintained, it will not be an efficient operating device. Under pressure, hot melt moves rapidly through the mold. Water or some other medium circulates in the mold to remove heat (for TPs) or add heat (for TSs). Air is released from cavities to eliminate melt burning and/or voids in the part. All kinds of action operate, including sliders and unscrewing devices [1]. Parts are ejected (knockout pins, air, etc.) at the proper time. These

Table 2.7 Examples of steels used in different parts of a mold

Type of steel	Typical uses in injection molds
4130/4140	General mold base plates
P-20	High-grade mold base plates, hot-runner manifolds, large cavities and cores, gibs, slides, interlocks
4414 SS, 420 SS (prehardened) P5, P6	Best grade mold base plates (no plating required), large cores, cavities and inserts; hobbed cavities
01	Gibs, slides, wear plates
06	Gibs, slides, wear plates, stripper rings
H-13	Cavities, cores, inserts, ejector pins and sleeves (nitrided)
S7	Cavities, cores, inserts, stripper rings
A2	Small inserts in high-wear areas
A6	Cavities, cores, inserts for high-wear areas
A10	Excellent for high-wear areas, gibs, interlocks, wedges
D2	Cavities, cores, runner and gate inserts for abrasive plastics
420 SS	Best all-around cavity, core, and insert steel; best polishability
440C SS	Small to medium-size cavities, cores, inserts, stripper rings
250, 350	Highest toughness for cavities, cores, small unsupported inserts
455M SS	High toughness for cavities, cores inserts
M2	Small core pins, ejector pins, ejector blades (up to $\frac{5}{8}$ in., 16 mm diam.)
ASP 30	Best high-strength steel for tall, unsupported cores and core pins

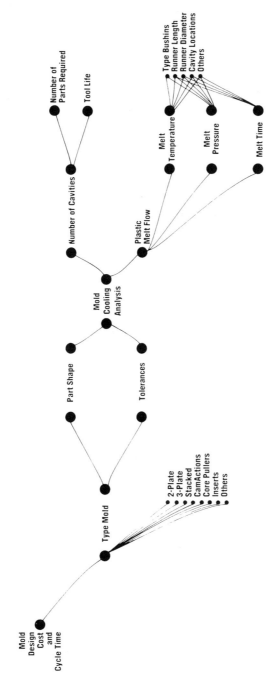

Figure 2.23 Mold operation: important control factors.

basic operations in turn require all kinds of interaction, including such parameters as fill-time, hold pressure, and other variables, as shown in Figs 2.10 and 2.23. Typical mold steels are shown in Table 2.7.

Each of the plastics used has special distinctive properties. Some are abrasive or corrosive; others require very tight heat control and pressure (Fig. 2.10, page 131). Settings that work for one resin will not probably work for another; a machine change to a duplicate will probably require different settings. Shrinkage requires special attention. Crystalline resins shrink more than amorphous resins (Chapter 1). Differential shrinkage can cause warpage. With tight part tolerances, it is necessary to leave more, rather than less, steel in the mold so that corrections requiring metal 'cutting' can correlate processing with tolerances.

Table 2.8 Protective coatings for molds/dies

Material	Method of application
Chromium	Plating
Nickel	Plating
Electroless nickel	Solution treatment
Nedox electroless nickel	Solution treatment followed by TFE impregnation; used on copper and ferrous alloys
Tufram TFE aluminum	Deep anodizing process followed by TFE impregnation; used on aluminum alloys
TFE ceramic	Spray and bake application; used for all die materials that can withstand 250 °C bake
Tungsten silicide	Solution treatment; used on steel and ferrous alloys
Tungsten carbide	Explosion impact or flame spray with plasma arc; used for all high-melting metals to improve abrasion resistance
Aluminum oxide	Plasma flame spray; used for extreme abrasion resistance; used on steel dies but usually limited to small dies because of expansion problems; works best on 18-8 stainless
PTFE	Spray and bake application; used for low-friction and low-adhesion application; poor abrasion resistance
Polyimide, aramid	Straight organic coatings with high softening points (450–550 °C), which are applied by spray and baked; low friction characteristics against some resins (for example, PVC); moderate abrasion resistance
Filled polyimide, aramid	Aramid and polyimide systems containing TFE and other fluorocarbon resins to improve the friction properties

Table 2.9 Guide to mold/die cleaning methods[a]

Method	Mold wear	Expense	Cleaning speed	Cleaning degree	Hazard waste	Disposal problem	Operating hazard	Damage caused
Chemical tank	2	4	5	1	1	Yes	Yes[b]	a
Ultrasonic chemical	2	10	3	5	1	Yes	Yes[b]	a
Wet-blast glass bead	5	7	8	1	10	No	No	b, c
Dry-blast glass bead	8	1	10	10	10	No	No	b, c, d
Plastic dry-blast	1	1	10	10	10	No	No	None
Hand cleaning	2	5	5	10	5	No	No	e
Dry-blast with sand or other abrasive	10	1	10	10	10	No	No	b, c, d

a = surface etching, b = surface pitted, c = round edge, d = removes chrome, e = filing/grinding premature surface/edge wear.

[a] 1 = low/slow, 10 = fast/best.
[b] Use proper precautions.

CAD and CAE programs are available that can aid in mold design and in setting up the complete process. These programs are concerned with melt flow to part solidification, and the meeting of performance requirements. Many different factors are incorporated, including heat transfer, thermal conductivity, thermal expansion, coefficients of friction, and machine and mold operating setup.

Some plastics, particularly filled plastics, can be very abrasive. This necessitates use of abrasion-resistant metals in the mold or the application of special coatings (Table 2.8). Some resins degrade during processing and are corrosive; so proper materials are required for construction and coatings. Mold cleaning keeps the molds operating properly. Certain plastics and mechanical mold operations may require daily cleaning (as recommended by resin suppliers) (Table 2.9).

Types

Although molds may look different on the outside, their structures are probably similar (Fig. 2.22, page 159), consisting of a frame and five plates. There is a top clamping plate; the A plate retains the mold female cavities; the B plate retains the mold male cores; then comes a support plate; finally

Table 2.10 Functions of an injection mold

Mold component	Function performed
Mold base	Hold cavity (cavities) in fixed, correct position relative to machine nozzle
Guide pins	Maintain proper alignment of the two halves of a mold
Sprue bushing (sprue)	Provide means of entry into mold interior
Runners	Convey molten plastic from sprue to cavities
Gates	Control flow into cavities
Cavity (female) and force (male)	Control size, shape, and surface texture of molded article
Water channels	Control temperature of mold surfaces, to chill plastic to rigid state
Side (actuated by cams, gears, or hydraulic cylinders)	Form side holes, slots, undercuts, threaded sections
Vents	Allow escape of trapped air and gas
Ejector mechanism (pins, blades, stripper plate)	Eject rigid molded article from cavity or force
Ejector return pins	Return ejector pins to retraced position as mold closes for next cycle

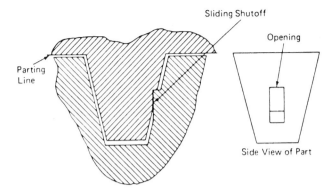

Figure 2.24 Aligning core and cavity to eliminate an undercut.

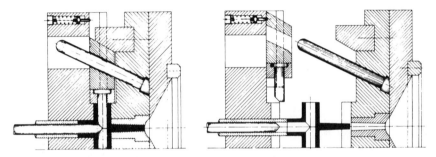

Figure 2.25 A mold with mechanically actuated side cores.

the ejector housing, which includes the knockout plate. The mold parting line is the area where the A and B plates meet (Table 2.10). The mold frame supports and aligns the cores and cavities when the mold opens and the knockout plate moves forward, ejecting the molded part and runner system if runners solidify (Fig. 2.24).

The cavity gives the part its outside shape and the core fits into the cavity to make the internal shape of the part. Shrinkage of the plastic occurs onto the core as the part cools. The runner system includes melt flow channels which guide and control the flow of melt from the plasticator nozzle to the cavities. A gate is at the end of the runner system, and meters the melt into the cavities. The gate is critical to the cycle time and most important to the quality of the part.

Cooling channels are located in the A plate, the B plate, and the support plate. They direct and hold the cooling fluid, which can be water or oil. The temperature for the cooling agent generally ranges from 60–350 °F (16–177 °C), depending on the melt requirements for TP materials; with TSs it is usually lower. The final element of an injection mold is the ejector

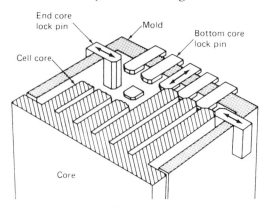

Figure 2.26 A mold with side-core action; it produces a battery case in polyproylene.

Figure 2.27 Stainless steel cavity blocks for PVC pipe elbows; the blocks have retractable cores.

system. This system includes such components as the knockout plate – which ejects the finished product from the core – the knockout pins, the knockout sleeve, and the stripper plate.

Injection mold manufacturers are responsible for parts of many shapes, sizes, materials, etc., so they specialize in the design and/or manufacture

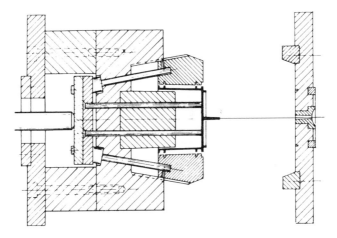

Figure 2.28 Mold with wedge-side core action.

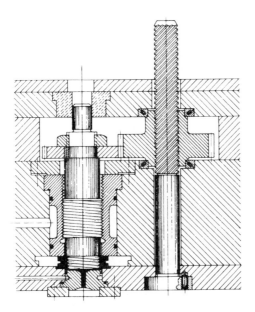

Figure 2.29 A mold with a rotating core that operates during mold opening and closing. The drive gear rotates via the worm shaft, which in turn transmits the rotation to the gear core. The core then unscrews the threaded molded part.

of specific types of mold. Examples of different molds with different actions are shown in Figs 2.25 to 2.31. Figure 2.31 shows a way to injection mold a telephone part.

Many parts of an injection mold will influence the final product's performance, dimensions, and other characteristics. These mold parts include the cavity shape, gating, parting line, vents, undercuts, ribs, and hinges. The mold designer must take all these factors into account. At times, to provide the best design, the product designer, processor, and mold designer may want to jointly review where compromises can be made to simplify meeting product requirements. With all this interaction, it should be clear why it takes a certain amount of time to ready a mold for production.

In the design of any IM part there are certain desirable goals that the designer should use. In meeting them, problems can unfortunately develop. For example, the most common mold design errors of a sort that can be eliminated usually occur in the following areas:

- Thick/thin sections, transitions, warp, and stress
- Multiple gates and weld lines
- Wrong gate locations
- Inadequate provision for cavity air venting
- Parts too thin to mold properly, as diaphragms
- Parts too thick to mold properly
- Plastic flow path too long and tortuous

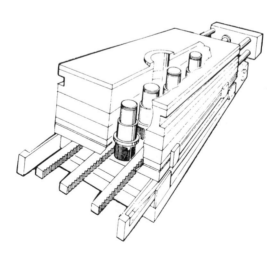

Figure 2.30 Cutaway view of a closed mold frame. Cores can be positioned in rows and each core resides within a gear. When engaged by one of the parallel racks, the gear causes the core to rotate, unscrewing the molded caps.

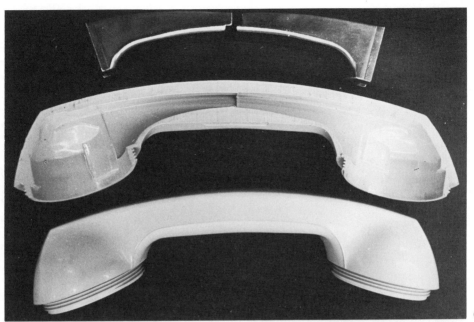

(a)

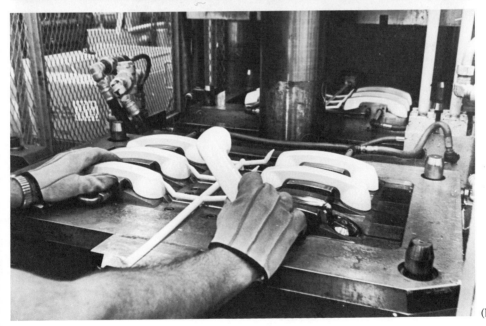

(b)

(c)

Figure 2.31 Molding a telephone part using a removable metal insert: (a) the complete molded part (bottom), a cross-section (middle) and the two-part metal insert to shape the interior, which after molding is shaken out; (b) the bottom half of the mold including two-part metal inserts placed on each mold cavity; and (c) the molded part containing the two-part metal inserts that are being removed and shaken to dislodge the metal inserts (other molding methods do not include such removable metal inserts or the use of cold runners).

- Runners too small
- Gates too small
- Poor temperature control
- Runner too long
- Part symmetry versus gate symmetry
- Orientation of polymer melt in flow direction
- Hiding gate stubs
- Stress relief for interference fits
- Living hinges
- Slender handles and bails
- Thread inserts
- Creep or fatigue over long-time stress (extremely important)

Prototype molds

For engineering evaluation, such as performance of the product in its intended use, a prototype mold must be designed with the essential features planned for the production mold, such as runners, gating and gate location, venting, and cooling. With little additional cost for ejection, such a mold, even though designed for prototyping, should perform in a manner similar to the production mold, but with one or two cavities only.

Production molds

Production molds generally differ only in the method of product and runner ejection. A semiautomatic (SA) mold requires an operator to remove the product from the mold and to ensure that the mold is clean before initiating the next cycle. Today, however, almost all molds are designed so that the assistance of an operator is not required to remove the product. Better ejection designs and mold finishes eliminate the need for mold spray, and automatic lubrication of the mold can be connected with the lubrication system of the machine.

Runner systems

Even though a molder has received a mold intended to meet a parts performance requirements at the lowest cost, it may not be cost-efficient because of a poor runner design. One can simulate the gate and runner design on commercial CAD programs before the runner system is built. Once it is built, very limited rework can be done.

Traditionally there have been several misconceptions about proper runner design. For example, the larger the runner in a cold-runner system, the faster the melt enters the cavity. Here *larger* actually means a longer chill-cycle time; a larger shot size and a larger machine capacity are needed, and more scrap and a higher reprocessing cost with potential increase of contamination; there is a greater projected area, requiring a higher clamping force, and so on. Impossible as it appears, with high production it may pay to have extensive rework done on the mold, runner, and cavity to obtain higher profits. It is easy to have the mold properly designed initially with all the tools available

Types of molds for TPs based on different runner systems are shown in Fig. 2.32. Hot-runner molds, although commercially available since the 1960s, have become dramatically more popular during the past decade. Hot-runner systems virtually eliminate runners, keeping the plastic from cooling as it goes from the nozzle to the cavity. One result is faster cycle time. Other advantages include increased capacity, automated parts handling, reduced injection pressures and clamping requirements, less

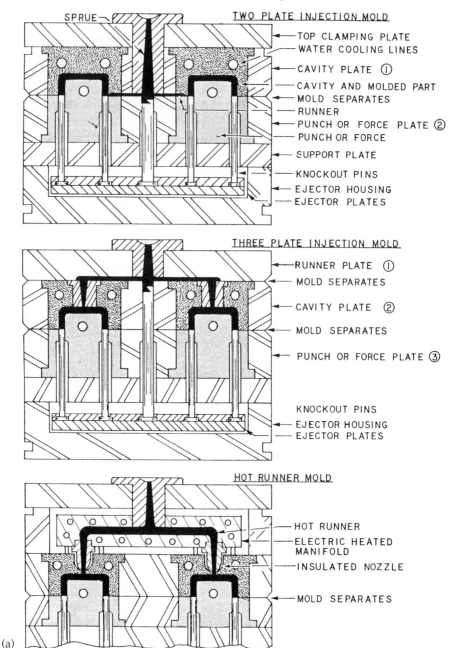

TWO PLATE INJECTION MOLD
SPRUE
TOP CLAMPING PLATE
WATER COOLING LINES
CAVITY PLATE ①
CAVITY AND MOLDED PART
MOLD SEPARATES
RUNNER
PUNCH OR FORCE PLATE ②
PUNCH OR FORCE
SUPPORT PLATE
KNOCKOUT PINS
EJECTOR HOUSING
EJECTOR PLATES

THREE PLATE INJECTION MOLD
RUNNER PLATE ①
MOLD SEPARATES
CAVITY PLATE ②
MOLD SEPARATES
PUNCH OR FORCE PLATE ③
KNOCKOUT PINS
EJECTOR HOUSING
EJECTOR PLATES

HOT RUNNER MOLD
HOT RUNNER
ELECTRIC HEATED MANIFOLD
INSULATED NOZZLE
MOLD SEPARATES

(a)

Figure 2.32(a) Injection molds: cold-runner two-plate mold, cold-runner three-plate mold, and hot-runner mold [1].

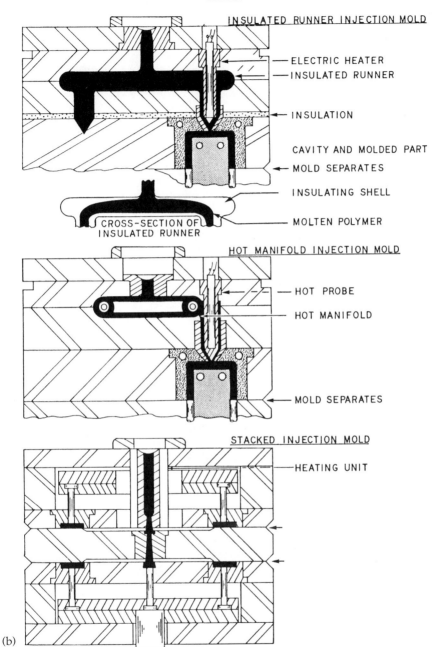

Figure 2.32(b) Injection molds: insulated-runner mold, hot-manifold mold, mold stacked mold [1].

material required, and improved gate cosmetics. But hot runners will not work with certain plastics and part configurations (Chapter 8).

Other considerations

Because parts that are made in multicavity molds tend to lose their dimensional accuracy, there is also a trend to move away from high-cavity molds. Molds with fewer cavities are particularly useful when manufacturing products with tight tolerances, and when quicker changeovers, reduced cycle times, and faster startups are required.

Venting

The air inside the mold must be released as the plastic enters the cavity space. Improvements in mold designs and the increased use of strippers or air ejection, combined with rapid injection, made the location and shape of vents more important. Vent design has become an important part of mold engineering, and vents are machined at the time the mold is built. The designer analyzes the probable flow path and provides as many vents as possible for fast, uniform filling of the cavity.

Mold heating and cooling

The processing temperature needs to be high enough for easy filling, but cold enough for rapid cooling (for TPs). Mold cooling occurs by normal heat transfer. Cooling efficiency depends on factors such as the temperature difference ΔT_1 between coolant entering and leaving, the coolant volume passing through the channels, its speed, the Reynolds number indicating laminar (poor) or turbulent (good) flow, and the chemistry of the coolant. By feeding all available data into the computer and varying the parameters under control of the designer, such as diameters, location, and lengths of bores, number of cooling circuits, distances from molding surfaces, and mold materials, the engineer can select the most advantageous design.

Alignment of mold halves

To ensure the mold halves meet without lateral shifting, leader pins and bushings have been used almost universally. They are still used extensively, but where high accuracy of alignment is required, leader pins are assisted or replaced by taper locks.

THE MOLDMAKERS DIVISION

THE SOCIETY OF THE PLASTICS INDUSTRY, INC.

3150 Des Plaines Avenue (River Road), Des Plaines, Ill. 60018, Telephone: 312/297-6150

TO _____ FROM _____ QUOTE NO. _____

_____ _____ DATE _____

_____ _____ DELIVERY REQ _____

Gentlemen:

Please submit your quotation for a mold as per following specifications and drawings:

COMPANY NAME _____

Name	1.	_____ B/P No. _____	Rev. No. _____ No. Cav. _____
of	2.	_____ B/P No. _____	Rev. No. _____ No. Cav. _____
Part/s	3.	_____ B/P No. _____	Rev. No. _____ No. Cav. _____

No. of Cavities: **Design Charges:**   **Price:**   **Delivery:**

Type of Mold: ☐ Injection   ☐ Compression   ☐ Transfer   ☐ Other (specify) _____

Mold Construction
☐ Standard
☐ 3 Plate
☐ Stripper
☐ Hot Runner
☐ Insulated Runner
☐ Other (Specify) _____

Mold Base Steel
☐ #1
☐ #2
☐ #3

Special Features
☐ Leader Pins & Bushings in K.O. Bar
☐ Spring Loaded K.O. Bar
☐ Inserts Molded in Place
☐ Spring Loaded Plate
☐ Knockout Bar on Stationary Side
☐ Accelerated K.O.
☐ Positive K.O. Return
☐ Hyd. Operated K.O. Bar
☐ Parting Line Locks
☐ Double Ejection
☐ Other (Specify) _____

Material

Cavities	Cores
☐ Tool Steel	☐
☐ Beryl. Copper	☐
☐ Steel Sinkings	☐
☐ Other (Specify) _____	

Press
Clamp Tons _____
Make/Model _____

Hardness

Cavities	Cores
☐ Hardened	☐
☐ Pre-Hard	☐
☐ Other (Specify) _____	

Finish

Cavities	Cores
☐ SPE/SPI	☐
☐ Mach. Finish	☐
☐ Chrome Plate	☐
☐ Texture	☐
☐ Other (Specify) _____	

Cooling

Cavities	Core
☐ Inserts	☐
☐ Retainer Plates	☐
☐ Other Plates	☐
☐ Bubblers	☐
☐ Other (Specify) _____	

Ejection

Cavities	Cores
☐ K.O. Pins	☐
☐ Blade K.O.	☐
☐ Sleeve	☐
☐ Stripper	☐
☐ Air	☐
☐ Special Lifts	☐
☐ Unscrewing (Auto)	☐
☐ Removable Inserts (Hand)	☐
☐ Other (Specify) _____	

Side Action

Cavities	Cores
☐ Angle Pin	☐
☐ Hydraulic Cyl.	☐
☐ Air Cyl.	☐
☐ Positive Lock	☐
☐ Cam	☐
☐ K.O. Activated Spring Ld.	☐
☐ Other (Specify) _____	☐

Type of Gate
☐ Edge
☐ Center Sprue
☐ Sub-Gate
☐ Pin Point
☐ Other (Specify) _____

Design by: ☐ Moldmaker   ☐ Customer
Type of Design: ☐ Detailed Design   ☐ Layout Only
Limit Switches: ☐ Supplied by _____   ☐ Mounted by Moldmaker
Engraving: ☐ Yes   ☐ No
Approximate Mold Size: _____
Heaters Supplied By: ☐ Moldmaker   ☐ Customer
Duplicating Casts By: ☐ Moldmaker   ☐ Customer
Mold Function Try-Out By: ☐ Moldmaker   ☐ Customer
Tooling Model/s or Master/s By: ☐ Moldmaker   ☐ Customer
Try-Out Material Supplied By: ☐ Moldmaker   ☐ Customer

Terms subject to Purchase Agreement. This quotation holds for 30 days.

Special Instructions: _____

The prices quoted are on the basis of piece part print, models or designs submitted or supplied. Should there be any change in the final design, prices are subject to change.

By _____ Title _____

Distribution: Use of this 3 part form is recommended as follows 1) White and yellow - sent with request to quote. Pink - maintained in active file. 2) White original - returned with quotation. Yellow - retained in Moldmaker's active file.

Figure 2.33 Guide for mold quotation.

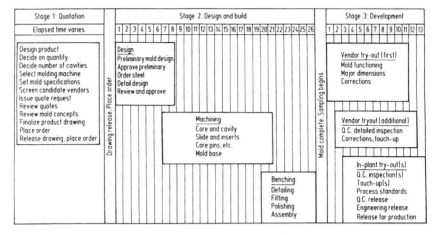

Figure 2.34 Time line: divided into weeks, it gives an idea of events that could occur; also called Gantt chart.

Special molds

Special molds include stack molds, insert molds, two-color molds, injection blow molds, and coinjection molds.

Mold guides

To ensure the mold meets design requirements, it is important to consider all the requirements and variables. Figure 2.33 was prepared by the Moldmakers Division of SPI (Society of Plastics Industry). People who work with ordering or manufacturing molds know that it takes time. Figure 2.34 provides a guide on time considerations that must be allowed to go from the design stage to producing a mold that will function properly.

PROCESS CONTROLS

Process controls can range from unsophisticated to very sophisticated devices. Their cost includes the equipment and using them correctly (they take time, patience, and a willingness to learn new ways of molding). Figure 2.35 provides a simplified approach to understanding controls, and Fig. 2.36 reviews some variables that influence part performance. Figures 2.10 and 2.23 (pages 131 and 161) show the many parameters that are interfaced to develop the most efficient machine operation. Trade-offs are inevitable in a complex operation such as IM. Many of these variables influence end results, and some of the variables interact.

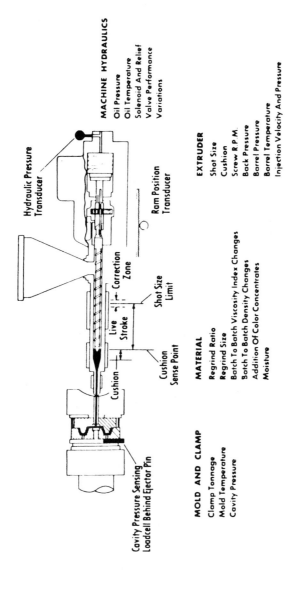

Figure 2.35 Process controls: reasons why they are needed.

MACHINE HYDRAULICS

Oil Pressure
Oil Temperature
Solenoid And Relief
Valve Performance
Variations

EXTRUDER

Shot Size
Cushion
Screw R.P.M.
Back Pressure
Barrel Pressure
Barrel Temperature
Injection Velocity And Pressure

MATERIAL

Regrind Ratio
Regrind Size
Batch To Batch Viscosity Index Changes
Batch To Batch Density Changes
Addition Of Color Concentrates
Moisture

MOLD AND CLAMP

Clamp Tonnage
Mold Temperature
Cavity Pressure

Hydraulic Pressure
Transducer

Ram Position
Transducer

Correction
Zone

Shot Size
Limit

Live
Stroke

Cushion
Sense Point

Cushion

Cavity Pressure Sensing
Loadcell Behind Ejector Pin

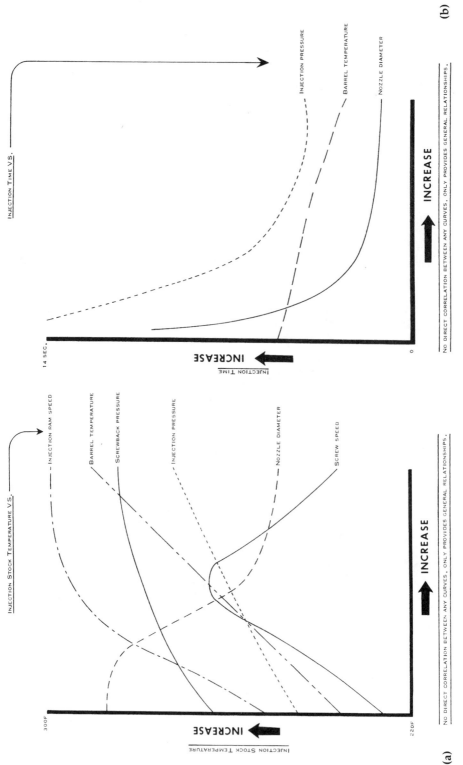

Figure 2.36 (a) Injection stock temperature versus injection ram speed, barrel temperature, screwback pressure, injection pressure, nozzle diameter, and screw speed; (b) Injection time versus injection pressure, barrel temperature, and nozzle diameter.

The development of powerful computer controls and programs has greatly accelerated integrating process variables with a goal of zero defects at the lowest cost. The computer simplifies the fine tuning of machine settings with molding variables. Examples include establishing melt conditions for mold filling and packing, which involves the simultaneous measurement and control of two or more critical variable parameters (Fig. 2.10, page 131). It is during this phase of operation that most variations make themselves evident and can be easily detected. A change in melt viscosity is reflected as a change in ram speed and can be detected by measuring the ram position with respect to time.

A change in resin viscosity is reflected as a change in melt pressure and can be detected by measuring mold or cavity pressure with respect to time. Other variations that similarly display themselves and can be detected include melt temperature, hydraulic pressure, and oil temperature.

COMPUTER-INTEGRATED INJECTION MOLDING

The ultimate result of computer-integrated injection molding (CIIM) in software packages is to translate the results of computer simulation of the molding of a specific part into machine settings for specific microprocessor-controlled machines (Figs 2.35 and 2.36). CIIM automates the entry of a large number of set points in microprocessor-controlled machines and maximizes their efficiency.

Microprocessor control systems

Microprocessor control systems (MCS) make it possible to completely automate an IM plant. They control machines, automatically, enabling them to achieve high quality and zero defects. These systems readily adapt to enhancing the ability of processing machines. There are many moldings that would be difficult, if not impossible, to produce at the desired quality level without this feature.

Once processing variables are optimized through computer simulation, these values are entered in computer programs in the form of a large number of machine settings. Establishing the initial settings during startup is inherently complex and time-consuming. The many benefits of these systems are well recognized and accepted, but it is evident that self-regulation of IM can be effective only when the design of the part and the mold are optimized, and when the correct processing conditions for the operation have been predetermined. Otherwise, a self-regulating machine is confused and can provide conflicting instructions. The results could be disastrous, damaging the machine or the mold. Therefore, the efficient utilization of microprocessor control systems depends on the success of utilizing correct and optimum programs.

Process simulations

The simulation approach replaces the traditional trial-and-error method. Programs are packaged for the complete molding process, including materials selection, molding and cost optimization, flow analysis, computerized shrinkage evaluation, and mold thermal analysis. The programs are mold filling, packing, and so forth, which accurately model the performance of microprocessor-programmed injection. Major 3D CAD systems for part and mold design, as well as for structural and flow analysis, are integrated with these systems.

Improving performance

Machine control coordinates functions of the molding machine. Control functions have evolved to advanced high-speed microprocessor-based systems. Surface-mount control-board technology is being used to reduce the size of machine control systems.

To complement the new controls, sophisticated hydraulics have been introduced. Servo control valves offer increased flexibility and accuracy, as well as shortened machine function response time. Microprocessor controls and servo proportional hydraulics provide dynamic response to achieve true closed-loop systems. Closed-loop systems maintain long-term repeatability of machine velocities and pressures independent of component wear and factors such as oil temperature, ambient temperature, and variations that occur in material viscosity.

MOLDING VARIABLES VERSUS PERFORMANCE

Melt flow behavior

There are variables during molding that influence part performance such as machine settings (Fig. 2.37 and Table 2.4 on page 144). The information presented here shows how melt flow variables behave to influence product properties. A flow analysis can be made to aid designers and moldmakers in obtaining a good mold. Of paramount importance is controlling the fill pattern of the molding so that parts can be produced reliably and economically. A good fill pattern for a molding is one that is usually unidirectional in nature, thus producing a unidirectional and consistent molecular orientation in the molded product. This approach helps to avoid warpage problems caused by a differential orientation, an effect best demonstrated by the warpage that occurs in thin center-gated disks. In this case all the radials are oriented parallel to the flow direction, with the circumferences transverse to the flow direction. The difference in the amounts of shrinkage manifests itself in terms of warpage of the disk.

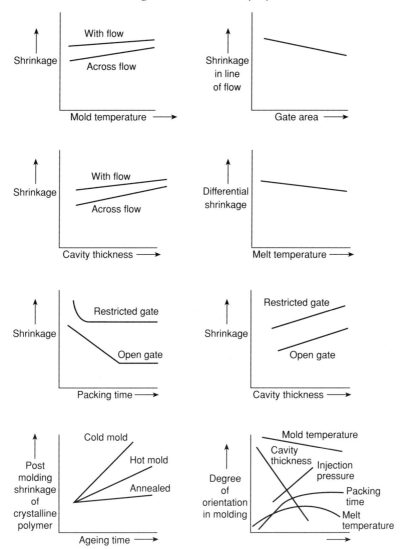

Figure 2.37 Machine settings: how they affect plastic properties, including shrinkage.

In order to achieve a controlled fill pattern, the mold designer must select the number and location of gates that will produce the desired pattern. Flow analysis can help by allowing the designer to try multiple options for gate locations and evaluate the impact on the molding process. This analysis can often be conducted with the product designer to achieve the best balance of gate locations for cosmetic impact and molding considerations. Figures 2.38 to 2.44 show various flow patterns, orientation patterns, and property performances.

In the practical world of mold design there are many instance where design trade-offs must be made in order to achieve a successful overall design. Although naturally balanced runner systems are certainly desirable, they may lead to problems in mold cooling or increased cost due to

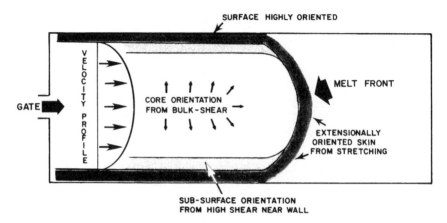

Figure 2.38 Cavity melt flow: looking at a part's thickness (fountain flow).

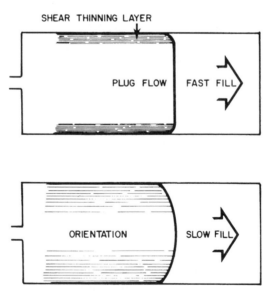

Figure 2.39 The effects of different fill rates.

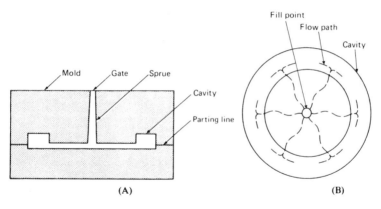

Figure 2.40 During the compensation phase, plastic melt does not flow uniformly through the diaphragm of the plate mold (a), but spreads in a branching pattern (b).

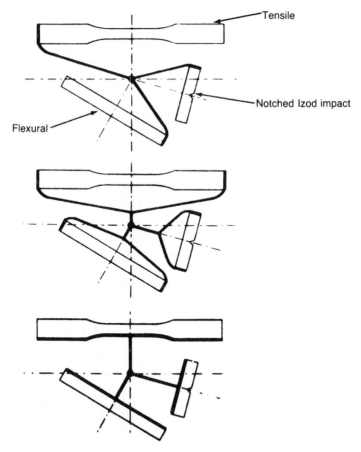

Figure 2.41 Test specimens with different ways of gating produce different flow directions and properties.

STRESS PARALLEL TO ORIENTATION

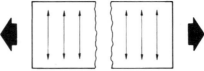

STRESS PERPENDICULAR TO ORIENTATION

Figure 2.42 Orientation affects strength: the highest tensile strength is in the direction parallel to the orientation.

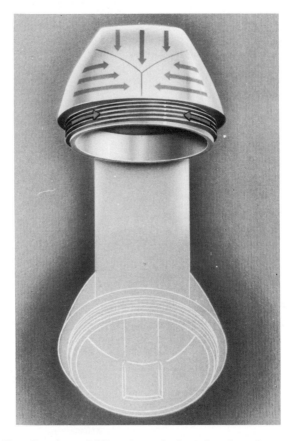

Figure 2.43 Flow lines or weld lines in a telephone handset: the gate was located at the top-center of the handle.

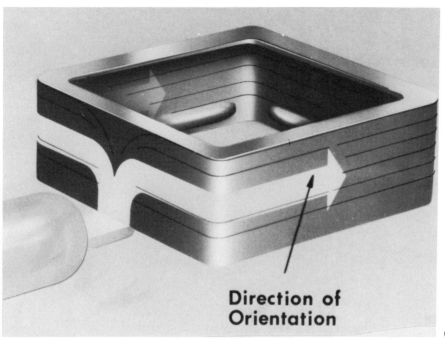

**Direction of
Orientation**

(a)

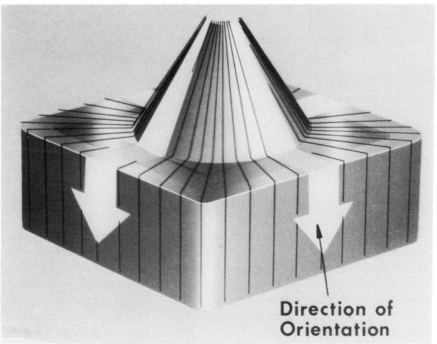

**Direction of
Orientation**

(b)

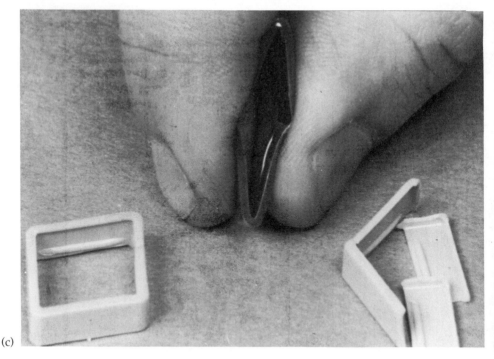

(c)

Figure 2.44 Locating a gate to obtain the required performance of a retainer product that is subject to being flexed in service: (a) retainer edge gated; (b) retainer center gated; and (c) left and center retainers (between fingers) that were edge gated, with the failed retainer on the right which was center gated.

excessive runner-to-part weights. And there are many cases, such as parts requiring multiple gates or family molds, in which balanced runners cannot be used. Flow analysis tools allow successful designs of runners to balance for pressure, temperature, or a combination of both.

Parting lines

The IM processing parting line technique controls the process by using the movement between the two halves as the plastic is injected into the mold as the feedback variable. This movement across the mold parting line is used to initiate the transfer from injection to holding pressure; it therefore performs as a transfer point controller (TPC). TPCs have been around for some time and are a common component of most process control packages for IM. Four strategies are included in the usual commercial transfer point packages; parting line adds a fifth. Parting line has a major advantage in that its sensor is simply added to the outside of the mold. This technique adds little or no machining cost. It may be an add-on to older machines without full control packages.

Back pressure

IM back pressure indicates resistance to the backward movement of the screw during preparation for a subsequent melt shot. This pressure is exerted by the plastic on the screw while it is being fed into the shot chamber (forward end of the barrel, in front of the screw). During rotation of the screw and the material under pressure, thorough mixing of the plastic is achieved, and some temperature increase also occurs. In dealing with heat-sensitive and shear rate insensitive plastics, care must be taken to keep this value within prescribed limits. The action reviewed concerns a conventional screw where back pressure is used to improve the melting characteristics of an otherwise marginally performing screw for the plastic being processed.

With a two-stage screw, the first stage is hydraulically isolated from the second-stage screw by the unfilled devolatization zone. Consequently, back pressure cannot be used to affect melting. Applying back pressure affects the second stage only, and serves to increase the reverse pressure flow component. This will necessitate a longer filled length of the second stage to produce adequate conveying, so the length of unfilled channel will be reduced and devolatilization impaired. In an extreme case, backfilling can progress to the vent port and vent bleed will occur. The only practical advantage lies in the additional mixing it induces in the second stage. However, the additional length of a two-stage screw is almost always sufficient to ensure adequate mixing without application of back pressure.

Screw bridging

When an empty hopper is not the cause of failure, plastic might have stopped flowing through the feed throat. An overheated feed throat, or startup followed with a long delay, could build up sticky plastics and stop flow in the hopper throat. Plastics can also stick to the screw at the feed throat or just forward from it. When this happens, plastic just turns around with the screw, effectively sealing off the screw channel from moving plastic forward. As a result, the screw is said to be bridged and it stops feeding the screw.

The common cure is to use a rod to break up the sticky plastic or to push it down through the hopper and into the screw, where its flight may take a piece of the rod and force it forward. The type of rod fed into the screw should be made of the plastics being processed. Other rods used could be of relatively soft material such as copper.

Weld/meld lines

Problems can develop when molding parts include openings and/or multiple gating (Fig. 2.45). In the process of filling a cavity the hot melt is obstructed by the core, and by the meeting of two or more melt streams. With a core the melt splits and surrounds the core. The split stream then reunites and continues flowing until the cavity is filled. The rejoining of the split streams forms a weld line that lacks the strength properties in an area without a weld line; this is because the flowing material tends to wipe air, moisture, and lubricant into the area where the joining of the stream takes place and introduces foreign substances into the welding surface. Furthermore, since the plastic material has lost some of its heat, the temperature for self-welding is not conducive to the most favorable results. A surface that is to be subjected to load-bearing should not contain weld lines. If this is not possible, the allowable working stress should be reduced by at least 15% for unreinforced plastics and 40–60% with RPs. The meld line is similar to a weld line except the flow fronts move parallel rather than meeting head on.

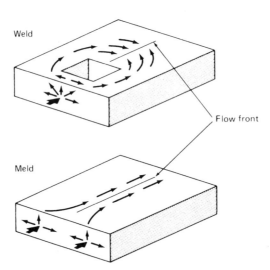

Figure 2.45 Flow paths are determined by part shape and gate location. Flow fronts that meet head-on will weld together, forming a weld line. But parallel fronts tend to blend, potentially producing a less distinct weld line but a stronger bond (called a meld line).

TOLERANCES AND SHRINKAGES

Certain IM parts can be molded to extremely close tolerances of less than a thousandth of an inch (25.4 µm), or down to 0.0%, particularly when TPs are filled with additives or TSs are used (Chapter 1). To eliminate shrink and to provide a very smooth and aesthetic surface, one should use a small amount of chemical blowing agent (<0.5 wt%) and a regular IM packing procedure (Chapter 9).

Table 2.11 provides a guide on shrinkage rates. Chapter 1 includes dimensional information pertaining to IM.

Table 2.11 Guidelines for nominal TP mold shrinkage rates using ASTM test specimens

	Average rate per ASTM D 955	
Material	*0.125 in. (3.18 mm)*	*0.250 in. (6.35 mm)*
ABS		
Unreinforced	0.004	0.007
30% glass fiber	0.001	0.0015
Acetal, copolymer		
Unreinforced	0.017	0.021
30% glass fiber	0.003	NA
HDPE, homo		
Unreinforced	0.015	0.030
30% glass fiber	0.003	0.004
Nylon-6		
Unreinforced	0.013	0.016
30% glass fiber	0.0035	0.0045
Nylon-6,6		
Unreinforced	0.016	0.022
15% glass fiber + 25% mineral	0.006	0.008
15% glass fiber + 25% beads	0.006	0.008
30% glass fiber	0.005	0.0055
PBT polyester		
Unreinforced	0.012	0.018
30% glass fiber	0.003	0.0045
Polycarbonate		
Unreinforced	0.005	0.007
10% glass fiber	0.003	0.004
30% glass fiber	0.001	0.002
Polyether sulfone		
Unreinforced	0.006	0.007
30% glass fiber	0.002	0.003
Polyether-etherketone		
Unreinforced	0.011	0.013
30% glass fiber	0.002	0.003

Table 2.11 *Continued*

Material	Average rate per ASTM D 955	
	0.125 in. (3.18 mm)	*0.250 in. (6.35 mm)*
Polyetherimide		
Unreinforced	0.005	0.007
30% glass fiber	0.002	0.004
Polyphenylene oxide/PS alloy		
Unreinforced	0.005	0.008
30% glass fiber	0.001	0.002
Polyphenylene sulfide		
Unreinforced	0.011	0.004
40% glass fiber	0.002	NA
Polypropylene, homo		
Unreinforced	0.015	0.025
30% glass fiber	0.0035	0.004
Polystyrene		
Unreinforced	0.004	0.006
30% glass fiber	0.0005	0.001

There are various methods of estimating shrinkages. An easy method for estimating shrink allowance is as follows:

$$M = (1 + S)L$$

where M = mold dimension, S = plastic shrinkage (in./in. or mm/mm), and L = part dimension.

If parts are small and have thin walls, this estimate is the best guide. If parts are larger (>10 in., 0.25 m) and/or use rather high-shrink plastics, consider using

$$LM = L/(1 - L)$$

where LM = largest mold dimension.

MOLDING TECHNIQUES

In addition to the conventional IM reviewed, specialized techniques are used to meet specific product requirements that generate cost reductions and reduce cycle time; coupled with this are the necessary molding capabilities to produce specific products. They include gas-assisted IM, coinjection, liquid IM, injection–compression molding (coining), continuous IM, fusible-core molding, multilive feed molding, reaction IM

(Chapter 11), reactive IM (Chapter 3), tandem IM, metal and ceramic IM, two-color IM, foam molding (Chapter 9), expandable polystyrene (Chapter 9), structural sandwich molding, parts consolidation molding, offset molding, jet molding, oscillatory molding, molding with rotation (stretch/orientation that differs from injection stretch blow molding; see Chapter 4), and others [1, 9]. Some of these methods are now reviewed.

Gas-assisted IM

A significant development in injection molding technology has been the introduction of gas assist. Nitrogen, an inexpensive inert gas, is introduced to the plastic melt through the injection nozzle, the mold runner, or directly into the mold cavity. The gas does not mix with the plastic, but takes the line of least resistance through the less viscous parts of the melt. The plastic is pushed against the mold and leaves hollow channels within the part.

Along with the ability to produce hollow parts, parts with heavy ribs and bosses can be achieved with low in-mold stresses, reduced part warpage, and the elimination of sinks. Along with the lowering of inmold stresses, gas-assisted injection offers material savings (since gas displaces resin and less plastic is used), lower clamp tonnage requirements, and reduced cooling/cycle times. The gas pressure is maintained through the cooling cycle. In effect, the gas packs the plastic into the mold without a second-stage high-pressure packing in the cycle as used in IM, which requires high tonnage to mold large parts [1, 14, 69].

Coinjection

Coinjection means that two or more different plastics are 'laminated' together. These plastics could be the same except for color. When different plastics are used, they must be compatible in that they provide proper adhesion (if required), melt at approximately the same temperature, and so on. Two or more injection units are required, with each material having its own injection unit. The materials can be injected into specially designed molds: rotary, shuttle table, etc. [1].

The term *coinjection* can denote different products, such as sandwich construction, double-shot injection, multiple-shot injection, structural foam construction, two-color molding, and inmolding. Whatever its designation, a 'sandwich' configuration has been made in which two or more plastics are laminated together to take advantage of the different properties each plastic contributes to the structure.

This form of injection has been in use since the early 1940s. Many different advantages exist: (1) it combines the performance of materials; (2) it permits the use of a low-cost plastic such as a regrind; (3) it provides

a decorative 'thin' surface of an expensive plastic; (4) it includes reinforcements; (5) it permits the use of barrier plastics (Chapter 4). Coinjection
molding is being redefined today in light of the approaches now available
for molding multicomponent parts such as automotive taillights, containers, and business machine housings.

Liquid IM

Liquid IM (LIM) has been in use longer than reaction IM (RIM), but the
processes are practically similar. The advantages it offers in the automated low-pressure processing of (usually) thermoset resins – fast cycles,
low labor cost, low capital investment, energy saving, and space saving –
may make LIM competitive to potting, encapsulating, compression transfer, and injection molding, particularly when insert molding is required
[1].

Different resins can be used, such as polyester, silicones, polyurethanes,
nylon, and acrylic. A major application for LIM with silicones is encapsulation of electrical and electronic devices.

LIM employs two or more pumps to move the components of the liquid
system (such as catalyst and resin) to a mixing head before they are forced
into a heated mold cavity. Screws or static mixers are used in some
systems. Only a single pump is required for a one-part resin, but systems
having two or more parts are normally used. Equipment is available to
process all types of resin systems, with unsophisticated or sophisticated
control systems. A very critical control involves precision mixing. If voids
or gaseous by-products develop, vacuum is used in the mold.

Injection–compression molding (coining)

Coining, also called injection stamping and more often injection–compression molding, is a variant of injection molding (Figs 2.46 and 2.47). The
essential difference lies in the manner in which the thermal contraction of
the molding during cooling (shrinkage) is compensated. With conventional injection molding, the reduction in material volume in the cavity
due to thermal contraction is compensated by forcing in more plastic melt
during the pressure-holding phase.

By contrast with injection–compression molding, the melt is injected
into a cavity that has a relatively short shot in a compression mold (male
plug fits into a female mold) rather than the usual flat surface matching
mold halves for injection molding. The melt injected into the cavity is
literally stress-free; it works without a holding pressure phase, and the
transport of plastic melt that accompanies this action avoids stresses in the
part, particularly in the gate area(s). The ICM process for thermoplastics
has been used for parts of different sizes, particularly thick-walled parts

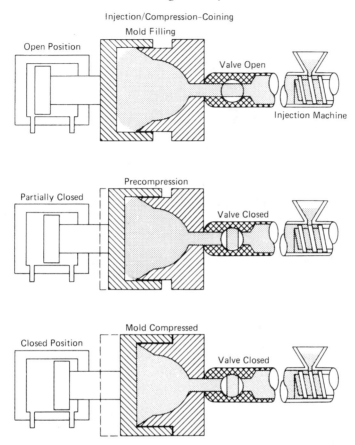

Figure 2.46 Coining combines injection molding and compression molding.

with tight dimensional requirements, such as optical lenses. When melt enters the mold, it is not completely closed. The short-shot melt literally flows unrestricted in the cavity and is basically stress-free. After injection is completed, the mold is closed, with the pressure on the melt very uniform.

Continuous IM

IMMs have been used to mold continuous all-plastic products or strips. An example is the Velcro strips that uses rotating mold halves with a constant flow of plastic melt from the injection unit to the mold [1]. There are systems where metal fiber, etc., are continuously fed through a multicavity mold and precision plastic parts molded around the metal.

Injection molding

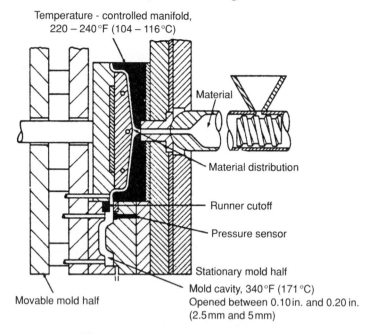

Temperature - controlled manifold,
220 – 240°F (104 – 116°C)

Material

Material distribution

Runner cutoff

Pressure sensor

Stationary mold half

Mold cavity, 340°F (171°C)
Opened between 0.10 in. and 0.20 in.
(2.5 mm and 5 mm)

Movable mold half

Figure 2.47 Close-up of a coining mold.

Figure 2.48 shows six copper wires being directed through the open mold halves. The IMM is on a movable platform; moving in a rectangular motion. The wires have gone through squeeze rolls, to produce the desired diameter, and move at a constant speed. With the mold closed, melt is injected into the multicavity mold (20 cavities around each wire).

The mold has recesses to accurately retain the wires. During mold filling, the mold and the IMM move at the constant speed of the wires. When the mold opens, the IMM moves sideways to reposition the mold away from the wires that have the plastic 'buttons'. The wires continue traveling while the IMM returns to the starting position. The platform moves sideways (back to its original position) and the mold closes, so it is ready for the next injection shot. These buttons are accurately molded (diameter and thickness) and accurately spaced about 1 in. (25.4 mm) apart. The accurate spacing is kept from shot to shot. Tolerances for all dimensions are in thousandths of an inch (tens of micrometers). The product was used in high-frequency electrical lines. Figure 2.48 shows the buttons around the wire exiting the IMM. The runners were cold runners, one of the three major types. Each runner feeds melt around two wires.

Figure 2.48 Coaxial cable cores produced by continuous IM using polystyrene buttons around copper wires.

Fusible-core molding

The use of fusible-core technology (FMCT), as well as soluble-core technology (SCT), to injection mold parts with cavities that could not otherwise be formed or released has been known in the plastics industry since at least the 1940s, but not frequently used (since it was more of a mystery in the past). Other forms, types, or terms include lost-core technology (LCT), soluble salt-core technology (SSCT), lost ice-core technology (LICT), and ceramic-core technology (CCT). LCT has been the most popular term, used since the 1940s and possibly earlier; it also pertains to all the other terms.

More recently, fusible cores and soluble cores have been used. Automobile engine intake manifolds molded of glass fiber reinforced nylon appear to be economical and technologically interesting. Use of a fusible core to mold the complex, curved part produced the sought-after properties of high quality and a smooth interior surface.

Multilive feed IM

The patented Scorim process is a molding method to improve the strength and stiffness of parts by eliminating weld lines and controlling the orientation of fibers. A conventional injection molding machine uses a special head that splits the melt flow into the mold into two streams. During the holding stage, two hydraulic cylinders alternately actuate pistons above and below the head, compressing the material in the mold in one direction

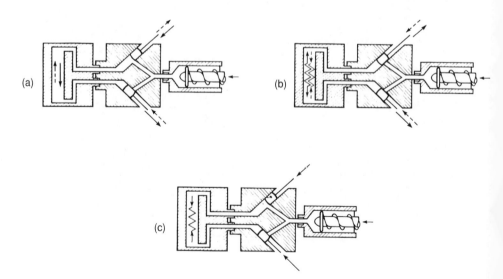

Figure 2.49 Multifeed molding, schematic.

then the other. This action aligns the fibers, removes weld lines, and induces orientation in liquid crystal polymers (LCPs). Figure 2.49 shows two packing pistons that oscillate 180° out of phase, two packing pistons that oscillate in phase, and two packing pistons that compress melt with equal constant pressure.

COSTING IMMS

A major investment is the purchase of IMMs. The cost of an IMM, in combination with the capability of that machine to repay the investment, can make the difference between success and failure of a business. Many molders make their purchasing decisions using empirical information based on hearsay or the performance of another machine they already own. This approach has its merits, but it could be disastrous for those with little knowledge of machines [1, 65, 69, 87, 88, 90].

Just like people, not all machines are created equal. Recognize that identical machine models, built and delivered with consecutive serial numbers to the same site can perform so differently as to make some completely unacceptable. There can be significant differences between machines, so the molder usually uses one machine for certain jobs and another for special precision jobs. Differences are due to factors such as hydraulic design, which affects long-term pressure drift. The consistency of the machine control affects the machine timing. Another area is the final calibration or tuning of a molded product during startup.

One cannot depend on identical calibration or identical performance from many sources. The machine to be purchased needs to perform as required. There is always new technology that can successfully differentiate good machines from those with poor expected performance.

To compare IMMs, you need to have done your homework; you need to find out what you need to monitor in the machine and how you desire it to operate. You also need to know the relative importance of each factor for the parts you intend to manufacture. You need to be able to compare a machine under test conditions to a common yardstick, and you need to know where flaws exist that might inhibit productivity [90].

The monitoring system needs to relate to the molded part requirements. This sets up a good set of parameter guides to be monitored to define the relationship between process deviation and part quality with the soundness of the machine design and construction of the machine. Factors to analyze include machine movements (clamping speed, injection ram time, back pressure holding capability, etc.), number of wires to the machine sequence control using quick-disconnect clips in an effort to synchronize the measurements with the machine cycle, and location of pressure transducer(s) connecting the injection ram cylinder to clamping speed.

Reviewing these data will show what can and cannot be met to operate the machine to a set of standards such as cycle deviation, clamping speed limitation, injection time, back pressure drifting, mold hold time, and plasticizing time. Some believe a machine runoff should be conducted with a mold that is representative of the type to be used in production. It is okay but not necessary. A simple molding block with a bleed hole that allows some material to escape during injection and hold will be sufficient. Thus, the repeatability of the machine is measured rather than the performance of the mold. The plastic to be used, however, should be the type that will be used in production.

TROUBLESHOOTING

All types of processing (IM, extrusion, etc.) have become more sophisticated, particularly with regard to process and power controls; so troubleshooting requires a thorough, logical understanding of the complete process (Fig. 1.1, page 2) and continues to be a very important function. Problems are presented throughout this book, with suggested approaches to solutions. One must assemble information of this type as the basis for a troubleshooting guide (Tables 2.12 and 2.13). Each problem will have its own solution or solutions (Fig. 2.50). Simplified guides to troubleshooting granulators, conveying equipment, metering/proportioning equipment, chillers, and dehumidifiers are available.

No two similar machines (from one or more suppliers) will operate in exactly the same manner, and plastics do not melt or soften as perfect blends, but they do all operate within certain limits.

A simplified approach to troubleshooting is to develop a checklist that incorporates the basic rules of problem solving: (1) have a plan and keep

Figure 2.50 Anticipate any problems.

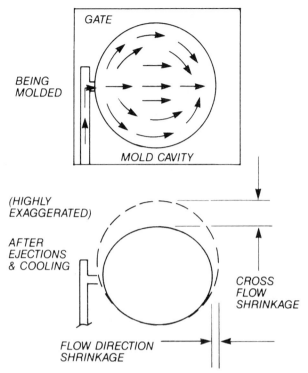

GATE

BEING
MOLDED

MOLD CAVITY

(HIGHLY
EXAGGERATED)

AFTER
EJECTIONS
& COOLING

CROSS
FLOW
SHRINKAGE

FLOW DIRECTION
SHRINKAGE

Figure 2.51 Directional shrinkage when processing crystalline TP.

updating it based on experience gained; (2) watch the processing conditions; (3) change one condition/control at a time; (4) allow sufficient time for each change, keeping an accurate log of each; (5) check housekeeping, storage areas, granulators, etc.; and (6) narrow the range of areas in which the problem belongs – machine, mold/dies, operating controls, material, part design, and management. To accomplish the last item, several steps may be taken:

(a) Change the resin. If the problem remains the same, it is probably not the resin.

(b) Change the type of resin used, as that may pinpoint the problem. Figure 2.51 is an example where shrinkage of crystalline plastics (Chapter 1) is not isotropic; even shrinkages in all directions occurs with amorphous plastics.

(c) If the trouble occurs at random, it is probably a function of the machine or the heat control system. Change the mold/die to another machine to determine if it is the machine. Also consider changing the operator.

Table 2.12 IM troubleshooting guide (Courtesy of RTP Co., Winona MN)

Try remedies in descending order	Problem												
	Blisters	Brittleness	Excessive flash	Gas burns	Oversized part	Poor surface finish	Poor weld lines	Short shots	Silver streaking	Sink marks	Undersized part	Voids	Warping
Change gate location						6	8						6
Clean mold faces	5		4	5									
Clean vents	6			2			5	11				12	
Check for material contamination		4							4				
Check for uneven mold temperature													1
Check mold faces for proper fit			5										
Dry material	1	6	6	6		7			1			11	
Increase amount of material							4	1		8	7	10	
Increase back pressure						5		6			6		
Increase clamp pressure			2										
Increase cooling time										10			9
Increase holding pressure						8		12		11	1		
Increase injection hold time							2	2		2	2	2	
Increase injection pressure						2	1	3		1		1	
Increase injection speed						3	9		2		8		2

Remedy	1	2	3	4	5	6	7	8	9	10	11	12	13
Increase injection time									12				
Increase mold temperature	7		5		1	3	7	6		5	9	8	3
Increase size of gates							8		4	10	4		
Increase size of runners							9		5	11	5		
Increase size of sprue							10		6		6		
Increase size of vent				4		6	4						
Locate gates near heavy cross sections			7								7		
Raise material temperature					4	10	5		7	4			
Redesign ejection mechansim													10
Reduce amount of regrind	5												
Reduce back pressure	2							7					
Reduce cylinder temperature	1	8	4	3				3					7
Reduce holding pressure		7	3										8
Reduce injection pressure		3	2					8					
Reduce injection speed	3	1	1	1					9		7		
Reduce mold temperature							3		9	3	9		4
Reduce molded stress	8												5
Reduce overall cycle time			6					5					
Reduce screw speed	4	3						5					

Table 2.13 Common molding faults

Defect	Possible cause	Suggested remedy
Short moldings	Insufficient feed	Adjust feed setting
	Insufficient pressure	Increase pressure
	Inadequate heating	Increase temperature or lengthen cycle
	Insufficient injection time	Increase injection time
	Cold mold	Increase mold temperature
	Back pressure due to entrapped air	Improve venting of mold
	Unbalanced cavity in a multicavity cavity mold	Check sizes of cavities
Flashing at mold parting lines	Insufficient locking force	Increase locking force
	Injection pressure too high	Reduce injection pressure
	Material too hot	Reduce cylinder temperature
	Mold faces out of line	Rebed mold faces
	Mold faces contaminated	Clean mold faces
	Flow restricted to one or more cavities (in multicavity mold)	Check and remove restriction
Surface sink marks	Material too hot when gate freezes	Reduce cylinder temperature or enlarge gate
	Insufficient dwell plunger forward time	Increase dwell time
	Insufficient material shot into cavity	Increase feed Increase cylinder temp. Increase mold temp.
	Insufficient pressure	Increase pressure
	Piece ejected too hot	Increase cooling time in the mold
Voids	Material too hot (gas formation)	Reduce cylinder temp.
	Condensation of moisture on polymer granules	Predry granules
	Condensation of moisture on the mold surface	Increase mold temp.
	Internal shrinkage after case-hardening of outer layer	Increase pressure Increase mold temp. Enlarge size of gates Lengthen dwell time

Table 2.13 *Continued*

Defect	Possible cause	Suggested remedy
Weld Lines	Material too cold	Increase cylinder temp.
	Mold too cold	Increase mold temp.
	Injection pressure too low	Increase injection pressure
	Gates wrongly located (including too big a distance from gate to weld joint) or designed	Relocate gates and/or redesign
Distortion of moldings	Ejection of molding at too high a temperature	Increase mold cooling time
	Ejection pin working unevenly	Correct or adjust ejection pins
	Existence of molded-in stresses due to material too cold, bad design, cavity overpacked in vicinity of gates	Increase cylinder temp. Redesign molding Check feed setting. Reduce injection pressure and cylinder temperature. Reduce injection time
Crazing and blistering	Excessive surface strain because of cold mold	Increase mold temperature
Surface streaks	Overheating of material	Reduce cylinder temperature
	Moisture in granules	Predry granules
Burn marks	Air trapped in mold cavities	Improve mold venting
Brittleness	Material too cold	Increase cylinder temp.
	Material has degraded	Decrease cylinder temp.
	Contamination with other material	Check the material for contamination Check cylinder and hopper
	Mold too cold	Increase mold temperature

(d) If the problem appears, disappears, or changes from one operator to another, observe the differences between their actions.

(e) If the problem always appears in about the same position of a single-cavity mold, it is probably a function of the flow pattern due to unsatisfactory cooling, and requires readjustments.

(f) If the problem appears in the same cavity or cavities of a multicavity mold, it is in the cavity or gate and runner system.

(g) If a machine operation malfunctions, check the hydraulic or electric circuits. As an example, a pump makes oil flow, but there must be resistance to flow to generate pressure. Determine where the fluid is going. If actuators fail to move or move slowly, the fluid must be bypassing them or going somewhere else. Trace it by disconnecting lines if necessary. No flow, or less than normal flow in the system, will indicate that a pump or pump drive is at fault. Details on correcting malfunctions are in the machine instruction manual.

(h) Check for hydraulic contamination. Too little attention is paid to the cleanliness required of the oil used. Dirt is responsible for the majority of malfunctions, unsatisfactory component performance, and machine degradation, particularly with the increased use of electro-hydraulic servosystems. Injection pressure, holding pressure, plasticating pressure, boost pressure, and boost cutoff are adversely affected by increased contamination levels in the fluid. Sources of contamination include new oil, a hydraulic system built with poor quality control, air from the environment, wear of hydraulic components, leaking or faulty seals, and shop maintenance activity. Contamination control is accomplished with the proper filters (such as 10 μm) (see suppliers), and with preventive maintenance procedures that are both correct and properly used.

(i) Set up a procedure to 'break in' the new mold/die.

The procedure for setting up a mold/die is as follows. (1) Obtain samples and molding cycle information if the mold was used by others. (2) Clean a used mold. (3) Visually inspect the mold and make corrections if required. (4) Check out, on a bench, the actions of the mold/die cams, slides, unscrewing devices, and so on. (5) Install safety devices. (6) Operate the mold/die in the machine, and move it very slowly under low pressure. (7) Open the mold/die and inspect it. (8) Dry-cycle the mold without injecting melt to check knockout stroke, speeds, cushions, and low-pressure closing [54]. (9) After the mold is at operating heat, dry-cycle it again; expansion or contraction of the mold parts may affect the fits. (10) Take a shot, using maximum mold lubrication and under conditions least likely to cause mold damage, usually low melt feed and pressure. (11) Build up slowly to operating conditions, and run the process until stabilized (usually 1–2 h). Record operating information. (13) Take the part to quality control for approval. (14) Make required changes. (15) Repeat the process until it is approved by the customer.

Faulty or unacceptable parts usually result from problems in one or more of these areas: (1) **premolding**, material handling and storage (Chapter 16); (2) **molding**, conditions in the processing cycle; and (3) **postmolding**, parts handling and finishing operations (Chapter 17).

Problems caused in premolding and postmolding may include those involving contamination, color, the static dust collector, and so on. In molding (item 2) the molder is required to produce a good-quality melt based on visual observation as it flows freely from the nozzle. Each mold is unique and each material is unique, so one cannot generalize about what makes a good melt. The experience of the molder and a knowledge of the process needs are the final determining factors.

There are several ways to determine the efficiency of the melt. One method is to observe the screw drive pressure; it should be about 75% of maximum. If it is less than that, lower the rear-zone heat until the drive pressure starts to rise. With melt quality changing, raise the center zone to restore quality to what is required. Heat changes should be accomplished in 10–15 °C increments, with 10–15 min of stabilization time allowed before the next change.

Once the rear zone is set, one should lower the front zones to whatever level will still give good molding conditions. With crystalline types, such as nylon, PP, and PE, the operator must watch the screw return. If the screw is moving backward in a jerky manner, there is insufficient heat in the rear zone; the unmelted resin is jamming or plugging the screw compression zone. The heat energy required to melt crystalline plastics is different from that needed for amorphous plastics (Chapter 1).

Wear

All screws, barrels, molds or dies, and any device that handles melt will wear, but hopefully by an insignificant amount that does not influence processability [54]. The wear of screws (particularly on the flight OD) and barrels is a function of (1) the screw–barrel–drive alignment; (2) the straightness of the screw and barrel; (3) the screw design; (4) the uniformity of barrel heating; (5) the material being processed; (6) abrasive fillers, reinforcing agents, pigments, and so on; (7) the screw surface material; (8) the barrel liner material; (9) a combination of the screw surface and the barrel liner; (10) improper support of the barrel; (11) excessive loads on the barrel discharge end and heavy molds or dies; (12) corrosion caused by additives such as flame retardants; (13) corrosion caused by certain polymer degradation; and (14) excessive back pressure on the injection recovery.

Screws are usually aligned properly by the supplier before shipment, but can become misaligned during shipment, during installation, and by accidental impacts and other aspects of their use. An angular misalignment will generally cause wear uniformly around the screw in a fairly localized area. In that vicinity the barrel will be worn around the entire ID. If the barrel is bent, the screw will be worn all around near the center and near the discharge, whereas the barrel is usually worn on one side near the center. Wear on screws and barrels generally falls into three categories.

Abrasive wear is caused by abrasive fillers such as calcium carbonate, talc, glass fibers, barium sulfate (used in magnetic tapes, etc.), and even the titanium dioxide pigments used in all white and pastel shades. Glass fibers tend to abrade the root of the screw at the leading edge, and in severe cases can undermine the screw flight completely, usually leaving no flight in the compression–transition zone. This action occurs extensively when partially melted or unmelted plastic pushes the glass against the screw or barrel.

Adhesive wear or galling is caused by metal-to-metal contact. Certain sensitive metals can momentarily weld to each other because of very high localized heating. As the screw rotates, the weld separates, and metal is pulled from the screw to the barrel or vice versa. Proper clearance usually eliminates this problem with proper alignment and hardness. With an improperly designed screw for a plastic operating at high output rates, an unmelted blockage will result, forcing the screw against the barrel and causing rapid adhesion wear.

Corrosion wear is caused by chemical attack in the melting of certain plastics, such as PVC, ABS, PC, and PUR, as well as flame-retardant compounds, fiber-sizing agents, and so on. Material suppliers can identify the offending agents. The wear usually shows a pitted appearance and is usually downstream, where it has a chance to overheat and degrade. This type of wear can be controlled by using proper operating procedures; do not let the machine stay at the operating heat for any length of time. Proper selection of the screw design and corrosion-resistant screw/barrel materials can help. Nonreturn valves and screw tips are also subject to wear, so it is important to use the best available material.

Different coatings such as chrome and nickel plating are used to protect the screw surface. Depending on the specific plastic being processed, a particular coating will be available. The wear surfaces, primarily of flight lands, are usually protected by welding special wear-resistant alloys over these surfaces. The most popular and familiar alloys are Stellite (trademark of Cabot Corp.) and Colmonoy (trademark of Wall Colomonoy Corp.); others are also used and are available from different suppliers. Different heat treatments are also used on the steels to increase wear resistance.

Inspection

Screws do not have the same outside continuous diameter. Upon receiving a machine or just a screw, it is a good idea to check its specified dimensions (diameters versus locations, channel depths, concentricity and straightness, hardness, spline/attachment dimensions, etc.) and make a proper visual inspection. This information should be recorded so that comparisons can be made following a later inspection. The initial check also guarantees proper delivery. Some special equipment should be

Table 2.14 Manufacturing tolerances on screws [54]

Diameters		Channel depths	
Outside diameter	±0.001 in.	*Depth*	*Tolerance*
Shank diameter	±0.005 in.	0.000–0.150 in.	±0.002 in.
Injection registers	±0.0005 in.	0.151–0.350 in.	±0.003 in.
Clearance diameters	+0.015 in.	0.351–0.750 in.	+0.005 in.
Lengths			
Overall length (OAL)	±$\frac{1}{64}$ in.	Hollow bore length:	±$\frac{1}{4}$
Transition zones	±$\frac{1}{10}$ dia.	Flight widths:	
Vent sections	±$\frac{1}{10}$ dia.	0–0.500 in.	±0.010 in.
Shank lengths	±$\frac{1}{32}$ in.	0.501–1.000 in.	±0.0515 in.
Ring valve location	+$\frac{1}{32}$ in.	>1.001 in.	+0.020 in.
Concentricity			
TIR of OD			
<100 in.	0.002 in.	Hollow bore to shank	0.015 in.
>100 in.	0.004 in.	Injection registers	0.001 in.
Hardness			
Base material 4140	28–32 Rc	Colmonoy[c] no. 5	36–40 Rc
Flame-hardened flights	48 Rc min.	Colmonoy[c] no. 56	46–50 Rc
Nitralloy[a] 135M		Colmonoy[c] no. 6	50–55 Rc
(or equivalent)	60–70 Rc	Colmonoy 84	36–42 Rc
Stellite[b] no. 6	38–42 Rc	N-45[d]	40–44 Rc
Stellite[b] no. 12	42–48 Rc	N-50[d]	44–48 Rc
		N-55[d]	46–50 Rc

Finish
Unplated screws   16 RMS max.
Plated screws
  Root   8 RMS max.
  Flight sides, OD, and shank   16 RMS max.

[a] Trademark of Joseph T, Ryerson & Son, Inc.
[b] Trademark of Cabot Corp.
[c] Trademark of Wall Colmonoy Corp.
[d] Trademark of Metallurgical Industries Inc.

used for inspection other than the usual methods (micrometer, etc.) to ensure that the inspection is reproduced accurately. Such equipment is available from suppliers [54] and actually simplifies testing and takes less time, particularly for roller and hardness testing. It is important that screws are manufactured to controlled tolerances such as those given in Table 2.14.

Rebuilding screws/barrels

In a properly designed plasticator, the majority of wear is concentrated on the screw because the screw can be replaced and built more easily than the barrel. The rebuilding of injection (also extrusion, blow molding, etc.) screws has become so common that the rebuilding business became a major segment of our industry. One reason for the popularity of screw building is that rebuilding is usually considerably less expensive than replacement of a new screw. Rebuilding is usually done with hardfacing materials. With the proper choice of hardfacing metallic material, the rebuilt screw can perform better than the original screw [54, 63, 65].

It is considerably more difficult and costly to repair a worn barrel than to rebuild a worn screw. If it is not greater than about 0.5 mm (0.02 in.), the whole barrel can be honed to a larger diameter and an oversized screw can be used. If the wear occurs near the end of the barrel, a sleeve can be placed inside. But despite these remedies, a worn barrel generally needs to be replaced.

3

Extrusion

BASIC PROCESS

The extruder, which offers the advantages of a completely versatile processing technique, is unsurpassed in economic importance by any other process. This continuously operating process, with its relatively low cost of operation, is predominant in the manufacture of shapes such as films, sheets, tapes, filaments, pipes, rods, and others. The basic processing concept is similar to that of injection molding (IM), in that material passes from a hopper into a cylinder in which it is melted and dragged forward by the movement of a screw. The screw compresses, melts, and homogenizes the material. When the melt reaches the end of the cylinder, it is usually forced through a screen pack prior to entering a die that gives the desired shape with no break in continuity (Fig. 3.1).

A major difference between extrusion and IM is that the extruder processes plastics at a lower pressure and operates continuously. Its pressure range is usually 200–1500 psi (1.4–10.4 MPa) and could go to 5000 or possibly 10 000 psi (34.5–69 MPa) (Table 3.1). IM pressures are 2000–3000 psi (14–210 MPa). However, the most important difference is that the IM melt is not continuous; it experiences repeatable abrupt changes when the melt is forced into a mold cavity (Chapter 2). With these significant differences, it is actually easier to theorize about extrusion and to process plastics through extruders, as many more controls are required in IM.

Good-quality plastic extrusions require homogeneity in terms of the melt heat profile and mix, accurate and sustained flow rates, a good die design, and accurately controlled downstream equipment for cooling and handling the product. Four principal factors determine a good die design: internal flow length, streamlining, the materials of construction, and uniformity of heat control. Heat profiles, such as those in Fig. 3.2, are preset via tight controls (Chapters 1 and 2). To accomplish this control, cooling systems are incorporated in addition to heater bands. Barrels use forced

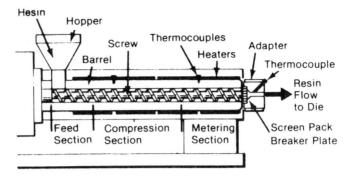

Figure 3.1 Cross section of a single-screw extruder.

Table 3.1 Typical extruder diehead pressures

	Melt pressure at the die	
Extruded shape	*psi*	*MPa*
Film, blown	1000–5000	6.9–34.5
Film, cast	200–1500	1.4–10.4
Sheet	200–1500	1.5–10.4
Pipe	400–1500	2.8–10.4
Wire coating	1000–5000	6.9–34.5
Filament	1000–3000	6.9–20.7

air and/or water jackets. Some machines have water bubbler channels located within the screws. (Table 2.1, pages 132–4, provides a guide for machine settings.)

It is important to realize that the barrel–plastic interface constitutes only about 50% of the total plastic–metal interface. Thus, with only barrel heating and/or cooling, only about 50% of the total surface area available for heat transfer is being utilized. The screw surface therefore constitutes a very important heat-transfer surface. Many extruders do not use screw cooling or heating; they run with what is called a neutral screw. If the external heating or cooling requirements are minor, then screw heating or cooling is generally not necessary. But if the external heating or cooling requirements are substantial, then screw heating or cooling can become very important and is usually essential [3, 9, 54, 60, 63].

On leaving the extruder, the product is drawn by a pulling device, and in this stage it is subject to cooling, usually by water and/or blown air. This is an important aspect of control if tight dimensional requirements exist and/or conservation of plastics is desired. Lines usually do not have adequate control of the pulling device. The processor's target is to deter-

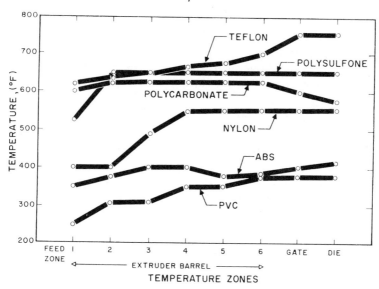

Figure 3.2 Temperature profiles of different plastics going through an extruder.

mine the tolerance required for the pull rate and to see that the device meets the requirements. One should check with a supplier on the speed tolerances available. Even if tight dimensional requirements do not exist, the probability is that better control of the pull speed will permit tighter tolerances and reduce the material output. One should check the cost of replacing the puller.

As the molecules of the melt flow are aligned in the direction of the output from the die, the strength of the plastic is characteristically greater in that direction than at right angles. Depending on the product use, this may or may not be favorable. The degree of orientation can be controlled (reviewed later and in Chapter 1).

Most plastics extruders incorporate a single screw rotating in a horizontal cylindrical barrel with an entry port mounted over one end and a shaping die mounted at the discharge end. A large-capacity, single-screw extruder may have a screw diameter of 600mm and be designed to extrude 29 ton h^{-1}; the smallest models, for torque rheometers, have a capacity of 5 ton h^{-1}. Very small extruders are used in laboratories and commercially for special applications such as extruding a narrow red stripe for decoration on polypropylene soda straws. Most US single-screw extruders are sized in diameter by inches 2–10 in. (51–254 mm).

The twin-screw extruders and other multiple-rotor devices are usually more expensive than conventional single-screw extruders. Their use is confined primarily to tasks that cannot be performed easily. For both single-screw and twin-screw extruders the screw design affects the performance capabilities of the extruder. A typical simple screw design for a solid-fed single-screw extruder must convey the plastic entering the screw into a heated, compacted environment, where the shear force developed by the rotating screw melts the plastic and mixes it to a reasonably uniform temperature while pressurizing the melt and pumping it uniformly through the die.

Advances in screw design and control features are making the single-screw extruder increasingly versatile. Nevertheless, new developments in combining plastics and incorporating various additives are placing more demands than ever on twin-screw extruders, thus assuring their expanded use. Both single-screw and twin-screw extruders can be designed with one or more vents to remove volatiles and with additional feedports downstream of the first feedport. Sometimes extruders are arranged in tandem or as coextruders to produce different layers of molten plastics through the same die. At the extruder exits, melt filtration may be provided, usually by screen packs. A static mixer is occasionally placed in an adapter tube leading to the die, and sometimes a gear pump may precede the die to assure a more consistent output rate. The possibilities for die design are almost unlimited and cover many applications, such as compounding (coloring and blending of melts, additives, and fillers), crosshead extrusion (wire, garden hose), sheet, film, coatings, pipe, rod, profiles, and coextruded products.

The success of any continuous extrusion process depends not only upon uniform quality and conditioning of the raw materials (Chapters 1, 2, and 16) but also upon the speed and continuity of the feed of additives or regrind along with the virgin resin. Practically only thermoplastics go through extruders; markets have not developed to date for extruded thermosets. As reviewed in Chapters 1 and 16, variations in the bulk density of materials exist in the hopper, requiring controllers such as weight feeders and perhaps requiring some type of packing feed, such as rams and screw packers.

Each line has interrelating operations, as well as specific line operations, to simplify processability. The extruder is usually followed by some kind of cooling system to remove heat at a controlled rate, causing plastic solidification. It can be as simple a system as air or water cooling, or a cooled roll contact can be used to accelerate the cooling process. Some type of takeoff at the end of the line usually requires an accurate speed control to ensure product precision and save on material costs by tightening thickness tolerances. The simplest device might be a pair of pinch rolls or a pair of opposed belts (a caterpillar takeoff). A variable-speed drive is usually desired to give the required precision.

Extruder film is the biggest output from extruders. Film is usually defined as material having a thickness of up to ten thousandths of an inch (250 μm), thicknesses above this are called **sheet**. There are fundamentally two different methods of extruding film, blow extrusion and cast or slot-die extrusion. Blow extrusion produces tubular film, which may be gussetted or layflat; cast or slot-die extrusion produces flat film.

Of the 38 wt% of all plastics going through extruders about 45% are PE, 25% PVC, 14% PP, 8% PS, and 8% others. Many plastics are first extruded in polymer manufacturing plants and in compounding operations before reaching final extrusion (injection molding and other) processes used to make fabricated products. Many other materials are formed through extruders: metals, clays, ceramics, foodstuffs, etc. The food industry extrudes noodles, sausages, snacks, cereals, and so on. This book confines itself to plastics, predominately thermoplastics.

PLASTICS HANDLING

Care should be taken to prevent conditions that promote surface condensation of moisture on the plastic and moisture absorption by any existing pigments in color concentrates. Surface condensation can be avoided by proper storage of the plastic and keeping it in an area at least as warm as the operating temperature for at least 24 h prior to its use. If moisture absorption by a color concentrate is suspected, heating in an oven for 8–24 h at 250–300 °F (120–150°C) should permit sufficient drying. With hygroscopic resins, special precautions and drying are required (Chapters 1 and 16). Heat from a hopper dryer can be used to improve melt performance and extruder output capacity (Fig. 3.3). When the dryer preheat is insufficient, heat can be applied in the screw's solids-conveying zone and/or the barrel feed throat (assuming the capability exists).

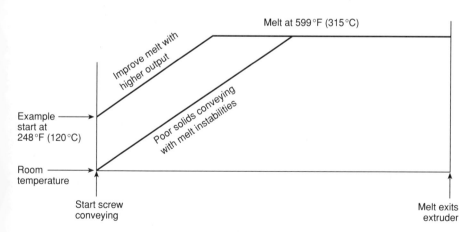

Figure 3.3 Improving material and/or machine performance by preheating 'solids' entering the extruder's feed throat.

Extrusion

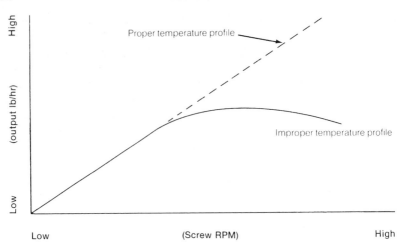

Figure 3.4 Plastics entering the barrel throat at a higher temperature will increase the extruder output capacity with an increase in RPM.

A typical heat profile that is adequate at low screw speed may be inadequate when speeds are increased, because of the greater rate of material flow through the extruder. (In fact, the heat profile may even be inadequate at low speed.) The output may not increase at a linear rate with increasing speed unless the heat profile increases (Fig. 3.4). With a proper heat profile, a linear relationship between speed and output can be maintained through a wider operating range. However, too high a heat at low screw speed is likely to cause melting all the way back to the barrel throat, causing resin bridging and degradation.

Different techniques are used to improve feeding from the hopper: a ram stuffer, a tapered (larger) screw in the throat section, a grooved feed section, starved feeding, etc. All have advantages and disadvantages, based on the capabilities of the machines and of the resin being processed. What may help with one machine in the plant may be useless in another machine using the same resin. Early melting can be avoided by using a barrel feed with an intensively water-cooled, grooved barrel section. The grooves impart better solids conveying for resins with a low coefficient of friction, such as UHMWPE, and other resins where additives make them extremely slippery. The grooves usually have a right-angled profile and axial arrangements, although helical grooves are used for certain materials. Optimum sizing and dimensioning of the feed section and the correct control of cooling are necessary, as well as proper design of the feed screw, to achieve a high-output revolution with a low-friction load.

If a resin exhibits a feeding inefficiency (as the screw speed is increased, the output of the extruder does not measurably increase), grooved feed sections should be considered. Most materials feed well, as seen through moderate or high pressure levels early in the barrel, and the state of screw design had led to efficient 'melting rate limited' extrusion. A grooved feed that is already limited by the melting rate will overdrive the screw and cause melt quality effects that must be corrected through the use of shallower screws. If a material is not melting rate limited, feeding reasonably well, a grooved feed will increase the output. Probably the next larger screw will accomplish the same result at a lower cost. One must consider economics, including the added maintenance and operational costs of grooves.

Starve or metered feeding devices do not allow equipment to run with feed flights completely filled, so this approach is best avoided, although some vented screws do use it. Machines will run most efficiently and with the best cleaning when feed flights are full. Starve feeding leads to flights that are partially empty for some distance, an undesirable situation (causing wear problems, etc.). Some two-stage screws that handle a variety of resins may not be versatile enough, then a feeder gives added control. Starving can help when machines are having a feed problem and no other approach is possible.

As the feeder is adjusted, the screw's performance will eventually deteriorate, so the output consistency or the melt efficiency is threatened. Adjusting the screw design to preclude the need to starve is a better alternative, and is usually economically feasible.

Twin-screw extruders producing profiles will often use starve feeders to reduce torque requirements. But when feasible, a better approach is to accomplish this via a screw design with adequate extruder torque power. Slight starving of twins may not seriously deteriorate the melt, but moderate to high levels of starving will reduce stability.

As explained in previous chapters and in Chapter 16, regrind can be a major problem. Many extruder lines, such as film trim and scrap, are reprocessed; the material is granulated and/or plasticized by a small extruder and fed back with virgin resin. The extrudability of the combination may be quite different from that of virgin resin. Problems can also develop in the blending of two or more different resins, such as LLDPE and LDPE, to improve specific performances. The LLDPE starts melting or breaking down at a lower heat than LDPE, so their blend is more difficult to process than virgin or regrind blends. New screw designs overcome such problems.

BARREL AND FEED UNITS

The feed throat and feedhopper units are important in ensuring that plastics are properly plasticized. The feed throat is the section in an extruder barrel (and other screw plasticator such as injection molding) where material is directed into the screw channel. It fits around the first few flights of the screw. Some extruders do not have a separate feed throat. In these extruders the feed throat is an integral part of the barrel, even though it may not be the best design approach. The feed-throat casting is generally water-cooled to prevent and early temperature rise of the plastics. If the temperature rises too high, it may cause the plastic to adhere (stick) to the surface of the feed opening, causing a material-conveying problem to the screw, usually called **bridging**. The problem can also develop on the screw, with plastic sticking to it, restricting forward movement of material.

Where the feed-throat casting connects with the barrel, a thermal barrier is usually included to prevent barrel heat from escaping through the feed throat. This action is not possible in a barrel with an integral feed opening.

The geometry of the feed section should be such that the plastic can flow into the extruder with minimum restrictions. The shape of the inlet is usually round or square. The smoothest transition from feedhopper to feed throat will occur if the cross-sectional shape of the hopper is the same as the shape of the feed opening; a circular hopper feeds into a circular feedport. Some machines have grooved barrel sections to aid in feeding the screw. The effective length of the grooves may be 3–5 diameters of the screw. Its depth varies with axial distance; maximum depth is at the start of the grooves and reduces to zero where the grooved section meets the smooth barrel. This approach produces excellent cooling. The result is a good thermal barrier between the feed section and the barrel. It also gives a high-pressure capability. Stress between the plastic and the grooves can be very high, so wear can be a problem, particularly when the plastic contains abrasive additives. To reduce or eliminate this problem, the grooved design is made with highly wear-resistant steel materials.

The feedhopper must ensure the plastics enter the screw plasticator at a controlled weight rate (not volume rate). The plastics flow by gravity in many machines. Although satisfactory for some materials, gravity flow may be unsuitable for bulky materials with very poor flow characteristics; additional devices are then used to ensure steady flow. A vibrating pad may be attached to the hopper to dislodge any bridges that may want to form. Stirrers are also used to mix materials (virgin, recycled, etc.), to provide an even distribution, or to prevent separation. The stirrers can be used to wipe material from the hopper wall if the bulk material tends to stick to it. Crammer feeders are used with certain bulky materials when other devices do not have the capability to provide the proper feeding action (Chapter 16).

Air entrapment can occur with certain plastics, particularly those having low bulk density. As explained in Chapter 1, if air cannot escape through the feedhopper, it will be carried with the plastic melt through the die, and problems can develop. One way to eliminate air entrapment is to use a vacuum feedhopper, but this is rarely practical. Problems can develop over loading the hopper without losing vacuum. Double-hopper vacuum systems are used where plastic is loaded into a top hopper and the air is removed before the plastic is dumped in the main hopper feeding the screw. This unit has merit when used very carefully but it can encounter problems such as leaks.

Another important aspect to the hopper is its sidewall angle from the horizontal. The angle should be larger than the angle that would cause internal friction between material and the sidewall. Some plastics slide easily but others want to stick. Experience or trial-and-error evaluations are needed. If the bulky material has a very large angle of internal friction, it will bridge in almost all hoppers; this situation may be resolved by using a force-feeding mechanism (Chapter 16) [1, 26, 32, 54, 63, 65, 69, 91].

TYPES OF EXTRUDER

There are two basic types of extruders: continuous and batch (discontinuous). Continuous extruders have the capability to develop a steady, continuous flow of material, whereas batch extruders operate in a repeat-cycle pattern. Continuous extruders use rotating-screw processes, processes reviewed in this chapter. Batch extruders generally have a reciprocating screw (like the injection molders reviewed in Chapter 2); they also use designs for preparing compound mixes, etc.

There are many different types of the basic continuous-screw extruder. The main types are classified as (1) screw extruders with single or multiple screws; (2) disk or drum extruders; (3) discontinuous, or batch, reciprocating extruders. Details will be presented regarding continuous-screw extruders, particularly single screws, since they represent the vast majority of extruders used to manufacture the products in this chapter [9].

The disk or drum types are represented by viscous drag and elastic – plastic melt plasticizing actions. The viscous types include the spiral disk, drum, dispact, and stepped-disk extruders. The elastic types include the Maxwell and screw/disk extruders. The discontinuous types comprise the ram extruders (melt feed and capillary rheometer types) and reciprocating extruders (injection molding and compounding extruders) [1, 6, 9, 26, 29, 39, 59, 60, 63–70].

Innovations in the design of screws and associated control systems will continue to keep pace with advances in technology and applications research. Among the likely future developments will be the screwless extruder, in which a smooth barrel with a dam, in conjunction with a rotating cylinder, will hasten the melting process. Such improvements will provide higher operating rates and product quality, along with lower processing costs [9].

SCREW/BARREL PERFORMANCE

Screw design is at the heart of the extrusion process, which suffers if the screw design is not well suited to the application. Before 1967 most screws were of conventional single-stage and two-stage design, and theoretical attention was directed to optimizing the length and depth of conventional feed, transition, and metering sections in relation to processing requirements. Some screws were built with different pitches and multiple flights, and occasionally mixing pins, ring barriers, undercut spiral barriers, and Dulmage sections (many parallel interrupted mixing flights), plus a few other novelties were employed. In a notable study on mixing screw design, a highly sensitive oscillograph recorder traced temperature and pressure. It was demonstrated that incorporation of a Union Carbide mixing section in a conventional screw extruding polyethylene at good output rates could reduce temperature and pressure fluctuations entering the die to about ±0.25%. Other mixing sections were developed, and since then, high-performance screws have been used more and more [63].

The extruder screw is the heart of the machine. Everything revolves around the extruder screw, literally and figuratively. The rotation of the screw causes forward transport, contributes to a large extent to the heating of the plastic, and causes homogenization of the material. In simple terms the screw is a cylindrical rod of varying diameter with helical flights wrapped around it. The outside diameter of the screw, from flight tip to flight tip, is constant on most extruders. The clearance between screw and barrel is usually small. The ratio of radial clearance to screw diameter is generally around 0.001, with a range of about 0.0005–0.0020.

In all these applications the design of the extruder screw is the most critical part of the machine. Although a single all-encompassing (or universal) design was once employed, today's screws are generally customized for a particular process (review Chapter 2). The specific screw type involves factors such as the type of polymer being processed, the temperature limitations, the degree of mixing, the amount of pressure required to move the polymer, and the form of extrudate and its level of homogeneity. Current screw designs are configured for such specific needs as high dispersion, high distribution (mixing), high temperature, low temperature, and good homogeneity.

Another factor in screw design involves removal of unwanted vapors during processing (which may be facilitated by use of vents at various points on the barrel) and the need for sufficient pressure to move the plastic through a die. Pressure requirements usually run from 1500 psi

Table 3.2 Guide to extruder plastics output rates in $lb\,h^{-1a}$

Material	Screw diameter (in.)					
	$1^1/_2$	$2^1/_2$	$3^1/_2$	$4^1/_2$	6	8
ABS	280	400	825	1350	2270	4100
Acrylic	320	470	900	1500	2700	4750
PC	210	320	680	1025	1850	3200
PP	280	400	825	1350	2270	4100
HIPS	340	560	1100	1800	3250	5750
PVC, flexible	300	450	900	1500	2700	4750
PVC, rigid	180	250	500	800	1450	2300
LDPE	310	525	1050	1750	3000	5500
LLDPE	200	300	600	1000	1800	3260
HDPE	215	325	725	1175	2150	3750

[a] Output deviation from average is ±10–15%; output rates are based on different processing machine settings and the general composition of the plastics (as reviewed in the chapter). To obtain the actual output rate, weigh the actual output based on machine settings and the specific plastic processed. A 'rough' estimate for output rate (OR) in $lb\,h^{-1}$ can be calculated by using the barrel's inside dimension (ID) in inches in the following equation: $OR = 16\,ID^{2.2}$. Pounds × 0.4536 = kilograms; see the appendix for metric conversion charts (page 642). Standard barrel inside diameters, in inches (mm in parentheses) are generally $1^1/_2$ (38), 2 (50), $2^1/_2$ (64), $3^1/_4$ (83), $3^1/_2$ (89), $4^1/_2$ (115), 6 (153), and 8 (204).

(103 MPa) for a compounding or reclaim material to 4000–5000 psi (276–345 MPa) for a blown-film extrusion operation.

Extruders are usually selected on the basis of their size. Machine sizes are classified by their screw diameters and lengths. Diameters may be 0.5–18 in. (10–460 mm, 46 cm). Lengths are measured in length-to-diameter ratios (*L/D*); these range from 6 to 48. Most polymer processing machines have *L/D* values of 24–36.

Rates of throughput, or the speed at which material is moved through the extruder, have continuously been pushed higher as a result of design advances. Throughput rates generally range from a few kilograms per hour to more than 5 tonnes per hour on single-screw machines (Table 3.2). Twin-screw extruder diameters are generally sized in millimeters, and range from 14 mm to 300 mm in diameter. Throughput rates on twin-screw machines vary from a few kilograms per hour to as much as to 30 tonnes per hour.

Single-screw extruders

The standard metering extrusion screw with its three zones (conveying, compression, and metering) operates rather like a conventional injection

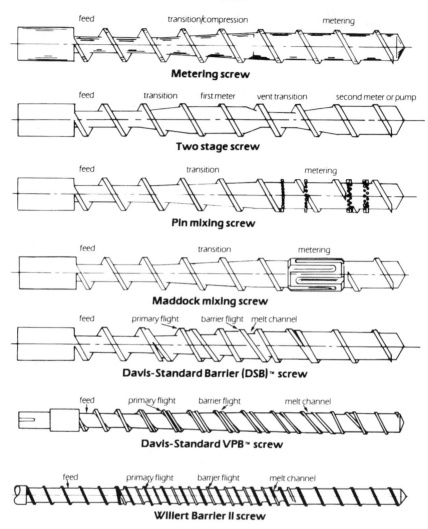

Figure 3.5 Screw designs with different mixing sections.

molding (IM) screw, as reviewed in Chapter 2. The nomenclature is the same for each (Fig. 2.11, page 136), except that no valve is used at the end of the extrusion screw (Fig. 3.5) Extrusion screws operate at lower pressures and in a continuous mode (IM is repeatable with abrupt, completely on/off pressure changes and very fast cycles). Even though many variables must be considered, extrusion requires fewer controls and presents fewer problems than IM.

Single-screw extruders have changed greatly over the years. Today's functional modular concept developed mainly for reasons of effectiveness and favorable cost comparisons. Their output rates have significantly surpassed those of older designs. The output rates in Table 3.2 can be used as a guide to predict the output rate of a process. The performance of all machines and production lines (film, profile, etc.) will depend on the many factors that have to be controlled and synchronized, going from upstream through the extruder and the downstream equipment. The type of screw has always been a major influence in the complete line.

The blown-film extruder has typified the new generation of extruders. The most effective screw design has usually been an L/D of 25. Longer machines with $L/D = 30–33$ are chosen for venting or special requirements. High outputs are obtained with LDPE blown-film or PP cast-film extrusion. The $L/D = 20$ machines are now almost always used only for heat-sensitive plastics. The $L/D = 25$ version offers exactly the right compromise for obtaining a high output and preventing overheating and damage of thermally sensitive plastics.

Even in today's hi-tech world, the art of screw design is still dominated by trial-and-error approaches. However, computer models (based on proper input and experience) play an important role. When new materials are developed or improvements in old materials are required, one must go to the laboratory to obtain rheological and thermal properties before computer modeling can be performed effectively (Chapter 1). New screws improve one or more of the basic screw functions of melt quality, mixing efficiency, melting performance along the screw, melt heat level, output rate, output stability, and power usage or energy efficiency.

Heating can be controlled by using different machine settings, which involve various trade-offs. For example, in choosing the optimum rotation speed, a slow speed places the melt in contact with the barrel and screw for a longer time via heat conduction, and the slower speed produces less shear, so that dissipative heating is reduced, and properties of the plastic

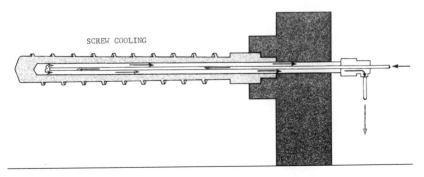

SCREW COOLING

Figure 3.6 Concept of screw with internal heat and/or cooling system.

(particularly of a film) are enhanced. An internal heat control is sometimes used with a screw (Fig. 3.6). This type of screw is characterized by deeper channels, steeper helical angles, and an internal heating element. Its internal heating lowers the amount of viscous heating needed to process the material. As a result, the melt heat can be reduced by 50 °F (30 °C).

Mixing and melting

In typical extrusion operations, mixing devices are used in the screws. Many dynamic mixers, such as those included in Fig. 3.5, are used to improve screw performance. Static mixers are sometimes inserted at the screw end (Fig. 3.7) or at the and of the barrel. There are also mixing devices that remain independent of the screw.

Proof of the success and reliability of dynamic online mixers is shown by their extensive use. Each mixer offers its own advantages and disadvantages (Chapter 2) with different machines and materials. Unfortunately there is no one system that solves all melting problems. The data available from the different equipment suppliers can be used in comparative studies.

Static mixers are successfully installed in the adapters between the barrel and the die for further thermal homogenization of the melt after it

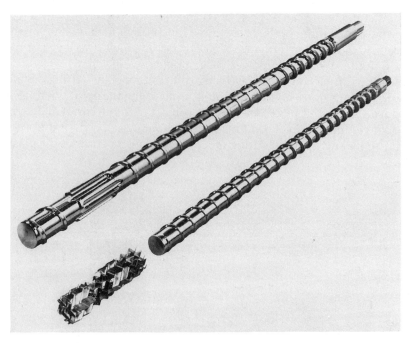

Figure 3.7 A static mixer located at the end of a screw.

leaves the screw. This causes transitory fluctuations and temperature variations in the melt to decrease considerably. As a result, an increased pressure drop occurs and the heat level increases a few degrees; temperature peaks disappear. Thus, in selecting a static mixer, one should be sure that its resistance to flow is as low as possible.

The dynamic mixing elements should be fitted as near as possible to the end of the metering zone. For maximum performance, a standard screw can have the following sectional dimensions in respect to its diameter (D): feed zone = $5D$, compression zone = $3D$, metering zone = $12–13D$, and mixing zone $2–3D$. In principle, any mixing zone, wherever it is located within the metering zone, is more effective in homogenizing than the screw threads normally present in that position. Where practical, mixers should be located in a region where the melt viscosity is not too low. Only in the metering zone can the flow phenomenon be explained theoretically. In this analysis the melt flow conveyed in the screw channel to discharge is visualized as the resultant of the positively directed drag flow and the negative, backward-directed pressure flow. Also directed backward, across the flight lands, is the leakage flow due to the pressure drop. Leakage flow, despite its significance with nonwetting materials, is rarely considered in a flow analysis. The extrusion process is best balanced when the initial pressures measured at the start of the metering zone and in the extruder die are about equal.

It is important to note that the backflow increases in proportion to the third power of the depth of the thread, so screws with deep channels are not the best choice for thermoplastics (but are okay with thermosets). If the speed of single-screw extruders is increased, especially in the processing of high molecular weight, viscous melts, the extrudate obtained may be rough and unattractive in appearance, making it unsalable. Such results can also occur with slower-running machines using relatively deep-cut screws in conjunction with extrusion dies of low resistance to flow.

This situation is due to an unstable combination of screw and die, and can be related to unsuitable pressure-melt variations. Approaches to improving the situation include longitudinally adjustable (regular mechanical) screws with tapered clearances as throttling sections, and independent throttling valves in different positions of the die. The goal is to eliminate dead spots in which stagnating melt could be thermally degraded; so streamlining is in order, such as that provided by conical screw tips (Figs 3.5 and 3.6).

Inline dynamic mixers, independently driven at optimum speeds, perform distributive mixing with moderate pressure losses, low power requirements, and small heat increases. The use of dynamic mixers mounted on the end of a screw may not be optimum because extruders may have to be driven at slower speeds to avoid problems such as surging; but independently driven mixers can be sized and run at optimum speeds to provide the best mixing. Other benefits of independently driven mixers involve feeding. For example, metering pumps can inject liquid additives directly into the mixer in the exact quantitites needed to modify a resin. In fact, an auxiliary extruder can add a secondary melt stream to the mixer, thereby allowing for such techniques as resin alloying while simultaneously coloring or stabilizing the resin.

A major and important use of mixers is to ensure minimum exposure of the melt to shear and high heat, so that less stabilizer is needed with heat-sensitive melts. With these devices, a moderate pressure drop and melt uniformity are achieved in a short process time. The mixers can be placed adjacent to a die, to provide the lowest flush time between material changes.

All the mixing devices discussed in this section make it possible to achieve better melts by literally breaking up the solid beds. Like an ice cube that more easily melts in water when it is cracked, the plastic 'solids' break up, producing a better melt.

Venting

During extrusion, as in IM (Chapters 1 and 2), melts must be freed of gaseous components (monomer, moisture, plasticizers, additives, etc.), so a vented screw is used (Figs 2.16, on page 145 and 3.8). It is very difficult to remove all the air from some powdered materials, unless the melt is exposed to vacuum venting (a vacuum is connected to the vent's exhaust). The standard machines operate on the principle of melt degassing. The degassing is assisted by a rise in the vapor pressure of volatile constituents, which results from the high melt heat. Only the free surface layer is degassed; the rest of the plastic can release its volatile content only

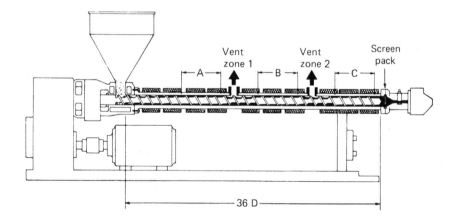

Figure 3.8 A double-vented extruder.

through diffusion. Diffusion in the nonvented screw is always time-dependent, and long residence times are not possible for melt moving through an extruder. Thus a vented extruder is used.

Most single-screw vented extruders have two stages (Fig. 2.16); a few have two vents and three stages (Fig. 3.8). The first stages of the transition and metering sections are often shorter than the sections of a single-stage conventional screw. The melt discharges at zero back pressure into the second stage, under vacuum instead of pressure. The first-stage extrudate must not be hot enough to become overheated in the second stage. And the first stage must not deliver more output per screw revolution at discharge pressure than the second stage can pump through the die under the maximum normal operating pressure, such as might occur just prior to a screen pack change. This usually means that the second-stage metering section must be at least 50% deeper than the first stage.

In practice the best metering-section depth ratio (pump ratio) is about 1.8:1. The best ratio depends on factors such as screw design, downstream equipment, feedstock performance, and operating conditions. There is likely to be melt flow through the vent if the compression ratio is high or the metering depth ratio is slightly too low. If the metering depth ratio is moderately high, there is gradual degradation of the output. If the screw channel in the vent area is not filled properly, the self-cleaning action is diminished, and the risk of plate-out increases. In any case, sticking or smearing of the melt must be avoided, or degradation will accelerate.

Screen packs

Melt from the screw is usually forced through a breaker plate with a screen pack. Extra heat develops when melt goes through the screens, so some heat-sensitive materials cannot use a screen pack. The function of a screen pack is initially to reduce rotary motion of the melt, remove large unmelted particles, and remove other contaminants. This situation can be related to improper screw design, contaminated feedstock, poor control of regrind, and so on. Sometimes screen packs are used to control the operating pressure of extruders. However, there are advantages in processing with matched and controlled back pressure, operating within the required melt pressure, as this can facilitate mixing, effectively balancing out melt heat.

In operation, the screen pack is backed up by a breaker plate that has a number of passages, usually many round holes ranging from $\frac{1}{8}$ to $\frac{3}{16}$ in. (3.2–4.8 mm) in diameter. One side of the plate is recessed to accommodate round disks of wire screen cloth, which make up the screen pack (Table 3.3). Pressure controls should be used on both sides of the breaker plate to ensure the pressure on the melt stays within the required limits.

Table 3.3 Screens used before the breaker plate to filter out contaminants in the melt[a]

		Wire mesh		Sintered
Contaminant	Metal fibers	Square weave	Dutch twill[b]	powder
Gel captured	5	1	2	3
Contaminant capacity	6	2	3	3
Permeability	4	4	1	2

[a] Range is from poorest (1) to best (6). Multiple screens are normally used; example screen pack has 20 mesh against breaker plate, followed by 40, 60, and 100 mesh (coarsest mesh has lowest mesh number).
[b] Woven in parallel diagonal lines.

Based on the processing requirements, the screen changers may be manual or highly sophisticated. Manual systems are used for limited runs or infrequent changes. The packs are usually mounted outside the extruder between the head clamp and the die; they can be changed via mechanical or hydraulic devices. Continuous screen changes also are used. The more sophisticated the system, the higher its costs. One should consult suppliers about screen capabilities, disadvantages, and so forth.

The commonly used square-weave wire mesh has poor filtering performance but good permeability. If filtering is really important, another filter media should be employed. Metal fibers stand out in their ability to capture gels and hold contaminants. Gel problems are particularly severe in small-gauge extrusion such as low-denier fibers and thin films. It is particularly in these applications that metal-fiber filters have been applied. If the plastic is heavily contaminated, the screen will clog rather quickly. If the screens have to be replaced frequently, an automatic screen changer is often employed. In these devices, the pressure drop across the screens is monitored continuously. If the pressure drop exceeds a certain value, a hydraulic piston moves the breaker plate with the screen pack out of the way; simultaneously a breaker plate with fresh screens in moved into position. These units are called slide-plate screen changers.

Screen changers are generally classified as manual; slide plate (discontinuous); continuous flow and constant pressure; and backflush. Each have performance variations to meet different requirements. Manual screen changers usually require shutdown of the extruder to change the filter pack.

The more common type is a manually shifted slide-plate screen changer. A screen change can be performed within seconds following extruder shutdown and depressurization of the melt. Shifting a slide-plate screen changer while the line is running may introduce air into the melt, causing a momentary disruption of the process. An array of continuous screen-changer technologies has evolved to eliminate this problem. These comprise two types: continuous flow and constant pressure.

No air is introduced with continuous-flow designs, but the melt pressure upstream of the breaker plate will still rise during a screen change. However, the variation in die pressure may be sufficiently minor as to have no major effect on a number of processes such as blown and cast film. Technologies include a design that splits the flow through two filters in a single slide plate. Another involves screen changers that split the flow through single or dual filters in two slide plates or sliding bolts, or a rotating wheel. Others offer prefilled filters that create a momentary disruption as they are hydraulically placed into the melt.

Constant-pressure machines provide an absolutely minimal change in die pressure while continuing to filter impurities. They have been applied on virtually every type of process. They are well suited for pipe, profiles, thin-gauge film, and foam production.

Meanwhile, recycling is exerting pressure upon screen changers. To address new demands on changing frequency, screen use, and operator intervention, backflush screen changers are being introduced. Most rely on downstream melt pressure that forces clean polymer to flush contamination off a screen pack in the screen changer, in an off-line position. Some backflush changers utilize a piston pump to meter and inject clean plastic upstream against a small portion of the online screen pack, which has been isolated from plastic flow. Backflush screen changers are claimed to extend the time between screen changes by a factor between 10 and 100.

These are also filtration devices that employ laser-drilled drums in place of typical screen packs. The holes frequently number 600000 to over 3 million and are equivalent in filtration to screens of 100–150 mesh. Contaminants trapped on the surface of the holes are wiped away by mechanical arms, and are removed.

Multiple-screw extruders

Regardless of their particular designs, all extruders have the function of conveying plastic and converting it into a melt. For this purpose, both single-and multiple-screw extruders are suitable; but they all have characteristic features. Practical and theoretical data show that each type has its place. The single-screw machine dominates and will be the focus of discussion in this chapter. However, other types are available, such as the twin-screw extruders shown in Fig. 3.9, and they are often used to achieve improved dispersion and mixing, as in the compounding of additives.

Other claims for multiscrews, most often twin-screw designs, include a high conveying capacity at low screw speed, a positive and controlled

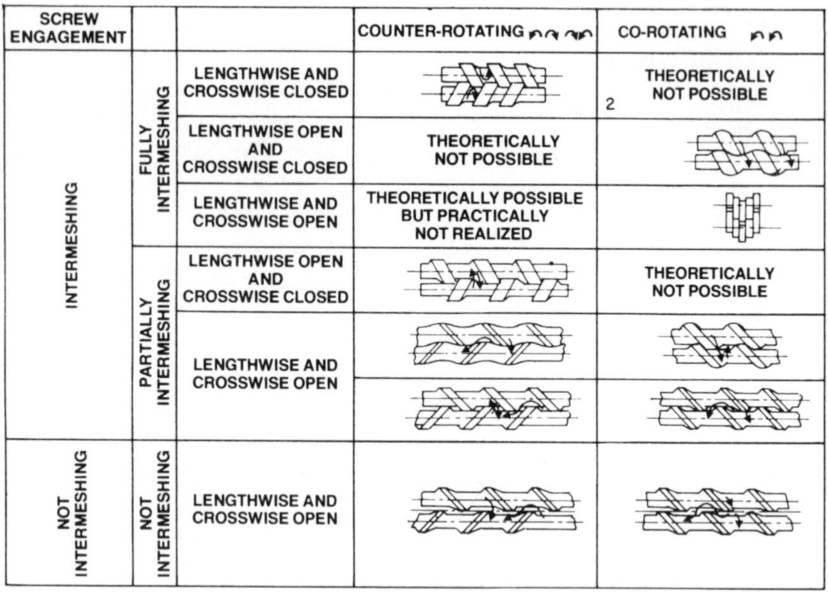

SCREW ENGAGEMENT			COUNTER-ROTATING	CO-ROTATING
INTERMESHING	FULLY INTERMESHING	LENGTHWISE AND CROSSWISE CLOSED	*(diagram)*	THEORETICALLY NOT POSSIBLE
		LENGTHWISE OPEN AND CROSSWISE CLOSED	THEORETICALLY NOT POSSIBLE	*(diagram)*
		LENGTHWISE AND CROSSWISE OPEN	THEORETICALLY POSSIBLE BUT PRACTICALLY NOT REALIZED	*(diagram)*
	PARTIALLY INTERMESHING	LENGTHWISE OPEN AND CROSSWISE CLOSED	*(diagram)*	THEORETICALLY NOT POSSIBLE
		LENGTHWISE AND CROSSWISE OPEN	*(diagram)*	*(diagram)*
			(diagram)	*(diagram)*
NOT INTERMESHING	NOT INTERMESHING	LENGTHWISE AND CROSSWISE OPEN	*(diagram)*	*(diagram)*

Figure 3.9 Different types of commercial twin-screw extruder mechanisms are used; they differ widely in operating principles and functions.

pumping rate over a wide range of temperature and coefficients of friction, low frictional heat generation which permits low-heat operation, low contact time in the extruder, relatively low motor-power requirements, and the ability to feed normally difficult feeding materials such as powders. Twin-screw types, because of their low-heat extrusion characteristics, have found increasingly wide usage in heat-sensitive PVC processing. The most popular and functional multiscrews are the twin-screw designs.

Twin-screw extruders with nonintermeshing counterrotating screws are mostly used for compounding by resin manufacturers, including situations where volatiles must be removed during extrusion. Meshing twin screws have found a substantial market in difficult compounding and devolatilization processes. To provide specialized compounding and mixing, particularly in the laboratory, different techniques are required, such as using interchangeable screw sections on a splined shaft (Fig. 3.10).

Most of the commercial machines on the market and in use today are intermeshing. One interesting feature of nonintermeshing twins is the possibility of running the two screws at different speeds, thus creating frictional relationships between them, which in some instances can be exploited for the rapid melting of powders. In some twins, one screw is significantly shorter than the other. This design is used for resins that may

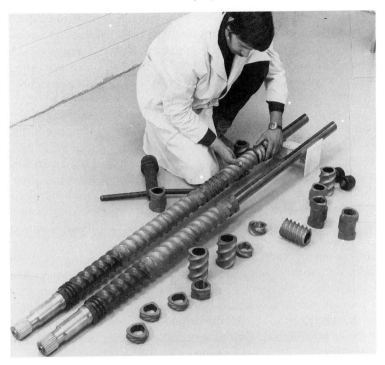

Figure 3.10 The Werner and Pfleiderer screw assembly offers specialized processing combinations.

be adequately conveyed by a single screw, once in the form of a melt, and which are difficult to feed into screw flights because of low bulk density or very low coefficients of friction of the solid against the surrounding walls. Thus, after melting by the twins, the resin moves through the single screw.

Constant screw diameters can be used, as in single designs, or the screws can be conical. A conical twin with a large-diameter, rear feed zone has a capacity greater than the compression and metering zones. This design allows a greater amount of powder to be moved to the feed zone per RPM while using the least possible clearance between flights, to provide a higher output rate with minimum shear and frictional heat accumulation. Different variables are involved in multiscrew machines compared to singles, so multiscrews are generally sized on the basis of $lb\,h^{-1}$ ($kg\,h^{-1}$) rather than L/D and/or diameters.

With intermeshing screws, the relative motion of the flight of one screw inside the channel of the other acts as a wedge that pushes material from the back to the front of the channel and from screw to screw. This pattern keeps the screw moving forward, almost as if the machine were a positive-displacement gear pump, which conveys material at rather low RPM with low compression and very little friction. The friction in a single screw that causes material to move forward also generates heat. Twin screws do not have the problem of frictional heat buildup because heat is not influenced by friction. Heat is controlled from an outside source (barrels) – an action that becomes very critical in the processing of heat-sensitive plastics, such as PVC. The multiple screws provide the advantages of higher output rates and very tight heat control, as required, e.g., to produce large PVC pipes. However, single screws are also used to produce PVC pipe.

Although multiple screws are more expensive than single screws, they do have some advantages. Twins can be used effectively in handling PVC dry blends that are compounded in a plant, potentially offering significant costs savings as compared to buying compounded PVC. With a multiple and its lower operating heat profile, lower levels of heat stabilizers can be used, with potential savings in material costs.

Other uses for twin screws include the processing of extruded expandable PS (EPS) sheet and high molecular weight polymers. For EPS there might be cost advantages in using one machine rather than the usual tandem single-screw setup. One machine mixes while the other extrudes, with the second machine providing the required cooling period. Products such as high molecular weight polyolefins or some of the TFE-fluoroplastics can be gently melted in twin screws, and these high-viscosity melts can be conveyed through a die without pulsations, a major advantage in processability.

In counterrotating systems, the basic advantage is that material which does pass through the nip of the two screws is subject to an extremely high degree of shear, just as if it were passing through the nip between two rolls of a two-roll mill. By varying the clearance (the free space left after the two screws have intermeshed), it is possible to vary the position of the material carried through axially. The narrower the clearance between the screws, the greater the shear force exerted, and the larger the proportion of material remaining in the 'bank'. With this action, it is easy to adjust the amount of shear to be applied.

In corotating twin-screw extrusion, one screw transports the material around and up to the point where the screws intermesh. Intermeshing creates two equal and opposite velocity gradients, so nearly everything that one screw has carried is taken over by the other screw. In this system, plastics will follow a figure eight path along the entire barrel length. Advantages of corotation are (1) chances are better statistically that all particles will be subjected to the same shear; (2) with the relatively long figure eight path, the melt heat has a good opportunity to influence the plastic; (3) at the deflection point, the shear energy introduced can be regulated within very wide limits by adjusting the depths of the screw flights; and (4) the system allows for a much greater degree of self-cleaning or self-wiping than in other designs, as one screw completely wipes the other screw. And more important, self-cleaning allows greater control over the residence time distribution, crucial for heat-sensitive plastics.

To summarize the comparison of screw types, in a single extruder the screw rotating inside the barrel is not able to push the material forward by itself. If the material filling the channels adheres to the screw, it becomes a rotating cylinder and provides no forward motion. Material pushed forward should not rotate, or at least should rotate at a slower rate than the screw. The only force that can keep the material from turning with the screw and make it advance along the barrel is the friction between the material and the inside surface of the barrel. The more friction, the less rotation of material with the screw; the less rotation, the more forward motion. To yield sufficient production with a low friction factor, the screw must have a large diameter and turn at high speed. However, a large-diameter screw rotating at high speed develops very high shear. The heat produced may increase the melt temperature over its limit, so cooling is necessary during the process (water jacket and/or blown air over the barrel). A nonintermeshing twin screw functions like a single-screw extruder, with friction as the prime mover. Twins with intermeshing screws operate on a completely different principle. The direction of rotation, whether corotating or counterrotating, has a great influence on the operation of the screws.

Counterrotating twins tend to accumulate and compress the material where the screws meet and remove material where they part, thereby creating zones of high and low pressure around the screws. Material that is forced by this pressure through small passages at high speeds is subjected to high localized shear. In corotating twins, material is transferred from one screw to the other without such high shear action.

In corotating twins, the screws act like a positive-displacement gear pump and do not depend on friction against the barrel to move material forward. As there is no relationship between the screw speed and the output rate, they can work with starve feeding. The channel depth is 3–4 times greater than in a single screw, and more important, the screw speed is much lower (use 15–20 RPM); so the total shear rate is much less (perhaps 80% less in twin screws). The low screw speed accounts for the low shear rate; this situation keeps the material more viscous and reduces the stock heat. Consequently, any additional heat required has to be applied through the barrel heaters.

Unfortunately, there are losses as well as gains. The higher-cost multiple-screw machines, with their more expensive and complicated drive systems, require constant preventive maintenance; otherwise, rather extensive downtime (and repair costs) may occur. Regardless of the disadvantages, multiscrews have an important role in processing plastics.

Their disadvantages should not influence their use if there are cost advantages.

Heating and cooling systems

There are three principal methods of heating extruders: electric heating, fluid heating, and steam heating. Electric heating is the most common because of its important advantages. It can cover a much larger temperature range, it is clean, inexpensive, and efficient. Electric heaters are generally placed along the barrel, grouped in zones. Each zone is usually controlled independently, so the temperature profile can be maintained along the barrel (Fig. 3.2, page 211).

Fluid heating allows even temperature over the entire heat-transfer area, avoiding local overheating. If the same heat-transfer fluid is used for cooling, an even reduction in temperature can be achieved. The maximum operating temperature of most fluids is relatively low for processing TPs, generally below 482 °F (250 °C). Because of the even temperature, fluid heating is used when processing TS plastics; even heating is required because accidental overheating could cause the TS to chemically react and solidify in the barrel. Even though very little TS is used in extrusion, the fluid system has been used with injection molding TSs (Chapter 2).

Steam heating was used in the past, particularly when processing rubber. Now rarely used, steam is a good heat-transfer fluid because of its high specific heat capacity, but it is difficult to get steam to the temperatures required for plastics processing, 392 °F (200 °C) and greater.

Cooling of extruder barrels is often an important aspect. The target is to minimize any cooling and, where practical, to eliminate it. In a sense, cooling is a waste of money. Any amount of cooling reduces the energy efficiency of the process, because cooling directly translates into lost energy; it contributes to the machine's power requirement. If an extruder requires a substantial amount of cooling, when compared to its past operating history or other machines, it is usually a strong indication of improper process control, improper screw design, excessive length-to-diameter ratio (L/D), and/or incorrect choice of extruder (single screw versus twin screw).

Cooling is usually required. Most extruders use forced-air cooling by blowers mounted underneath the extruder barrel. The external surface of the heaters or the spacers between the heaters is often made with cooling ribs to increase the heat-transfer area (ribbed surfaces will have a larger area than flat surfaces), which significantly increases cooling efficiency. Forced air is not required with small-diameter extruders because their barrel surface area is rather large compared to the channel/rib volume, providing a relatively large amount of radiant heat losses.

Fluid cooling is used when substantial or intensive cooling is required. Air cooling is a rather gentle type of cooling because its heat-transfer rates are rather small compared to water cooling. But it does have an advantage in that, when the air cooling is turned on, the change in temperature occurs gradually. Water cooling produces a rapid and steep change in temperature. This faster action requires much more accurate control and is harder to handle than air cooling. Control equipment is used to provide the proper control.

Water cooling is mostly used with grooved barrels and the feed-throat castings because they require intense cooling action. However, problems could develop. If the water temperature exceeds its boiling temperature, evaporation will occur. The water system is an effective way to extract heat, but causes a sudden increase in cooling rate. This produces a nonlinear control problem, so it is more difficult to regulate the extruder temperature. Nevertheless, the water-cooling approach is used with adequate control and, most important, proper startup procedures. Note that the cooling efficiency of air can be increased by 'wetting' it with water. However, this approach requires cooling channels made out of corrosion-resistant material [63–67].

Barrel and screw materials

The majority of barrels and screws are made from special steels (Table 3.4), which are nitrided to a minimum depth by special techniques. Low-alloy steels are sometimes used with wear-resistant liners. Bimetallic cylinders usually experience almost three times as much wear as the others. In the processing of abrasive materials, feed sections are sometimes finished in hard metal or other special materials, and matched with the screws. If there is wear in the extruder, the greatest damage is always on

Table 3.4 Materials of screw construction[a]

Material	Strength	Corrosion resistance	Cost
4140 tool steel	10	1	3
17-4PH stainless steel	8	4	3
Hastalloy C-276	1	10	10
4140 chrome-plated	10	5	3
4140 electroless nickel-plated	7	8	3
4140 with hardfacing	10	5	6
Stellite 6			
Colmonoy 56			

[a] Ratings: 1 = poorest, 10 = best.

the screw. The screw is often the only part to be replaced as it is assumed the barrel is not damaged. But this assumption is usually incorrect. If the screw is worn out, the barrel has been affected to some extent. It may well need complete replacement.

The rate of wear is increased considerably when the feed contains fillers such as titanium dioxide and glass fibers. As reviewed in Chapter 2, there are many variables that cause damage to the barrel and the screw. If a problem is likely to occur frequently, protect the screw and consider using barrels with replaceable inner liners.

GEAR PUMPS

Gear pumps, also called melt or metering pumps, have been standard equipment for decades in textile fiber production and in postreactor polymer finishing. In the 1980s they established themselves in all kinds of extrusion lines. They consist of a pump, a drive for the pump, and pump controls, located between the screen pack (or screw) and the die. Two counterrotating gears will transport a melt from the pump inlet (extruder output) to the pump discharge outlet (die) (Fig. 3.11). Gear rotation creates a suction that draws the melt into a gap between one tooth and the next. This continuous action, from tooth to tooth, develops surface drag that resists flow, so some inlet pressure is required to fill the cavity.

Strictly speaking, the gear pump is a closely intermeshing, counterrotating, twin-screw extruder. However, since gear pumps are solely used to generate pressure, they are seldom called an extruder, even though the gear pump is an extruder.

The inlet pressure requirements vary with material viscosity, pump speed, and mixing requirements. These pressures are usually less than 1000 psi (69 MPa) but cannot go below certain specified pressures such as 300 psi (21 MPa). An extruder specifically designed for use with a pump only has to 'mix', with no need to operate at high pressures to move the melt. It only has to generate the low pump-inlet pressure; thus it can deliver melt at a lower than usual heat, requiring less energy and often yielding a higher output rate. **The positive-displacement gear device pumps the melt at a constant rate.** It delivers the melt to the die with a very high metering accuracy and efficiency. It is common to have pressure differentials as high as 4000 psi (276 MPa) between the pump inlet and the discharge.

The pump's volumetric efficiency is 85–98%. Some melt is deliberately routed across the pump to provide lubrication, some slips past the gears. An incomplete fill on the inlet side will show up as a fast change in output and pressure at the exit. The extended loss of inlet pressure can damage the pump by allowing it to run dry. Overpressurization at the inlet, caused by the extruder's sudden surge, will at least change the melt

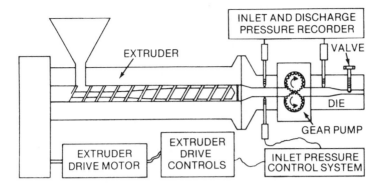

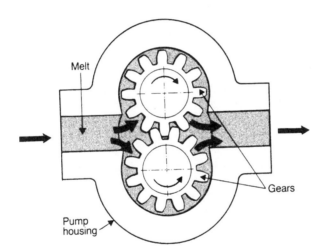

Figure 3.11 A gear pump used in a typical extrusion system.

conditions and, in extreme cases, can be dangerous to both equipment and operator. The problem can be eliminated by closed-loop pressure controls at the inlet and exit. To prevent overfeeding and overpressure, the metering section of the screw should have a barrel clearance larger than normal.

Melt pumps are most appropriate when the characteristics of the screw and die combine to give a relatively poor pumping performance by the total system. This can happen when die pressures are low but more often occurs when they are extremely high, or when the melt viscosity is extremely low. When pumps are used to increase the production rate by reducing the extruder head pressure without a corresponding increase in the screw speed, it often increases the solids content of the extrudate, creating an inferior product. This problem often necessitates additional filtration, which serves to increase pressure and may counteract many of the benfits expected from the pump, as well as increasing the financial investment even further.

Depending on the screw design, the extruder often creates pulses, causing the production rate to fluctuate. Some products can seldom tolerate even minor fluctuations, and a pump can often assist in removing these minor product nonuniformities. A pump can generally provide output uniformity of ±0.5% or better. Products include films (down to 0.75 mil, 19 μm, thickness), precision medical tubing, HIPs with 3500 lb h^{-1} (1600 kg h^{-1}) output, fiber-optic sheathing, fibers, PET magnetic tape, PE cable jacketing (weight per unit length variation) reduced from 14 to 2.7%), and so on.

Pumps are very helpful to sheet extruders who also do in-house thermoforming, as they often run up to 50% regrind mixes. Normally having a variable particle size, this mix promotes surging and up to 2% gauge variation. Pumps practically eliminate the problem and make cross-web gauge adjustments much easier. Pumps are recommended in (1) most two-stage vented barrels where output has been a problem, such as ABS sheet; (2) extrusions with extremely critical tolerances, such as CATV cable, where slight cyclic variations can cause severe electrical problems; (3) coextrusion, where precise metering of layers is necessary and low pressure differentials in the pump provide fairly linear outputs; and (4) twin-screw extruders, where pumps permit long wear life of bearings and other components, thus helping to reduce their high operating costs.

Besides improving gauge uniformity, a pump can contribute to product quality by reducing the resin's heat history. This heat reduction can help blown-film extruders, particularly those running high-viscosity melts such as LLDPE and heat-sensitive melts such as PVC. Heat drops of at least 20–30 °F (11–17 °C) will occur. In PS foam sheet extrusion, a cooling of 10–15 °F (6–8 °C) occurs in the second extruder as well as a 60% reduction in gauge variation by relief of back pressure. One must be aware that all melts require a minimum heat and back pressure for effective processing.

Pumps cannot develop pressure without imparting some energy or heat. The heat increase of the melt depends on its viscosity and the pressure differential between the inlet and the outlet (or ΔP). The rise can be 5 °F (3 °C) at low viscosity and low ΔP, and up to 30 °F (17 °C) when both these factors are higher. By lowering the melt heat in the extruder, there is practically no heat increase in the pump when ΔP is low. The result is a more stable process and a higher output rate. This approach can produce precision profiles with a 50% closer tolerance and boost output rates by 40%. Better control of PVC melt heat could increase the output up to 100%. In one case the output of totally unstabilized, clear PVC blown-film meat wrap went from 600 to over $1000 \, lb \, h^{-1}$ (from 270 to $450 \, kg \, h^{-1}$) with the use of the gear pump.

With pump use, potential energy savings amount to 10–20%. Pumps are 50–75% energy efficient, whereas single-screw extruders are about 5–20% efficient.

Although they can eliminate or significantly improve many processing problems, gear pumps cannot be considered a panacea. However, they are worth examining and could boost productivity and profits very significantly. Their major gains tend to be in (1) melt stability, (2) temperature reduction in the melt, and (3) increased throughput with tighter tolerances for dimensions and weights. They can cause problems when the plastic contains abrasive additives (the small clearances make the gear pump very susceptible to wear) and when the plastic is susceptible to degradation (gear pumps are not self-cleaning; combined with high temperatures, this will cause degraded plastic to be pumped).

DIES

Overview

The function of the die is to accept the available melt (extrudate) from an extruder and deliver it to takeoff equipment as a shaped product (film, sheet, pipe, profile, filament, etc.) with minimum deviation in the cross-sectional dimensions and a uniform output by weight, at the fastest possible rate. A well-designed die should permit quick changes of color and compatible plastics without producing large quantities of off-grade material. It will distribute the melt in the die's flow channels so that it exits with uniform velocity and uniform density (crosswise and lengthwise). Examples of different dies are shown in Figs 3.12 to 3.16.

Figure 3.13 shows the blown-film, spiral-groove die system used as the best method for even distribution of melt flow; distribution can be improved by lengthening the spirals and/or increasing the number of distribution points. The end sections of the sheet die (Fig. 3.14) have a higher temperature than the center area. This profile equalizes the cooling rate of the melt as it passes through the die. The target is to equilibrate the rate of melt flow and the stock temperature across the width of the sheet as it exits the die.

Examples of profile dies are shown in Fig. 3.15. They show how the shape of the die opening and the length of the land are related to the shape within the die for extruding different profiles. Examples of successful profile shapes with unbalanced walls are shown in Fig. 3.16.

If the exit opening of the extruder barrel does not match up with the entry opening of the die, an adapter is used between barrel and die. Dies specifically designed for a certain extruder will usually not require an adapter. However, since there is little standardization in extruder design and die design, the use of adapters is quite common.

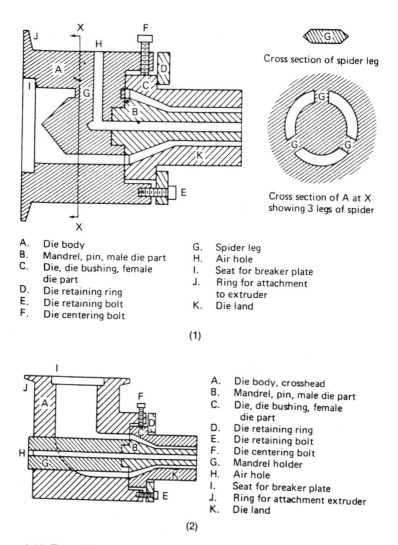

Cross section of spider leg

Cross section of A at X
showing 3 legs of spider

A. Die body
B. Mandrel, pin, male die part
C. Die, die bushing, female die part
D. Die retaining ring
E. Die retaining bolt
F. Die centering bolt

G. Spider leg
H. Air hole
I. Seat for breaker plate
J. Ring for attachment to extruder
K. Die land

(1)

A. Die body, crosshead
B. Mandrel, pin, male die part
C. Die, die bushing, female die part
D. Die retaining ring
E. Die retaining bolt
F. Die centering bolt
G. Mandrel holder
H. Air hole
I. Seat for breaker plate
J. Ring for attachment extruder
K. Die land

(2)

Figure 3.12 Examples of dies with nomenclature: (1) pipe or tubing die for inline extrusion; (2) pipe or tubing die for crosshead extrusion; (3) cast-film die; (4) sheet extrusion die.

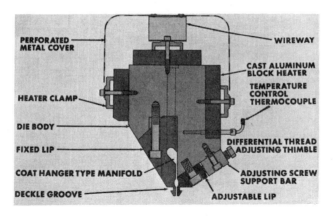

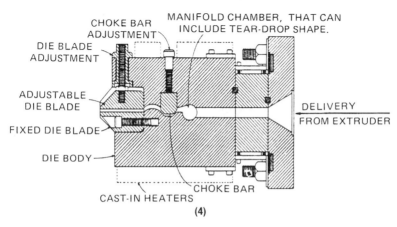

Figure 3.12 *Continued*

The die is a very critical part of the extruder. It is here that the forming of the plastic takes place. The rest of the extruder has only one task: to deliver the plastic melt to the die at the required pressure and consistency. Thus, the die-forming function is a very important part of the entire extrusion process.

The flow rate is influenced by all the variables that can exist in preparing the melt during extrusion – die heat and pressure with time in the die. Unfortunately, in spite of all the sophisticated polymer-flow analysis and the rather mechanical computer-aided design (CAD) capabilities, it is very difficult to design a die. An empirical approach must be used, as it is quite difficult to determine the optimum flow-channel geometry from

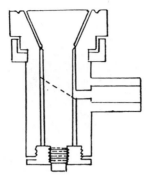

(1) *SIDE FEED DIE*
Advantages:
1. Low initial cost
2. Adjustable die opening
3. Will handle low flow
   materials
Disadvantages:
1. Mandrel deflects with
   extrusion rate, necessitating
   die adjustment
2. Die opening changes with
   pressure
3. Non-uniform melt flow
4. Cannot be rotated
5. One weld line in film

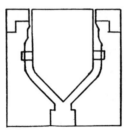

(2) *BOTTOM FEED SPIRAL DIE*
Advantages:
1. Positive die opening
2. Can be rotated
3. Will handle low flow resins
Disadvantages:
1. High initial cost
2. Very hard to clean
3. Two or more weld lines in
   film

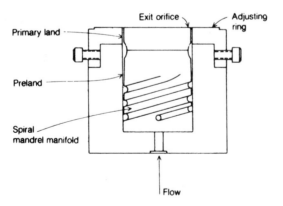

(3) *SPIRAL FEED ROLE*
Advantages:
1. No weld line in film
2. Positive die opening
3. Easy to clean
4. Can be rotated
5. Improved Film Optics
Disadvantages:
1. High head pressure
2. Will not handle low flow
   resins without modification

Figure 3.13 Blown-film dies.

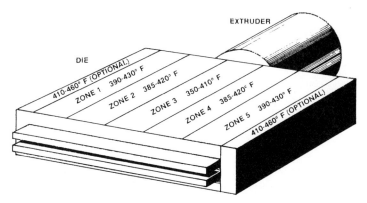

End sections of melt die are higher than center area equalizing the cooling rate of melt as it passes through the die, thus equalizing the flow rate and stock temperature of the sheet across the width of die.

Figure 3.14 Sheet die using temperature control pattern.

engineering calculations. It is important to employ rheological flow properties and other melt behavior (Chapter 1) via the applicable CAD programs for the type of die required. The most important ingredient is experience, and novices can find this within the computer program. Nevertheless, die design has remained more of an art than any other aspect of process design. Design experience can work only if the operator of the processing line has developed the important ability to debug it.

The example presented below, using a simplified equation (G. P. Lahti) obtained through a high-speed computer study during the early 1960s, continues to be extremely useful in CAD programs. It provides an excellent empirical approach that pertains to extrusion channels and dies of several shapes. Flow equations for dies of simple shapes, such as circular or rectangular channels, were known in the last century, and were first developed by M. J. Boussinesq in 1868 (*Journal de Mathematiques Pures et Appliquees*, **2**(13), 377–424, 27 July 1868). Formulas for pressure drop through more complex channels had not yet developed; this was because they required extremely complicated mathematics. As shown in Fig. 3.17, the following equations can be used:

$$Q = \frac{1}{\mu} \frac{\Delta P}{L} \frac{BH^3}{12} F$$

or

$$\Delta P = \frac{12 \mu QL}{BH^3} \frac{1}{F}$$

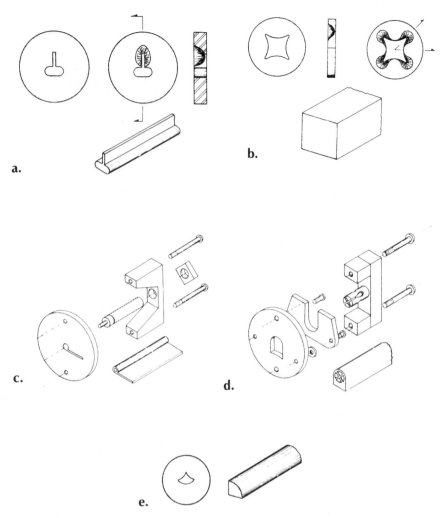

Figure 3.15 Die designs to produce different profiles. (a) The method of balancing flow to produce this shape requires having a short land where the thin leg is extruded. This design provides the same rate of flow for the thin section as for the heavy one. (b) This die for making square extrusions uses convex sides on the die opening so that straight sides are formed upon melt exiting; the corners have a slight radius to help obtain smooth corners. The rear and sectional views show how part of the die has been machined away to provide short lands at the corners to balance the melt flow. (c) In this die for a P shape, the hole in the P is formed by a pin mounted on the die bridge. The rate of flow in thick and thin sections is balanced by the shoulder dam behind the small-diameter section of the pin. The pin can be positioned along its axis to adjust the rate of flow to meet the melt characteristics. (d) In this die to extrude a rather complicated, nonuniform shape, a dam or baffle plate restricts the flow at the heavy section of the extrudate to obtain uniform flow for all sections. The melt flows between the die plate and the dam to fill the heavy section. The clearance between the dam and the die plate can be adjusted as required for different plastics with different melt behaviors. (e) In this die for extruding a quarter-round profile the die opening has convex sides to give straight sides on the right-angled portion, and the corners have a slight radius to aid in obtaining smooth corners on the extrusion.

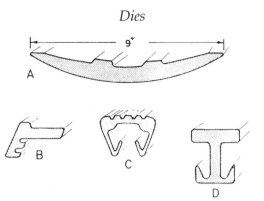

Figure 3.16 Examples of successful profile shapes with unbalanced walls: (a) a rigid PVC shoe; (b) an ABS house-trailer trim section; (c) a flexible PVC armrest for a bus; and (d) a rigid PVC insulator for an electrical bus bar.

where

H = minimum dimension of cross section, in. (mm)
B = maximum dimension of cross section, $B \geqq H$, in. (mm)
ΔP = pressure drop
μ = viscosity
Q = volumetric flow rate
L = length of channel
F = flow coefficient

To account for the entrance effect when a melt is forced from a large reservoir, the channel length (L) must be corrected or the apparent viscosity must be used, once it has been obtained from shear rate – shear stress curves for the L/H value of the existing channel. The entrance effect becomes negligible for $L/H > 16$. This single equation can be used for a variety of flow channels, as shown in Fig. 3.17.

A well-built die with adjustments – temperature changes (Fig. 3.12 [3]), restricter/choker bars, valves, and/or other devices – may be used with a particular group of materials. A die is usually designed for a specific resin. For example, conventional LDPE blown-film dies with 0.030 in. (0.8 mm) die gaps will not process LLDPE satisfactorily at high output rates. The higher-viscosity LLDPE increases back pressure significantly, thereby decreasing the throughput. And there may be melt fracture (shark-skinning), which produces a rough surface finish. With its extensional rheology, processors of LLDPE can overcome these problems with wide die gaps of about 0.090 in. (2.3 mm). With an increase in the die gap, the head pressure decreases, allowing significant increase in output.

LLDPE can be drawn or stretched in the melt with low induced orientation (Chapter 1), so a wide die gap does not add undesirable film stresses for LLDPE, as with LDPE. The optimum die gap for each applica-

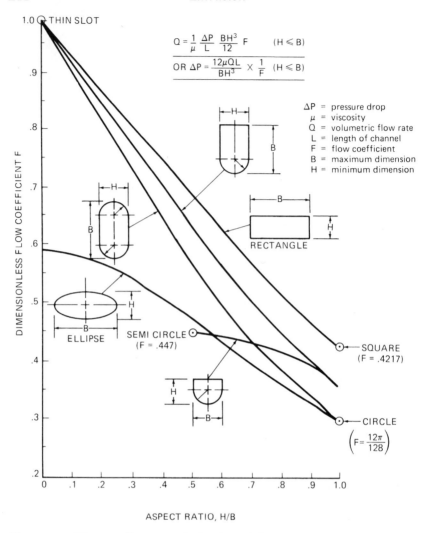

$$Q = \frac{1}{\mu} \frac{\Delta P}{L} \frac{BH^3}{12} F \qquad (H \leqslant B)$$

$$\text{OR } \Delta P = \frac{12\mu QL}{BH^3} \times \frac{1}{F} \quad (H \leqslant B)$$

ΔP = pressure drop
μ = viscosity
Q = volumetric flow rate
L = length of channel
F = flow coefficient
B = maximum dimension
H = minimum dimension

THIN SLOT

RECTANGLE

ELLIPSE

SEMI CIRCLE
(F = .447)

SQUARE
(F = .4217)

CIRCLE
$\left(F = \frac{12\pi}{128} \right)$

DIMENSIONLESS FLOW COEFFICIENT F

ASPECT RATIO, H/B

Figure 3.17 Flow coefficients calculated at different aspect ratios for various shapes using the same equation.

tion will vary according to resin grade, melt heat, and output rates. Only the die mandrel requires modification for conversion to LLDPE. If only LLDPE is to be run, the existing mandrel can be machined down. If both resins are to be run, two separate mandrels are required. As cast film is processed at considerably higher heats than blowm film, shear stresses are minimized and no die modifications are required.

A good PS-sheet die can usually run some other resins. The gauge-control capability of the die for PS or other resins is largely determined by the flow adjustments. Heavy-gauge dies might have a lip land length of 3–4 in. (75–100 mm) and a relatively coarse method of adjusting the massive lips. In contrast, a film die would have a lip land length of 0.75 in. (20 mm) and a lip adjustment capable of extremely fine adjustments. Many film dies and thin-gauge sheet dies utilize a flexible lip for extremely close gauge control.

Any analysis of die efficiency must include a careful examination of the compatibility of the die with the products to be extruded. If a die is

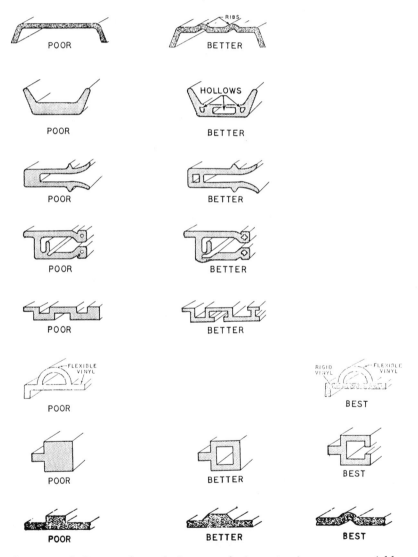

Figure 3.18 Influence of part design on reducing extrusion process variables.

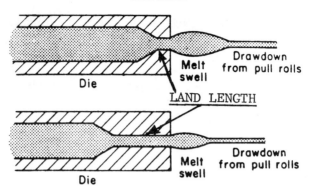

Figure 3.19 Effect of land length on swell.

designed for sheet thicknesses of 0.150–0.375 in. (3.8–9.5 mm), it is extremely difficult to extrude 5 mil (125 μm) film. As there is no die design that could be called a universal die, it is very inefficient to expect an operator to run a die beyond its capabilities. The result would be poor gauge control, and so forth. If the geometry of the flow channel is optimized for a resin under a particular set of conditions (heat, flow rate, etc.), a simple change in flow rate or in heat can make the geometry very inefficient. Except for circular dies, it is essentially impossible to obtain a channel geometry that can be used for a relatively wide range of resins and a wide range of operating conditions, such as those reviewed for LLDPE and LDPE.

Adjustment capabilities are therefore provided in the die to permit heat and pressure changes. Some dies require the heat profile to be across one direction or in different directions, using individual heating pads, and so on, with appropriate controls. The following general classification may be helpful as a guide to film and sheet selection for a die: (1) film dies are generally applicable for thicknesses of 0.010 in. (250 μm) or less; (2) thin-gauge sheet dies are normally designed for thicknesses up to 0.060 in. (1.5 mm); (3) intermediate sheet dies may cover a thickness range of 0.040–0.250 in. (1.0–6.4 mm); and (4) heavy-gauge sheet dies extrude thicknesses of 0.080–0.500 in. (2.0–12.7 mm).

To simplify the processing operation, the die design should consider certain factors if possible. The goals are to have extrudate (product) of uniform wall thickness (otherwise the heat transfer problem is magnified); to minimize the use of hollow sections; to minimize narrow or small channels; and to use generous radii on all corners, such as a minimum of 0.02 in. (0.5 mm). An 'impossible' or difficult process can be designed, but it requires extensive experience (both practical and theoretical), with trial-and-error runs, to make it practical (Figs 3.18 to 3.22).

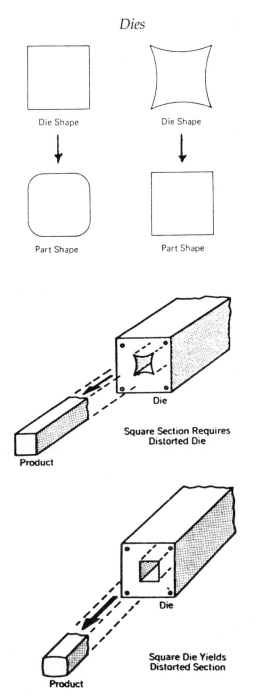

Die Shape

Die Shape

Part Shape

Part Shape

Die

**Square Section Requires
Distorted Die**

Product

Die

**Square Die Yields
Distorted Section**

Product

Figure 3.20 Effect of die orifice shape on square extrudate.

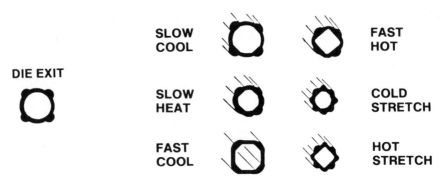

Figure 3.21 Examples of temperature, pressure, and takeoff speed (time) variation that can potentially influence the shape of the extrudate.

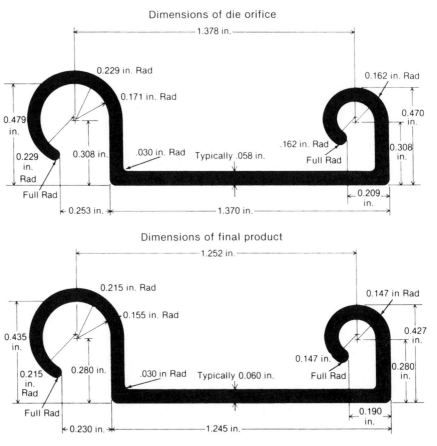

Figure 3.22 Examples of changes in dimensions of a PVC profile shape from the die orifice to the product.

Basics of flow

The non-Newtonian behavior of a plastic (Chapter 1) makes its flow through a die somewhat complicated. When a plastic melt is extruded from the die, there is some swelling (Fig. 3.19 and Table 3.5). After exiting the die, it is usually stretched or drawn down to a size equal to or smaller than the die opening. The dimensions are reduced proportionally so that, in an ideal resin, the drawn-down section is the same as the original section but smaller proportionally in each dimension. The effects of melt elasticity mean that the material does not draw down in a simple proportional manner; thus the drawdown process is a source of errors in the profile. The errors are significantly reduced in a circular extrudate, such as wire coating (Fig. 3.23). These errors must be corrected by modifying the die and takeoff equipment (Fig. 3.17).

There are substantial influences on the material due to the flow orientation of the molecules, so there are different properties parallel and perpendicular to the flow direction. These differences have a significant effect on the performance of the part (Chapter 1).

Another important characteristic, melts are affected by the orifice shape (Fig. 3.20). The effect of the orifice is related to the melt condition and the die design (land length, etc.), but a slow cooling rate can have a significant influence, especially with thick parts. Cooling is more rapid at the corners; in fact, a hot center section could cause a part to 'blow' outward and/or include visible or invisible vacuum bubbles. The popular coat hanger die, used for flat sheet and similar products, illustrates an important principle in die design. The melt at the edges of the sheet must travel farther through the die than the melt that goes through the center of the sheet.

Table 3.5 General effect of shear rate on die swell of various thermoplastics

Plastic	Die swell ratio at 392 °F (200 °C) for the following shear rates			
	$10\,s^{-1}$	$100\,s^{-1}$	$400\,s^{-1}$	$700\,s^{-1}$
PMMA-HI	1.17	1.27	1.35	–
LDPE	1.45	1.58	1.71	1.90
HDPE	1.49	1.92	2.15	–
PP, copolymer	1.52	1.84	2.1	–
PP, homopolymer	1.61	1.9	2.05	–
HIPS	1.22	1.4	–	–
HIPVC	1.35	1.5	1.52	1.53

(1) DDR in a circular die is the ratio of the cross sectional area of the die orifice/opening to the final extruded shape.

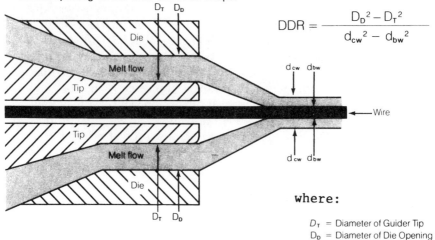

$$DDR = \frac{D_D{}^2 - D_T{}^2}{d_{cw}{}^2 - d_{bw}{}^2}$$

where:

D_T = Diameter of Guider Tip
D_D = Diameter of Die Opening
d_{bw} = Diameter of Bare Wire
d_{cw} = Diameter of Coated Wire

(2) DRB aids in determining minimum and maximum values that can be used for different plastics. Outside these limits can cause at least out of round and melt degradations.

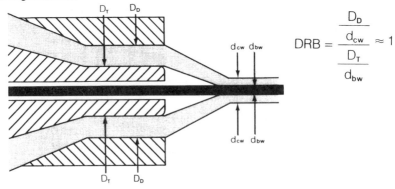

$$DRB = \frac{\dfrac{D_D}{d_{cw}}}{\dfrac{D_T}{d_{bw}}} \approx 1$$

Definition of draw ratio balance = 1

Figure 3.23 Drawdown ratio (DDR) and draw ratio balance (DRB): applicable to different products, as shown here for wire coating. Plastics have different DDRs and DRBs which can be used as guides to processability and to help establish the various melt characteristics.

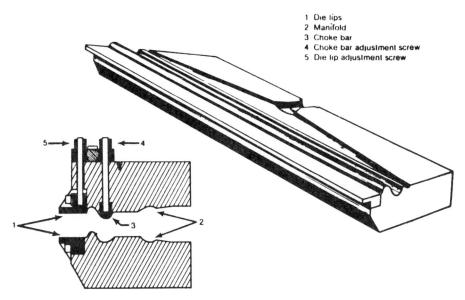

1 Die lips
2 Manifold
3 Choke bar
4 Choke bar adjustment screw
5 Die lip adjustment screw

Figure 3.24 A coat-hanger die: lower half and cross section.

Thus, a diagonal melt channel with a triangular dam in the center is used to restrict the direct flow to some degree. The principle of built-in restrictions is used to adjust the flow in many dies (Fig. 3.24). With blow-molding dies (Chapter 4) and profile dies, the openings require special attention to provide the proper product shape (Figs 3.21 and 3.22).

Special dies

Some special dies are shown in Fig. 3.25; they produce interesting flow patterns and products such as tubular-to-flat netting dies. For a circular output, a counterrotating mandrel and orifice have semicircular slits through which the melt emerges. If one part is held stationary, it forms a rhomboid or elongated pattern; if both parts rotate, it forms a true rhombic mesh. When the slits overlap, a crossing point is formed where the emerging threads are 'welded'. For flat netting, the slide is in opposite directions.

Materials of construction

Flat film and sheet dies are usually constructed of medium-carbon alloy steels. The flow surfaces of the die are chrome-plated to provide corrosion resistance. The exterior of the die is usually flash chrome-plated to prevent

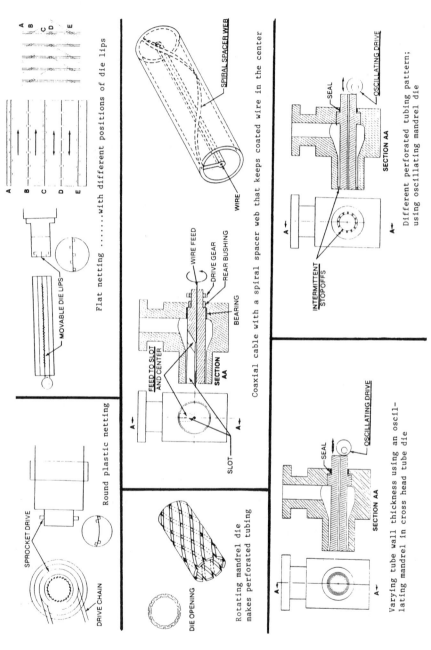

A B C D E

Flat netting ......with different positions of die lips

MOVABLE DIE LIPS

Round plastic netting

SPROCKET DRIVE

DRIVE CHAIN

DIE OPENING

Rotating mandrel die makes perforated tubing

SPIRAL SPACER WEB

WIRE

Coaxial cable with a spiral spacer web that keeps coated wire in the center

WIRE FEED

DRIVE GEAR

REAR BUSHING

BEARING

FEED TO SLOT AND CENTER

SECTION AA

SLOT

A

A

SEAL

OSCILLATING DRIVE

INTERMITTENT STOPOFFS

SECTION AA

A

A

Different perforated tubing pattern; using oscillating mandrel die

SEAL

OSCILLATING DRIVE

SECTION AA

A

A

Varying tube wall thickness using an oscil-lating mandrel in cross head tube die

Figure 3.25 Special-action dies that produce round and flat products.

rusting. Where chemical attack can be a severe problem (with PVC, etc.), various grades of stainless steel are used.

Profile, pipe, blown-film, and wire-coating dies are generally constructed of hot-rolled steel for low-pressure melt applications. High-pressure dies are made of 4140 steel and chrome plating is generally applied to the flow surfaces, particularly with EVA. Stainless steel is used for any die subject to corrosion.

Maintenance

The die is an expensive and delicate portion of any extrusion line. Great care should be taken in the disassembly and cleaning of components. Disassembly should be attempted only when the die has had sufficient time to heat-soak or at the end of a run. Experience has shown a temperature of 450 °F (232 °C) to be adequate for cleaning most nondegradable resins. For degradables, cleanup should begin immediately after shutdown to prevent corrosive action on the flow surfaces. While the heat is left on, all die bolts should be broken loose. The heat should then be turned off, and all electrical and thermocouple connections removed – carefully; then while it is still hot, the equipment is disassembled and thoroughly cleaned with 'soft' brass and copper tools.

If the extruded materials tend to cling to the flow surfaces, it is usually best to purge the die prior to cleanup, with a purging compound (Chapter 2). During assembly, the die bolts should be just snugged tight until the die heat is in the normal operating range. Once this heat is reached and a sufficient heat soak has been allowed (which could take at least 15–30 min), all bolts should be tightened to the manufacturer's recommended sequence and torque levels. If the die is stored disassembled, care should be taken in its handling to prevent damage to individual components and to flow surfaces, which can include storage in a vacuum-sealed container.

COEXTRUSION

Coextrusion provides multiple molten layers – usually using one or more extruders with melts going through one die – that are bonded together. This technique permits the use of melt heat to bond the various plastics (Table 3.6 and Figs 3.26 and 3.27) or using the center layer as an adhesive. Coextrusion is an economical competitor to conventional laminating processes by virtue of reduced materials-handling costs, raw materials costs, and machine-time costs. Pinholing is also reduced with coextrusion, even when it uses one extruder and divides the melt into at least a two-layer structure. Other gains include elimination or reduction of delamination and air entrapment.

Table 3.6 Examples of compatibility between plastics for coextrusion[a]

	LDPE	HDPE	PP	Ionomer	Nylon	EVA
LDPE	3	3	2	3	1	3
HDPE	3	3	2	3	1	3
PP	2	2	3	2	1	3
Ionomer	3	3	2	3	3	3
Nylon	1	1	1	3	3	1
EVA	3	3	3	3	1	3

[a] 1 = layers easy to separate, 2 = layers can be separated with moderate effort, 3 = layers difficult to separate.

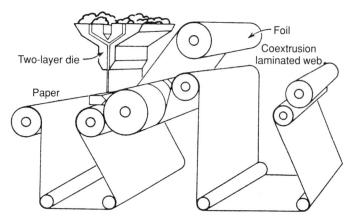

Figure 3.26 A typical coextrusion lamination process.

A processor desiring to enter the field once had little choice of equipment, but the increased interest in coextrusion has led to a proliferation. With rapidly changing market conditions and the endless introduction of useful materials, the design of machines has become much more involved. It is important that the processor has flexibility in making selections, but not at the expense of performance, dependability, or ease of operation. One should provide for the material or layer thickness necessary in product changeover without high scrap rates. The target is to incorporate scrap regrind in the layered construction.

It is very important to be able to control the individual layer distribution across the width of the die. As the viscosity ratio, or the thickness ratio, of the polymers being combined increases, it is normal for the individual layer distribution(s) of the composite film to become displaced. Viscosity differences influence reduction and saving of materials.

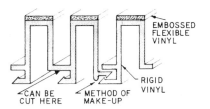

a.

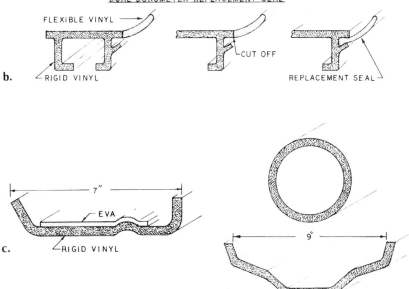

Figure 3.27 Design tips for coextrusion: (a) a dual extrusion for a modular cabinet wall panel; (b) if the flexible sealing portion wears out from abrasion, a replacement flexible insert can be slid into the slot in a rigid portion; (c) a cross section of a dual extrusion (a ball-return trough for a billiard table); (d) a bowling-ball return trough made from a 6 in. (15 cm) diameter extruded tube with one or more layers. The tube is slit while still workable and guided over a forming die; (e) typical dual extrusions of rigid and flexible PVCs; (f) typical extrusions of rigid and flexible PVCs showing different applications; (g) a cross section of a window frame with a metal embedment; (h) nonbondable plastic can be joined by keying or fitting; (i) noncircular hollows are easier to form if each part of the surrounding wall is made from the same family of plastic: (A) the rigid PVC base will remain flat and not bulge, (B) the air pressure inside the hollow will cause the flexible base section to bulge; (j) different applications for metal-embedment extrusions.

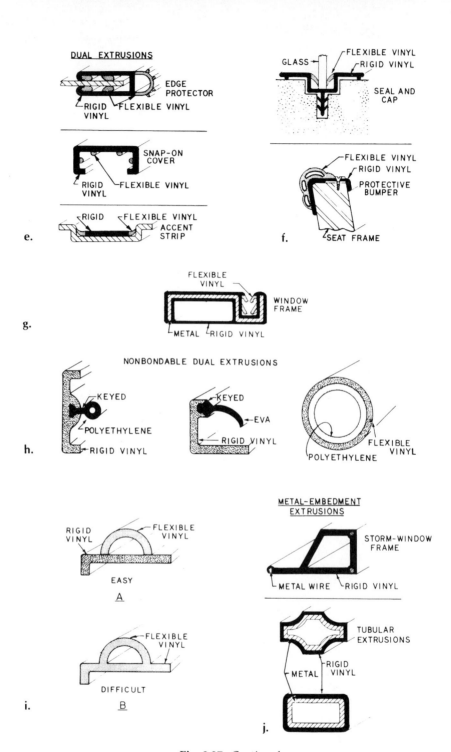

Fig. 3.27. *Continued*

Table 3.7 Comparison of feedblock and multimanifold coextrusion dies

Characteristic	Feedblock	Multimanifold
Basic difference	Melt streams brought together outside die body (between extruder and die) and flow through the die as a composite	Each melt stream has a separate manifold; each polymer spreads independently of others; they meet at die preland to die exit
Cost	Lower	Higher
Operation	Simplest	–
Number of layers	Not restricted; seven- and eight-layer systems are commercial	Generally restricted to three or four layers
Complexity	Simpler construction; no adjustments	More complex
Control flow	Contains adjustable matching inserts, no restrictor bar	Has restrictor bar or flow dividers in each polymer channel; but with blown-film dies, control is by individual extruder speed or gearboxes
Layer uniformity	Individual layer thickness correction of ±10%	Restrictors and manifold can meet ±5%
Thin skins	Better on dies >40 in. (1 m)	Better on dies <40 in. (1 m)
Viscosity range	Usually limited to 2/1 or 3/1 viscosity range of materials	Range usually much greater than 3/1
Degradable core material	Usually better	–
Heat sensitivity	More	Less
Bonding	Potentially better; layers are in contact longer in die	–

Several techniques are available for coextrusion, some of them patented and available only under license. Three types exist: feedblock, multiple-manifold, and a combination of the two (Table 3.7 and Fig. 3.28).

With the feedblock die, different melts are combined just upstream of the die, before they enter it via a special adapter. Laminar flow keeps the layers from mixing together, so the layup exits the die as an integral construction.

A multiple-manifold die involves the combination of melts within the die (Figs 3.29 and 3.30). Each inlet port leads to a separate manifold for the

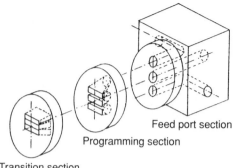

Feed port section

Programming section

Transition section

Figure 3.28 Schematic of a feedblock sheet die.

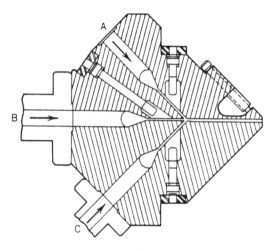

Figure 3.29 Multimanifold three-layer flat-film die.

individual layers involved. The layers are combined at or close to the final land of the die, and they exist as an integral construction through a single lip. Although the multimanifold die can be more costly than the feedblock type, it has the advantage of more precise control of individual layer thickness.

A third approach combines the feedblock and multimanifold dies to provide further processing alternatives as the complexities of coextrusion increase. This approach has been used successfully in barrier-sheet coextrusions where requirements preclude other alternatives. The feedblock is placed on the manifold or manifolds, permitting the combination of materials of similar flow characteristics in the feedblock while feeding dissimilar materials directly into the die.

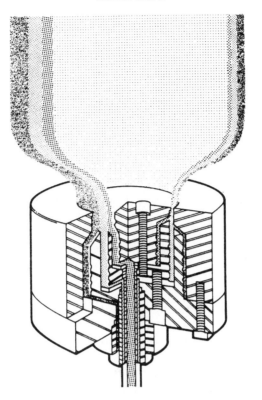

Figure 3.30 A three-layer blown-film die.

Assuming both types have a good manifold design, the multimanifold can process a broader range of flow resins than the feedblock. No matter which method is used, it is important to maintain the melt heat of each layer above the 'freezing' temperature of all layers, else the adhesion will be poor. The heat has to be the most important operating requirement for all resins. Figure 3.27 shows various design tips for coextruded profiles.

It is also desirable to be able to vary and control the thickness of the plies. As there are two or more extruders feeding the separate inlet channels, the first step is to calibrate them in terms of output rate versus screw speed; gear pumps can be used. The next step is to establish the required width-to-thickness and the takeoff speed, and to convert them into an output rate for each resin. These output rates and calibration curves are then used to establish the screw speeds. When a feedblock is not designed properly, problems such as excessive pressure drops, shear rates, or residence time can affect the quality of the extrudate. The teardrop manifold has proved beneficial for coextrusion (Fig. 3.12, part 4).

All extrusion processes require some form of melt transfer from the extruder to the feedblock or die. The transfer device can be short, such as an extruder adapter, or longer, as required in coextrusion piping. The pipe design is integral to coextrusion success; in general, the inside diameter should be large enough to avoid excessive pressure drops, but not so large as to cause extended residence time. Its heavy wall provides maximum thermal stability and heat distribution efficiency. The heat should be uniform to avoid hot and cold spots. Low-voltage heaters are desirable, but heater tapes are dangerous because of the possibility of nonuniform heat distribution.

Safety consideration should be the determining factor in the design of piping. Pipes have been known to cold-pack, depending on shutdown procedures. The location of the control thermometer is important. Over-heating of a zone, particularly on startup, can cause degradation of the plastic into a gas, creating extremely high pressures.

Tie-layers

Choosing an adhesive layer is by no means a simple operation; there are many different types, each with specific capabilities, with EVAs forming the bulk. Selection of a material is based on its providing good adhesion and surviving the process. For example, high melt strength in a blow film improves bubble stability. At temperatures above 460 °F (238 °C), EVAs could suffer from gel formation and decomposition. High melt strength can also help in cast extrusion and thermoforming processes, and the melt draw is important in coextrusion of cast film/sheet or a coating. Good melt draw is required to run higher take-up speeds and/or thinner adhesive layers without causing flow-distribution or edge-weave problems. Effects such as neck-in and edge bead are also minimized by choosing adhesives with a good draw.

Various processing conditions can require the tie resin to fall into a particular melt index (MI) classification. MI is inversely related to molecular weight (MW) (Chapter 1); a high-MW adhesive will have a low MI. Most adhesive are available in a range of MIs to meet different requirements.

The melt stability or flow is easily influenced by regrind. It is important that the regrind is compatible with the adhesive.

ORIENTATION

Orientation consists of a controlled system of stretching plastic molecules to improve their strength, stiffness, optical, electrical, and other properties. Used for almost a century, it became prominent during the 1930s for stretching fibers up to 10 times. Later it was adapted to stretching film,

Table 3.8 Example of orientation used to fabricate different types of TP film tapes

Ranges of application	Demands made	Rate of stretching	Thermoplastic
Carpet basic weave	Low shrinkage	1:7	PP
	High strength	1:5	PETP
	Temperature stability		
	Specific splicing tendency		
	Matt surface		
Tarpaulins	High strength	1:7	PP
			PE
Sacks	High strength	1:7	PP
	High friction value		PE
	Specific elongation		
	Weather resistant		
Ropes	High tensile strength	1:9 to 1:11	PP
	Specific elongation	(15)	
	Good tendency to splicing		
Twine	High tensile strength	1:9 to 1:11	PP
	High knotting strength		PP/PE
Separating weave	High strength	1:7	PP
Filter weave	Low shrinkage	1:7	PP
	Abrasion resistant	1:5	PETP
Reinforcing weave	Low shrinkage	1:7	PP
	Specific elongation	1:5	PETP
	Temperature resistance		
Tapestry and home textiles	UV resistance	1:7	PE
	Low static charge		
	Uniform coloration		
	Textile-type handle		
Outdoor carpets	Low shrinkage	1:7	PP
	Wear resistance	1:5	PETP
	Weather resistance		
	Elastic recovery		
	Uniform coloration		
	Defined splicing		
Decorative tapes	Effective surface	1:6	PP with blowing agent
	Low specific gravity		
Knitted tapes, sacks, and other packagings, seed and harvest protective nets	High knotting strength	1:6.5	PP
	Low splicing tendency		PE
	Suppleness		
	UV resistance		
Packaging tapes	High strength	1:9	PP
	Low splicing tendency	1:7	PETP
Fleeces	Fiber properties	1:7	PP and blends

sheet and more recently blow-molded containers. Many other products take advantage of its benefits (tape, pipe, profile, thermoformed parts, etc.) (Table 3.8). Practically all plastics can undergo orientation, although certain types find it particularly advantageous (PET, PP, PVC, PE, PS, PVDC, PVA, and PC). Of the 12 million tons of plastic film sales world-wide, about 20% are sales of oriented material (Chapter 1).

In extrusion the most important orienting processes are used for blown film, flat film/sheet, and blow molding (Chapter 4). During blown-film processing the blowup ratio determines the degree of circumferential orientation, and the pull rate of the bubble determines longitudinal orientation (Fig. 3.31). The optimum stretching heat for amorphous resins (PVC, etc.) is just above the glass transition temperature; for crystalline types (PET, PE, etc.) it is just below the melting point (Table 3.9). During

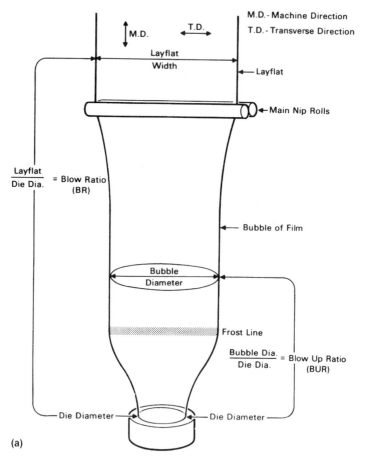

(a)

Figure 3.31 Blown film: (a) terminology, (b) orientation.

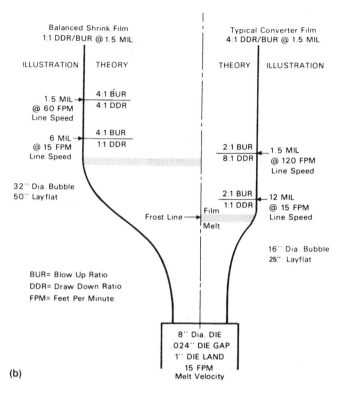

(b)

Figure 3.31 *Continued*

Table 3.9 Examples of orienting conditions for plastics

			Modulus of elasticity (Gpa)					
		Glass	Biaxially oriented film			Oriented fiber		
	Melting	transition			Unoriented	Melt	Gel	Density
	temperature,	temperature,			film	spun	spun	(g cm⁻³)
Plastic	T_m (°C)	T_g (°C)	Longitudinal	Transverse				
PVC	–	70 to 90	2.5–2.7	3–3.5	2.2	5.5	–	1.35
HDPE	138	–70 to –110	3–4	3–4	1.2	5	170	0.96
PP	134	–5 to –20	3–4	2–3	0.9	5	18	0.90
PA	260	50 to 75	2–2.5	2–2.5	0.5	4.5	19	1.13
PET	250	70 to 110	4–5.7	4.5–8.5	1.5	15	28	1.35

the stretching process, the structure changes because of crystallization, usually necessitating an increase in heat if further deformation is planned. Afterward, the orientation is 'frozen in' by lowering the heat or, with crystalline types, set by increasing the crystalline portion.

With orientation, the thickness is reduced and the surface enlarged. If film is longitudinally stretched in the elastic state, the thickness and the width are reduced in the same ratio. If lateral contraction is prevented, stretching reduces the thickness only.

The direct injection of liquid additives, such as polyisobutylene (PIB), to produce stretched film prevents difficulties in extruding and offers a processor a wider range of materials from which to select. It provides cost reductions due to the use of more economical formulations. This method is suitable for the injection of cross-linking agents, liquid colors, and the like, via the extruder or a gear pump.

In orienting film or sheet the processor uses a tentering frame (typically used for many decades in textile weaving), which is enclosed in a heat-controlled oven, with a very accurate and gentle airflow used to hold the oven at the required orienting heat (Fig. 3.32). The frame has continuous speed control and diverging tracks with holding clamps. As the clamps move apart at prescribed diverging angles, the hot plastic is stretched in the transverse direction, producing single orientation (O). To obtain

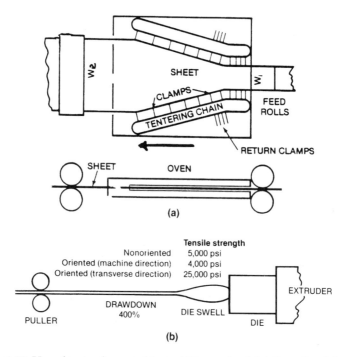

Figure 3.32 Use of tenter frame to biorient film or sheet: (a) the ratio of the feeder-roll speed to the puller-roll speed is 4:1 (ratio of width W_2 to width W_1); (b) the drawdown phenomenon with die swell to produce orientation in the longitudinal (machine) direction.

bidirectional orientation (BO) an inline series of heat-controlled rolls are located between the extruder and tenter frame. The rotation of each succeeding roll is increased, based on the desired longitudinal stretch properties.

Each line has interrelating operations, as well as specific line operations, to simplify processability. The extruder is usually followed by some kind of cooling system to remove heat at a controlled rate, to cause plastic solidification. It can be as simple as an air and/or water cooling system, or a cooled roll contact can be used to accelerate the cooling process. Some type of takeoff at the end of the line usually requires an accurate speed control to ensure product precision and/or save on material costs by tightening thickness tolerances. The simplest device might be a pair of pinch rolls or a pair of opposed belts (caterpillar takeoff). A variable-speed drive is usually desired to give the required precision.

Orienting plastics occurs in many different products. An example is an **oriented solid rod**. A solid rod (billet) is directed through a reducing conical die, usually with heat. This technique is a takeoff used in the metal industry with rods, wires, etc. The amount of deformation is based on a deformation ratio; the diameter before and after going through the die, as well as travel speed and temperature. The compressing action orients the plastic molecules and increases performance.

Oriented film (single or coextruded) has its major market in all types of packaging bags. They attract wide interest and are used for diverse applications. As an example, shrinkbands are an important product used by different industries (medical, electronic, etc.).

Most shrinkbands are made of oriented plastic films that shrink around a container when heat is applied. They are used as labels, tamper-evident neckbands, combination label/neck bands, and devices for promotional packaging. These are called dry bands, in contrast to cellulose bands that are applied wet and shrink to a tight fit as they dry. Dry bands are much more versatile than wet bands, and they have become the predominant form by a wide margin.

Most dry bands are made of PVC or vinyl copolymers. They can be made of flat film that is stretched (oriented) and seamed, or from tubular film that is stretched in the blowing process. The degree to which the band shrinks is determined when the film is stretched. To obtain a band that will eventually shrink 60%, the film must be stretched 60%. In effect, stretching the film programs **memory** into the material. That memory is recalled by applying heat after the band is placed on the container, and the band shrinks to its original dimensions. The degree of shrink required is determined by the shape of the container and the amount of container coverage required; for example, to label a tapered container, with a wider radius at the bottom than at the top, bands with relatively high shrink capability accommodate the drastic difference in radius. Depending on the grade of film, the material can yield a controlled accurate shrink of 65% or more with current technology.

The most common film thicknesses for most packaging applications are 1.5, 2, and 3 mil (38, 51, and 76 µm). Uniaxially oriented (preferential) film shrinks in height or width; biaxially oriented film shrinks in both directions. Uniaxial orientation is preferred for printed bands because they do not wrinkle and graphics are not distorted.

BLOWN FILM

More plastics go through blown-film lines than other extrusion lines. The process can vary in direction (up, down, or horizontal) and in the method of flattening the film prior to windup (Fig. 3.33). Developments in these lines relate to the extruder, dies, takeoff systems, and automation components. The development of high-speed extruders with a grooved feed zone and barrier screws makes it possible to increase output while providing greater processing flexibility; this makes it unnecessary to change the screws, particularly in coextrusion. Blown-film dies have been developed with the goals of low-pressure consumption, good self-cleaning, material changes, and ease of maintenance. The automation of blown-film plants to reduce film thickness tolerances involves the increased use of linear weight-control systems (upstream and downstream), as well as greater opportunities to influence profile thickness via suitable control elements on the die and cooling systems.

Regarding the film direction, horizontal operation entails no overhead installation and a low building height, but requires a larger floor space with probable adverse effects of gravity and uneven cooling. Vertical-down operation has the advantage of startup without flooding of the annular die gap by exiting hot melt. However, vertical-up operation is the usual method, provided sufficient melt strength exists for an upward startup, and so on. Special die blowheads are designed, usually with a multiple threaded helical mandrel discharging into an expansion space. The tubular melt assumes its final shape in a smoothing-out zone, which in all heads is a cylindrical land in a parallel position between the mandrel and the orifice. Its length is about 10–15 times the annular gap width (the lower value applies to thin film and the higher to thick film). The gap width is generally 0.5–2.0 mm.

Different methods of bubble cooling exist, each with advantages and disadvantages. For example, because of their different extensional rheologies (flow), LLDPE bubbles are less stable than those of LDPE. Proper cooling is very important in obtaining good gauge uniformity. Gentle, very cold air has a better cooling effect than high-velocity cool air; the gentle air helps to minimize bubble instability (usual pressure is 150–600 mm water column). Although single-lip air rings have proved

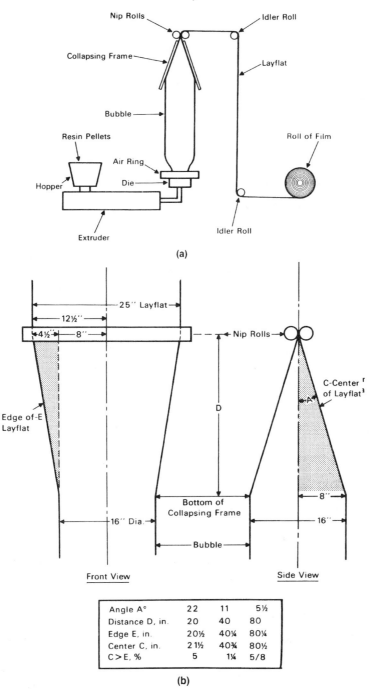

Angle A°	22	11	5½
Distance D, in.	20	40	80
Edge E, in.	20½	40¼	80¼
Center C, in.	2 1½	40¾	80½
C > E, %	5	1¼	5/8

(b)

Figure 3.33 A basic (vertical-up) blown-film line; geometry of collapsing bubble.

adequate for some applications, dual-orifice designs provide enhanced cooling, which effectively stabilizes the bubble and speeds up the line (Fig. 3.34). Internal bubble cooling (IBC) with a dual-lip air ring is also effective in increasing bubble stability at high production rates. However, improperly arranged IBC configurations can cause melt fracture due to chilling of the die lip.

Heat between the die and the pinch rolls influences the hauloff rate. LDPE, for example, leaves the die at 150–170 °C. On its arrival at the pinch rolls, the temperature should have fallen to 40 °C. The film should be wound up at as low a heat as possible in order to prevent excessive shrinkage on the roll, which causes blocking. Thin-walled film can be taken off at speeds of at least 20–50 m min^{-1}. With film of 150–300 µm thickness, rates of at least 10–20 m min^{-1} are achieved.

The blow ratio (Fig. 3.31(a)) is usually 1.5–4.0, depending on the material being processed and the thickness required. With crystalline types, melt leaving the die changes from a transparent (amorphous) condition to hazy. The level at which this transition occurs is called the frost line (Fig. 3.31(b)). The visual appearance (whether it is level/straight or shows a varying line and height) of the melt exiting the die can be related to processing conditions (uneven melt flow, heat variations, degree of orientation, etc.). For polyolefins (PE, PP, etc.) to be printed, a corona discharge pretreatment is usually given following the layflat operation. This has the effect of oxidizing and activating the surface. The weldability may be impaired and the blocking tendency increased if the treatment is too intensive.

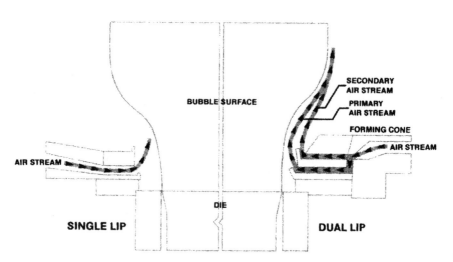

Figure 3.34 Comparison of conventional single-lip and dual-lip air rings for cooling blown film.

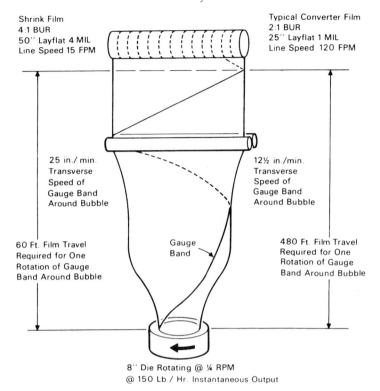

Figure 3.35 Averaging out thickness changes in blown film by rotating the diehead.

A certain degree of variation in thickness is unavoidable. When caused by the tubular film die, these variations always occur at the same position. A local film excess usually appears as a line. This can be countered by die head, hauloff and windup gear, or, with a vertical extruder, the extruder barrel is rotated or moved from side-to-side at regular intervals (Fig. 3.35).

Film gauge is affected by several factors, including takeoff speed, extruder output, barrel and die temperatures, blowup ratio, die-ring adjustment, and the rate of cooling of the bubble. Among the surface defects, fish eyes are due to imperfect mixing in the extruder or to contamination. Both these factors are controlled by the screen pack that creates a back pressure which improves homogenization.

The advantages of blown-film extrusion over flat extrusion include the ability to produce film with a more uniform strength in both the machine and transverse directions. In flat-film extrusion (particularly at high take-off rates), there is a relatively high orientation of the film in the machine direction and a very low orientation in the transverse direction. In blown-film extrusion by balancing blowup ratios against takeoff rate, it is possible to achieve physical properties which are very nearly equal in both directions, and this gives a film of maximum toughness. Another advantage of blown film is in bag making, where the only seal necessary is one across the bottom of the bag, whereas with flat film either one or two longitudinal seals are also necessary. The main disadvantage is the lower clarity of the blown film but this can be improved by more efficient cooling. Blown film is often slit, after extrusion, and is then wound up as flat film. Blown diameters can be produced, giving flat film widths that are much wider than anything produced by slot-die extrusion. Such large widths of polyethylene film find extensive uses in agriculture, horticulture, and building.

FLAT FILM

Flat films processed through slit dies are cooled principally by using chilled rolls. Many different resins are used, with thicknesses of 15–200 µm (Figs 3.36 and 3.37). Alternatively, certain plastics go directly into a water tank, but that creates many technical difficulties in production. Thus, the chill-roll process is preferred; and film up to 3 m in width will have output rates of at least 120 m min^{-1}.

In this process the melt film contacts (as quickly as possible, vertically or at an angle) the first water-cooled highly polished (to 1 µm) chrome-plated roll. An air knife can be used: its placement parallel to the die makes it possible to press the film smoothly onto the first cooling roll by means of a cold airstream. Lubricant plate-out on the cooling rolls is avoided by operation with contact rolls. At hauloff rates of up to 60 m min^{-1}, reel

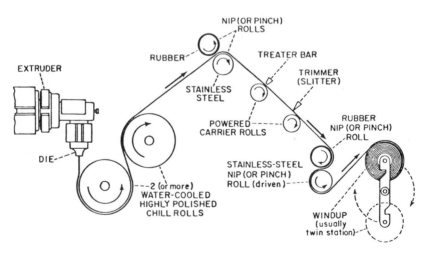

Figure 3.36 Chill-roll system for flat-film extrusion line.

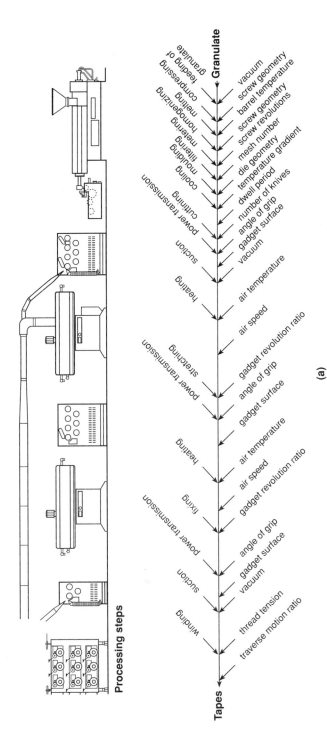

Processing steps

Granulate

feeding of granulate — vacuum
compressing — screw geometry
melting — barrel temperature
homogenizing — screw geometry
metering — screw revolutions
filtering — mesh number
moulding — die geometry
cooling — temperature gradient
— dwell period
cutting — number of knives
power transmission — angle of grip
— gadget surface
suction — vacuum
heating — air temperature
— air speed
stretching — gadget revolution ratio
power transmission — angle of grip
— gadget surface
heating — air temperature
— air speed
fixing — gadget revolution ratio
power transmission — angle of grip
— gadget surface
suction — vacuum
winding — thread tension
— traverse motion ratio

Tapes

(a)

Figure 3.37 (a) Chill-roll process used in oriented film tape line; (b) performance of oriented PP based on orienting heat and stretch ratio.

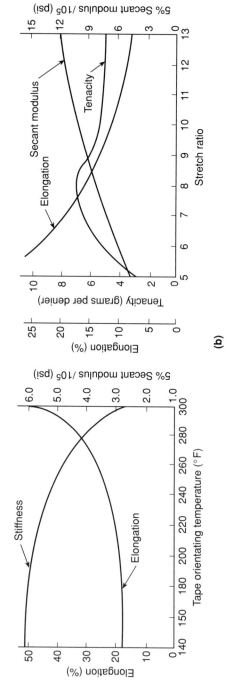

Figure 3.37 *Continued*

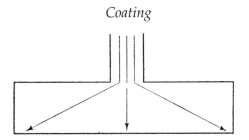

Figure 3.38 Varying flow paths in simple slot die.

change is carried out by hand. Automatic changeover equipment is required at higher hauloff rates.

Advantages of the chill-roll process over the blown-film process include preparing almost transparent film from crystalline resins (the frost line forms about 50 mm above the contact line with the chill roll); no risk of blocking; a simple crease-free windup; continuous film thickness control; high output; relatively small space requirement; and the fact that pretreatment for printing can be applied simultaneously to both sides of the film. Disadvantages are the limitation on maximum width of about 3 m (blown-film layflat is at least 12 m); loss through edge trimming; and only uniaxial orientation.

Since the width of a slot die for flat film is large compared with the diameter of the extruder head, the flow path to the extreme edges of the die is longer than to the center (Fig. 3.38). Flow compensation is usually obtained, in the case of thin film, by a manifold die. It consists of a lateral channel (or manifold) of a diameter such that its flow resistance is small compared with that offered by the die lips.

The inside surface of the die should be precision machined and well polished since the slightest surface irregularities will create striations or variations in gauge. A manifold is efficient in its task of flow compensation only if the viscosity of the melt is fairly low, so higher temperatures are necessary for flat-film extrusion. This in turn limits the use of the manifold die to materials of good thermal stability. Another consequence of the higher extrusion temperature is that a heavier screen pack is necessary in order to maintain satisfactory back pressures.

COATING

With extrusion coating an extruder forces melted TP through a horizontal slot die onto a moving substrate or web of material. The rate of application controls the thickness of the continuous film deposited on the web. The melt stream, extruded in one layer or coextruded layers, can be used as a coating or as an adhesive to laminate or sandwich two or more materials together, such as plastic film, foil, and paper.

There are three types of lines: thin-film or low-tension applications at operating web tension levels of 8–80 lbf (35.6–356 N); paper and its combinations in the middle range of 20–200 lbf (89–887 N); and high-tension applications for paperboard applications at 150–1500 lbf (667–6672 N).

As the melt leaves the die lips, the thin molten film is pulled down into the nip between two rolls situated directly below the die. The substrate, which is moving at a speed faster than the extruded film, draws the hot film to the required thickness, while the pressure between the chill roll and the pressure roll forces the film onto the substrate. Meanwhile, the molten film is being cooled by the water-cooled, chromium-plated chill roll. The pressure roll is also metal but is covered with a rubber sleeve, usually neoprene or silicone rubber. After leaving the chill roll, the coated material is drawn over a driven slitter roll where the edges are trimmed to remove the uncoated edge and the small beads of plastic which builds up at the edges of the film. After trimming, the coated material is wound up on conventional windup equipment.

In this continuous operation, rolls of material are unwound, new rolls are automatically spliced on the fly, and, when required, the surface of the substrate is prepared by chemical priming or other surface treatment to make it receptive to the extrusion coating, and to help develop adhesion between the two materials (Figs 3.39 and 3.40). Pressure and temperature on the web and extrudate combine to produce adhesion. The substrate

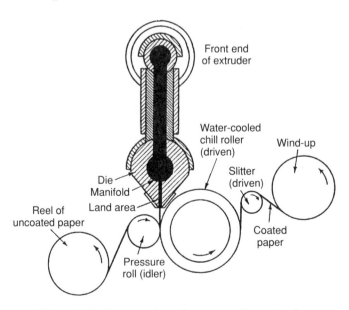

Figure 3.39 Cross section of paper-coating extruder.

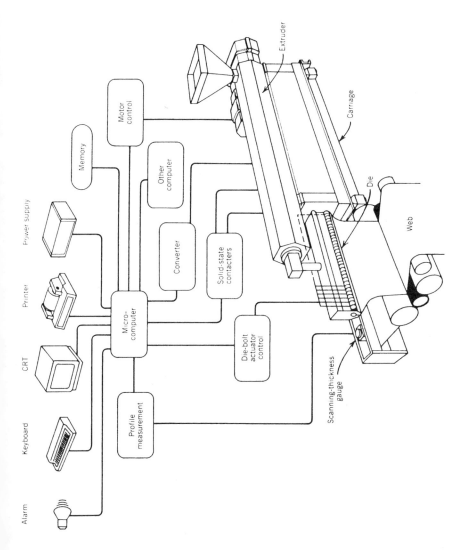

Figure 3.40 Controls on an extrusion coating line. (Courtesy Egan machinery)

normally provides the mechanical strength to the resultant structure, and the plastic provides a gas, moisture, or grease barrier.

It is not economical to coat narrow substrate widths, since the cost of equipment is high, so die widths are large with a consequent need for large extruders. Higher temperatures are required for extrusion coating than for unsupported film extrusion. The usual form of heating employed in coating extruders and dies is high-capacity electrical heaters.

The weight of the coating deposited on the substrate is governed by the speed of the chill roll. For any given extrusion rate, the higher the chill-roll speed, the greater the drawdown, hence the lower the coating weight. The chill roll has two other functions, one of which is to cool the coating to a temperature low enough to enable it to be stripped from the roll. This cooling has to be accomplished in about half a revolution of the roll thus necessitating very efficient water cooling of the roll. The chill roll also determines the surface finish of the film. For high gloss the surface of the chill roll must be very highly polished; where dull film surfaces are required, the chill roll may be etched or sandblasted.

Adhesion of the coating to the substrate is a matter of prime importance and is affected by several factors. High melt temperatures are essential; other factors include coating thickness (thick coatings adhere better than thin ones) and drawdown rate (high drawdown rates have an adverse effect on adhesion). Different substrates also have their own individual problems as in the case of paper and board, where a low moisture content is required for good adhesion. Adhesion of polyethylene to aluminum foil is not an easy matter to achieve and the problems are sometimes overcome by an etching treatment of the foil surface. The coating of regenerated cellulose film also presents a problem.

One of the problems encountered in extrusion coating is that of 'beading' at the edges of the hot film where 'necking in' occurs. This local thickening of the film has to be removed during the trimming operation otherwise, when the roll is wound up, the beads at either end support the weight of the roll, leaving a loosely wound and sagging roll in the middle. The extra trimming required means a reduction in the usable web widths, thus increasing the coating costs. In the case of polyethylene the incidence of neck-in can be reduced by decreasing the melt index or the density, but the risk of voids in the hot film at low coating speeds increases because of imperfect melt flow, so an optimum combination of these two factors has to be found.

The distance between the die lips and the chill roll should also be at a minimum for minimum neck-in; increasing the screw speed will also reduce neck-in. The balancing of requirements such as minimum neck-in, high coating speeds, absence of voids, etc., is a matter of experience. One should determine the minimum neck-in and bead size based on processing conditions, and use those observations as control parameters. Sensors

Table 3.10 Example of a surface coverage of polyethylene coating (average density 0.920 g cm^{-2})

Thickness (mm)	Mass (g) required by 1 m² of substrate	Length (m) covered by 0.45 kg PE
0.001	5.8	175
0.002	11.6	85
0.004	23.2	42.5
0.008	46.4	21

of neck-in and beads can be used in automatic process control. An example of typical surface coverage using PE coating resin is given in Table 3.10. With 3½ in. (89 mm) extruders, coating widths may be 600–1200 mm; with 4½ in. (114 mm) 900–2500 mm; with 6 in. (152 mm) 1000–4000 mm; and with 8 in. (203 mm) 3000–5000 mm.

Plastic coatings are applied in different forms and shapes on many different products, such as wire, cable, profiles (plastics, wood, aluminum, etc.), films/foils (plastics, aluminum, steel, paper, etc.), rope, and so on. Certain coatings only require snug fits, whereas others require excellent adhesion, usually necessitating cleaning, priming, and/or heating substrates. Wire and cable coating is reviewed later in this chapter. Examples of other coating lines include coating wood, film, and structural reinforcement wire.

Coating wood profiles

Wood, as well as other materials (plastic, aluminum, steel, etc.), can be easily coated in profile shape. Procedures used would be similar to wire coating, including possible preheating and cleaning. For noncontinuous profiles, special equipment is available; wood (or other material) is fed automatically from a storage/feeding magazine (when required preheated and/or cleaned), through a crosshead die, and finally through a cooling medium of air, water, gas, and so on (Fig. 3.41).

Coating films/foils

Extrusion coatings, using many different plastics, are applied to film, foil, or sheet substrates of plastic, wood, aluminum, steel, paper, cardboard, and so on. A 'curtain' of very hot melt is extruded downward from a slit die (similar to flat-film slit dies). Preheating and/or cleaning operations may be required. Hot melt contacts the substrate, which is supported by a large, highly polished chill roll with a small rubber or metal nip roll. The nip roll applies the required pressure to ensure proper air-free adhesion. As the melt is usually at its maximum heat and pressure, 'delicate' opera-

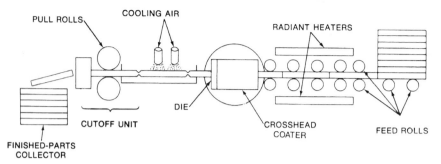

PULL ROLLS

COOLING AIR

RADIANT HEATERS

DIE

CUTOFF UNIT

CROSSHEAD COATER

FEED ROLLS

FINISHED-PARTS COLLECTOR

Figure 3.41 Noncontinuous profiles being automatically coated with plastics.

tion is required in the equipment and the surrounding area. Changes in air currents and moisture can cause immediate downtime; tighter controls may not be necessary if all that is required is not to open a 'door' – particularly a large garage door.

Products from extrusion coating/laminating lines have six main market classifications: liquid packaging; flexible packaging; board packaging; industrial wraps; industrial products; and sacks.

COMPOUNDING

Of all the 17000 plastics available worldwide, most must undergo compounding (the incorporation of additional ingredients needed for processing) in order to have desired properties. These ingredients may include additives to improve a polymer's physical properties, stability, or processability (Chapter 1). The method used to compound these materials can be either dry blending or melt mixing. In many cases both types of mixing are required. Mixing can be continuous or in batches. Batch systems tend to be more labor-intensive and less complicated. Continuous systems are more consistent and easier to instrument for statistical process control [9].

Batch mixers are still a very important part of every compounding operation, primarily in the premixing and postcompounding areas. There are three intensities of batch mixing: low, medium, and high. Low-intensity mixers are the most commonly used because they are the least expensive, are relatively easy to clean, and can be employed for a multitude of tasks where time and dispersion are not the prime consideration.

Medium-intensity mixers usually consist of low-intensity mixers fitted with attachments called intensifiers. The intensifiers are normally a high-speed mixing head added to a low-intensity mixer.

The high-intensity mixer is a high-speed unit with specialized blade designs. (Mixing times are typically measured in minutes, as compared to half an hour or more for other types of mixers.) High-intensity mixers are used most often in the areas of pigment dispersion and premixing of the compound.

Most continuous-melt mixer/extruders and single-screw extruders are simple in design and reasonably priced. Processing with single-screw extruders is normally limited to the barrel surface and conveying is only by drag flow. This minimizes the use of single screws for low-viscosity mixing. Heat transfer is difficult and is effective only on the barrels. Many of the shortcomings of the single screw can be overcome by grooved barrel walls, barrier screws, and shortened screw design. Twin-screw extruders provide much higher torque and volume than single-screw models.

The nonintermeshing twin-screw extruder has many advantages over the single screw. These include better distributive mixing and feeding because of the presence of higher volumes and the downward motion of the counterrotating screw.

The fully intermeshing twin screw is by far the most versatile and the most popular design currently used for mixing. The intermeshing screw allows the mixer to approach plug flow, creating uniform heat and shear mixing. This type of twin-screw extruder is a high-shear machine, permitting both distributive and dispersive mixing.

Distributive mixing is the homogenization of the ingredients without particle-size reduction. Dispersive mixing is a high-shear operation that reduces particle sizes.

Improvement in compounding has been through process control that typically monitors the temperature of the motor, lubrication system, and barrels; also the feed rates, shaft speed, torque, oil level, die pressure, pelletizing speed, etc.

With data acquisition, the compounder collects data and trend variables over the entire run to insure a consistent product. Statistical process control can be added to the system to inform the operator when the processing variables are out of specification.

Most compounding systems are starve-fed, so accurate control of the feed streams is required. An accurate feeder is the loss-in-weight feeder. In these systems the entire feeder, including the hopper, is placed on a load cell or is sometimes suspended from load cells. The weight can be measured as often as 10 times per second. This information is then processed to increase or decrease the feed rates, based on the set points.

FIBERS

So-called manmade fibers, which include those made from natural organic polymers as well as synthetic organic polymers, are mostly produced by extrusion through fine holes, a process known as melt spinning.

The term derives from the formation of fibers into yarns; it is still applied to these processes today. Since the natural polymers and their derivatives undergo decomposition at or below their melting points, they can be converted into a form suitable for spinning only by dissolution. Thus, the two processes for extruding polymer solutions, known as dry and wet spinning, were developed well before the third process, melt spinning.

The spinneret is a type of extrusion die; it is usually a metal plate with many small round or oval holes, through which a melt is forced and/or pulled (rayon, nylon, glass, etc.). For glass, the plate is made of precious metals such as gold, but principally platinum; they provide control of hole size and wear resistance against the rubbing action of the glass. Filaments may be hardened by cooling in air and/or water, or by chemical action. Spinnerets enable extrusion of filaments of 1 denier or less (1 denier = 40 µm). For commercial work 12–15 denier fiber is generally used [9].

The product obtained from the spinning is usually relatively weak, not yet commercially suitable. Filaments are usually subjected to orientation after leaving the die to increase their properties. Thermal setting and thermal relaxation processes provide dimensional stability; twisting and interlacing provide interfilament cohesion; texturing provides a voluminous yarn; crimping and cutting provide staple products similar in length and processing behavior to natural fibers.

Continuous-filament yarn, which consists of a small number of roughly parallel, continuous, individual filaments of unlimited length held together by a slight twist or by intermingling, is usually packaged as a tube or cone. Staple fiber is made up of a very large number of discontinuous, randomly oriented, individual fibers normally shipped in a box or bale. It is usually subjected to a series of processes, culminating in textile spinning to yarn. The precursor of staple fiber is tow, which consists of a large number of roughly parallel, continuous filaments. It is converted by cutting or breaking into staple fiber or directly into a top or sliver, intermediate stages between staple fiber and yarn. In the latter case the filaments remain parallel.

Substantial amounts of fiber are also sold as monofilament, which is a single-filament yarn of substantially greater diameter than those present in continuous-filament and staple yarn, and as tapes, fibrillated tapes, and slit-film products. In addition, certain types of so-called nonwoven fabric are directly formed from continuous filaments without isolation of a yarn. Products of this type include melded and spun-bonded fabrics.

The three most common spinning processes are melt spinning, dry spinning and wet spinning. Other types of fiber-forming process include reaction spinning; dispersion, emulsion, and suspension spinning; fusion-melt spinning; phase-separation spinning; and gel spinning. Numerous techniques for producing fibers without using a spinneret have been

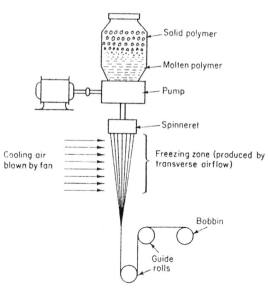

Figure 3.42 Melt spinning of fibers.

devised, including centrifugal spinning, electrostatic spinning, and tack spinning. Also under investigation is solid-state extrusion (SSE), extruding through a capillary rheometer with a conical die; the processing temperature is close to the melting temperature.

Melt spinning was developed in the late 1930s for nylon-6 and nylon-6, 6 and is now used for several fibers, including polyamides, polyesters, and polyolefins. Molten polymer is forced by an extruder or gear pump through the fine holes in a spinneret and solidified by passage through cool air to a driven roll that draws the fibers away from the face of the spinneret (Fig. 3.42).

In dry spinning, a polymer solution is extruded from a spinneret through a zone in which the solvent is rapidly evaporated, leaving filaments that are wound up at speeds up to about $1000\,\mathrm{m\,min^{-1}}$.

In wet spinning, a polymer solution is extruded from a spinneret into a nonsolvent that coagulates (i.e., precipitates) the polymer. In some cases, such as the viscose process, both the solution and the precipitation stages involve chemical reaction. Conventional wet spinning is by far the slowest in linear velocity of the principal processes, but since it permits very short distances between the holes in the spinneret face, a single spinneret may carry a very large number of holes. Hence, the productivity from a single spinning position may be very high.

NETTING

Plastic netting is produced through couterrotating die lips as a feedstock with strands at a nominal 45° to machine direction then oriented into a lightweight tubular netting (Fig. 3.43). A variety of mesh sizes, diameters and colors are available, usually made from PE. Different shapes are shown in Fig. 3.25 (page 252). Most of the netting produced for flexible packaging started during the mid 1950s (patent) and is now marked throughout the world in forms such as (1) tubular netting on traverse-wound rolls (commonly called rope by the industry); (2) header-label bags converted from rope, heat-set into open tubes (usually gussetted) and sealed by the manufacturer with a folded and sewn printed label; (3) G-bags, which are lengths of material cut with a hot knife to produce a gathered heat seal on one end; and (4) sleeving or cartridges, where netting is shirred onto a collapsed corrugated board. The board can be opened into a square or rectangle for transfer onto a tube or funnel.

The largest market for plastic netting is consumer-sized packaging in the produce industry. Nearly complete ventilation reduces spoilage of many fruits and vegetables such as citrus and onions. Automatic packaging equipment using rope shirred onto tubes offer labor and material costs competitive with other consumer-size packages. Premade header-label bags are also widely used by produce packers and by supermarkets that

Figure 3.43 Netting exits the die and undergoes postextrusion stretching.

package in-store. Other markets include pallet stretch-wrap, tools, and toys.

PIPE

A typical pipe line consists of a single- or a twin-screw extruder, a die, equipment for inside and outside calibration, a cooling tank, a wall-thickness measuring device, marking equipment, hauloff and automatic cutting and pallet equipment, or a windup unit for self-supporting pipe coils or lengths that are coiled on a drum (Fig. 3.44). Figure 3.12 on pages 238–9, provides examples of pipe dies.

Single-screw extruders are generally used when processing PVC compound in granule form; twins handle powders of PVC. The adjustment and control of back pressure are very critical. PVC pipe is a big and very competitive market, so quality and profitability have been the most important requirements for years. Improving the equipment is almost of secondary importance because the equipment for good-quality products is already available.

The consumption of material, the main cost factor, can be gradually minimized by the use of measuring and control systems. The goal is always tighter tolerance control to save material. Calibrating disks and pressure calibration methods of many different designs are used to meet various requirements. Figure 3.45 provides calibration examples.

The operator's expertise in using these calibration systems is as important as controlling the complete line. A system of feedback and control of the extruder by a microprocessor is used to control wall thickness, combining ultrasonic gauging with gravimetric proportioning. Such new technologies are available but have to be debugged. Perhaps it could be said that any equipment from the past is definitely noncompetitive, based on all the new equipment that has been made available from upstream, through the extruder, and downstream. In the past few years, all the equipment has been significantly altered to increase profits.

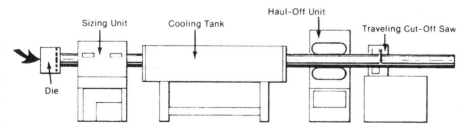

Figure 3.44 Important downstream equipment used in pipe and profile extrusion.

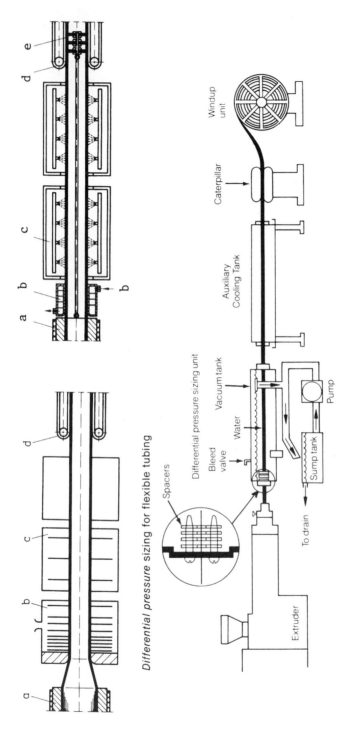

Differential pressure sizing for flexible tubing

Figure 3.45 Calibration systems for pipe/tube extrusion lines. Top left schematic shows a vacuum tank calibration of rigid pipe with cascading, temperature-controlled water baths: (a) pipe die, (b) vacuum with disks, (c) heated-zone water baths, and (d) caterpillar takeoff puller. Top right schematic shows pressure calibration of rigid pipe using plug insert with water-spray cooling: (a) pipe die, (b) pressure calibration, (c) water-spray cooling, (d) caterpillar conveyor-belt takeoff puller, and (e) plug insert to retain internal pressure, helping to control inside pipe diameter. The bottom schematic is an example of differential pressure sizing for flexible tubing.

POSTFORMING

Inline postforming, or postextrusion processing, refers to the special processing that may be done to the extrudate, usually just after it emerges from the die but before the material has a chance to cool. It provides performance and cost advantages, principally for long production runs. The process is used with different products such as sheet, film, rod, profile, and tubes. Upon leaving the die, and retaining heat, the plastic is continuously postformed (Figs 3.46 to 3.50). As an example, Fig. 3.48 is a system that can be used with different profiles, such as small or large extruded tubes producing corrugated tube/pipe. Moving molds would use corrugated tubular cavities with vacuum, pressure, or water cooling lines.

With this type of inline system, the hot plastic is reduced only to the desired heat of forming. All it may require is a fixed distance from the die opening. Cooling can be accelerated with blown air, a water spray, a water bath, or combinations thereof. This equipment, like others, requires precision tooling with perfect registration.

When the material is worked in such a state it is known as inline processing, as opposed to cutting, forming, or other processing, which might be done on the cold extrusion. Inline processing is usually done close to the extruder and is done automatically, with little or no extra labor

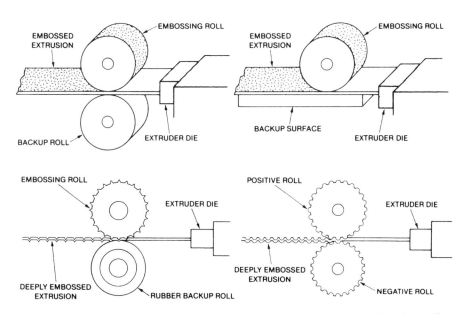

Figure 3.46 Inline postforming with extruder: embossing one or both sides with shallow or deep patterns.

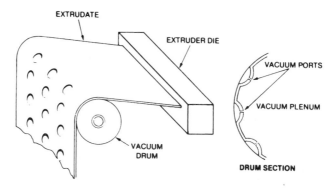

Figure 3.47 Inline vacuum-forming embossing roll with water-cooled temperature control.

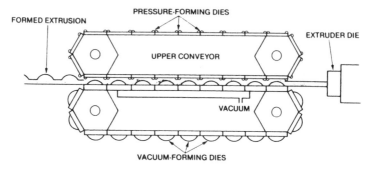

Figure 3.48 Inline vacuum/pressure former for plastic sheet with matched, water-cooled, forming molds on a continuous conveyor system.

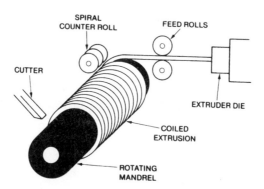

Figure 3.49 An inline coil former can produce telephone cords, springs, etc., using extruded round, square, hexagonal, and other shapes.

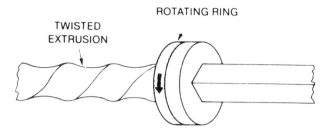

Figure 3.50 Inline fixed/rotating rings used to twist extrudate.

on the part of the machine operator. The extra processing, which may involve shaping, cutting, reforming, or a surface modification of the extrudate, can considerably increase the value of the extrusion without materially increasing its cost, but it may also be done to enable the use of a lower-cost die, perhaps for flattening a tubular extrusion into an oval so that a much lower-cost circular die can be used (Chapter 10).

PROFILE

Profile extrusion is similar to pipe extrusion. A die plate, in which an orifice has been cut, is bolted to the face of the normal die assembly so that it is then possible to change from one profile to another without the necessity of a major die change. To maintain a smooth flow, the die-plate orifice is blended to match the bore of the die body.

There is a surface drag on the molten plastic as it passes through the die and this has the effect of lessening the flow through the thinner sections of the orifice. To counteract this tendency, it is necessary to alter the shape of the orifice, and there is often a wide difference in the orifice shape and the desired extrusion profile. Some examples of sections and the die orifices necessary to produce them are shown in Figs 3.15, 3.16, 3.18, 3.21, 3.22, and 3.27. Table 3.11 includes examples of dimensional tolerances.

For complex profiles it is impossible to design a die plate from first principles. It is usual to produce a first approximation in an unfinished form then to arrive at the correct orifice by trial and error on the extruder.

As with pipe, the profile market is largely dominated by PVC and is highly competitive. Automation at the processor level has reached a very advanced stage. The operator's capability is needed to ensure maximum product efficiency by controlling die swell, rate of pull, and so on. High performance lines operate at over $2\,\mathrm{m\,min^{-1}}$.

As changes have been made in PVC compounding to optimize processing, there has been a considerable change in the type of impact modifier used. The use of acrylic, with butyl acrylate, has almost replaced

Table 3.11 Some dimensional tolerances for plastic profile extrusions

Dimension	Rigid vinyl (PVC)	Polystyrene	ABS	Polypropylene	Flexible vinyl (PVC)	Polyethylene
Wall thickness	± 8%	± 8%	± 8%	± 8%	± 10%	± 10%
Angles	± 2°	± 2°	± 3°	± 3°	± 5°	± 5°
Profile dimensions. ± mm (in.)						
0–3 (0–$\frac{1}{8}$)	0.18 mm (0.007 in.)	0.18 mm (0.007 in.)	0.25 mm (0.010 in.)	0.25 mm (0.010 in.)	0.25 mm (0.010 in.)	0.30 mm (0.012 in.)
3–13 ($\frac{1}{8}$–$\frac{1}{2}$)	0.25 mm (0.010 in.)	0.30 mm (0.012 in.)	0.50 mm (0.020 in.)	0.38 mm (0.015 in.)	0.38 mm (0.015 in.)	0.63 mm (0.025 in.)
13–25 ($\frac{1}{2}$–1)	0.38 mm (0.015 in.)	0.43 mm (0.017 in.)	0.63 mm (0.025 in.)	0.50 mm (0.020 in.)	0.50 mm (0.020 in.)	0.75 mm (0.030 in.)
25–38 (1–1$\frac{1}{2}$)	0.50 mm (0.020 in.)	0.63 mm (0.025 in.)	0.68 mm (0.027 in.)	0.68 mm (0.027 in.)	0.75 mm (0.030 in.)	0.90 mm (0.035 in.)
38–50 (1$\frac{1}{2}$–2)	0.63 mm (0.025 in.)	0.75 mm (0.030 in.)	0.90 mm (0.035 in.)	0.90 mm (0.035 in.)	0.90 mm (0.035 in.)	1.0 mm (0.040 in.)
50–75 (2–3)	0.75 mm (0.030 in.)	0.90 mm (0.035 in.)	0.94 mm (0.037 in.)	0.94 mm (0.037 in.)	1.0 mm (0.040 in.)	1.1 mm (0.045 in.)
75–100 (3–4)	1.1 mm (0.045 in.)	1.3 mm (0.050 in.)	1.3 mm (0.050 in.)	1.3 mm (0.050 in.)	1.7 mm (0.065 in.)	1.7 mm (0.065 in.)
100–125 (4–5)	1.5 mm (0.060 in.)	1.7 mm (0.065 in.)	1.7 mm (0.065 in.)	1.7 mm (0.065 in.)	2.4 mm (0.093 in.)	2.4 mm (0.093 in.)
125–180 (5–7)	1.9 mm (0.075 in.)	2.4 mm (0.093 in.)	2.4 mm (0.093 in.)	2.4 mm (0.093 in.)	3.0 mm (0.125 in.)	3.0 mm (0.125 in.)
180–250 (7–10)	2.4 mm (0.093 in.)	3.0 mm (0.125 in.)	3.0 mm (0.125 in.)	3.0 mm (0.125 in.)	3.8 mm (0.150 in.)	3.8 mm (0.150 in.)

modification with EVA. When modification is carried out with acrylate, the elastomer phase is embedded in the form of beads in a continuous PVC matrix. During processing it is retained to the decomposition range. This wide processing latitude has at least made it easier to achieve the present high outputs in profile extrusion. It has also provided low shrinkage, high heat distortion, and good weather resistance.

REACTIVE

Reactive extrusion is one of the rare occasions when an extruder is used as a chemical reactor instead of merely as a melt processor. Reactive extrusion, also called reactive processing or reactive compounding, refers to the performance of chemical reactions during extrusion processing of polymers (Chapter 1).

An extruder may be considered as a horizontal reactor with one or two internal screws for conveying reactant polymer or monomer in the form of a solid or slurry, melt, or liquid. The most common reactants are polymer or prepolymer melts and gaseous, liquid, or molten low molecular weight compounds. A particular advantage of the extruder as a chemical reactor is the absence of solvent as the reaction medium. No solvent-stripping or recovery process is required, and product contamination by solvent or solvent impurities is avoided.

The chemical reaction may take place in the melt phase or, less commonly, in the liquid phase, as when bulk polymerization of monomers is performed in an extruder, or in the solid phase when the polymer is conveyed through the extruder in a solvent slurry. The types of reactions developed include bulk polymerization, graft reaction, interchain coploymer formation, coupling or branching reaction, controlled molecular weight degradation and functionalization or functional group modification (Chapter 1) [9, 63–65, 68–70]. The extrusion device as a reactor combines several chemical process operations into a single piece of equipment with accompanying high spacetime yields of product. An extruder reactor is ideally suited for continuous production of material after equilibrium is established in the extruder barrel for the desired chemical processes.

Because of their versatility, most extruder reactors are twin-screw extruders that possess a segmented barrel, each segment of which can be individually cooled or heated externally. In addition to external heating, a molten material may be shear heated by the resistance of viscous material to the conveying motion of the screw; these processes provide energy for chemical reaction. Extruder screws often have specialized sections or configurations, e.g., high-shear mixing sections. Twin-screw extruder screws may be equipped with interchangeable screw elements that provide different degrees of mixing and surface area exposure by varying the depth between screw flights, the individual flight thicknesses, and direction and degree of flight pitch.

In a typical reactive extrusion process, the reactants are fed into the extruder feed throat where the material is usually heated to initiate reaction or increase the reaction rate. The reactant mixture is conveyed through sequential barrel segments where degree of mixing and specific energy input bring the reaction to the desired degree of completion within the limits of residence time in the extruder. At this stage the reaction may be quenched by cooling or addition of a catalyst quencher where applicable, and volatile by-products or excess reactants may be removed. Melt is forced from the extruder through a die with one or more openings. The design of the die openings is one factor determining the pressure against which the extruder has to pump by the conveving motion of the internal screw as is the case with conventional extrusion processing. Melt from the die is usual rapidly cooled by contact with a fluid medium such as water.

SHEET

Sheet is usually defined as being thicker than film, or thicker than 1–4 mm (0.04–0.15 in.). However, certain products of the industry consider 0.254 mm (0.010 in.) the separation thickness.

Sheet thickness can be at least 2 mm (0.5 in.), and widths can be up to 30 m (10 ft). Hot melt from a slit die is directed to a combination of an air knife with two cooling rolls, or a more popular choice, to a stand of three cooling rolls (Fig. 3.51), which cools, calibrates, and produces a smooth sheet. To aid the chill rolls, end sections of the die are operated at a higher heat than the center (Fig. 3.14, page 241). Cooling rolls require this type of heat control from their ends to the center.

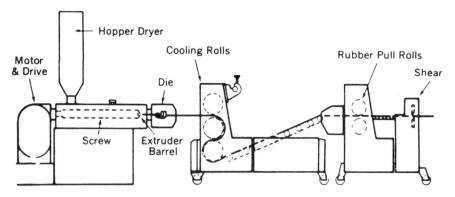

Figure 3.51 Sheet line using a stand of three cooling rolls.

Table 3.12 Important review of sheet problems

Defect	Probable cause
Continuous lines in direction of extrusion	Die contamination or scored rollers or moisture in the polymer
Continuous lines across the sheet	Jerky operation; maybe too low a temperature
Curved lines across a coloured sheet	Colour not mixed well enough
Discoloration	Too high a mass temperature or too high a percentage of reground material added
Bad gloss	Roll temperature too low or defective rolls
Dull strip	Die set too narrow at this point

The operation (as well as design) of a slit die, particularly for wide sheets, requires extensive experience. Its rather high melt pressure can deform the die.

The heated rollers are highly polished to give a good surface to the semimolten sheet. This final polishing process cannot be used to cover up defects in the sheet caused by faulty extrusion, and great attention must be paid to eliminating surface defects before the sheet leaves the die. A brief note on sheet faults and their probable causes is given in Table 3.12.

TUBING

Tubings are defined as long, hollow, flexible cylinders used to transport fluids or solids. Pipe is rigid and larger in size. Tubing is flexible with outside diameters up to 150 mm. A flexible material can be stressed or bent without breaking. The extent of flexibility is important in selecting a tubing material. Associated with flexibility is kinking. Kinks are caused by reducing the bend radius to a point where the tubing wall collapses.

Tube processing is similar to pipe processing where the relatively big-tank pipe technoogy is transformed into tubing such as very small medical tubing. The tubing extrusion process requires high precision and tight process control to meet very tight dimensions and performances. Die configurations are shown in Fig. 3.12 (pages 238–9). Vacuum-type calibrating/sizing units can be used with very strict control on extruder output rate and downstream pullers.

Plastic tubings were introduced in the late 1930s as replacements for rubber tubing and hose. Today cured elastomer tubings of natural or synthetic rubber, silicone rubber, and fluoroelastomer materials are used, but the growth of the tubing market has principally been in thermoplastic materials that can be extruded into long lengths without the need for postcuring. Because virtually all thermoplastics can be extruded, the choice depends on the application. Unlike pipe, used mostly for plumbing, tubing is used in many different applications.

Requirements of the end-use application should be used to specify tubing properties. Properties to consider include flexibility (bend radius), hardness, chemical resistance, environmental stress-crack resistance, permeability and chemical absorption, burst pressure and working pressure, recommended use temperatures, flammability, electrical conductivity, peristaltic pumpability, surface wettability, abrasion resistance, dielectric strength, fatigue resistance, and appearance.

WIRE AND CABLE

The demand for covered wire and cable in the various communication industries is very large and steadily increasing, so this extrusion process is a very important one. The types of crosshead die used are shown in Figs 3.23 and 3.52.

Coating is performed by extruding plastic. This may be accomplished by feeding the wire directly through a hole in the center of the feed screw, but by far the more popular method is to use a crosshead die (similar to Fig. 3.12, part 2) through which the wire is fed. Hot melt is extruded over a preheated wire to improve adhesion and reduce shrinkage stresses. Before the wire enters the die, preheating is done by an electric current, radiant heaters, and so on; thick wires or cables can be heated by a gas flame or hot gas. Wire may travel at rates of $1300 \, \text{m} \, \text{min}^{-1}$ ($4000 \, \text{ft} \, \text{min}^{-1}$). Regardless of the speed, the rate of movement has to be held extremely uniform (or perfect). In order to achieve uniformity, all peripheral (expensive) equipment is carefully controlled and monitored, from the input drum to the output drum.

There are two basic types of die, called high- and low-pressure coaters. With high pressure, the melt meets and coats the wire between the die lips prior to exiting the die. The result is good contact of plastic to wire, tight control of the plastic OD, and the ability to handle plastics that require tight melt control, particularly with operation at 'peak' heats and pressures. In the low-pressure type, the melt makes contact with the wire after they both exit the die. Plastic hugs the wire, with formation of a loose jacket which facilitates removal of plastic insulation. If spiders are used to support the central mandrel, they are usually thin and streamlined to minimize disruption of the velocity. Adjustment of the wall thickness distribution and concentricity via die-centering bolts can be manual or automatic (Fig. 3.12, part 2). Automatic control can be achieved with inline

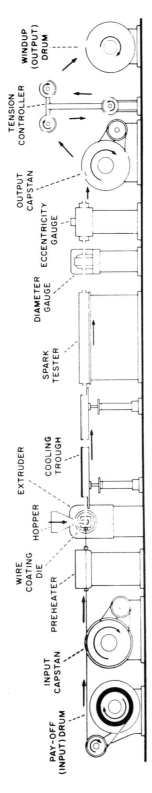

Figure 3.52 General layout of a wire-coating extrusion line using a crosshead die.

wall-thickness measurement probes, and so forth. With high-pressure dies, a vacuum can be used in the die just before the wire goes through its snug central support, to obtain an air-free and better bond. With a low-pressure die, low air pressure is applied through the center of the mandrel tube to prevent collapse of the melt tubing on exiting (eliminating melt adherence on the front of the die, which requires downtime and cleanup) and to aid in maintaining the plastic ID.

Cooling of the thermoplastic coated wire usually occurs as soon as it leaves the die through water-cooling troughs, which may have cascading heated sections. The troughs are usually 20–100 ft (6–31 m) long. With thermoplastics or elastomers and natural rubber, the required higher heat of melt solidification is added via hot-gas systems, vulcanization cures, and so forth.

PROCESS CONTROL

Controlling extrusion processes takes two types of systems, one on the extruder and another on the finished extruded product. In turn, these controls have to be interfaced. Extensive efforts are continually being made to achieve the maximum in speed, precision, and output (Chapter 1).

For products with specific requirements, typically the best approach has always been to control the weight per unit area or the length of output that is directly related to a constant melt throughput, by weight. As reviewed in this chapter (and others), many machine controls and designs are required, such as feed rate of loading and screw design. Thickness gauges were once very popular for adjusting film takeoff speeds. However, better control is achieved with weight control systems. Depending on the given weights per unit length or area and the speed of the line, the melt throughput rates of the extruder and precalculated accurately enough to ensure, via a melt throughput control system, that the weight can be kept constant within a tolerance range of at least ±0.5%. Whereas measurement of takeoff speed using a wheel pulse counter is relatively easy, measurement of melt throughput involves considerable expense.

From a practical standpoint, extrusion lines appear to be relatively simple and compact, with a long operating life; their controls should provide ease of startup and shutdown, and must be easy to service. But the real world is not that simple; the various materials and types of equipment in the lines are not stable, and they all work within rather tight limits. Although all types of very tightly controllable equipment are available, the operator or processor (and others) must recognize that all equipment and all materials have limits, optimum operating conditions, and so on. To benefit from setting up or operating a line, one has to determine what is required, determine the limits of operating materials and equipment, and establish quality control and other requirements as summarized in Fig. 1.1 (page 2).

Acquiring all this knowledge takes time and experimentation. One must be aware of potential problems or limitations. As new ones develop, one should accommodate them logically in processes. This approach provides a means of determining which materials and equipment to purchase or upgrade. One could thus set up an ideal line with so-called interchangeable and controllable features. Given a specification that must be met on a complete line, one must meet the target (or no payment is made). Few processors desire tight specifications, but such responsibility can be managed, one step at a time. The processor should determine as well as is possible what is required to meet product performance (color, dimensions, strength, etc.), and relate the requirements to material and equipment currently in use. It would be unfortunate if one were to purchase the 'best' material and equipment available at a particular time, only to find that it did not perform as needed and that another piece of equipment was required. This situation happens all too often, as people hastily, or without full knowledge of an operation, make foolish mistakes.

A processor does not have control of all the steps from basic material to finished product. Material-processing capability is limited by what is received and by when inspection is made or required; so selection tests are important and must be subject to change (Chapter 16). In turn, the process line has many variables that must be coordinated. A practical procedure today, with all closed-loop process control systems, is to subdivide the controls into distinct subsystems (Fig. 3.53). They can then be controlled within single control loops or by simple intermeshed circuits. A single-loop feedback circuit has one input and one output signal; disturbances that affect the process are registered by the controller directly, by an additive term in the value of the controlled parameter. A disturbance is registered by its own sensor and interacts with the control signal with an open-loop controller. When the effect of a disturbing factor on a process is well understood, disturbance control provides a rapid reaction to unpredictable influences.

Additional process control signals can be achieved in a single close-loop system by having it cascade the signal. Cascade control offers the advantages of faster reaction time, reduced susceptibility to disturbance, and less effect from incorrect settings. As this chapter's review of all the variables that can exist and are controllable would suggest, interrelating them with multiperforming controllers is desirable. Computer-integrated microprocessor control systems can sometimes be used; they offer some benefits when properly installed with matching hardware, but cannot do the complete job. A complete package that would properly include all

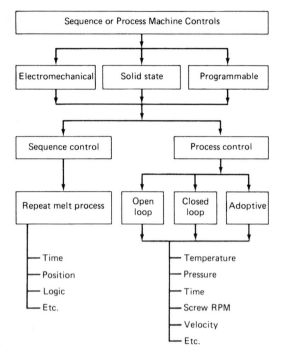

Figure 3.53 Simplified overview of process controls for an extrusion machine.

parameters is desirable but at present not very realistic. Computerized integrated controllers are available and used, each type for its unique capability. Better complete multiparameter controllers still need to be designed. One must appreciate that use of a controller, whether simple or highly computerized, requires the operator to recognize which parameters need to be controlled and to what degree; the operator then has to correlate these factors. For example, if the output rate has to be increased, is a screw RPM increase satisfactory, or should there be a change in back pressure, resin feed rate, and so on, or combinations of certain controls, to provide a product that is acceptable (according to various criteria)? The possibilities of measuring back pressure in the melt and the melt heat at the screw tip are of particular value in processing. In this manner it is possible to control the reproducibility of the two most important parameters in every processing method: pressure and heat.

From the processor's point of view, the extruder has the following objectives: (1) high throughput proportional to screw speed and independent of back pressure; (2) uniform, pulsation-free delivery, with optimum melt heat (with respect to power consumption and quality of extrudate); (3) uniform melt heat locally and also throughout the run; (4) delivery of an orientation-free, relaxed melt; (5) homogeneous mixing of the resin with all its additives; and (6) a pore-free extrudate, free of volatiles, at high output rates for both granular and powder feeds.

With all the varieties of materials available, and their individual grades with many different formulations, it has not yet been possible to design a screw in advance, based on the melt or the rheological/flow physical relationships involved in plastification and conveying. Trial and error, with an observant processor using reliable and reproducible controls, makes the screw perform to its maximum efficiency.

As reviewed up to now, controls on extruders can consist of anything from a discrete instrument panel to a fully integrated system of microprocessors. Assuring the correct temperature of the extruder barrel is crucial to performance. Consequently, this function is maintained with extreme accuracy by the control system. The speed at which the screw rotates is also important and can be controlled by using electronic circuits in conjunction with speed tachometers.

An extruder's head pressure is often monitored to ensure a steady state of operation. Some processors, however, prefer to control the head pressure by use of a feedback loop to the drive controller. This is quite common in the fibers and film industry, where a constant pressure at the inlet to the gear pump is required to give precise throughput.

The melt temperature is another variable than must be continuously monitored to optimize extruder performance. The temperature can be ascertained by inserting a thermocouple at some point just downstream of the screw. The thermocouple is designed to be positioned in various areas of the machine, including the adapter flange, a connecting pipe, at a die inlet, or in the body of a die.

A common problem in measuring melt temperature is verifying the accuracy of the readings. Many times, a melt temperature probe is likely to be affected by the metal temperature of the surrounding steel rather than the actual polymer melt. For this reason, variable-depth thermocouples are often employed because they allow the tip of the thermocouple to be moved inward through the melt. This placement gives a truer indication of melt temperature homogeneity, as well as a good average melt temperature.

Although pressure and temperature monitoring capabilities are present in all control units, the more sophisticated microprocessor systems include extra features such as alarms, interlocks, and screens. Printers linked to these systems can copy all the data on demand or at set time intervals; they can even print out each change that is made by the operator. Such a report is known as a tattle tale because it is almost impossible to change the operating conditions without detection and reporting by the control system. The result is a steadily monitored state of operation, hence consistent quality.

Other computerized enhancements include software which can simulate machine performance. These programs, which are now offered by most equipment manufacturers, have been upgraded over the years by the use of data generated at test facilities. As a result, they can be highly effective tools in predicting extruder behavior during operation.

Simulation programs can greatly simplify screw design for any application once the rheological and physical properties of the plastic being processed are known.

Downstream controls

Downstream product controls can be interconnected with process controllers so that any variation in the product is immediately reflected as a corrective change. A decision is derived from the feedback signal about the kind and extent of adjustment that should be made to a control variable; in other words, the system has to be matched to the process. A few systems permit the properties of the extrudate to be controlled: (1) pipe-thickness distribution is controlled by centering of the die either mechanically or thermally, and wall thickness by the hauloff speed; (2) the average thickness of blown and flat film is controlled by extruder speed and hauloff speed, the thickness profile by the die-gap restricter-bar height (thermal or piezoelectric), the film width by the diameter of the calibrator, the transverse thickness tolerances by cooling air, and the edge shift by side-gusseted triangles; (3) a profile's dimensional stability (weight per unit volume) is controlled by hauloff speed, screw speed, and recorded melt throughput or a product dimension; and (4) all product throughput is controlled by screw speed, and the running length by hauloff speed and screw (gravimetric metering).

Transverse direction (TD) gauge control of cast film and sheet has been successfully used for many years worldwide, by automatic dies employing thermal bolts to locally adjust the die gap. TD control of blown film is now being used in a system that takes a different approach; it cools segments of the die lip, in turn cooling the melt in contact with it, reducing the melt's drawdown capability and thus increasing the thickness. The TD gauge profile can be stabilized to a tolerance of $\pm 1.7\%$. In addition to minimizing resin consumption, the system also reduces film curvature, which increases as TD tolerance increases.

To monitor this gravimetric control system in the machine direction, an electrical sensor is used. The transverse profile of the bubble is read by a capacitance gauge that makes a complete circuit of the bubble in 2–3 min. The system's microprocessor takes the data and displays the profile on a color monitor. An additional display indicates the centering status of the die prior to startup, guiding the material adjustment. The system is then activated. Valves arrayed around the die at a spacing of about 0.75 in. (19 mm) direct cold, compressed air through cooling channels onto segments of the die lip, with the microprocessor controlling the operation. In running 53 μm LDPE film, the thickness tolerance is reduced from 3.3% to 2.0% after 15 min. After another 22 min the tolerance is reduced to 1.7%, which is maintained as long as the control is on.

There are different pulling and cutting units for film, sheet, pipe, and so forth. Tube pullers are driven by various methods. Electric drives are available with tachometer feedback accuracy ±0.5–1.0%. Digital drives can hold to ±0.01%.

Table 3.13 compares cutting equipment capabilities. Automatic slitter/ rewinders, for handling blown or cast films such as rigid PVC, PS, and nonuniform-thickness laminates, operate at speeds of 300–500 m min^{-1}, producing reels up to 3000 mm in diameter and slit reels up to 1000 mm in diameter. Film is drawn from an unwind to the slitting/rewinding stations through a translation bridge equipped with idler rollers which facilitate web feed and prevent the deposition of dust due to static. Rewinding/slitting machines are available that can handle highly tension-sensitive materials such as PP and PET, video film, magnetic tape, and heavy-gauge laminates, with working widths of 2500–3000 mm and finished-reel diameters of up to 600 mm for PP and PET. They operate at 600 m min^{-1} and produce slit widths of 100 mm and above. Rewind units are also available that operate on single- and twin-spindle roll-slitting principles. Units incorporate high-accuracy scissor slitting, crush cutting for nonbrittle materials, and razor-blade slitting for materials such as cellophane, PE, PP, PVC, and thin aluminum foils.

Machines for automatic winding of blown and cast film, which includes winding and slitting, are available; they handle all types, widths, and thicknesses, and different winding characteristics, produce different reel diameters on various core sizes, and divide film webs into different lane widths during production. Reel changes are automatic, and reel production can continue for one or two shifts without operator intervention, because of the incorporation of a reel-bar storage system.

Lines produce reels in widths of 900, 1200, and 1600 mm and permit winding of all film types on various cores with reel diameters of up to 600 mm. They also provide automatic shaft changes at a predetermined film length and gap winding that is steplessly adjustable from 1 to 10 mm. Jumbo reels run up to 2000 mm in diameter and weigh up to 3 tons in other systems.

Table 3.13 Examples of cutting equipment capabilities

Cutter	Line speed ($m\,min^{-1}$)	Cuts (min^{-1}) maximum	Accuracy ($\pm mm$)	Advantages	Disadvantages
Saws	150	30	0.015	Easy setup and large capacity	Requires cleanup via air systems, etc. and uses clamping/travel table units
Guillotine	90	50	0.015	Large capacity and angle cuts	Slow blade speed; rigids need profiled bushing and/or blade; high air consumption and few cuts per minute
Flywheel	4500	12 000	0.004	High cut rates and high accuracy	Not good under 300 cuts per minute; must adjust blade RPM for speed changes; profile bushings may be needed for rigids; limited to small-angle cuts
Die-set stationary traveling	–	90	0.00	Inline finishing and high accuracy	Price, slow line speed, long setup time, and long runs only

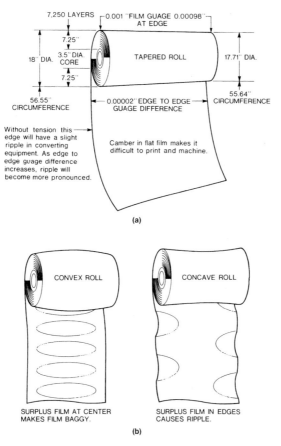

Figure 3.54 Influences on film performance during windup: (a) effect of tension; (b) effect of uneven thickness.

Developments in web tension control systems are providing increased capability and function to eliminate problems (Fig. 3.54). They included ultrasonic roll-diameter sensors, pneumatic pressure gauge tension monitors, capston–Mt. Hope tension systems, and so on. As an example, replacement of a web-tensioning system's conventional electromechanical drive with an ordinary AC motor enables processors to lower system cost and improve web consistency, as has been done for many years. A vector control system uses a belt and pulley arrangement to remotely couple an encoder to the shaft of the AC induction motor. This approach provides closed-loop feedback, without requiring that one modify the relatively inexpensive motor by installing a special feedback device.

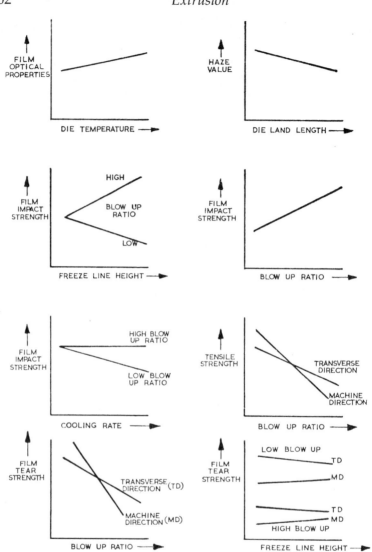

Figure 3.55 How blown-film extrusion machine settings can affect the properties of plastics.

EXTRUDER OPERATION

This section summarizes what has been reviewed in this chapter. Machine operation takes place in three stages. The first stage covers the running of a machine and its peripheral equipment. The next involves setting processing conditions to a prescribed number of parameters for a specific material, with a specific die in a specific processing line; to meet product performance requirements. Figure 3.55 relates to blown-film settings. The final stage is devoted to problem solving and fine tuning of the operation, which will lead to meeting product performance requirements at the lowest cost of operation. A successful operation requires close attention to many details, such as the quality and flow of feed material, a heat profile adequate to melt but not degrade the material, and a startup and shutdown that will not degrade the plastic. Processors must become familiar with a troubleshooting guide.

Care should be taken to prevent conditions that promote surface condensation of moisture on the resin and moisture absorption by the pigments in color concentrates. Processors must avoid contamination from other plastics, dust, paper clips, and so on, and take special care in cleaning feedhoppers, hopper dryers, blenders, scrap granulators, and other material-handling equipment. Resin silos, containers and hoppers should be kept covered to prevent contamination. Also, certain established procedures for startup should be followed to prevent contamination, overheating, and excessive pressures.

It is important to provide safe operating conditions for personnel and equipment. One must realize that high production rates cannot be achieved until all parts of the extrusion system reach the optimum operating conditions. This condition is best achieved by gradually increasing heat and rates until desired optimum conditions are reached.

Prior to startup one must take certain precautions. (1) Unless the same resin is already in the machine from a previous run, the entire machine should be cleaned or purged, including the hopper, barrel, screw, breaker plate, die, and downstream equipment. If a resin was left in the barrel for a while, with heats off, the processor must determine if the material is subject to shrink and could cause moisture entrapment from the surrounding area, producing contamination that would require cleanup (this situation could also be a source of corrosion in the barrel/screw). (2) One must check heater bands and electrical connections, handling electrical connections carefully. (3) Check thermocouples, pressure transducers, and their connections carefully. (4) Be sure the flow path through the extruder is not blocked. (5) Have a bucket or drum, half-filled with water, to catch extrudate whenever purging or processing plastics that have contaminating gaseous by-products. (6) Inspect all machine ventilation systems to ensure adequate airflow.

Startup procedures also involve precautions. (1) Starting with the front and rear zones (die end and feed section), one should set heat controllers slightly above the resin melt point and turn on the heaters. Heatup should be gradual from the ends into the center of the barrel to prevent pressure buildup from possible melt degradation. (2) Increase all heaters gradually, checking for deviations that might indicate burned-out or runaway heaters by slightly raising and lowering the controller set point to check if power goes on and off. (3) After the controllers show that all heaters are slightly above the melt point, adjust the settings to the desired operating heats, based on experience and/or the resin manufacturer's recommendation, checking to ensure that the heat increase is gradual, particularly in the front/crosshead. (4) The time required to reach temperature equilibrium may be 30–120 min, depending on the size of the extruder. Overshooting is usually observed with on/off controllers. (5) Hot melts can behave many different ways, so no one should stand in front of the extruder during startup, and one should never look into the feed hopper because of the potential for blowback due to previous melt degrading, and so on. (6) After set heats have been reached, one puts the resin in the hopper and starts the screw at a low speed such as 2–5 RPM. (7) The processor should observe the amperage required to turn the screw, stop the screw if the amperage is too high, and wait a few minutes before restart. (8) In working with a melt requiring high pressure, the extruder barrel pressure should not exceed 7 MPa (1000 psi) during the startup period. (9) One should let the machine run for a few minutes, and purge until a good-quality extrudate is attained visually (experience teaches what it should look like; a certain size and amount of bubble or fumes may be optimum for a particular melt, based on one's experience after setting up all controls). If resin was left in the extruder, a longer purging time may be required to remove any slightly degraded resin. (10) For uniform output, all of the resin needs to be melted before it enters its metering zone, which needs to be run full. When time permits, after running for a while, the processor should stop the machine, let it start cooling, and remove the screw to evaluate how the resin performed from the start of feeding to the end of metering. Thus one can see if the melt is progressive and can relate it to screw performance and product performance. (11) Turn up the screw to the required RPM, checking to see that maximum pressure and amps are not exceeded. (12) Adjust the die with the controls it contains, if required, at the desired running speed. Once the extruder is running at maximum performance, set up controls for takeoff equipment, which may require more precision settings, if required. One may get into a balancing act of interrelating extruder and downstream equipment.

Shutdown procedures vary slightly, depending on whether or not the machine is to be cleaned out or just stopped briefly. If the same type of resin is to be run again, cleanout is generally not required. The goal is to avoid degradation by reducing exposure of the resin to high heat. If cleanout is required, because a different type of resin is to be processed, it is necessary to disassemble the equipment at a heat high enough to allow cleanout before material solidification.

The procedure for shutdown without cleanout is as follows: (1) one should empty the hopper as well as possible; (2) if coating, one should take the action required to remove the substrate (with wire coating, one removes the wire from the crosshead die; in coating paper, film, etc., the extruder is usually on a track and can be withdraw from the substrate, etc.); (3) one reduces all heat settings to the melt heat; (4) one reduces the screw speed to 2–5 RPM, purging the resin, if required, into a water bucket or drum prior to reducing the heat to melt heat; (5) when the screw appears to be empty, one stops the screw, and shuts off the heaters and the main power switch (however, steps 4 and 5 must be completed before the melt heat drops significantly, or a premature shutdown will occur with material remaining in the barrel); and (6) if a screen pack with breaker plate is used, one disconnects the crosshead from the extruder and removes the breaker plate and screen. If necessary, appropriate action is taken to clean them.

For cleanout of the extruder at shutdown, the first three steps are the same as steps 1, 2, and 4 in the preceding paragraph. Then the procedure is as follows. (4) When the extruder appears to be empty, one stops the screw. (5) Shut off and disconnect the crosshead heaters. Reduce other heaters to about 170–330 °C (400–625 °F), depending on the resin temperature at the melt point. (6) Disassemble the crosshead and clean it while still hot. Remove the die, and the gear pump if used, and remove as much resin as possible by scraping with a copper spatula or brushing with a copper wire brush. Remove all heaters, thermocouples, pressure transducers, and so on. Consider using an exhaust duct system (elephant trunk) above the disassembly and cleaning area, even if the resin is not a contaminating type; it keeps the area clean and safe. (7) Push the screw out gradually while cleaning with a copper wire brush and copper wool. Care should be exercised if a torch is used to burn and remove resin; tempered steel, such as Hastelloy, may be altered, and the screw distorted or weakened, and subjected to excessive wear, corrosion, or even failure (broken). (8) After screw removal, continue the cleaning, if necessary. (9) Turn off heaters and the main power switch.

Final cleaning of parts, particularly disassembled parts, is best done manually, or much better, in ventilated burnout ovens, if available, at 540 °C (1000 °F) for about 90 min. For certain parts, with certain resins, the useful life could be shortened by corrosion; check with the part manufacturer. After burnout, remove any grit that is present with a soft, clean cloth. If water is used, one can air-blast to dry. With precision machined parts, water cleaning could be damaging, because of potential corrosion when certain metals are used.

COSTING

By providing tight controls on products, one can extrude to tight tolerance and significantly reduce the amount of plastic consumed, thus reducing cost. Even if the products do not have to meet very tight tolerances, the payoffs can be significant. The biggest cost in production (and other high-production automated lines, such as injection molding and blow molding, is the cost of plastics that may be 40–60 wt%.

Another important area for cost reductions is in energy consumption. Like the output capacity, the energy efficiency of an extruder depends on the torque available on the screw, screw RPM, heat control, and material being processed. Unfortunately, costly energy losses may be 3–20% and due to various factors, with the major loss occurring in the drive mechanism. The power for screw rotation is supplied by a variable-speed motor drive, and is transmitted through a gear-reduction unit, a coupling, and a thrust bearing. Gear reducers impart the final speed and torque to the screw. Most gear reducers use double-reaction helical or herringbone gears for their ruggedness and to hold noise levels within acceptable limits. Worm and pinion gear combinations have been used on smaller extruders. The efficiency of the power transmission gear with the worm is a maximum of 85%; helical gears reach 95% efficiency, and herringbone gears 97%.

Thrust bearings absorb the thrust force exerted by the screw as it turns against the material in the barrel. The size or rating of the thrust bearing provides an anticipated number of operating hours (if kept clean). The operating pressure, size of extruder, and operating speed are important factors in selection of a thrust bearing. The motor size must be sufficient to allow for the energy required to melt the plastic and to pump the melt at the desired output condition and rate. For example, higher outputs at higher screw speeds require more power than low outputs. The specific heat of the resin (Table 1.28, page 86) will also be a determining factor. As a guide, simple extrusions usually require 1 hp (750 W) for every 10–15 lb h^{-1} (4.5–6.8 kg h^{-1}) output. For high energy or where high heat at low speeds is required, 1 hp may be required for every 3–5 lb h^{-1} (1.4–2.3 kg h^{-1}). Plastics require different power or running torque on screws, based on the type of plastic and the screw size. Extrusion of rigid PVC requires about twice the power needed for LDPE. Increasing screw diameters could at least double torque requirements for every inch (25 mm) of diameter increase. In comparing single extruders with intermeshing, corotating, twin-screw extruders, the twins provide significant energy conservation and efficiency.

Energy consumption is a major factor in production costs, as well as all equipment efficiency. Many extruders, as well as other equipment, are usually overpowered. This situation may be better than using underpowered equipment; but processes should not waste energy. In an extruder, as in any other machine, the energy output is always equal to the energy input, regardless of the forms into which it may be converted. The energy may be furnished by mechanical or thermal sources. Mechanical energy is supplied by the drive motor; thermal energy is supplied positively by electrical heater bands or negatively by cooling devices. It is important to evaluate and compare equipment performance in order to minimize energy waste. Some machines (of the same type) are more wasteful, whereas others are more efficient. One should check the power consumption on incoming electrical lines going into equipment that operates from minimum to maximum load conditions. Perhaps all that is needed is an ammeter or wattmeter physically placed on the line. The results will be obvious: use more efficient units, charge for operating costs based on efficiencies, and/or replace certain units.

TROUBLESHOOTING

Throughout this book there are discussions of why problems develop and how they can be eliminated or kept to a controllable minimum. To do the best job of eliminating or reducing problems, it is very important to be able to diagnose the extruder quickly and accurately in order to minimize downtime or off-quality products. Two important requirements for efficient troubleshooting are good instrumentation and good understanding of the extrusion process. Instrumentation is very important in process control, but it is absolutely essential in troubleshooting. Without good instrumentation, troubleshooting tends to be a useless approach at best, no matter how well one understands the process. Lack of instrumentation can prove very costly if it causes a certain problem to remain unsolved for even a short time. Typical problems can be summarized as follows:

- *Nonuniform feedstock*
  - variations within lot, between lots, and between sources
  - variations in percentage and nature of regrind components
  - variations in moisture content, additives, premixing, predrying, etc.
- *Solids conveying*
  - bridging in hopper causing erratic feed
  - density variations or low-density feed
  - poor filling of feed flights for powder feed at higher screw speeds

Table 3.14 Troubleshooting guide: common extrusion problems and how to solve them

Problem	Causes(s)	Solution(s)
General considerations		
Surging	Resin bridging in hopper	Eliminate bridging
	Incorrect melt temperature	Correct melt temperature
	Improper screw design	Check design
	Rear barrel temperature too low or too high	Increase or decrease rear temperature
	Low back pressure	Increase screen pack
	Improper metering length	Use proper screw design
Gels	Melt temperature too high	Lower melt temperature
(Contaminants that look like	Not enough progression in screw	Use new screw
small specks	Bad resin	Check resin quality
or bubbles)	Melt temperature too high	Check melt temperature
Melt fracture	Melt temperature too low	Increase melt temperature
(Rough surface	Die gaps too narrow	Heat die lips
finish. Also		Increase die gaps
called		Use processing aids
'sharkskin')		
Bad color	Color concentrate incompatible with resin	Ensure melt index of concentrate base material is close to melt index of resin
Bubbles	Wet material	Dry thoroughly
	Overheating	Decrease temperature; check thermocouples
	Shallow metering section	Use proper compression-ratio screw
Overheating	Improper screw design	Use lower-compression screw
	Restriction to flow	Check die for restrictions
	Barrel temperature too low	Increase temperature
Die lines	Scratched die	Refinish die surface
	Contamination	Clean head and die
	Cold polymer	Check for dead spots in head; adjust barrel and head temperature to prevent freezing
Flow lines	Overheated material	Decrease temperature
	Poor mixing	Use correct screw design
	Contamination	Clean system
	Improper temperature profile	Adjust profile

Table 3.14 *Continued*

Problem	Causes(s)	Solution(s)
	Blown film	
Wrinkles	Dirty collapsing frame	Clean frame
	Too much web tension	Adjust tension
	Improperly designed air ring	Use new air ring
	Gauge variations	See gauge variations
	Insufficient cooling	Use refrigerated air
		Increase flow of air
		Reduce output
	Misalignment between nip rolls and die	Check alignment
Folds, creases	Excessive stretching between nip and roller	Reduce winding speed
	Nip assembly drive not constant	Adjust or replace drive
Blocking	Inadequate cooling	Use better cooling method
	Excessive winding tension	Adjust tension
	Excessive pressure on nip rolls	Adjust pressure
	Bad resin	Check resin
Port lines	Melt temperature too low	Increase melt temperature
	Die too cold or too hot in relation to melt temperature	Adjust die temperature
Splitting	Excessive orientation in machine direction	Increase blowup ratio
	Die lines	See die lines
	Degraded resin	Reduce melt temperature
	Poor resin choice	Ensure resin is suitable
Die lines	Nick on die lip	Change die
	Dirty die	Clean die
	Inadequate purging	Increase purging time between resin changes
Gauge variations (machine direction)	Surging	Check temperature
		Check hopper for bridging
	Inconsistent take-up speed	Check take-up speeds
Gauge variations (transverse direction)	Nonuniform die gap	Adjust gap
		Center air ring on gap

Table 3.14 *Continued*

Problem	Causes(s)	Solution(s)
Printing problems	Insufficient treatment	Use properly treated film
	Additives interfering with ink	Use resins with no interfering additives
		Erratic treatment
		Reduce slip levels to about 600 ppm for water-based inks

Sheet[a]

Problem	Causes(s)	Solution(s)
Poor gauge uniformity	Melt flow is not stable	Use gear pump to stabilize flow
Viscosity not stable	Poor mixing	Use static mixer
Streaks	Contaminated system	Clean hopper
		Check screw and die; clean if necessary
Total discoloration	Excessive regrind	Check amount of regrind used
Discontinuous lines	Too much moisture	Increase resin drying
		Use hot regrind

Pipe and tubing[b]

Problem	Causes(s)	Solution(s)
Poor output	Improper die or screw design	Ensure die and screw are designed for desired output
ID blisters	Insufficient vacuum	Increase vacuum
	Excessive moisture	Maintain normal percentage of moisture in compound
	Gases entrapped	Reduce temperature
	Water inside pipe	Stop water access
ID burn streaks	Mandrel heat too high	Check mandrel heat
	Stock temperature too high	Reduce temperature slowly
ID grooves	Mandrel is coated with material	Clean mandrel
ID wavy surface	Screw clearance set improperly	Adjust clearance
	Puller drive slipping	Adjust or replace puller drive
OD burn streaks	Material hung up on die	Clean die
	Temperatures too high	Reduce temperatures slowly
OD uneven circumference	Too much air pressure on puller	Reduce air pressure
	Insufficient air pressure	Check air pressure and all connections

Table 3.14 *Continued*

Problem	Causes(s)	Solution(s)
OD discolored	Stabilizer level too low	Check stabilizer level
OD pock marks	Air bubbles adhering to pipe in flotation tank	Install wiper in tank
	Improper adjustment of spray rings that surround water tank	Readjust spray rings
OD oversized	Air supply too high	Adjust air supply
	Insufficient water supply	Increase water supply
	Pipe hot when measured	Allow pipe to cool before measuring
Wall too thick	Misadjusted die bushing	Adjust die bushing to achieve uniform thickness
	Wrong die setup	Use correct setup

[a] Most of the problems covered under blown film are also relevant to sheet extrusion.
[b] These data pertain to extrusion lines using water-filled vacuum-sizing tanks.

- poor frictional characteristic in feed section
- poor design of feeding system or screw feed section
- *Melting and metering*
  - poor screw design for the application
  - poor barrel temperature profile
  - screw RPM exceeding extruder melting capacity
  - worn screw or barrel
  - fouled cooling system, burned-out heaters
  - inadequate extruder size, drive power, or RPM range
  - inadequate or poorly maintained machine controls
  - inadequate or excessive pressure drop through screen pack and die
  - vent bleed
- *Melt quality*
  - melt pressure and temperature fluctuations
  - output surge product dimensional fluctuations
  - incomplete melting, excessive melt temperature
  - poor distributive or dispersive mixing
  - contaminants, degradation, plugged screen, etc.
  - poor die design, adjustment, or temperature settings

There can be plastic material and/or machine-related problems. Examining material can eliminate the problem, such as checking bulk properties, melt flow and the thermal properties (Chapter 1). If a material problem is suspected, one should first examine the quality control (QC) records

Table 3.15 Troubleshooting guide for screen changers

Possible causes	Remedy
Seal leakage	
Head pressure above maximum set pressure	Reduce head pressure
Excessive overhung load	Support downstream die adapter, etc.
Seals improperly installed	Remove, clean, and reinstall
Excessively scored seals or slide plate	Correct cause of damage (see below) or replace
Disk spring relaxed	Replace
Initial pressure setting of upstream body too low	See procedure for setup in manual
Excessively scored seals	
Maximum set pressure rating too high	See procedure for setting U/S body
Wire strands protruding above face of slide plate	Ensure correct screen size and installation
Screens tearing out or folding on shift	Utilize positive screen retention
Tramp metal caught by screens	Utilize hopper magnet or other metal-catching device
No slide-plate movement	
Guards not closed	Close guards
Pressure ready light not on (w/opt. solenoid valve)	Check hydraulic power unit
Limit switches or interlock wiring faulty	Repair or replace
Direction control valve not operating	Repair or replace
Erratic or slow slide-plate movement	
Air in system	Purge air
Low precharge in accumulator	Recharge accumulator
Oil viscosity too high	Replace with oil of correct viscosity
Pressure drop across breaker plate too high	Reduce pressure drop
Polymer viscosity too high	Decrease viscosity
Initial pressure setting of upstream body too high	Procedure for setting U/S body
Leakage past the direction control valve or changer cylinder	Repair or replace
No pressure from hydraulic power unit	
Motor not running	Check disconnects, motor overloads, motor
Unload solenoid valve not closed	Replace
Pump/shaft damaged	Replace pump
Low pressure, motor stops	
Faulty pressure switch	Replace switch

Table 3.15 *Continued*

Possible causes	*Remedy*
Low pressure, motor continues to run	
Low fluid level	Fill reservoir
Suction strainer blocked	Clean or replace
Oil overheated	See below
Air entrained in fluid	Allow bubbles to subside; use antifoam
Leakage past direction control valve	Repair or replace
Leakage past screen-changer cylinder	Repair or replace
Pump damaged/worn	Replace pump
Normal pressure, does not stop	
Faulty pressure switch	Replace switch
Rapid pump/motor cycling	
Faulty pressure switch	Replace switch
Leakage past check valve	Replace pump
Pump damaged or worn	Replace pump
Leaking past direction control valve or screen-changer cylinder	Repair or replace
Oil overheating	
Rapid screen-changer cycling	Shift less frequently
Restricted air movement or high temperature around reservoir	Assure free movement of air around hydraulic unit
Fluid bypass through system causing excessive pump-up time or rapid pump cycling	See above

on incoming material to see if a change in feedstock properties occurred. Unfortunately, the only QC test on incoming material is often a melt index (MI) test [1, 3, 9]. This test is only able to detect a very limited number of material-related extrusion problems. Thus, in many cases, material testing may have to be more extensive than the regular QC testing [3].

Guides for troubleshooting have been included in different sections of this chapter where problems or defects were associated with probable cause. Guides for components, such as electric and hydraulic motors, barrel heaters, and screen changers, are readily available by equipment manufacturers. Table 3.14 provides a general extrusion guide; Table 3.15 is a guide concerning screen changers. One must be aware that causes of problems are not always obvious; for example, Fig. 3.56 shows contamination in the die presenting a problem that is almost impossible to resolve by equipment controls.

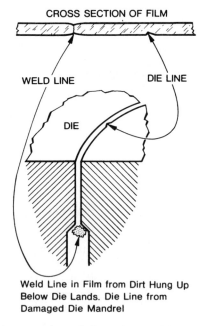

CROSS SECTION OF FILM

WELD LINE   DIE LINE

DIE

Weld Line in Film from Dirt Hung Up
Below Die Lands. Die Line from
Damaged Die Mandrel

Figure 3.56 A contaminated die and a troubleshooting problem.

To determine the source of a problem, understand the basics of a process and apply them to problem solution. For example, with film and sheet dies, tip adjustments only control the uniformity of the transverse thickness. To control the average thickness a proper relationship between the extruder pumping rate and the speed of rollup is required. Closing the die-lip opening has very little effect on the extrusion rate and does not make the entire sheet thinner.

Regarding surges, any variation of the extrusion rate with time originates in the extruder and thus must be corrected in the extruder. This could be the most common and most difficult problem to correct, as it has many interrelated sources, none of which are in the die. A surge can be due to inconsistent feed from the hopper to the barrel, varying frictional forces in the barrel feed zone, improper screw design, and other causes. The gear pump may be the solution.

Melts move in laminar flow in the die, where no mixing action takes place; so poor mixing and incomplete thermal homogenization must be corrected prior to melt flow in the die. Because they are generally very efficient heat insulators, melts are in the dies for only a few seconds. It is possible to alter frictional effects on the melt skin through changes in the die heat, but there is usually a better chance to produce an even heat in the extruder.

Die adjustments do not significantly influence the extrusion rate; the lip opening, restrictor bar clearance, die heat, and overall die design have small to moderate effects on the back pressure. However, the extrusion rate is only slightly affected by substantial changes in back pressure. Thus attempts to open the lip or increase the die heat to increase the extrusion output rate can be exercises in futility, as well as the source of many additional problems.

Extrusion instabilities or variabilities

In extrusion, as in all other processes, an extensive theoretical analysis has been applied to facilitate understanding and to maximize the manufacturing operation. However, the 'real world' must be understood and appreciated as well. The operator has to work within the many limitations of the materials and equipment (the basic extruder and all auxiliary upstream and downstream equipment). The interplay and interchange of process controls can help to eliminate problems and/or aid operation with the variables that exist. The greatest degree of instability is due to improper screw design (or using the wrong screw). Proper instrumentation, particularly barrel heat, is important to diagnosis of the problem(s). For uniform/ stable extrusion, it is important to check the drive system periodically, the take-up device, and other equipment, and compare it to its original performance. If variations are excessive, all kinds of problems will develop. An elaborate process control system can help, but it is best to improve stability in all facets of the extrusion line. Examples of instabilities and problem areas are (1) nonuniform plastics flow in the hopper; (2) troublesome bridging, with excessive barrel heat that melts the solidified plastic in the hopper and feed section and stops plastic flow; (3) variations in (a) barrel heat, (b) screw heat, (c) screw speed, (d) screw power drive, (e) die heat, (f) diehead pressure, and (g) the take-up device; (4) insufficient melting and/or mixing capacity; (5) insufficient pressure-generating capacity; (6) wear and/or damage of the screw/barrel; (7) melt fracture/ sharkskin (Chapter 1).

One must also check the proper alignment of the extruder and the downstream equipment. Proper alignment and isolation of the vibrators is a must for high-quality, high-speed output. Proper instrumentation is vitally important, for quick and accurate diagnosis. A prerequisite for stable extrusion is a good extruder drive, a good temperature control system, a good take-up device and, most important, a good screw design. The extruder should be equipped with some type of proportioning temperature control, preferably a PID-type control or better [9].

4

Blow molding

BASIC PROCESS

Blow molding (BM), the third most popular method of plastics processing, offers the advantage of manufacturing molded parts economically, in unlimited quantities, with little or virtually no finishing required. It is principally a mass production method. The surfaces of the moldings are as smooth and bright, or as grained and engraved, as the surfaces of the mold cavity in which they were processed. Among the special techniques available are stretch blow molding and coextrusion. One can improve the cooling efficiency and reduce cycle time with gases (CO_2, etc.). Other developments include shuttle postcooling, insertion of printed film in the mold to avoid the need for subsequent decorating, and so on.

Blow-molded parts demonstrate that, from technical and cost standpoints, BM offers a promising alternative to other processes, particularly injection molding (IM) and thermoforming. The technical evolution of BM, plus accompanying improvements and new developments in plastics, has led to new BM parts. With the coextrusion technology and the hardware in place, the variety of achievable properties can readily be extended by the correct combination of different materials (Chapter 3). The expertise and economics of the method have reached the point where many ideas once deemed futuristic have arrived.

Blow molding is versatile. No longer confined to the very popular production of bottles and other containers [3], is offers several processing advantages, such as molding extremely irregular (reentrant) curves, low stresses, the possibility of variable wall thicknesses, the use of plastics with high chemical resistance, and favorable processing costs. Reentrant curves are the most prominent features, so much so that it is difficult to find examples without them.

They combine esthetics with strength and cost benefits (Figs 4.1 to 4.6 and Table 4.1).

Table 4.1 Hollow and structural blow-molded shapes

Industry	Application	Required properties
Automotive	Spoilers	Low temperature, impact, cost
	Seat backs	Heat distortion, strength/weight
	Bumpers	Low temperature, impact dimensional stability
	Underhood tubing	Chemical resistance, heat
Furniture	Workstations	Flame retardance, appearance
	Hospital furniture	Flame retardance, cleanability
	Office furniture	Flame retardance, cost
	Outdoor furniture	Weatherability, cost
Appliance	Air-handling equipment	Flame retardance, hollow
	Air-conditioning housings	Heat distortion, cost
Business machine	Housings	Flame retardance, cost
	Ductwork	Cost
Construction	Exterior panels	Weatherability, cost
Leisure	Flotation devices	Low temperature, impact strength cost, weatherability
	Marine buoys	Low temperature, impact strength cost, weatherability
	Sailboards	Low temperature, impact strength cost, weatherability
	Toys	Low temperature, impact strength cost, weatherability
	Canoes/kayaks	Low temperature, impact strength cost, weatherability
Industrial	Tool boxes, ice chests	Low temperature, impact strength, cost
	Trash containers, drums	Low temperature, impact strength, cost
	Hot-water tanks	Low temperature, impact strength, cost

As the final mold equipment for BM consists of female molds only, simply by changing machine parts or melt conditions, it is possible to vary the wall thickness and the weight of the finished part. If the exact thickness required in the finished product cannot be accurately calculated in advance, this flexibility is a great advantage from the standpoints of time and cost. With BM it is possible to produce walls that are almost paper-thin. Such thicknesses cannot be achieved by conventional IM, but with certain limitations, they can be produced by thermoforming. Both BM and IM can be succesfully used for very thick walls. The final choice of process

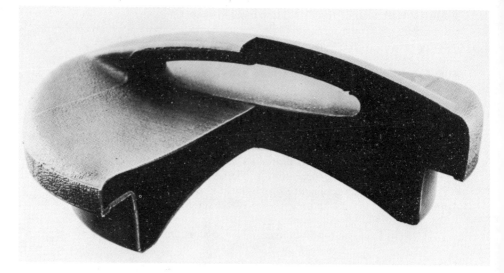

Figure 4.1 Integral handle for a container lid: double-walled, blow-molded HDPE.

Figure 4.2 Aquacycle: its blow-molded PP wheels incorporate paddles and fins on their sides; they operate most efficiently.

for a specific wall section is strongly influenced by such factors as tolerances, reentrant curved shapes, and costs.

BM can be used with plastics such as PE that have a much higher molecular weight than is permissible in IM (Chapter 1). For this reason

Figure 4.3 Truck fascia: multiple-extrusion blow-molded PP.

items can be blown that utilize the higher permeability, oxidation resistance, UV resistance, and so on, of the high-MW plastics. This feature is very important in providing resistance to environmental stress cracking. This extra resistance is necessary for plastic bottles used in contact with the many industrial chemicals which promote stress cracking.

With BM the tight tolerances achievable with IM are not obtainable. However, in order to produce reentrant curved or irregularly shaped IM products, different parts can be molded and in turn assembled (snap-fit, solvent-bonded, ultrasonically bonded, etc.). In BM of a complete irregular or complex product, even though IM tolerances cannot be equaled, the cost of the container is usually less. No secondary operations such as assembly (adhesives, etc.) are required. Other advantages are also achieved, such as significantly reducing (if not eliminating) leaks, reducing total production time, and so forth.

BM can be divided into three major processing categories: (1) extrusion BM (EBM), which principally uses an unsupported parison; (2) injection BM (IBM), which principally uses a preform supported by a metal core pin; and (3) stretch BM, for EBM or IBM, to obtain bioriented products,

Figure 4.4 Extrusion blow molded automotive panels have generous radii at their corners and edges.

providing significantly improved cost-to-performance advantages. Almost 75% of processes are EBM, almost 25% are IBM, and about 1% use other techniques such as dip BM. About 75% of all IBM products are bioriented. These BM processes offer different advantages in producing different types of products, based on the materials to be used, performance requirements, production quantity, and costs.

BM requires an understanding of every element of the process, starting with the basic extruder used in conventional extrusion and IM machines. (See Chapters 2 and 3 for information on the machines used to plasticate/melt materials for BM.)

With EBM the advantages include high rates of production, low tooling costs, incorporation of blown handleware, and a wide selection of machine builders. Disadvantages are a usually higher scrap rate, the use of

Figure 4.5 A 52 gallon (200 dm³) electric hot-water heater tank produced by blow molding.

Table 4.2 Injection versus extrusion blow molding

Injection blow molding	*Extrusion blow molding*
Use for smaller parts	Used for larger parts, typically ≥8 fl oz (237 cm³)
Best process for GPPS and PP; most resins can be and are used	
Scrap-free: no flash to recycle, no pinchoff scars, no postmold trimming	Best process for polyvinyl chloride; many resins can be used provided adequate melt strength is available
Injection-molded neck provides more accurate neck-finish dimensions and permits special shapes for complicated safety and tamper-evident closures	Much fewer limitations on part proportions, permitting extreme dimensional ratios: long and narrow, flat and wide, double-walled, offset necks, molded-in handles, odd shapes
Accurate and repeatable part weight and thickness control	Low-cost tooling often made of aluminum; ideal for short-run or long-run production
Excellent surface finish or texture	Adjustable weight control; ideal for prototyping

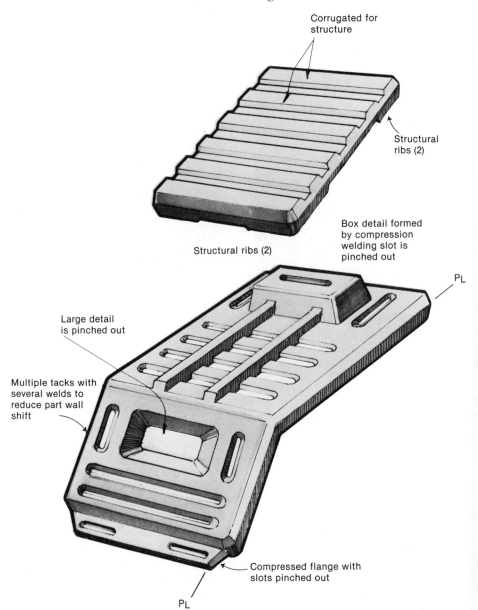

Corrugated for
structure

Structural
ribs (2)

Box detail formed
by compression
welding slot is
pinched out

Structural ribs (2)

PL

Large detail
is pinched out

Multiple tacks with
several welds to
reduce part wall
shift

Compressed flange with
slots pinched out

PL

Figure 4.6 A single blow-molded part can often replace several different injection-molded parts; the blow molding depicted is a structural multilayer coextrusion.

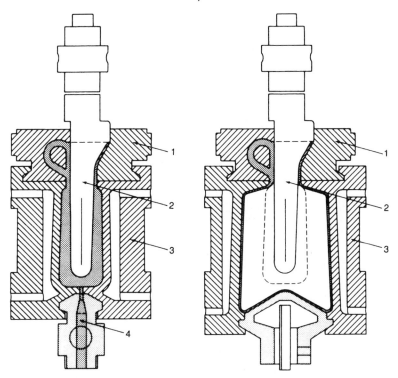

Figure 4.7 An integral carrying handle for an injection blow molded product: (1) the precision 'neck' mold, including a solid handle; (2) the preformed core and blow pin; (3) the basic water-cooled bottle female mold; and (4) the injection nozzle of the injection molding machine.

recycled scrap, and limited wall thickness control or resin distribution. Trimming can be accomplished in the mold for certain mold types, or secondary trimming operations have to be included in the production lines, and so forth.

With IBM, the major advantages are that no flash or scrap occurs during processing, it gives the best of all thickness and material distribution control, critical neck finishes are molded to a high accuracy, it provides the best surface finish, low-volume quantities are economically feasible, and so on. Disadvantages are its high tooling costs, the current lack of blown handleware (there is only solid handleware Fig. 4.7), it is somewhat limited to relatively smaller blown parts (whereas EBM can easily blow extremely large parts), and so forth. Similar comparisons exist with biaxially orienting EBM or IBM. With respect to coextrusion, the two methods also have similar advantages and disadvantages, but mainly major advantages. IBM can process PET (mono- or multilayer), stretching it into the popular two- and three-liter carbonated beverage bottles (Table 4.2).

Auxiliary equipment used for molding support functions is no different from the equipment used by other processes. However, the nature of the process requires the supply of clean compressed air to 'blow' the hot melt located within a mold. The air usually requires a pressure of 80–145 psi (0.55–1 MPa) for IBM. EBM requires 30–90 psi (0.21–0.62 MPa). However, stretch blow molding often requires a pressure up to 580 psi (4 MPa). The lower pressures generally create lower internal stresses in the solidified plastics and a more proportional stress distribution; the higher pressures are low when compared to conventional injection molding. The result of the lower stresses is improved resistance to all types of strain (tensile, impact, bending, environment, etc.).

EXTRUSION BLOW MOLDING

Continuous extrusion

In extrusion blow molding the melted plastic is continuously extruded as a tube (called a parison) into free air. It is located between the two halves of the mold (Figs 4.8 and 4.9). The melt flowing through the die can form different cross sections with or without changes in the parison's wall thickness as it exits the BM die (Table 4.3). When the parison has reached the required length, a mold-mounted or shuttled device below the die closes around the parison and air forces the plastic melt against the cavity walls. A blow pin is inserted through the parison, permitting air to enter. The parison then cools against the mold cavity.

Unlike injection blow molding (IBM), flash is a by-product of the process which must be trimmed and reclaimed. This 'excess' is formed when the parison is pinched together and sealed by the two halves of the mold.

In addition to continuous extrusion, there is also the intermittent extrusion, also called discontinuous extrusion. Both can have a diehead with one or more openings; so one or more parisons can be extruded into single- or multiple-cavity molds. For increased production output, multiple extrusion heads are used (Fig. 4.10).

With continuous extrusion, the extruded parison is continuously formed at a rate equal to the rate of part molding and moving the mold (Figs 4.11 and 4.12). The mold is moved away from the parison drop, and when more than one mold is used, another mold moves into position. To avoid interference with the parison formation, the mold-clamping mechanism must move quickly to capture the parison and return to the blowing station where the blow pin enters. More than half of all blow-molded parts are made using the continuous extrusion method. This process has

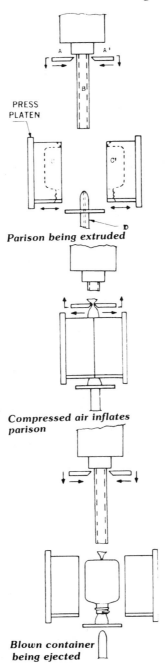

Figure 4.8 Basic continuous extrusion blow molding process: A = parison cutter, B = parison, C = blow mold cavity, D = blow pin.

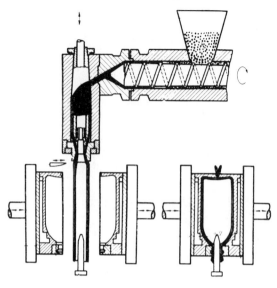

Figure 4.9 Typical phases in intermittent extrusion blow molding with accumulator in the die.

three different modes of operation: the rising method, the rotary wheel, and the shuttle.

With the rising method, the parison is continuously extruded above the mold. When it reaches the proper length, the open mold rises quickly to clamp the parison and returns downward to the blow station so that the parison continues to extrude. After the part is blown and cooled, the mold opens, the part is removed, and the process repeats.

The rotary wheel uses 3–20 clamping stations, possibly more. The stations are mounted to a vertical or horizontal wheel. A parison is pinched in the mold, parts are molded and cooled, and a cooled part is removed, all simultaneously as the wheel rotates past the extruder. This method provides high production rates.

With the shuttle method, two molds are located under the extruder. As the parison reaches the proper length, one mold located under the parison is clamped quickly, the parison is cut and shuttled quickly to its blowing station. With this dual-sided machine, the other (opened) mold moves under the parison drop. Thus, the molds alternately shuttle sideways (Fig. 4.10).

Intermittent extrusion

With an accumulator, the flow of the parison through the die is cyclical, permitting intermittent or discontinuous EBM (Figs 4.13 and 4.14). The

Table 4.3 Examples of differently performing extrusion blow molding dies

Die type	Feature	Advantage/disadvantage
Simple die	Fixed die gap	Simple; inexpensive; no adjustment facility
Die profiling	Premanently profiled; preferred in die land area	Fixed circumferential wall-thickness change; time-consuming; complex
Die centering	Can be permanently shifted laterally to correct parison drop path	Compromise between required drop path and equal wall thickness
Open-loop axial die-gap control	Can be axially shifted during extrusion	Equal circumferential wall-thickness change possible; no feedback
Servohydraulic closed-loop axial die-gap control	As above, with greater speed, accuracy, and flexibility	Equal circumferential wall-thickness change possible, with feedback
Stroke-dependent die profiling	Permanently ovalized die gap	Fixed, unequal circumferential wall-thickness change possible affects entire parison length
Die/mandrel adjustable profiling	Settable adjustment of die-gap profile	Settable, unequal circumferential wall-thickness change possible; rapid optimization
Servohydraulic closed-loop radial die-gap control	Programmable ovalization and shifting of die gap	Programmable circumferential wall-thickness change possible, independent of parison length

connecting channels between the extruder and accumulator, as well as the accumulator itself, are designed to prevent restrictions that might impede flow or cause the melt to hang up (Chapter 3). Flow paths should have low resistance to melt flow to avoid placing an unnecessary load on the melt.

To ensure that the least heat history (Chapter 1) or residence time is developed during processing, the design of the accumulator ensures the first melt into the accumulator is the first to go out when the 'ram' literally empties the accumulator chamber; the goal is to have the chamber totally emptied on each stroke. Plastics that are not heat-sensitive permit some relaxation in their heat history.

Figure 4.10 A multiple continuous extrusion head (three parisons) blow molding three containers simultaneously in a shuttle clamping system.

Intermittent systems fall into three classes: reciprocating screw, ram, and accumulator. The reciprocating screw units act much like the injection molding machines (Chapter 2). Plastic is conveyed and melted by the screw turning. As melt accumulates at the front of the barrel, the screw is pushed backward. Once a sufficient quantity of melt is in the barrel, the screw stops rotating and pushes forward (ram action) forcing the melt through a die to form a parison (Fig. 2.5, page 124).

The ram-type machine incorporates a screw that delivers melt into a chamber. An accompanying piston then forces the molten material from the chamber and through the die (Fig. 2.3, page 123).

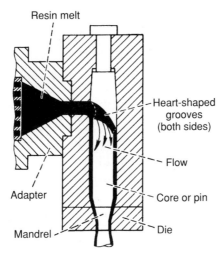

Figure 4.11 Side-fed or radial-flow head around the core; die-fed die with heart-shaped grooves.

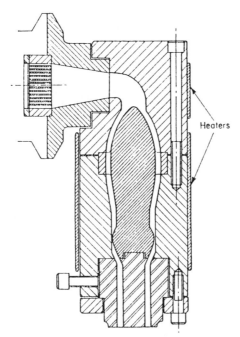

Figure 4.12 Continuous blow molding head having a spider-supported core.

Accumulator heads, the most common, are used particularly for molding large parts. Accumulator heads, attached to the end of extruders, are designed to collect and eject a measured amount of plastic (Figs 4.9 and 4.13).

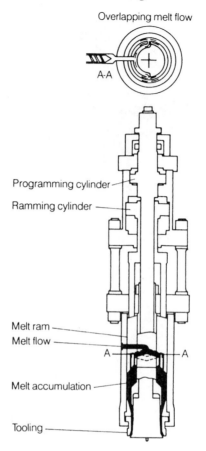

Figure 4.13 An accumulator melt flow head.

Melt behavior

Melt properties are of critical importance to BM; more so than for conventional extrusion. To a large extent they determine the quality achieved. The melt viscosity decides whether sagging or lengthening of a parison during extrusion from the die can be compensated, particularly in noncircular parisons (Fig. 4.15). Because engineering resins have so far been used mainly with IM, most processors attempt to use easy-flowing, low molecular weight IM-grade resins. But in BM, particularly EBM, the objective is very different; the melt should be viscous and of high molecular weight (high melt strength) (Chapter 1). This requirement also generally ensures another important feature, better impact strength. The melt viscosity should be nearly independent of the shear rate and the processing heat.

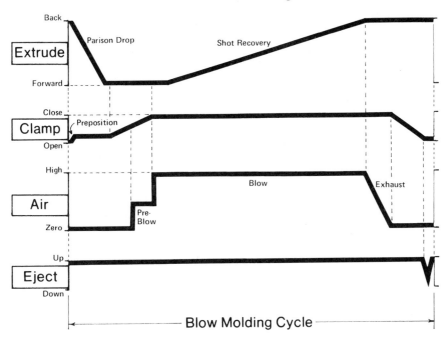

Figure 4.14 Blow molder using an accumulator head.

In raw-material data sheets there are usually no characteristics that indicate whether the resin is suitable for BM, or up to what limit an application is sensible. With increasing markets for these materials, more useful data sheets for all types of resins are becoming available.

Another important melt condition, extensibility, relates to how successfully large blowup ratios can be met, and whether edges and corners can be properly blown without thin spots. For ease of processing, there should be a large difference between the processing heat and solidification heat. This characteristic has a particularly advantageous effect on the quality of the blown surface. The greater the difference, the longer the time the melt has to be shaped, resulting in better surface definition. With a large difference, the contact of the parison with the surface of the mold is not influenced (or is affected very little). With little difference in the heating profile, uneven contacting of the mold can cause an uneven surface appearance, uneven stress, and so on. Increasing the mold heat can offset some of the negative effects. Unfortunately a large difference in the heat profile involves increasing the cycle time, but this increase can be minimized to meet quality production requirements with no rejects.

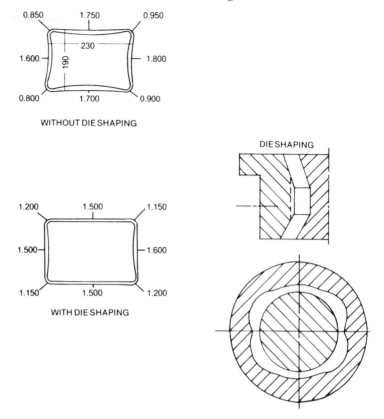

Figure 4.15 Noncircular blow mold die, with and without wall thickness die shape (dimensions in millimeters).

Parison thickness control

The control and monitoring functions range from extremely simple ones to expensive complete microprocessor systems. Some machines use electrical relays that permit molding with inexpensive machines which do not require a very sophisticated maintenance shop or skilled operators. However, to produce good-quality parts at the lowest cost, machine systems generally have to be more sophisticated, to meet any degree of performance.

Electronic parison programming is an effective way to control material usage and to improve quality and productivity. The most common method is orifice modulation (Fig. 4.16). The die is fitted with a hydraulic positioner that allows positioning of the inside die diameter during the parison drop. The OD to ID relationshing of the tapered die orifice open-

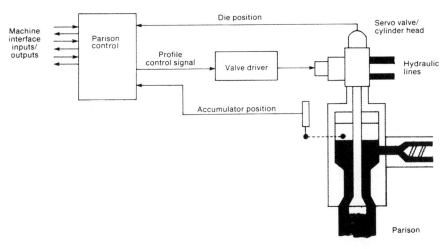

Figure 4.16 An accumulator head with programmable process controller; it controls the melt characteristics (interrelated with extruder performance), the rate of melt flow to form the parison, and the profiling thickness of the parison as it extrudes from the die.

ing is varied in a programmed manner to increase or decrease the parison wall thickness.

Electronic parison programming utilizes an electronic unit, commonly called a programmer, that uses a closed-loop servosystem supplying proper signals to control the amount, direction, and velocity of the movement of the hydraulic positioner. Programmers are designated by the number of program points available, usually 5–100.

In extrusion of a parison, especially a large parison, the wall thickness will vary as the weight of the resin increases, and it sags. Parison control may be helpful, but another method of minimizing wall-thickness variations is to increase the pressure of the melt in the die, either by regulation of the barrel back pressure or possibly by pressure variations via a ram when an accumulator is used. But an even wall-thickness distribution on the parison circumference and wall-thickness control in the longitudinal direction are seldom sufficient to meet today's quality specifications. Increasing requirements and ever more geometric shapes necessitate a partial wall-thickness adjustment, realized by shaping the flow channels in the die. The wall thickness along the length can be controlled by the conventional programmers reviewed. Circumferential distribution is controlled by patented predeformation of the die ring.

Parison control requires a compromise between the desired net weight and the need to maintain a sufficient safety margin over a set of minimum specifications, which include minimum wall thickness, drop speed, drop strength, dimensional stability, and fluctuations in net weight. Most of these parameters can be directly affected by the molder's ability to control the parison wall thickness. The most common and practical way of doing this has been to adjust the gap between the die and mandrel (Table 4.3).

For some resins, it is advantageous to maintain a constant shear rate at the die orifice. Constant shear ensures the melt is uniformly stressed and produces a uniform surface texture. To provide constant shear, an accumulator ram speed is controlled according to a predetermined profile. As the parison drop increases, the ram stroke increases. Electronic and mechanical systems are being used with adjustments that are reproducible at speeds of $20\,mm/s^{-1}$. This action can be applied only when a torpedo head is used. An adjustable core 'plug' at the center of the torpedo moves longitudinally and controls the orifice opening without changing the volume of melt in the accumulator (Fig. 4.16). In a central feedhead with the outer die ring moving, the volume of the melt in the flow channel varies, making it difficult to control the speed of the flow.

Processing capability developments in the past involved commodity resins, as they were predominantly used. Now there is more focus on engineering resins. Overlapping the melt streams in accumulator heads to obtain uniform wall thickness distribution is now a more fundamental requirement. The available accumulator heads differ in their feed-channel designs and are frequently protected by patents; so one must be careful when purchasing them. In a parison head, the melt is divided into separate streams by the mandrel or spiders. Weld lines form where the flow fronts reunite. As the parison is deformed differentially in BM, these weld lines are potential weak sections in areas of extreme deformation (Chapter 2).

If it is necessary to minimize streaks, another factor must be considered. No oxygen should be permitted to reach the melt and cause oxidation during downtime. A special die with an automatic shut off prevents this action. During downtime, the barrel heat should not drop lower than $20\,°C$ above the softening point, in order to prevent excessive volume contraction of the melt.

Accumulators can provide special products, an example being clear stripping systems. Two separately extruded melts enter the head and fuse together prior to exiting. Like overlapping layer formation of the parison, this feature also provides a positive way to strengthen weld lines.

Parison swell

A very important factor in EBM is the effective diameter swell of the parison. Ideally the diameter swell would be directly related to the weight swell and would require no further consideration. In practice the existence

Table 4.4 Average parison swell for some commonly used plastics

Plastics	Swell, present
HDPE (Phillips)	15–40
HDPE (Ziegler)	25–65
LDPE	30–65
PVC (rigid)	30–35
PS	10–20
PC	5–10

of gravity, the finite parison drop time, and the anisotropic aspects (the parison has directional properties) of the BM operation prevent reliable prediction of parison diameter swell directly with the weight swell. After leaving the die, the melt – which has been under shear pressure – undergoes relaxation that causes cross-sectional deformation or swell.

Parison swell tends to be the most difficult property to control in efforts to produce low-cost, lightweight products. Diameter swell is easy to see; with the parison dropping, one may be able to see it actually shrink even after it stretches. If it is shrinking in length, the wall must be thickening, and the parison is heavier per unit length, a behavior known as weight swell. Table 4.4 gives swell action for some common plastics.

For a given die cross section, the weight varies in proportion to the deformation. The time dependence and the increasing effect of weight with increasing parison length cause a narrowing of the diameter and a reduction in wall thickness from the lower end toward the die. Excessive length and prolonged hanging may lead to collapse, even with the usual procedure of providing some air pressure flow during parison formation. With this section, unlike the other processing methods, the characteristics of the melt within and after it exits the die are extremely critical in designing the die to suit the blow mold. A high melt strength is desirable to minimize this problem.

Diameter swell is an important processing parameter that must be controlled for EBM of unsymmetrical containers and particularly with blown side-handle products. Resins with an excessively high diameter swell could experience curtaining and stripping problems (adherence to mold cavity). An inadequate diameter swell contributes to an increase in rejects for defective handles.

Pinchoff

The pinchoff is a very critical part of the EBM mold, where the parison is squeezed and welded together, requiring good thermal conductivity for rapid cooling and good toughness to ensure long production runs. The pinchoff must have structural soundness to withstand the resin pressure and repeated closing cycles of the mold. It must usually push a small amount of resin into the interior of the part to slightly thicken the weld area. It also provides a cut through the parison to provide a clean break later when flash is removed.

Most molds use a double-angle pinchoff with 45° angles and a 10 mil (250 μm) land (Fig. 4.17(a)). When the blown part is large relative to the parison diameter, the plastic will thin down and even leave holes on the weld line, requiring pinchoff (Fig. 4.17(b)). Using shallow angles (15°) has a tendency to force the plastic to the inside of the blown part, thereby increasing the thickness at the weld line. A pinchoff with a dam (Fig. 4.17(c)) also helps to solve problems. The depth of the flash pocket is related to the pinchoff and is very important for proper molding and automatic trimming.

A gross miscalculation of pocket depth (which must be learned through experience) can cause severe problems. For example, if the pocket depth is too shallow, the flash will be squeezed with too much pressure, putting undue strain on the mold, mold pinchoff areas, and machine-clamp press section. The molds will be held open, leaving a relatively thick pinchoff, which is difficult to trim properly. If the pocket is too deep, the flash will not contact the mold surface for proper cooling. In fact, between molding and automatic trimming, heat from the uncooled flash will migrate into the cool pinchoff and cause it to heat up, creating problems like sticking to the trimmer. During trimming it can stretch instead of breaking free.

The knife-edge cutter width of the pinchoff depends on the resin used, the wall thickness of the blown part, the size of the relief angle, the closing speed, and the time when blowing starts. As a general guide for small parts up to 10 mil (250 μm), the width is 4–12 mil (100–300 μm). When processing LDPE, one uses the narrowest edge. Edge tapering can help, starting from the center. Its internal edge should be machined smooth and possibly broken slightly to prevent surface damage of the parison or the molded part during closing and opening. Such surface damage could include dull patches next to the weld line or scratches and scoring during ejection.

Deflashing is not as easy with engineering resin products as with PEs. Even if the processor maximizes the clamping force and the design of pinchoff edges, conventional methods such as knockingoff, punching, or cutting can be a problem for most production operations. For these plastics, other cutting methods are generally used, such as mechanical separation, water-jet cutting, or laser-beam cutting. Robots are used to handle these types of cutters.

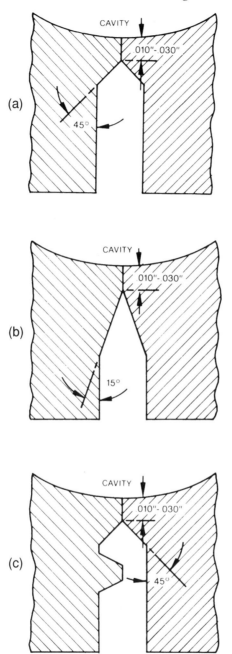

Figure 4.17 Typical pinchoff double-angle design.

Blowing the parison

The air used for blowing serves to expand the parison tube against the walls of the mold cavity, forcing the melt to assume the shape of the mold and forcing it into the surface details, such as raised letters and surface designs (Table 4.5). The air performs three functions: it expands the parison against the mold, exerts pressure on the expanded parison to produce surface details, and aids in cooling the blown parison. During the expansion phase of the blowing process, it is desirable to use the largest available volume of air, so the parison is expanded against the mold walls in the shortest possible time. A maximum volumetric flow rate into the cavity at a low linear velocity can be achieved by making the air-inlet orifice as large as possible.

A blow pin is usually located opposite the pinched end of the parison. However, the pins can be located in any position and usually around the mold parting line. Various techniques are used. Air can enter through the extrusion die mandrel (as with most pipe lines, Chapter 3), through a blow pin over which the end of a parison has dropped (Fig. 4.8), or through blowing needles that pierce the parison. It is possible to avoid the blow-pin mark when using EBM by employing hypodermic needles and pulling them out before the material solidifies.

Blowing inside the neck is sometimes difficult. Small orifices may create a venturi effect, producing a partial vacuum in the tube and causing it to collapse. If the linear velocity of the incoming blow air is too high, its force can actually draw the parison away from the extrusion head end of the mold, producing an unblown parison. The air velocity must be carefully regulated by control valves placed as close as possible to the outlet of the blow tube. The gauge pressure of the air used to inflate commodity and engineering resin parisons is normally about 90 psi (0.62 MPa) with range 30–300 psi (0.21–2.1 MPa). Too high a blow pressure will often 'blow out'

Table 4.5 Guide for air blowing pressure

Plastic	Pressure (psi)
Acetal	100–150
PMMA	50–80
PC	70–150
LDPE	20–60
HDPE	60–100
PP	75–100
PS	40–100
PVC (rigid)	75–100
ABS	50–150

the parison. Too little pressure will yield end products lacking adequate surface details. As high a blowing pressure as possible is desirable to give both minimum blow time (therefore higher production rates) and finished parts that reproduce the mold surface. The optimum blowing pressure is generally found by experimentation on the machine that will be used in production. The blow pin should not be so long that air is blown against the hot plastic opposite the air outlet. That action can create freeze-off and stresses in the container at that point.

Air is a fluid, just like the parison, and as such it has a limited ability to blow through an orifice. If the air-entrance channel is too small, the required blow time will be excessive, or the pressure exerted on the parison will not be adequate to reproduce the surface details in the mold. General guidelines for determining the optimum diameter of the air-entrance orifice during blowing are (1) up to 1 quart ($0.95\,dm^{-3}$) use $\frac{1}{16}$ in. (1.6 mm); (2) for 1 quart to 1 gallon (0.95–$3.8\,dm^{-3}$), use 0.25 in. (6.4 mm); and (3) for 1–54 gallons (3.8–$205\,dm^{-3}$) use 0.5 in. (12.7 mm).

The pressure of the blowing air will cause variations in the surface detail of the molded items. Some PEs with heavy walls can be blown with air pressure as low as 30–40 psi (210–280 kPa). Low pressure can be used because items with heavy walls cool slowly, giving the resin more time, at a lowered viscosity, to flow into the indentations of the mold surface. Thin walls cool rapidly; so the plastic reaching the mold surface will have a high melt viscosity, and higher pressures will be required, 50–100 psi (0.3–0.7 MPa). Larger items such as 1 gallon containers require a higher air pressure, 100–150 psi (0.7–1.0 MPa). The plastic has to expand farther and takes longer to get to the mold surface in larger items. During this time the melt heat will drop slightly, producing a more viscous mass, which requires more air pressure to reproduce the details of the mold.

A high volumeteric airflow at a low linear velocity is desired. A high volumetric flow gives the parison a minimum time to cool before coming in contact with the mold, and provides a more uniform rate of expansion. A low linear velocity is desirable to prevent a venturi effect (see above). Volumetric flow is controlled by the line pressure and the orifice diameter. Linear velocity is controlled by flow-control valves close to the orifice.

The blowing time differs from the cooling time, being much shorter than the time required to cool the thickest section to prevent distortion on ejection. The blow time for an item may be computed from Table 4.6 and the following formula:

$$\text{Blow time (s)} = \frac{\text{Mold volume}\left(m^{3}\right)}{m^{3}\,s^{-1}} \times \frac{\text{Final mold pressure (kPa)} - 101\,kPa}{101\,kPa}$$

Blow molding

Table 4.6 Discharge of air at 14.7 psi (101 kPa) and 70 °F (21 °C)

Gauge pressure (psi)	Discharge of air (ft³ s⁻¹) for specified orifice diameter			
	$^1/_{16}$ in. (1.6 mm)	$^1/_8$ in. (3.2 mm)	$^1/_4$ in. (6.4 mm)	$^1/_2$ in. (12.7 mm)
5	0.993	3.97	15.9	73.5
15	1.68	6.72	26.9	107
30	2.53	10.1	40.4	162
40	3.10	12.4	49.6	198
50	3.66	14.7	58.8	235
80	5.36	21.4	85.6	342
100	6.49	26.8	107.4	429

This is for free air; but there will be a pressure buildup as the parison is inflated, so the blow rate has to be adjusted. The value of $m^3 s^{-1}$ is obtained from Table 4.5, according to the line pressure and the orifice diameter. The final mold pressure is assumed to be the line pressure for purposes of calculation. Actually, the blow air is heated by the mold, raising its pressure. Calculations ignoring this heat effect will be satisfactory when blow times are under one second (for small to medium-size parts); but if blow times are longer, the air will have time to pick up heat, causing a more rapid pressure buildup and blow times shorter than calculated.

Cooling

Processors often employ methods to speed up cooling. For example, postcooling of blow-molded parts can shorten the blow cycle. Shuttle machines, which maximize production in continuous EBM, are preferred by many molders because they can produce finished containers in the machine; but trimming cannot proceed until the scrap areas, the thickest areas of the part, have been cooled sufficiently, so the cycle depends on getting parts cool enough to trim. If trimming is not done in the machine, the parts can be demolded when they are cool enough to maintain their shape. Wheel machines are sized to do the job by mounting a sufficient number of mold stations to provide adequate cooling for each blown container as it takes its turn around the wheel.

Resins vary in cooling requirements. It is not usually necessary to postcool PVC; it gives up its heat much more readily than the polyolefins (making it more appropriate for a dedicated operation than for custom blow molding). Also the bigger the part, the more cost-effective its cooling.

Postcooling can be extremely beneficial when directed at the thickest areas, especially at pinchoff, neck finish, and thick handle areas.

Prepostcooling is very important in developing any 'fast' cooling control. It originates with the proper placement of cooling lines in the molds. Unfortunately most blow molds are inefficiently designed for maximizing cooling rates. Improvements are gradually developing, similar to those available for injection molds [1]. In fact, a few computer-aided design programs have been produced just for blow-mold cooling analysis.

Efficient mold cooling also depends on accurate venting. Air trapped between the mold cavity and the plastic will significantly slow down cooling and also could cause problems with the part. Certain molds incorporate vacuum lines to ensure proper contact.

Unfortunately, it is not always possible to run the mold as cold as possible because of a high ambient dew point. With high humidity, the mold sweats, causing the surface of a blown part to be damaged. Mold sweating is easily solved by increasing the mold heat, which in turn increases the cycle time. To eliminate the problem and keep the cycle time short, as in other processes, the mold is enclosed in a dry air curtain. One should also consider dehumidifying the room, or just the molding machine in an enclosure or tent, if practical.

As much as 80% of the blow molding cycle is cooling time. Several methods are used to reduce cycle time. A part is normally cooled externally by the mold cavity, forcing heat to travel through the entire wall thickness. With the relatively poor thermal conductivity (Chapter 1) of most plastics, molding cycle times of thick parts can be considerable. Internal cooling systems are used to reduce the cooling time of the mold, thus reducing costs by removing some of the heat from the inside.

Shuttle and rotational systems are used. By moving the molds to a location other than under the parison, the molds themselves with blown parts are cooled by water sprays, heat-absorbing liquids, etc. Another system uses air chillers which reduce the temperature of the blown air to about $-95\,°F$ ($-70\,°C$) and blow pins that permit heated air in the blown part to exit. This means that a continuous flow of fresh, cool air enters the part. With such systems, the output can increase by 10–30%.

With the liquefied gas system, immediately after the blowing action, liquid carbon dioxide (CO_2) or nitrogen (N_2) is atomized through a nozzle in the blow pin into the interior of the blown part. The liquid quickly vaporizes. This action removes heat and is exhausted at the end of the blowing cycle with production rates of 25–35%. The cost of this system requires high production. Another potential disadvantage could be that precise control is required.

A similar system uses supercold air with water vapor; very dry, subzero, blown air is used to expand the parison. The expanding air circulates and is exhausted. Immediately after the parison has expanded, a fine mist of water is injected into the cold airstream. As it flows, the water mist turns into snow that circulates within the blown part. The snow melts and vaporizes. At the end of the molding cycle, the water mist is stopped and air is purged, permitting the interior to become dry before the mold opens and the part is removed. This system has improved production up to 50%.

An air-exchange system may be easier to operate. After expanding the parison, plant air is allowed to circulate internally and exhausted. Differential pressure inside the part is kept at about 80 psi (550 kPa) keeping the plastic in contact with the cavity wall. Production increases of 10–15% can be obtained.

Clamping

The mold-clamping methods are hydraulic and/or toggle, similar to, but less sophisticated than those used with injection molding (Chapter 2). Clamping systems vary depending on machine operation (Fig. 4.18), part configuration, and the location of the parting line. Sufficient daylight is needed in the mold platen area to accommodate parison systems, ejection of blown parts, unscrewing or insertion equipment, and/or other special equipment.

The improvements made in clamping units provide a great variety of movement and action, particularly in the larger BM machines. With advanced operating technology, a machine runs more smoothly and more exactly to permit a wide variety of action in the mold, as well as accurate control of the closing speed (Fig. 4.19).

The delayed closure action in the final phase of mold closing determines pinchoff weld formation, and the reproducibility of this delayed closure phase ensures the uniformity of BM parts. A disadvantage, still not resolved, is that it is often necessary to work with scale settings and not actual physical values, in recording and setting the closing speed as well as in other aspects of BM. However, proportional valve technology offers

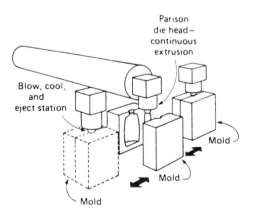

Figure 4.18 Shuttle continuous extrusion blow molding machine. Molds on this dual-sided system move alternately to close on the parison.

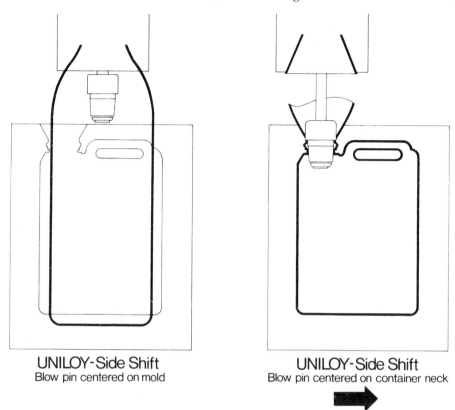

UNILOY-Side Shift
Blow pin centered on mold

UNILOY-Side Shift
Blow pin centered on container neck

Figure 4.19 Mold movement can locate the blow pin in any required horizontal position.

the great advantage that the full clamping force is available within a very short time through suitable control of the valves during mold closure (this is similar to the clamp action of injection molding [1]). This action facilitates easy pinchoff removal. Flash removal is more difficult with engineering resins because they freeze rapidly. Thus, to facilitate flash removal with these resins, more care (control) is required in holding down the BM pinchoff edge during mold closure.

The closing speed and clamping force of the mold have an important influence on product quality. If closing speed is increased, the lengthening of the parison due to low melt viscosity is minimized; formation of the edge layers of the parison, which freeze very rapidly, is better controlled.

For resins with low viscosity and/or for BM when there is a large opening in the delay position, a closing speed of $100\,\text{in.}\,\text{s}^{-1}$ ($40\,\text{cm}\,\text{s}^{-1}$) helps to resolve some of the problems. Closing speeds such as these can be operated repeatedly only with a speed control system. The required clamping force for the part is determined less by the blowing pressure acting on the part surface than by the length of the flash to be removed. For engineering resins, a clamping pressure of 1000–2000 psi (7–14 MPa) at the pinchoff edge should be considered. Closing movements in which the flash is separated before the melt has started to solidify are also used. Most of these systems are patented.

Many articles require molds with large thickness and shape variations between the two mold halves because of the particular contours of the part. For such molds, particularly those having very complex (reentrant) shapes, clamping platens and molds can be sectionalized, with each section separately controlled. Thus, sectionalizing can permit blowing complex parts quickly and repeatedly. The hydraulic systems permit sections to operate for different time periods and at different speeds. When only two mold halves are used, they can meet at the prescribed center point at the same time.

Shrinkage

The shrinkage behavior of different resins and the part geometry must be considered. Shrinkage is generally the difference between the dimension of the mold at room temperature (22 °C) and the dimensions of the cold, blown part, usually checked 24 h after production. The elapsed time is necessary to allow the part to shrink. Trial and error determines how much time is required to ensure complete shrinkage. Coefficients of expansion and the different shrinkage behaviors depend on whether plastic materials are crystalline or amorphous (Chapter 1).

Longitudinal shrinkage tends to be slightly greater than transverse shrinkage. Most of the horizontal shrinkage occurs in the wall thickness rather than a body dimension. Higher shrinkage occurs with higher-density PE and thicker walls. Longitudinal shrinkage is due to the greater crystallinity of the more linear plastics. Transverse shrinkage is due to slower cooling rates, which produce more orderly crystalline growth.

Mold shrinkage depends on many factors, such as resin density, melt heat, mold heat, part thickness, and pressure of blown air. Typical PE shrinkages are as follows: LDPE at a thickness up to 0.075 in. (1.9 mm) has a tolerance of 0.010–0.015, and at a thickness over 0.075 in. (1.9 mm) has a tolerance of 0.015–0.030; whereas HDPE at a thickness up to 0.075 in. (1.9 mm) has a tolerance of 0.020–0.035, and at a thickness over 0.075 in. (1.9 mm) has a tolerance of 0.035–0.055. Once the operating conditions are established, tolerances of ±5% may be expected. When fillers are included in these resins, as well as others, their shrinkage behavior changes, even to the extent of their having very little shrinkage when certain fillers are used.

The cavity defines the shape of a part. As with other processes, the cavity dimensions in BM are enlarged slightly to compensate for part shrinkage. For polyolefins, particularly in the neck section, slightly higher shrinkage rates are used for extrusion BM than for injection BM. The amount of shrinkage is 2% for the body and as high as 3.5% for the neck finish. However, rigid resins are relatively unchanged.

The most common special feature of the mold is the quick-change volume-control insert. Rigid volume control is necessary for certain products, such as dairy containers. Here HDPE is used (for many excellent reasons), and the container slowly shrinks and changes in size for many hours after molding. Because of production 'volume control' requirements, some dairies must fill containers molded half an hour before fill then switch to filling containers molded several days previously. Volume-control inserts that displace the difference in volume between the two types are added to the mold, usually as a disk in the sidewall, to ensure the volume and fill levels are the same in both containers at the time of filling. The device works because HDPE shrinkage is reduced virtually to zero for the life of the container when filled with milk or juice and stored at cold temperatures.

With most BM, shrinkage may be reduced by raising the blowing pressure and lowering the mold heat. Rapid cooling is desired, but cooling too rapidly can cause surface imperfections, distortion, and frozen-in stresses. Raising the stock heat may not appreciably affect the outside dimensions, but it causes more shrinkage to occur in the wall thickness, as higher heats lessen strain recovery and reduce blowing stresses.

Close tolerances on the capacity of the blown part also influence mold construction. Where multiple molds are required, the capacity must be closely regulated from one mold to another. Some of the casting processes have generally been more successful in reducing capacity variation than duplicating or hobbing. Final capacity adjustment uses an inserted bottom plate or other adjustable insert. Slight changes can then be made on each mold insert without noticeable variation in the finished parts.

Plastics will sometimes thin out at an insert or parting line. This phenomenon is of little importance with resins such as LDPE, but it can be a serious problem with HDPE and similar resins. Any inserts should be fitted tightly. The parting line should be mitered as close as possible, and sufficient clamping force should be provided to prevent partial opening of the mold. Solid molds have a slight technical advantage over molds with inserts with respect to the thinning action.

To aid in controlling shrinkage, vacuum-assisted molding ensures closer contact of the parison with the mold wall. The rate of heat transfer is effectively accelerated, usually with cycle reductions of up to 10%. Table 4.7 on page 323 shows the effective cycle gains made by using a vacuum in conjunction with various wall thicknesses in an HDPE container. The advantage of vacuum assistance is further enhanced by utilizing maximum blow pressures. With parts trimmed in the mold, the problem of deflashing is significantly alleviated when reduced blow cycles and vacuum assist are used. This improvement is due to the more rapid rate and degree of part cooling.

With either EBM or IBM, the requirement for a thermally controlled parison at the lowest heat increases the demand on internal pressure/ volume to ensure adequate movement or expansion of the parison in the transverse direction. It is possible that vacuum-aided stretch blowing can provide faster expansion rates of the parison and improve performance at the higher stretch/blow ratios. With PVC, PS, and other low-shrink alloyed resins, the vacuum assist may need little or no pressure beyond 1 atm (100 kPa). The function of vacuum (−100 kPa) has the basic effect of −(−100 kPa) + (+100 kPa) = +200 kPa. The normal shrink allowances can usually be reduced by one-third. If a vacuum retained for in-mold labeling is used, it effectively provides a simultaneous vacuum-assist system to the blowing cycle.

Surface treatment

There are surface treatments that allow low-cost blow-molded PE bottles, gasoline fuel tanks, and other containers that provide barriers to gases,

Table 4.7 Cycle time reduction using vacuum-assisted blow molding of HDPE[a]

Wall thickness, in. (mm)	Blow pressure (psi)	Normal cycle time (s)	Cycle time with vacuum assistance (s)	Cycle time vacuum only (s)
0.040 (1.0)	40	14	12	12
	60	12	11	
	80	11	10	
	100	10	9.2	
	125	10	9	
0.050 (1.21)	40	16	14	15
	60	15	13	
	80	14.5	11	
	100	14.5	10	
	125	14	10	
0.060 (1.52)	40	20	16	18
	60	18	15	
	80	18	15	
	100	17	14	
	125	16.5	14	

[a]Coolant input kept constant at 28 °F (−2 °C); coolant flow kept at a constant 31 gallons per minute (2 dm³ s⁻¹) set at 27 in. Hg gauge; surge tank capacity is 10 ft³ at 27 in. Hg.

moisture, etc. These include containers for paints, paint thinners, lighter fluids, polishes, cleaning solvents, cosmetics, and toiletries. An example is the fluorination surface treatment to provide a gasoline barrier when blow molding with PE – a barrier to nonpolar solvents. A barrier is created by the chemical reaction of the fluorine and the polyethylene, which form a thin (20–40 nm) fluorocarbon layer on the bottle surface. Two systems are available for creating the layer. The in-process system uses fluorine as a part of the parison-expand gas in the blowing operation. With it, a barrier layer is created only on the inside. The posttreatment system requires bottles to be placed in an enclosed chamber filled with fluorine gas. The method forms a barrier layer on the inner and outer surfaces.

INJECTION BLOW MOLDING

IBM has three stages (Figs 4.20 to 4.22). The first stage injects hot melt through the nozzle of an injection molding machine into a manifold and into one or more preform (test-tube shaped) cavities. An exact amount of resin is injected around the core pins. Hot liquid from a heat control unit is directed by hoses through mold-heating channels around the preform cavity; these channels have been predesigned to provide the correct heat control on the melt within the mold cavity. The melt heat is decreased to the required amount.

The two-part mold opens, and the core pins carry the hot plastic pre-form to the second stage, the blow molding station (counterclockwise). Upon mold closure, air is introduced via the core pin, and the plastic blows out and contacts the surface of the mold cavity. Controlled chilled water circulates through predesigned mold channels around the mold cavity (usually 4–10 °C), and solidifies the hot plastic.

The two-part blow mold opens, and the core pin carries the complete blown container to the third stage, which ejects the part. Ejection can be done by using a stripper plate, air, a combination of stripper plate and air, robots, and so forth.

IBM can have four or more stations (stages). A station can be located between the preform stage and the blow-mold stage to provide extra heat-conditioning time. A station between the blow-mold stage and the ejection stage can provide additional cooling and/or secondary operations, such as hot stamping and labeling. A station between the ejection stage and the preform stage can be used to see whether or not ejection has occurred, to add an insert to the core pin, and so forth.

The process parameters in preform production that determine the qual-ity of the part are the injection speed, injection pressure, hold-on (packing) pressure, heat control of the preform, and melt mix (Chapter 2). The process permits the use of resins that are unsuitable for EBM, specifically those with no controllable melt strength such as PET, which is used in

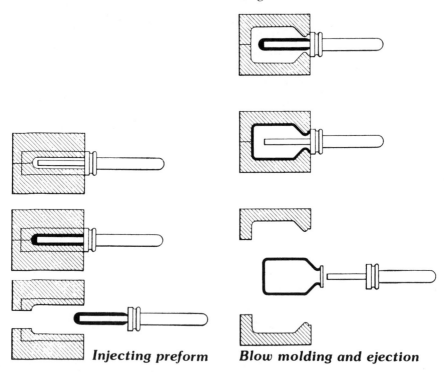

Injecting preform   **Blow molding and ejection**

Figure 4.20 Basic injection blow molding process.

carbonated beverage bottles (stretch blown). The information on blowing parisons, cooling, clamping, and shrinkage that was presented for EBM is also applicable to IBM.

Several different methods of IBM are available, each with different means of transporting the core rods from one station to another. These methods include the shuttle, two-parison rotary, axial movement, and rotary with three or more stations used in conventional IM clamping units.

Precise dimensional control offers several important advantages. It produces scrap-free, close tolerance, completely finished bottles or containers that require no secondary operations. Neck shapes and finishes, internally and externally, can be molded with an accuracy of at least ±4 mil (± 100 μm). It also offers positive weight control in the finished product accurate to +/−0.1 g. These close size and weight tolerances are important for compatibility with downstream operations such as filling bottles.

Multistation machines are used. As an example, with a four-station system, the conditioning station enables production of parts with better dimensional and physical properties, giving better utilization of machine time.

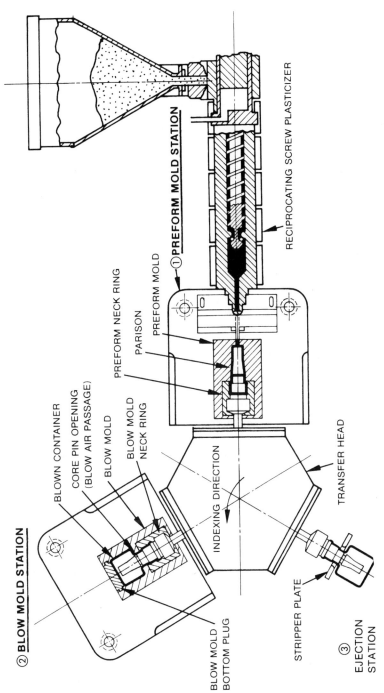

Figure 4.21 A three-station injection blow molder.

① **PREFORM MOLD STATION**

RECIPROCATING SCREW PLASTICIZER

PREFORM NECK RING

PARISON

PREFORM MOLD

② **BLOW MOLD STATION**

BLOWN CONTAINER

CORE PIN OPENING
(BLOW AIR PASSAGE)

BLOW MOLD

BLOW MOLD
NECK RING

INDEXING DIRECTION

TRANSFER HEAD

BLOW MOLD
BOTTOM PLUG

STRIPPER PLATE

③ EJECTION
STATION

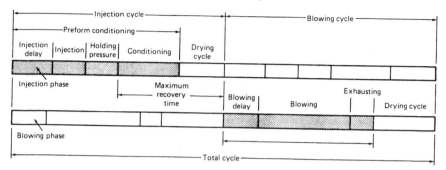

Figure 4.22 Injection blow molding: a complete cycle begins with injection molding of the preform followed by the blowing cycle.

In the past few years, new markets have been established by new materials developed especially for IBM. Cough-medicine bottles, once made of glass, have virtually become all PP. Other specially formulated and compounded plastics include PVC homopolymers and PC.

Early IBM techniques, all two-position methods, were adaptations of standard injection molding machines fitted with special tooling. The primary difficulty with all of the two-position methods was that the injection mold and blow mold stations had to stand idle during part removal. This led to the 1961 invention in Italy by Gussoni of the three-position method, which uses a third station for part removal. By the late 1960s, the concept had been developed by commercial machinery builders, and it forms the basis for virtually all IBM today (Fig. 4.21). One or more additional stations can be added for various purposes: detection of a bottle not stripped or temperature conditioning of the core rod, decoration of the bottle between bottle blow mold and stripping stations, or temperature conditioning or preblowing of the parison between parison injection and bottle blow mold stations. These high-speed machines mold to tight tolerances. Machines are available that have outputs of 50 000 PET bottles per hour ($8\frac{1}{2}$–$20\frac{2}{5}$ floz, 240–580 cm³ bottles).

STRETCHING/ORIENTING BM

High-speed EBM and IBM take an extra step in stretching or orienting. Figure 4.23 shows stretched IBM; with EBM the stretching action is similar to this, as it occurs during compressed air inflation (Fig. 4.8, page 325). In EBM the parison, which is mechanically held at both ends, is stretched rather than just blown. Stretching can include the use of an expanding rod within the IBM preform or an external gripper with EBM.

By biaxially stretching the extrudate before it is chilled, significant improvements can be obtained in the finished containers, as described in Chapters 1 and 3. This technique allows the use of lower-grade resins or thinner walls with no decrease in strength; both approaches reduce material costs. Stretched BM gives many resins (mono- or bioriented) improved physical and barrier properties (Tables 4.8 and 4.9). The process allows wall thicknesses to be more accurately controlled and also allows weights to be reduced.

Draw ratios used to achieve the best properties in PET bottles (typical two- and three-liter carbonated beverage bottles) are 3.8 in the hoop direction and 2.8 in the axial direction; they will yield a bottle with a hoop tensile strength of about 29 000 psi (200 MPa) and an axial tensile strength of 15 000 psi (104 MPa).

Stretch blow is extensively used with PET, PVC, ABS, PS, AN, PP, and acetal, although most TPs can be used. The amorphous types, with a wide range of thermoplasticity, are easier to process than the crystalline types such as PP. If PP crystallizes too rapidly, the bottle is virtually destroyed during the stretching. Clarified grades of PP have virtually zero crystallinity and overcome this problem.

The stretch process takes advantage of the crystallization behavior of

Table 4.8 Volume shrinkage of stretch blow molded bottles[a]

Type of bottle	Percent
Extrusion blow molded PVC	–
Impact-modified PVC (high orientation)	4.2
Impact-modified PVC (medium orientation)	2.4
Impact-modified PVC (low orientation)	1.6
Nonimpact-modified PVC (high orientation)	1.9
Nonimpact-modified PVC (medium orientation)	1.2
Nonimpact-modified PVC (low orientation)	0.9
PET	1.2

[a] Seven days at 80 °F (27 °C).

Table 4.9 Gas barrier transmission comparisons for a 24 fl.oz (680 cm³) container weighing 40 g.[a]

Type of bottle	Rate ($m^2 day^{-1}$)	
	Oxygen (cm^3)	Water vapor (g)
PET (oriented)	10.2	1.10
Extrusion blow molded PVC	16.4	2.01
Stretch blow molded PVC (impact-modified)	11.9	1.8
Stretch blow molded PVC (nonimpact-modified)	8.8	1.3

[a] At 100 °F (38 °C).

Blow molding

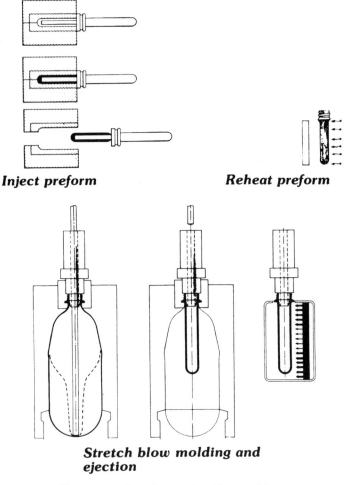

Inject preform   **Reheat preform**

**Stretch blow molding and
ejection**

Figure 4.23 Stretch injection blow molding.

the resin and requires the preform to be temperature-conditioned then
rapidly stretched and cooled into the bottle shape.

There are inline and two-stage processes. Inline processing is done on a
single machine, whereas two-stage processing requires either an extrusion or an injection line to produce the solid parisons or preforms. With
either type of process, a specific-reheat blow machine is used to produce
the bottle. With inline systems, the hot, firm plastic preform or parison
passes through conditioning stations that bring it down from the 'melt'
heat to the proper orientation heat (Table 4.10). A rather tight heat profile is maintained in the axial direction. Advantages of this approach
are that the heat history is minimized (crucial for heat-sensitive resins),
and the preform or parison can be programmed for optimum material
distribution.

With the two-stage process, cooled preforms or parisons are conveyed through an oven (usually using quartz lamps) which reheats them to the proper orientation heat profile. Advantages include minimization of scrap (for EBM, there being no scrap with IBM using either the inline or the two-stage process), higher output rates, and the capability to stockpile preforms or profiles and improve thread finishes (EBM). Inline stretch IBM, which offers more flexibility from a material view, does not give the degree of parison programming available in EBM stretch. An extruded parison can be heat-stabilized, then the parison is held externally and pulled to give an axial orientation while the bottle is blown radially. Another technique stretches and preblows the parison before completing the blowing operation. In both operations, scrap is produced at the neck and the base of the bottle. The processes completely orient the whole bottle, including the threaded end; this neck orientation does not occur with stretch IBM. With EBM, threads are finished postmolding.

Extrusion processes can also use programmed parisons and two sets of molds. The first mold is used to blow and heat-condition the preform, actually to cool it to the orientation heat. The second mold uses an internal rod to stretch and blow the bottle. With two-stage EBM, the preforms or parisons are usually open-ended. They are reheated then stretched by pulling one end while the other end is clamped in the blow mold. Minimum scrap can be produced, and the compression-molded threads provide a good neck finish. This system utilizes a conventional extrusion line and a reheat blow machine whose oven contains quartz lamps to provide the proper heat. (When used to heat the preform or parison, they rotate in a heat-profile-controlled oven to provide uniform heat circumferentially.)

MOLDS

As blow molds do not have to withstand high pressure, a wide selection of construction materials is available. The ultimate selection will depend on a balance of the following factors: cost, thermal conductivity, and required service life. The more commonly employed materials for small parts are aluminum and aluminum alloys, steel, beryllium copper

Table 4.10 Stretch blow molding processing characteristics

Plastic	Melt temperature		Stretch orientation temperature		Maximum stretch ratio
	°F	°C	°F	°C	
PET	490	250	190–240	88–116	16
PVC	390	199	210–240	99–116	7
PAN	410	210	220–260	104–127	9
PP	334	168	250–280	121–136	6

(Be/Cu), and cast zinc alloys (Kirksite, etc.). Aluminum molds are excellent heat conductors, are easy to machine, can be cast, and are reasonably durable, particularly when fitted with harder pinch blades and neck inserts (Table 4.11).

Molds with aluminum pinchoff areas could last 1–2 million cycles if properly set up and maintained. But most molds use Be/Cu and steel pinchoffs, providing more durability, and they can easily be repaired. Be/Cu is preferred because of its high thermal conductivity; steel is predominantly used because of its wear resistance and toughness.

Although Be/Cu and Kirksite are better conductors of heat, aluminum is by far the most popular material for molds, because of the high cost of Be/Cu and Kirksite's short life (soft zinc alloy). Aluminum is light in weight, a relatively good conductor of heat, very easy to machine, and low in cost. A major disadvantage is that it is easily damaged, particularly if it is abused. Its potential porosity may easily be eliminated by coating the inside cavity with a sealer such as automotive radiator sealant.

Aluminum is used for single molds, molds for prototypes, and large numbers of identical molds, as found on wheel-type blowing equipment or equipment with multiple-die arrangements. (It is best to prototype using a material that the production mold will use to duplicate heat transfer conditions.) Aluminum may tend to distort after prolonged use. Thin areas, like pinchoff areas, can wear in aluminum.

Be/Cu molds are corrosion resistant and very hard when compared to

Table 4.11 Materials used in the construction of blow molds[a]

| Material | Hardness[b] | Tensile strength | | Thermal conductivity $(Btu\,in.\,ft^{-2}\,h^{-1}\,°F^{-1})$ |
		psi	MPa	
Aluminum				
A356	BHN-80	36 975	255	1047
6061	BHN-95	39 875	275	1165
7075	BHN-150	66 700	460	905
Beryllium copper				
23	RC-30	134 850	930	728
165	(BHN-285)			
Steel				
0-1	RC 52-60	290 000	2000	
A-2	(BHN-530-650)			
P-20	RC-32	145 000	1000	257
	(BHN-298)			

[a] BHN = Brinell hardness; RC = Rockwell hardness (C scale).
[b] Specific gravities (lb in.$^{-3}$) Al = 0.097, Be/Cu = 0.129–0.316, steel = 0.24–0.29.

aluminum, making Be/Cu the choice for PVC. However, it is about three times as heavy as aluminum, costs about six times as much per volume, has a lower thermal conductivity, and requires about one-third more time to machine.

Thermal conductivity of the Be/Cu alloy is slightly poorer as well. For polyolefin blow molding, some mold makers have combined the materials by inserting Be/Cu into the pinchoff area of an aluminum cavity, thereby obtaining a lightweight, easy-to-manufacture mold with excellent thermal conductivity and hard pinchoff areas.

Unlike injection blow molds that are mounted onto a die set, all extrusion blow molds are fitted with hardened steel guide pins and bushings. These guide pins ensure the two mold halves are perfectly matched. The remaining hardware in the process, dies, mandrels, blow-pin cutting sleeves, and neck-ring striker plates, are all made from tool steel hardened to 56–58 Rc.

In blow molding plastics that produce corrosive volatiles (PVC, nylon, acetal, etc.), it is necessary to employ corrosion-resistant steels or to use plating, even gold plating. However, platings can lead to reductions of strength and hardness, and damage to edges and surfaces can easily occur. With proper care, lifespans of 10^5 to 10^6 cycles can be expected. Steel molds are used for continuous production and longer runs.

Hardening is not necessary, and chrome plating is not customary. For large-volume parts with a content of 4yd^3 (3m^3) and upwards, welded machined steel plate construction can be considered. Cast metal is of no value for the mold bases. Given the right care, the life of steel molds is the longest, at over 10^8 parts per mold. Inserts subject to wear and tear have to be refurbished or exchanged at intervals of 10^5 or 10^6 cycles (Table 4.12).

Unfortunately, most BM molds do not yet provide the high level of cooling that has been achieved for decades with IM molds. A greater effort is now being made to properly incorporate cooling channels in BM molds, to provide necessary and controllable heat transfer. CAD systems now are available for proper design of the molds. Internal cooling with liquid N_2, CO_2, or refrigeration and ice crystals can improve efficiency by up to 50%.

PLASTICS

Nearly all BM resins used to be commodity types (PE, PP, PVC, and PS), but engineering types have been used more recently. Typical heats used for blow molding some resins are given in Table 4.13. The polyolefins (PE and PP) and rigid PVC have proved to be the most suitable materials for blow molding. Its heat control and rheology allow PE to be processed relatively easily. The thermal sensitivity of PVC and the reprocessing of

Table 4.12 Guide to selecting construction materials for blow molds[a]

| Property | Steel | Machined | | Cast | | |
		Aluminum	Be/Cu	Aluminum	Kirksite	Be/Cu
Pinch life	4	3	2	2	1	3
Cavity life	4	3	4	2	1	3
Surface finish	4	3	4	2	1	3
Heat control	2	4	4	2	1	3
Mold modifications	2	4	2	1	1	2
High volume	4	3	4	2	1	2
Mold lead time	2	3	2	4	4	3
Low cost	2	3	1	4	4	3
Prototype cost	1	3	2	3	4	3
Complex shapes	3	4	3	3	2	2
Moving mold parts	4	3	3	3	1	1

[a] 4 = best, 1 = poorest.

the flash can cause several difficulties if not handled properly. In retrospect, PVC in particular, which imposed important and high performance requirements on the processing operation, imparted major impulses to the further development of EBM technology and provided the impetus for a thorough engineering analysis of melt flow through the extruder and blow heads.

The wide diversity of products being manufactured makes it clear that processing plants will have to solve logistical problems such as the separation of material cycles and the drying of materials, as has been done in the standard extrusion and injection molding plants.

Most engineering resins are hygroscopic and therefore must be properly dried before processing. Moisture generally has an adverse effect on the melt viscosity, surface, and physical properties of the product. To avoid these problems, resins must be dried, usually to a residual moisture content of <0.02 wt%. The optimum drying heat and time are given in the manufacturer's data sheet (Chapters 1, 2, and 3).

Blow molds for engineering resins are made from the same materials that are normally used for processing HDPE, principally steel and aluminum. The pinchoff edges should always be made of steel. With commodity resins, such as PE and PP, a sandblasted cavity surface can be used to aid in venting and also to provide a smooth surface on the blown part (a characteristic of the melt that prevents penetration of the 'rough' surface). With engineering resins, the surface of the cavity is reproduced precisely; so sandblasting does not aid venting. Conventional methods of mold venting through slits in the parting line are of minor importance or

Table 4.13 Guide to processing temperatures of plastics for blow molding

Plastic	Temperature (°C)
LDPE	130–180
MDPE	150–200
HDPE	160–220
HMWPE	180–230
PVC	190–205
PP	200–220
PS	280–300
PA	240–270
POM	150–280
SB	170–210
ASA	200–230
ABS	180–230
ABS/PC	230–250
PPE	240–250
PBT	245–260
PBT/PC	240–260
PUR	180–190

could even cause problems with these resins, but appropriate venting measures can be taken.

It is necessary to provide a heat control system for the mold to obtain the required part finish (Table 4.14). The mold surface heat depends on the resin being processed and is usually about 40–50 °C below the softening temperature. A higher mold heat means a longer cooling time, although engineering resins may require the higher heat to provide their highest-quality performance. But the effect of this heat control is not great enough to compensate for extrusion defects (Chapter 3). The shrinkage of engineering resins averages 0.7% and, because of their low melt viscosity in comparison with HDPE, is not directionally dependent.

Any scrap (flash, rejects, etc.) can be recycled. It is vital to granulate the material properly and prevent contamination. Processing of up to 100% dry regrind is possible, as well as blending with virgin resins. However, with increasing regrind content, melt viscosity is reduced, parison swell worsens, and the performance properties of the blown container may be reduced or unacceptable (Chapters 1, 2, 3, and 16).

The continuous EBM process, although well suited for most plastics, is the best process for PVCs. PVC can degrade rapidly if overheated slightly. The relatively slow uninterrupted flow of material in this process reduces the tendency for hot spots to occur, which would damage the material.

Table 4.14 Examples of recommended temperatures for cavities in blow molds

Plastic	Temperature	
	°C	°F
PE and PVC	15–30	59–85
PC	50–70	122–160
PP	30–60	85–140
PS	40–65	105–150
PMMA	40–60	105–140

IBM is heavily used for pharmaceutical and cosmetic bottles because the bottles are frequently small and precise neck finishes are frequently important; here IBM is more efficient than EBM. The plastics most commonly used are HDPE (very inert, low cost, forgiving material), PP, and PS. PS receives a degree of orientation which enhances impact resistance. IBM has been used for decades to fabricate these and other plastics; it was not used with PVC until the late 1970s. The use of PVC had to await the development of plastics that could take the heat of the process without degradation. Development of machinery was also a factor. New plastics and improved machinery allowed PET to grow in importance and replace other materials, particularly for packaging carbonated beverages. Depending on the type of production and output requirements, a one-step or two-step process for PET is available. The one-step process is used for lower production output, the two-step process for higher output. With careful temperature conditioning of an IM preform, PET can easily be given a biaxial orientation or stretched using IBM equipment.

Although PET usually lacks the required melt strength for EBM (Chapter 1), it can be processed when coextruded with other materials. Basic EBM applications use PETG, a noncrystalline glycol-modified polyester, but so far its noncrystalline molecular structure has not permitted biaxial orientation.

HDPE is the dominant material for EBM and IBM. PP was the preferred material for many applications. It is expected that PP will regain some market share as a result of new two-step EBM technology. With this process, it can produce environmentally friendly, transparent PP products that are cost-competitive with PET packaging.

LDPE is also easily processed by both methods, but LDPE applications are not as common. Ultrahigh molecular weight polyethylene (UHMWPE) is processed by EBM, especially where environmental stress-crack resistance is important. Like PVC, its heat sensitivity indicates continuous EBM, not intermittent EBM (rapid parison formation of this approach adds considerable shear heat; Chapter 1). Special PVC heat-stable compounds are routinely IBM, although the resultant cavitation and bottle size are limited. Rigid PVC can be stretched biaxially during EBM.

Nylons are available for IBM and EBM. They are also used as barrier layers in coextrusion BM. IBM and EBM can also be used on other materials. In a few years' time, automotive under-the-hood temperatures are expected to reach 400 °F (204 °C), so equipment manufacturers and resin suppliers are experimenting to blow mold high-temperature engineering resins that will overcome their rather poor melt strengths.

Coextrusion/coinjection

As explained in the first three chapters, multilayer constructions provided advantages in combining different materials to meet BM product requirements at low cost. Two or more plastics may be combined to produce packaging superior than each individually (Fig. 4.24).

OTHER PROCESSES

Another process somewhat related to IBM is displacement blow molding or dip blow molding. Normally used for small containers, it provides some of the advantages of IBM with lower molded-in stresses. A premeasured amount of melted plastic is deposited into a cupel the shape of a parison preform. A core rod is inserted into the cupel, displacing the melt and packing it into the neck-finish area [3].

Many related operations have been used to improve blow molding. They range from in-mold labeling to very complex and irregular shapes (Figs 4.25 to 4.27).

The most advanced forms of EBM technology use special or different mold designs. Called by different names, such as three-dimensional (3D) EBM, they can fabricate complex geometric parts that even have acute angles. Three-dimensional EBM has been used in the past, but only recently have products become marketable. It is an intermittent process where the parison is dropped from an accumulator head and deformed or manipulated to fit the contours of the mold cavity; the parison does not make the conventional straight drop. The parison can follow horizontal as well as vertical positions – at any angle to the parison. Little or none of the parison lies across the mold face, so parts have little or no flash.

Manipulation of the parison is accomplished by different techniques. One method has the accumulator head moving to follow the contours of an almost horizontal mold. Another method has the clamping platens moving to position the mold under the parison. There is a technique

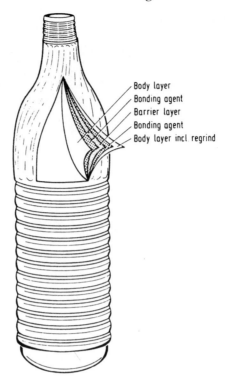

Body layer
Bonding agent
Barrier layer
Bonding agent
Body layer incl regrind

Figure 4.24 Coextrusion blow molding can provide flash-free multiple layers with easy, high-speed production; six or more layers can be produced at a time.

where the vertical parison is gripped by successive robotically controlled mold sections starting at the bottom and moving up.

This 3D technique has been used successfully in molding very complex automotive air ducts because it saves materials and is also capable of coextrusion; it combines parts that have hard and soft/flexible sections. It can contain flexible bellows, etc., and permits the use of inserts.

Some formed plastics have been produced by IBM, e.g., PS containing CO_2. Parts are blow molded and expanded simultaneously.

CONTROLS

Different types of microprocessor-based modules control BM machines and melt parameters, ranging from single to multiple functions. The modules interact at high speeds, coordinating process variables such as heat, timing, parison or preform molding speed and pressure, and melt wall thickness. The principles are similar to controlling injection and extrusion machines (Chapters 2 and 3).

Control technology is used to improve machine cycle rates, as in employing proportional hydraulics to safely speed up mold movements. In

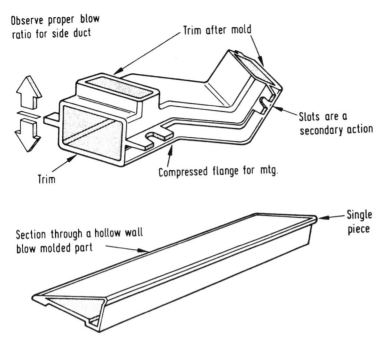

Figure 4.25 A blow-molded air duct for an auto spoiler.

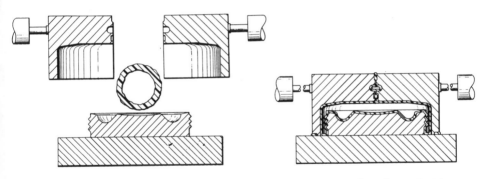

Figure 4.26 A mold used in extrusion blow molding a complex shape that includes a threaded forming core. It shows the three-part mold in the open and closed positions, with the blow pin located in the top two sections of the mold.

addition, production monitoring systems have become part of some BM plants, helping managers make effective decisions. These improvements in monitoring and controlling have contributed significantly to the manufacture of products with zero defects and to profits.

BM controls closely regulate many process variables such as viscosity, stretch, orientation, and swell. Additional productivity in a BM operation can arise from the coordination of various functions in a single control system. Controls clearly play an important role, but they are not the

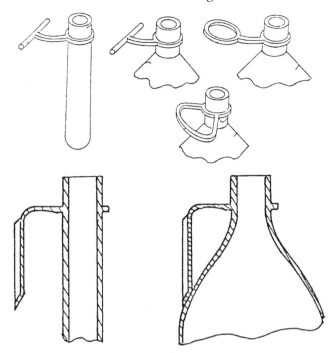

Figure 4.27 Ring, strap, and integral pouring handles.

answer to many processing problems, as explained in the first three chapters. They have to be properly integrated with the complete process, as summarized in Fig. 1.1 (on page 3) and Table 4.15.

COSTING

Table 4.16 provides a cost comparison of BM techniques for PVC and PET materials. Cost advantages can only be obtained by periodic updating of BM factors for processing and operating characteristics. Any guide should include factors such as plastic flow (mainly determined by basic properties), parison or preform sag and swell, molding cycle time, most favorable melt temperature, and blowing pressure. Upgrading blow molding machinery, particularly auxiliary equipment (Chapter 16), is important for producing thinner-walled products that meet their performance requirements at lower molding costs.

The molding properties to evaluate should include stiffness or flexibility, resistance to environmental stress cracking (for certain plastics such as PE), resistance to attack and penetration by chemicals, appearance (especially gloss and surface defects), uniformity of wall thickness, distortion (shrinkage and warpage), strength of the parting line, shape and thickness of the weld line (for EBM), part weight, discoloration, deteriora-

Table 4.15 Factors to consider when reviewing the complete blow molding process

Part design	Part configuration (size/shape) relate shape to flow of melt in mold to meet performance requirements, which should at least include tolerances
Material	Chemical structure; molecular weight amount and type of fillers/additives; heat history; storage; handling
Mold design	Number of cavities; layout and size of cavities cooling lines, side actions, knockout pins, etc., relate layout to maximize proper performance of melt and cooling flow patterns to meet part performance requirements, preengineer design to minimize wear and deformation of mold (use proper steels, aluminum, etc.), and layout cooling lines to meet temperature-to-time cooling rate of plastics (particularly crystalline types)
Machine capability	Accuracy and repeatability of temperature, time, velocity and pressure controls of extrusion injection unit; accuracy and repeatability of clampling force; flatness and parallelism of platens; even distribution of clamping; repeatability of controlling pressure and temperature of oil; minimizing oil temperature variation; no oil contamination (by the time you see oil contamination, damage to the hydraulic system could have already occurred); machine properly leveled
Molding cycle	Setting up the complete molding cycle to repeatedly meet performance at the lowest cost by interrelating material, machine, and mold controls such as those listed in this chapter

tion due to temperature fluctuations or UV radiation, odor, taste and/or toxicity.

TROUBLESHOOTING

There is a logical approach to setting up a troubleshooting guide. The troubleshooting sections of the first three chapters are equally relevant to blow molding, particularly Chapters 2 and 3. Table 4.17 lists some of the common BM problems with information on causes and solutions.

In setting up and operating a machine, one should not make too many adjustments or changes at the same time. The goal is to make them one at a time and to wait for any change to occur, allowing at least 15 min. Record the action taken and the result, as well as the machine used, the resin used, other machine/control settings, and the operator. This type of log helps one to study and evaluate the operating performance. It is not a new approach, except perhaps in its departure from any formal record

Table 4.16 Manufacturing cost comparison example of 16 fl. oz (454 g) blow-molded bottles[a]

	Standard Extrusion blow molding: two-parison head, fourfold	Stretch blow molding PVC: two single-parison heads, fourfold	Stretch blow molding PET
1.0 Machine cost ($)			
Including head, molds, ancillaries (license fee, stretch PVC and PET)	270 000	450 000	850 000
2.0 Hourly machine costs ($ h^{-1})			
Five-year depreciation (30 000 h)	9.00	14.85	28.33
Five-year financing, cost at 12.5%	2.80	4.65	10.20
Labor (1 worker)	13.00	13.00	13.00
Energy at $0.06 per kWh	2.50	5.35	11.00
Floor space	1.50	2.00	4.00
Maintenance and consumables	2.25	3.75	4.50
Total	31.05	43.60	71.03
3.0 Bottle specs (hourly/annual production)			
3.1 16 fl.oz finish weight (454 g)			
Regular 37 g (1.3 oz)			
Stretch PVC 20 g (0.7 oz)			
Stretch PET 20 g (0.7 oz)			
Cycle time (Bottles per hour)	8.4 s (1714)	7.5 s (1920)	(4000)
Bottles per year (millions)	10 286	11 520	24 000
4.0 Annual costs ($ y^{-1})			
4.1 16 fl.oz (454 g)			
Resin 37 g	585 200		
$0.70 lb^{-1} ($1.54 kg^{-1})			
20 g		334 950	
$0.66 lb^{-1} ($1.46 kg^{-1})			
20 g			634 360
$0.60 lb^{-1} ($1.32 kg^{-1})			
Machine costs	186 300	261 600	426 180
Total	771 500	596 550	1 060 540
Annual royalty to Du Pont (PET)			
Cost per thousand	75.00	51.78	45.44

[a] Figures are not be to considered as absolute costs, but rather **reflect comparisons** between various machine options. All calculations are based upon 100% efficiency. All bottle weights are finish weights (flash being considered as 100% reusable).

Table 4.17 Guide to common blow molding problems

Problem	Cause(s)	Solution(s)
Rough parison; orange peel	Melt fracture; melt temperature too low	Polish all tooling Raise melt temperature
Poor gloss	Mold too cold	Increase die surface temperature
Black specks in part	Contamination from degraded material	Purge to clean system Keep materials clean
Gels in parison	Excessive fines in regrind Moisture in resin Screw too deep	Screen out regrind fines Dry material before use Use higher-shear screw and lower barrel temperatures
Bubbles in wall	Moisture in trapped air	Increase extrusion pressure If moisture, lower screw speed; reduce feed-zone temperature
Uneven wall thickness circumferentially	Pin not centered in die ring	Adjust die-pin position
Parison hooking	Head temperature not uniform	Stagger heater-band gaps on head
Incomplete blow	Extrusion rate too high Blowup air pressure Blowup time too short Parison is cut at pinchoff	Reduce screw speed Increase blow-air pressure Reduce mold-closing speed
Holes in parison and/or bottles	Contaiminated or degraded resin Trapped air Moisture in resin	Purge and clean tooling and screw Let extruder run for a few minutes Dry the resin
Parison stretches	Resin melt index too high Melt temperature too high	Use lower melt index Reduce melt temperature Increase screw speed Boost extrusion rate
Parison blowout	Blowup too rapid Melt temperature too high Pinchoff too sharp Blowup ratio too high	Program blowup start with low air pressure and increase Align molds Use larger parison

Table 4.17 *Continued*

Problem	Cause(s)	Solution(s)
Die, weld, and spider lines in parison	Damaged die ring Mandrel spider legs cause improper knitting	Repair or replace die tooling Streamline spider legs Reduce die temperature to increase back pressure
	Contamination from material	Clean diehead
Webbing in handle	Parison walls touch when mold closes Wrong parison diameter	Align parison closer to handle side of mold Increase die diameter Reduce melt temperature
Rocker bottoms	Blowing air not vented before mold opens Insufficient cooling	Increase air exhaust time Clean cooling channels of mold Increase blow time
Tails not pulled	Parison is too short Plastic or foreign matter holding mold	Lengthen the parison by increasing extruder speed Clean mold parting surfaces
Bottles thin in various areas	Parison curling Parison too long or short	Adjust die ring concentricity Increase/decrease extruder speed and adjust parison temperature Reduce head temperature
Molds not separating from neck finish	Cutting ring is dull Poor contact between cutter ring and striker plate	Sharpen or replace cutting sleeve Increase overstroke and downward pressure of blow pin
Weak shoulders on bottles	Parison sag Parisons too long or short Container to light	Reduce melt temperature and decrease/increase extrusion rate Program increased weight
Slanted neck finish	Blow pin/cutter entry too deep Parison folding over	Raise blow pin until it just cuts Replace dull knife blade Adjust knife-cut delay timer
Parts sticking in mold	Mold too hot Cycle too short	Improve mold cooling Lengthen cycle

Table 4.17 *Continued*

Problem	Cause(s)	Solution(s)
Mold parting line indented in part	Blowup air introduced prematurely	Delay blowup
	Hooking parison	Reduce mold temperature
Handle missing	Insufficient die swell	Position parison closer to handle
		Use larger tooling
Sink marks	Air trapped in mold	Improve venting
		Lower mold temperature
Parison tails	Parison is too long	Reduce extruder speed
	Pinchoff improperly designed	Design pinchoff to compression cool tail
Poor detail definition	Blow-air pressure too low	Increase blow-air pressure and blow time
	Poor mold venting	Improve venting
	Cold mold	Increase mold temperature

Coextrusion blow molding
Most of the above tips also apply to blow molding multilayer containers

Skips in barrier layer	Temperature of barrier material too high	Reduce barrier material temperature
	Pressure fluctuations at extruder	Maintain constant pressure at extruder screw tip
	Degraded material in head	Purge head and/or extruder
Barrier integrity of handle breached	Too little material in handle	Program more material into handle and pinchoff area
	Poor pinchoff	
Layer separation, blistering or bubbles in container	Adhesive layer too cold, did not flow around structure; adhesive too hot to stick to adjacent layer	Adjust temperature of adhesive material up or down
	Adhesive layer cooled too fast	Raise mold temperature to prevent fast cooldown
	Moisture in materials	Dry materials

keeping. It helps one to eliminate duplication and unnecessary performances, providing factual information for a troubleshooting guide. Some processors have successfully reduced downtime by employing a specially trained person on each shift to help the operator troubleshoot processing problems.

5

Calendering

___

INTRODUCTION

The calendering process is used to produce plastic films and sheets. It melts the plastic then passes the pastelike melt through the nips of two or more precision-heated, counterrotating, speed-controlled rolls into webs of specific thickness and width, as shown in Figs 5.1 and 5.2. The parallel rolls have extremely flat surfaces and rotate at the same speed or at slightly different speeds. The web may be polished or embossed (depending on roll surface), rigid or flexible (a range of material properties are used). Although plastic forming occurs in the calender itself, downstream equipment is needed to produce the TP film or sheet. Upstream of the calender, a mixer blends the raw material, usually in powdered form, with the desired additives, such as plasticizers or fillers (Chapter 1). After plasticizing, the pastelike melt passes through the multiple-roll mill.

Heavy-gauge products, 0.02 in. (0.6 mm) off the rolls, are drawn down or stretched at takeoff as little as consistent with good stripping from the final calender roll. Lighter-gauge film, 0.007 in. (0.18 mm) off the calender, is typically drawn down at a ratio of 1.2:1 and wound up at a thickness of 0.006 in. (0.15 mm).

Calenders may consist of between two and seven rolls, possibly more. They are characterized by the number of rolls and their arrangement: I, Z, or inverted L. Most popular are the four-roll inverted-L calender and the Z calender. Z calenders have the advantages of lower heat loss in the sheet (because of shorter travel) and a simpler construction. They are simpler to construct because they need less compensation for roll bending; this is because there are no more than two rolls in any vertical line (as opposed to three in a four-roll inverted-L calender).

Other variations in these massive, multimillion-dollar calender lines are dictated by the very high forces exerted on the rolls to squeeze the plastic melt into a thin web. High forces can bend the rolls, producing a web

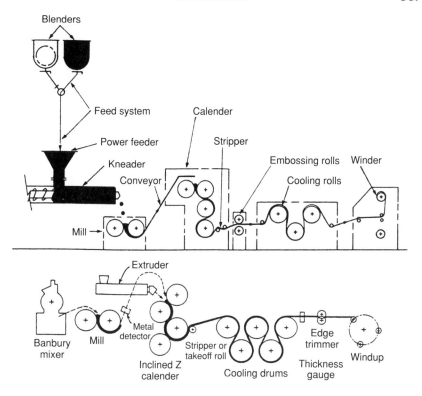

Figure 5.1 Calendering lines.

thicker in the middle than at the edges. This can be counteracted by different methods: (1) crowned rolls, which have a greater diameter in the middle than at the edges; (2) crossing the rolls slightly, thus increasing the nip opening at either end of the rolls; and (3) roll bending, where a bending moment is applied to the end of each roll by having a second bearing on each roll neck, which is then loaded by a hydraulic cylinder. Calenders call for high temperatures, with little variation across the rolls during the application of high pressures.

The high forces are necessary to squeeze the plastic mass into a very thin film; any unevenness in the forces along the roll is reflected as variations in the film thickness. One cause of pressure fluctuations is too much clearance in the bearings, which can be resolved either by pulling back the rolls against one side of the bearing or by using tapered roller bearings.

After forming, through the multiple rolls of the calender, the film or sheet is cooled by passing through precision-surfaced cooling rolls at precisely controlled temperatures and/or a cooling tower. Thickness gauges are usually located within this cooling section of the production

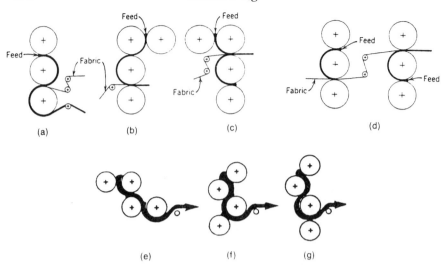

Figure 5.2 Arrangements of parallel rolls for calender coating: (a) one-sided coating with vertical three-roll; (b) one-sided coating with inverted-L four-roll; (c) double-sided coating with inverted-L four-roll; (d) double-sided coating with two three-rolls; (e) one-sided coating with Z-configuration four-roll; (f) one-sided coating with S-configuration four-roll; and (g) one-sided coating with modified-S four-roll.

line; the feedback they provide makes thickness corrections automatically. Gauge controls can also be used in the neckdown section, where the hot plastic leaves the calender rolls and transfers to the cooling rolls. After cooling, the plastic is trimmed at the edges and wound. Trim may account for up to 5% of the width, depending on the line's operating efficiency, and most trim is immediately directed back through a granulator and blended with the virgin material (Chapter 16).

This is the typical sequence for a calender line, but there are many variations to suit the end product. Auxiliary equipment can be included, such as annealing, quench tank, decorating, slitting and heat sealing, and orienting (uniaxial or biaxial stretching).

Calendering was developed over a century ago to produce natural rubber products; it is now used principally with TPs. A major product is elastomeric tire-fabric coating. Calendering is a highly developed art, recently elucidated by combining the complexity of melt behavior with the mechanics of rotating machines. With this understanding comes the ability to make calenders more productive by increasing their speed, to produce films with tighter thickness tolerances and greater uniformity, and to handle thicker sheets more effectively.

MATERIALS

A wide variety of plastics can be used; about 80 wt% is PVC but others are ABS, PE, PP, and styrenes. When calendering other plastics, there is a trade-off in economy and quality. The basic plastic limitation of the calendering process is the need to have a sufficiently broad melt index to allow a heat range for the process. This permits the material to have a relatively high viscosity in the banks of the calender (banks indicating where two rolls meet, or the nip of the rolls). As a result of the viscosity, a shear effect can be developed throughout the process, especially between the calender rolls. Thus, the calender forms the web as a continuous 'extrusion' between the rolls. Unlike the process in an extruder (Chapter 3) or an injection molder (Chapter 2), the plastic mass cannot be confined when being calendered. Because of the lack of confinement, the shear effect and a broad melt band are essential aspects of calendering.

The blending or compounding of the plastic with different additives and fillers is a critical part of the process. The blending must produce a uniformly colored and stabilized product, in powder form. After blending, the rate of consumption dictates the temperature of the melt. Because the plastic is processed between the process heat and its critical heat of degradation, the time at heat becomes extremely critical and an important part of the process. For example, the processor will minimize the amount of melt in the bank (nip) of the rolls. The residence time of the plastic flux at high heat must be limited. PVC is especially heat sensitive.

The plastic mix fed to the calender may be a simple hot melt, as with PE. For PVC it is obtained by premixing the plastic, stabilizers, plasticizers, etc., in ribbon blenders, then passing the blend to a Banbury mixer where the mass is gelled for about 5–10 min at 120–160 °C (248–320 °F). The gelated lumps are made into a rough web on a two-roll mill and the web is fed to the calender. If fabric, paper, or other material is fed through the rolls, plastic can be pressed into the surface of the web; then the calender becomes a coating machine (Fig. 5.2).

Ingredients for PVC film can include solid and liquid raw materials. Solids are weighed in batches and dropped into a high-speed mixer. Liquids are metered into the mixer by piston pumps or other volumetric metering devices. A batch mixing system is generally inline with a continuous fluxing system, starting at the holding bin. Some operations do not use batch mixers for blending. Continuous metering systems, using loss-in-weight feeders and pumps, meter directly into the fluxing machine, a continuous mixer.

Films

The PVC film industry distinguishes between two calender types and roll arrangements that are rigid and flexible PVC. Rigid PVC manufacturers prefer the L configuration with four to seven rolls being fed from the floor level. Since there are no disturbing vapors from lower calender rolls within the pickoff area, i.e., condensation on rolls and film, it is preferable to have the pickoff rolls on an elevated level.

Flexible PVC is commonly fabricated using a four-roll inverted-L calender or an F calender. These systems enable the plasticizer-saturated vapors to escape via the suction hood above the calender, where they are filtered before being released into the atmosphere.

A universal five-roll L calender is used for rigid and flexible PVC film. It provides heat stability and superior film control for good surface appearance. The major difference between this universal calender and the other calenders is in mounting and placement of the first roll. Although the first roll is generally mounted horizontally beside the second roll, the latest design still has the first roll mounted horizontally but now it is next to the third roll, resembling a four-roll L calender. The second roll remains in action underneath the third roll, so four gaps are still available. Roll/film wrap length, however, is reduced so dwell time is 40% less, decreasing PVC heat loss and shrinkage.

Modern designs for rigid or flexible PVC lines currently show few significant differences. Both consist of the same roll groups, but arranged at different elevation levels. The pickoff section has up to 15 rolls, each with an OD of 6 in. (15 cm), driven and temperature controlled in sections. This roll arrangement drops film vertically into an embosser with three rolls (the embossing roll itself, a cooled rubber roll, and a contact cooling roll to the rubber roll). For tempering and cooling the film, another set of up to 19 roll is arranged following the embosser; they have diameters of 6–10 in. (15–25 cm). These rolls are driven and temperature controlled in sections of between two and five rolls. Temperature accuracy can be controlled within ±2 °F (±1 °C). A beta-gauge controls film or sheet thickness.

Recycled

Handling scrap and cold trim from the product line can pose a very difficult problem. The scrap and trim could represent 10–40 wt% of the mix with virgin material. The actual amount depends on the width of the calender in relation to the sheet width. The flux rate and energy required to remelt the scrap are considerably less than required to flux the virgin plastic. With this differential, there is the potential danger of material decomposition whenever scrap is processed. But optimum uniformity and product standardization require a blend of new and old material.

Reprocessed material is best added to the blender when a standard can be established, although some plants feed it directly to the fluxing equipment. Careful control of the scrap percentage in the total mix is essential to obtain a quality-controlled product. These lines are expensive to set up and operate. They are most economically operated on long runs, where they compete with film and sheet extruders (Chapter 3). Extreme care is required to ensure the material or the processing area is not contaminated. Any foreign matter can severely damage the expensive and highly polished rolls as well as causing other serious damage to the line.

Contamination

Vital to this process is the removal of any metal, including microscopic metallic particles that usually occur in granulated scrap. These contaminants would destroy the very expensive precision surfaces of the calender rolls. Damage would also occur to the downstream cooling rolls. All kinds of precautions are taken to remove any contaminants, starting with the incoming raw materials, the fillers, and the additives. Materials are passed through an extensive battery of screens and metal detectors.

The compound is usually fed through an extruder with multiple screens. Metal detectors tend to be used along the flow line of the plastic on its way from the compounder to the first nip rolls. Detectors are included under the belt that conveys the final melt, perhaps in a ropelike form of diameter 2–4 in. (5–10 cm).

FLUXING AND FEEDING

Fluxing is a plastic composition to improve flow; fusion is the heating of the vinyl compound to produce a homogeneous mixture. Fluxing units used in calender lines for preparing thermoplastics include batch-type Banbury mixers and Farrel continuous mixers (FCMs), Buss Ko-Kneaders (BKKs) and planetary gear extruders (PGE). The dry blend is fed into the mixer (extruder). Excellent mixing within a short dwell time and heat transfer control contribute to an improved product. During fluxing, each particle receives the same 'gentle' treatment, generating less heat history (Chapter 1) and producing more uniform feed rate, color, gauge, and surface. The feed can discharge onto a two-roll mill. Operating this way, it provides for a second fluxing device, mainly for working in scrap or for convenience as a buffer.

The stock delivered to the first calender nip needs to be well fused, homogeneous in composition, and relatively uniform in temperature. The optimum average temperature for good fusion depends on the formulation. A rigid PVC formula based on medium molecular weight resin (intrinsic viscosity of 0.90–1.15 [9]) has a typical optimum temperature of 356–374 °F (180–190 °C) at the first calender nip. For best calendering there should be no cold volume elements below 347 °F (175 °C) and no hot spots

Table 5.1 Examples of flexible PVC calendering conditions

Roll no.	Heavy-gauge product				Light-gauge product			
	Roll temp.		Roll speed		Roll temp.		Roll speed	
	°F	°C	ft min^{-1}	m min^{-1}	°F	°C	ft min^{-1}	m min^{-1}
1	347	175	125	38	338	170	263	80
2	352	178	128	39	343	173	269	82
3	353	181	138	42	349	176	279	85
4	363	184	148	45	354	179	289	88

above 383 °F (195 °C). These typical temperature readings are necessary for close control of fusion and mixing conditions; they are based on studies of flow effects and their relationship with volume and element size. This interaction depends on stock temperature and in turn on the performance of PVC melts.

Flexible PVC is normally calendered at temperatures 50–68 °F (10–20 °C) lower than for rigid PVC. Typical calendering conditions for unsupported flexible PVC film and sheet are shown in Table 5.1. In flexible PVC production, a short single-screw extruder, acting as a strainer, filters out contaminants from stock before reaching the calender. This method is not applicable for rigid PVC because it drastically increases the head pressure and the consequent overheating would cause the stock to decompose.

PROCESSING CONDITIONS

Preparation of stock for calendering, conditions on the calender, the take-off, thickness measurements, and control to windup must all be adapted to the plastic system being calendered. Other considerations include whether the plastic is laminated to a fabric, etc., on the calender; whether the web is embossed or slit inline; what finish is required (glossy, matt or semimatt); and whether the product needs special properties such as optical clarity, biaxial orientation (Chapters 1 and 3) or very low strain recovery.

General processing considerations include temperature control, calender speed, finish, gauge, orientation, and embossing. Significant progress in calendering is due to improvements in control of the overall heating and cooling processes involved in the complete calender lines (trains), as well as the speed and tension of the web at each stage of processing.

Design aspects

Design improvements in calenders have advanced the art of calendering. A preloading device eliminates the main bearing clearance (play) of the penultimate calender roll. A four-roll L calender also eliminates the third roll. Another aid is an air knife to firmly hold down the film on the first section of pickoff rolls and to minimize neck-in, shrinkage and plate-out. Neck-in is the difference between the width of the hot web as it leaves the hot rolls and the width of the cooler web downstream. This 'shrinkage' dimension is used as a guide to ensure the processing conditions are correct. Plate-out is an objectionable coating gradually formed on calendering rolls during processing. It is caused by the extraction and deposition of certain components in the compound such as pigments, lubricants, plasticizers, and stabilizers; it requires immediate cleanup.

Another improvement, the compact universal embosser, is used for horizontal or vertical film drop, embossing on the top or bottom side. The air knife was developed to meet rising quality demands and increased production speeds. At high speeds, more air is drawn between the film and the pickoff roll, preventing the film from making contact with the roll. Since pickoff roll to temperatures are lower than calender roll temperatures, liquids in the compound tend to condense and transfer from the roll to the film, causing surface deficiencies. The air knife ensures contact between the film and the pickoff roll. Air blown over the film creates pressure, preventing air entrapment between the film and the pickoff roll.

Surface finishing

Calenders provide styling or finishing on films and sheets. PVC surface calendered goods require depths of matt or provide an even roughness, which can be measured with a device such as a profilometer. Different markets use different terms to describe their finishes, e.g., satin, Sheffield, and light matt. A one-sided matt finish may be applied with an embossing roll. A sharper, deeper, and more precise matt finish can be applied to film or sheet on the calender. To make two-sided matt products, such as the big market for two-sided credit card stock, the last two calender rolls are matt. Only the last calender roll is matt for one-sided matt goods meeting critical specifications, such as special print film and one-sided credit card stock. To produce these products, the last two calender rolls are sandblasted with aluminum oxide grit of controlled particle size, such as 120–180 mesh. The heavier matts, coarser grit, and more passes of the sandblasting action on the rolls are required. Proper selection of metal for the calender rolls is also necessary. Chilled cast-iron rolls take a sharp matt, which wears off slowly.

The principle used in embossing PVC film and sheet is to deliver a relatively hot web to a relatively cold embossing roll. This action imparts the desired surface finish to the web and freezes it by rapidly cooling the embossment pattern. Development continues for embossing a wide range of rigid, semirigid, and flexible PVC compounds. Film temperature, roll temperature, and pressure are optimized empirically for each compound formulation and pattern to be embossed. The steel embossing roll and the rubber pickup roll are internally cooled by a liquid medium; both of them require a very uniform surface temperature to develop the embossing pattern. But the rubber pickup roll tends to stick to a hot PVC film unless it is cooled externally as well as internally. External cooling in a wet trough is required because most rubbers have poor heat-transfer properties, so poor that a nonexternally cooled backup roll surface would rise almost to the processing temperature of the PVC film.

The surface of the sheet is determined on the last roll and may be glossy, matt, or embossed. After leaving the calender, the sheet is cooled by passing it over cooling rolls then through a thickness gauge (e.g., a beta-ray device) prior to being wound up. Quick roll-change is used for the last two calender rolls. Changes from high gloss to matt, or to different profiles, can now be made in about 10 h or less; in the past it took at least 80 h. Embosser rolls can be changed in minutes instead of hours.

Coating

One special and important application of the calender is the coating of paper, woven and nonwoven textiles, plastic film and sheet (Fig. 5.2). A calender with three rolls is usually sufficient for one-sided coating, but four rolls are used for extremely thin coatings. Double-sided coating can be applied simultaneously on both sides of a fabric, using a four-roll calender, or sequentially by two three-roll calenders.

Roll configurations are shown in Fig. 5.2. The most popular configuration is the inverted L. However, these and other configurations are used to meet different product requirements.

So-called frictional calender coating takes a material such as an elastomer and forces it into the interstices of the woven or cord fabrics while passing through the rolls of the calender.

CONTROLS

Many control features in calenders provide versatility, better quality, and higher operating rates. Complete control of friction ratios gives good tracking of stock via individual drives on all calender rolls. Close tolerances on film profiles are obtained from axis crossing, roll straightening, and gap profiling of calender rolls; greater profiling accuracy is obtained by hot grinding of calender rolls at operating temperatures. Better control over stock temperature is achieved by using drilled rolls with heat-transfer fluid circulated through them at high rates and by using accurate (±3–4 °F, ±2 °C) systems to control the fluid temperature.

Automatic web-thickness profile control, well known since the 1940s in extrusion, can be used in calendering plastics such as PVC. As an example, perhaps 64–76 individually temperature-controlled air nozzles, arranged in a row, create a series of zones across the width of the sheet. The web thickness can be controlled by adjusting the coating thickness applied to each zone.

Calender bowl deflection can occur. It is the distortion suffered by calender rolls resulting from the pressure of the plastic or elastomer running between them. If not corrected, the deflection produces film or sheets thicker in the middle than at the edges. The machines are set up so the required adjustments can be made to correct this distortion. As an example, less deflection at high operating conditions can be achieved by the use of stiffer rolls, based on higher-modulus steels or dual-metal construction.

In older calenders severe cases of 'roll float' caused by changes in the balance of forces on a roll were sometimes encountered. Preloading devices reduce roll floating and thus provide better gap control between pairs of rolls, especially during startup. Film thickness can be measured automatically with improved feedback control to the calender and drawdown system. Programming temperature changes at rates of 4.5 °F min^{-1} (2.5 °C min^{-1}) or less will avoid roll distortion caused by thermal shock.

With vertical stacks of rolls, the separating forces exerted by the stock at the nips may be adjusted by the operating conditions, so the middle rolls defect very little. As a first approximation, it may be assumed that the third roll on a four-roll L calender or an inverted-L calender does not deflect in balanced operation. Deflection of the fourth roll primarily determines gauge variation across the width of the web. For thin plastic webs, roll-separating forces in the final nip may be as high as 6000 lb in.$^{-1}$ (1072 kg cm^{-1}), which may cause considerable deflection of the final calender roll. Unless this is compensated, gauge variation will occur across the web.

Good drawdown and pickoff control is achieved by using tightly spaced multiroll pickoffs with grouped roll-speed controls. Helical cavities in each pickoff roll allow high circulation rates of heat-transfer fluid. Quick-opening devices on all nips reduce damage to rolls from running together in the event of power failure or stock runouts. Mechanical, square-cut devices with automated tension control and roll change as the roll reaches a preset diameter ensure better windup.

Compensation for deflection of calender rolls is normally achieved by combining the three methods of axis crossing and skewing, roll straightening (also called roll counterbending), and roll crown (also called roll profile). Roll crown may be ground or blocked.

Axis crossing increases the roll separation at the ends but does not change it in the middle. Roll straightening involves the application of external bending moments to both ends of the shaft of a calender roll in order to compensate for deflection (+) or excessive roll crown (−). With the usual approach, positive straightening is used to compensate for deflection caused by roll-separating forces in the nip of a calender; negative straightening is used to reduce excessive fixed crown on a roll, in which it has been ground.

Radiation film-thickness gauges are now used with computer systems that process the raw data and print out detailed production reports. They also control the gap in the last calender nip by varying the opening of the rolls, the axis crossing, and the roll straightening on the final calender roll. In order for these thickness gauges (also called beta-gauges) to work effectively, the web composition must be exactly known and accurately controlled. The gauges measure attenuation of radiation, which is proportional to the quantity of matter in the measured area. For known formula compositions, this can be translated into web thickness. Although good results have been reported, these gauges cannot compensate for deficiencies in older calenders or for improper control of the compounding formulations.

ADVANTAGES OF CALENDERING OVER EXTRUSION

Calendered sheet is usually less glossy than extruded material (Chapter 3). Calendering may be preferable for certain applications requiring its higher tensile properties and unusually close gauge control. Extrusion of colored films or sheets requires the extruder to be cleaned and purged when changing colors. A calender requires a minimum of cleaning between color changes. Calendering is definitely used for long production runs to be economically profitable, producing smooth and other finishes at higher speeds. Applications include window-shade stock, electrical tape, PVC tile, PE-coated fabric, tire-fabric coating, and swimming-pool liner.

COSTING

Most output in plastics is with PVC. Flexible film and sheet accounts for the greatest volume. The capital cost for a new calendering line will average $4–5 million, not including building costs. The sum of the fixed costs and the variable costs for operating a new line in the United States

Table 5.2 Examples for producing PVC film and sheet by different manufacturing methods

	Calender	Extruder–calender	Blown film	Flex-lip extruder	Plastisol cast	Melt roll
Lines installed, USA	155	2	90	40	60	5
Relative resin cost	lowest	low	higher	higher	higest	low
Machine cost ($ million)	1–6	1–2.5	0.3–0.6	0.3–0.6	0.3–0.7	0.3–1.3
Rate and range (lb h⁻¹)	800–8000	500–1500	600 (4½ in.)	750 (4½ in.)	750	100–1000
Product gauge range (in.)	0.002–0.050	0.002–0.005	0.001–0.003	0.001–0.125	0.001–0.012	0.0015–0.020
Sheet accuracy (%)	3 (1–5)	3 (1–5)	10	10	7	5 (2–10)
Time to heat (hr)	6	5	3	3	½	3
Time for 'on stream'	2–5 min	10 min	2 h	5 h	10 min	2–5 min
Gauge adjust time	seconds	seconds	5–30 min	5–30 min	seconds	1 min
Autogauging capability	yes	yes	no	no	no	no
Color or product change time	5–30 min	10–40 min	30–60 min	30–60 min	15 min	30–60 min
Windup speed (yd min⁻¹): average (max.)	80 (150)	60 (80)	15 (20)	15 (30)	20 (40)	20 (30)
Limitations	High capital cost, heat time	Lower rate, versatility problem	Poor accuracy, long on stream time, low rate, degradation, reduced versatility		Fumes, inefficiency, high energy cost, resin cost, release paper cost	Reduced rate and range, soft materials only slow manual gauge change
Applications and advantages	Versatility, high rate, accuracy, ease and, adjustment ease at reprocess	Accuracy, gauge adjust, reduced cost	Low investment, multiplant capability, thin gauge (0.003 in. and under) and heavy gauge (0.050–0.125 in.)		Grain retention (pattern cast in), soft hand and drape	Good on wall covering, thin material, coated fabric, accuracy, reduced investment

amount to $450–550 per hour. Taking PVC as an example, these lines have a throughput of 90–6000 lb h^{-1} (400–2700 kg h^{-1}); the average range of componding costs for calendering PVC is about $1.08–2.47 per pound ($0.49–1.12 per kilogram).

Not including material costs, the expense of operating a large calender line may be $350–560 per hour, depending on the degree of automation, investment, labor, power costs, and overheads. For PVC sheet 7.1 ft (2.13 m) wide and 0.02 in. (0.51 mm) thick traveling at 91 ft min^{-1} (27.4 m min^{-1}) and wound up at a rate of about 1076 lb h^{-1} (489 kg h^{-1}), the conversion cost is about $0.46 per pound ($0.21 per kilogram). These costs are based on PVC resin at $1.70 per pound ($0.77 per kilogram) and plasticizer at $2.42 per pound ($1.10 per kilogram). Table 5.2 provides a general manufacturing comparison of calendering with combination extruder/calender, extrusion blown film, extruder with flexible-lip die, plastisol-cast, and melt roll.

Calendering costs that will affect product costs include (1) bank marks and shiny patches on the surface; (2) cold marks (crow's feet) caused by the stock being too cold; (3) blistering due to high temperature or a large bank of rolls; thicker sheets are more vulnerable to blistering; (4) pinholes caused by foreign matter or by the presence of unplasticized plastic particles; and (5) watermarking usually due to lubrication contamination.

To reduce the cost of products, aim to calender under conditions that yield the highest windup speed consistent with good quality. However, there are numerous factors which limit calender speed. At a given speed, more frictional heat is generated by the more highly filled and rigid stocks, and by reducing web thickness. Excessive speed produces degradation or sticking of many products on which the stock temperature is limited. Optimal speeds of the final calender roll for double-polished, clear melt viscosities at a given temperature degrade at about the same maximum temperatures as the rigid stocks. Air occlusion limits the speed for some heavy-gauge products, whereas the matt quality or micro-roughness of the film, or melt fracture in the last calender nip, limits the speed on others.

Bank marks are minimized by optimizing formulations, calendering speeds, and roll temperatures so as to obtain the most orderly behavior of the rolling banks of stock at the calender-nip entrances. Proper use of drawdown permits windups to be run substantially faster than the final calender roll on many thin, unsupported film products. Calenders and takeoffs are run almost synchronously on heavy-gauge products.

Films and sheets with a high gloss taken off a highly polished final calender roll tend to stick to the roll more than their matt counterparts. Very soft webs also tend to stick to the final calender roll. The fastest calender speeds are generally obtained using medium stiffness, moderately filled, matt products in a median thickness range.

The amount of edge cutting influences costs. Edge cuts in excess of 3 in. (76 mm) are normal to avoid undesirable edge effects. On critical products it is realized that close control of the rework ratio is necessary for uniformity of output. Excess rework from critical products is used in those whose specifications permit this practice [9].

TROUBLESHOOTING

Earlier sections of this chapter cover an extensive range of calendering problems and their remedies; collectively they should provide an adequate troubleshooting guide.

6

Casting, encapsulation, and potting

OVERVIEW

Casting, encapsulation, and potting, these terms are often interchangeable; they interrelate very closely to describe processes and performance. However, there are differences. Even though TSs or TPs may be used, it is TSs that predominate, usually in liquid form. These resins, reactive TS liquids, are often used to form solid shapes. Such resin systems harden or cure at room temperature or at elevated temperatures because of the irreversible cross-linking of rather complex molecular structures (Chapter 1). This is different from the hardening of resins in solution, which harden when the solvent is evaporated. The hardening of the reactive resins produces no by-products, such as gases, water, or solvents. When reactive resins are used as impregnants, they are sometimes called solventless systems. However, there are resins and certain additives which release gases and may require degassing during processing; they are reviewed later in this chapter.

The reaction of resin systems is initiated when two or more components, usually a resin and a catalyst, are combined by mixing [3]. The individual components are reasonably stable at room temperatures, but once mixed they start to form long polymer chains, ultimately gelling or losing all their fluidity – they become rigid. The reaction is exothermic (gives off heat) until cross-linking is complete. However, the reaction may be accelerated by external heat. Depending on the mix proportions (including special catalysts, initiators, accelerators, etc.) the time to cure or completely cure into a solid form can be an instant, a very long period, or somewhere between. The required time cycle is based on the time required to handle pouring and/or assembling/encasing devices such as delicate electronic parts, medical parts, and ornaments. Most of the devices encased in plastics are usually being protected. Chapter 7 looks at protective coatings.

CASTING

Casting is the process whereby a liquid is poured into an open mold and allowed to react, cure, or harden to form a rigid part that reproduces the mold cavity (Fig. 6.1). The choice of casting material, the type of mold, and the method of fabrication often depend on the application. Production is rarely automated, but automation may be used when the economic benefits are considerable.

Casting differs from many of the other processes in that it seldom involves pressure or vacuum, although certain plastics and complex parts may require pressure and vacuum casting. Free-flowing plastics with low surface tensions and low viscosities are generally used for castings of intricate shapes with fine details in their design, and low-viscosity plastics are more suitable for producing bubble-free castings.

High-viscosity systems usually produce castings with better physical properties than low-viscosity plastics. However, handling of the high-viscosity plastics is usually more difficult and/or expensive. Most plastics suitable for castings are two-component systems. Mixing is mechanical or by hand. When the compound is 'gently' poured into a mold cavity, usually coated with a mold release agent, air in the cavity is naturally removed and the plastic allowed to set (harden).

Setting takes place at either room temperature or elevated temperature. During the chemical reaction there is usually the liberation of heat. The quantity of heat evolved depends on the cast shape and thickness. The rate of heat dissipation is important and leads to differing approaches in casting thin or heavy sections. In thin sections, where a large area is exposed in relation to the total volume of the plastic, the heat of the exothermic reaction is dissipated rapidly and the temperature of the casting is not very high. Thin sections can therefore be cast at room temperature with no danger of cracking. When the loss of heat is excessive, such as with thick castings, proper application or rate of heating may be necessary to accomplish the cure without any damage. The rate of heat development can be controlled by the compund ingredients and their

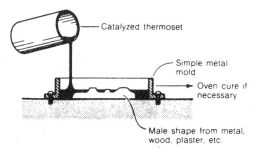

Figure 6.1 Casting of thermosets.

proportions. Heat sinks (or heat pipes) can be used for the absorption or transfer of heat away from a critical element or part. Different liquids, bulk graphite, etc., are used as heat sinks. The heat pipe can transmit thermal energy across a small temperature difference and almost at the speed of sound. The heat pipe consists of a tubular structure completely closed and containing a working fluid. For heat to be transferred from one end of the structure (e.g., in the casting) to the other, the working liquid is vaporized. The vapors travel and condense at the opposite end and the condensate returns as a working liquid to the other end of the pipe (tube). Saturated vapor transfers heat much more effectively than solid metals. Heat pipes can be used to remove heat or to add it.

Air bubbles are present in any casting operation; sometimes they are invisible, other times they are visible but not materially damaging. Removal of air bubbles, whether microscopic or larger, may be accomplished by a variety of methods. Air is present in the plastic and the additives, such as hardeners or fillers; air is also introduced when pouring the compound into the mold. The number of air bubbles that form depends on the viscosity and surface tension of the plastic– hardener system, the solubility of air in the system, and the characteristics of the mold surface. Although fillers extend the plastic and lower the bulk cost, they are a major cause of retained mixed-in air. Air entrapment can often be reduced by proper design of the mold. Elimination of sharp corners and provision of an adequate number of release openings or sprues, to facilitate air movement, considerably minimizes the amount of air in the casting. Plastic and hardener should be deaerated when possible; material handling and mixing should be carefully controlled to minimize the amount of air 'encased'. Use of high-speed mixers and vortexing of material should be avoided. Vacuum equipment used during mixing and pouring eliminates almost all air.

Casting can also form a film or sheet. Liquid plastic is poured onto a moving belt or precipitated in a chemical bath. Controls are required to obtain the proper rate of pouring and the moving belt to form cast film. The terms *cast film* or *film casting* are also used for the process of producing extruder-cast film (Chapter 3).

Acrylic casting

High molecular weight, cast acrylic (PMMA) sheet is a major product; it can be manufactured into rods, tubes, spheres, lenses, and other intricate shapes. Acrylic sheet may be manufactured in a batch-casting process by using individual glass or metal cells (forming a mold) or in a continuous operation, usually between polished metal belts. Rods, tubes, spheres, and lenses are usually cast in metal molds using conventional or rotational casting techniques. Cast acrylic products provide high transparency, a wide range of colors, rigidity, moderate impact strength, thermal as well as chemical resistance, and excellent weatherability.

The bulk, solvent-free polymerization of polymethyl methacrylate (PMMA) monomer is normally conducted during casting by means of a peroxide-initated free-radical process. During initiation, free radicals are normally generated by thermal decomposition of a peroxide catalyst added to the monomer. The free radicals attack the double bond of the vinyl monomer and generate the radical, which rapidly adds another molecule, thus extending the chain by one unit and regenerating a new active end site. The reaction from the monomer to the acrylic plastic is shown in Fig. 1.3 on page 4.

Transparent acrylic (PMMA) sheet is normally cast between individual plates of glass or highly polished metal, although other molds or mold surfaces can be used. Filled or opaque acrylic products may be processed in plaster, reinforced plastic, metal, wood, or concrete molds that provide adequate heat-transfer and surface-release characteristics.

Standard cell-casting operations utilize two flat glass plates (larger than the finished casting) separated by an elastomeric gasket that is unaffected by the monomer reaction. This gasket provides a pouring cavity for the monomer syrup. Cast acrylic sheeting has been made this way since the 1930s. Many equipment and material-handling changes have been made to provide quality products. The casting process requires heating for a few hours until polymerization is complete, followed by a gradual and controlled cooling period to produce sheets with excellent optical properties and other characteristics.

Sheeting can be fabricated by continuous casting of acrylic using two parallel, continuously conveying, endless, highly polished, stainless steel belts traveling through a temperature-controlled environment. These castings are usually made from solutions or dispersions of the monomeric syrup.

Polymerization rates may be increased and shrinkage reduced by adding fine PMMA powder or a prepolymerized acrylic syrup to the monomer. The powder may be produced by bulk, solvent, or emulsion polymerization (Chapter 1). The syrup may be produced in a batch or a continuous process with solids usually limited to 15.0–40.0%.

Continuous casting and batch casting do not produce identical products. Continuous casting does not match the greater precision obtained from batch casting, especially for thickness tolerance, surface finish, and transparency. The major advantage of continuous casting is its lower cost.

EMBEDDING

Embedment is a useful casting process for the protection or decoration of a device or assembly. Components and electronic circuits may be embedded in various types of materials, including different forms of plastic. Applications range from electrical insulation and protective packaging to mechanical devices. The largest use is for electrical and electronic devices. Since the start of the electrical industry, circuit components have been coated, buried, or otherwise encased in dielectric materials for protection from oxygen, moisture, temperature, electrical flashover, current leakage, salt spray, ocean environment, radiation, solvents, chemicals, micro-organisms, mechanical shock, vibration, gases, abuse or damage, and so on.

Embedding generally implies complete encasement in some uniform external shape; most of the package consists of the embedment material. The casting, mold, and assembly are designed to provide minimum internal stress as the plastic shrinks during curing. Encapsulation is a coating, and the part is normally dipped in a high-viscosity plastic (Chapter 7).

Components and circuits may be embedded in various types of plastics such as liquids, granular solids, or powdered solids. Materials include epoxies, silicones, polyurethanes, polyesters (TSs), polysulfides, and allylic plastics. These plastics incorporate many different fillers, extenders, flexibilizers or plasticizers, and foams such as expandable polystyrene, polyurethanes, and syntactic [3]. Decorative and other embedments are used to meet different applications and service requirements.

Embedment materials are selected for their resistance to specific conditions such as solvents, moisture and abrasive handling, repair, quality control, the recovery of metals and devices, etc. Analytical investigation in the case of failure frequently requires removal of the embedded device from the plastic. Methods include mechanical, thermal, plasma, and chemical treatment.

Characteristics during the embedding that can influence part performance are the stresses that develop between the components and the plastic during curing of the castings. Differentials between thermal expansion coefficients are a major source of design and processing problems [3]. Procedures can be used to eliminate problems, such as use of flexible plastic systems, controlling plastic viscosity to permit ease of mixing and pouring, curing heat control (very important), etc. The embedment of large components has the risk of runaway exothermic reactions. Temperature variations may produce excessive stresses and cracking throughout the casting, causing localized shrinkage. High temperatures generated by the exothermic reaction may cause volatilization of the curing agents and modifiers, leading to bubble entrapment.

ENCAPSULATION

Encapsulation is an offshoot of embedding. The part to be encapsulated is normally dipped into a high-viscosity or thixotropic plastic material to obtain a thin coating usually 0.01–0.50 in. (0.25–12.7 mm) thick. The primary protection is provided by a seal, which often imparts mechanical strength. The casting, mold, and assembly are designed to provide minimum internal stress as the plastic shrinks during curing; due to thin coating.

POTTING

Potting is similar to encapsulation except that the shell (container or mold) is not separated from the finished part. It is an embedding technique in which the device or part is encased in a plastic and in turn the mold (shell).

IMPREGNATION

This is a specialized method of embedding used mainly for electrical coils and transformers in which a liquid plastic is forced into the interstices of the component. **Trickle impregnation**, a related process, uses reactive plastics with a low viscosity, first catalyzing them then dripping them onto a transformer coil or similar device with small openings. Capillary action draws the liquid into the openings at a rate slow enough to allow escape of the air displaced by the liquid. When the device is fully impregnated, the plastic system is cured by exposing it to heat.

MOLDING

Molding is usually related to injection molding, compression molding, reaction injection molding, and other pressure molding processes (Chapters 2, 8, 11, 12, etc.). Thus, molding is a technique of embedment in which an object or device is encased in a plastic that flows into the mold under pressure. Encased molding is also known as insert molding.

PROCESSING CHARACTERISTICS

Processing techniques range from the unsophisticated (high labor costs but low capital costs) to sophisticated (almost zero labor costs but very high capital costs). Decisions on the appropriate technique are governed by production quantity, the material being processed, the available equipment, and the total cost. Small quantities are usually produced with an unsophisticated approach. Perhaps the material is mixed or prepared with catalyst by hand, using a 'dixie cup', before atmospheric pouring into open molds. Such a method usually involves inaccurate weighing (depending on labor and the specification). Waste can occur with too much mixing or preparation and from careless pouring. It affords a quick and easy evaluation for a given application or an inexpensive means for prototyping an small production runs.

Cleanliness is important when handling materials. When plastics are cured, especially TS plastics, they tend to adhere to whatever they contact and can rarely be dislodged by solvents. So it is well worth the effort to use disposable containers to clean up any spilled or unwanted plastic before it cures. Good housekeeping is also important where the process takes place.

Many pieces of equipment are available. Any selection needs to address the economics of the product being fabricated, the degree of precision necessary, and the production rate required. The mix can be manually proportioned using balancing and weighing scales. Proportioning pumps proportion according to a desired ratio; displacement cylinders are also used. Diaphragm pumps may be used with adjustable displacement to draw the plastic components from reservoirs and to meter the correct proportions into a mixing chamber. A popular approach is the use of gear pumps with drives ratioed to deliver the correct portions; they are used only when the mix is relatively free of abrasive fillers or additives.

Vacuum systems

The process may require a vacuum degassing of each plastic system component under heat and agitation prior to mixing. A vacuum chamber may need to be used for castings where the molds have been baked or cured under vacuum. Vacuum potters are also available as integrated systems which provide for degassing of materials, including heat and/or agitation, mixing under vacuum, baking out the mold and parts under vacuum prior to casting, and casting under vacuum. The material, molds, and parts to be potted are kept under vacuum throughout the entire process. These potters are also available if positive pressure (versus vacuum) is required during the process. Stream degassers are available for continuous degassing of materials; liquid flows through the degasser in a thin stream surrounded by vacuum. This approach enables air and volatiles to escape from a relatively fine stream without impeding plastic melt flow.

Care should be taken when using a vacuum for **degassing**. All components of a plastic system have a unique vapor-pressure curve with respect to temperature. At a given temperature there is a critical pressure (vacuum) where the plastic system component will vaporize. If vacuum degassing takes place (such as during casting) at too low a pressure (actually too high a vacuum), one or more parts of the material system will boil off (usually the catalyst). The result is a product, if cured, with properties different from those required. The appropriate vacuum levels for the material system may have to be determined experimentally or the material supplier may make a suggestion. Pressures of 3–5 mm Hg are usually sufficient for degassing room temperature processes. Higher pressures are required if higher temperatures are used.

Mixing

Improper mixing can produce resin-rich and/or catalyst-rich compounds. They become sticky when no curing occurs, or there are dry areas where the finished part has weak or powdery sections. The mixing is related to the miscibility of the resin components and often to the viscosity ratios. As an example, it is difficult to mix very low-viscosity catalyst into a very high-viscosity resin. Raising the temperature of highly viscous resin systems generally lowers the viscosities of their components, making mixing easier. But it may also accelerate the cure, so the material gels before the melt has finished flowing.

Cycle times

It is very important to determine the proper curing cycle for the resin system. Some systems cure at room temperature, usually requiring several hours or more to develop maximum strength, etc. Others need to be heated to a specific temperature to initiate the reaction, then they must be held at that temperature for a specified length of time. More critical resin systems may need programmed heating followed by a postcure cycle that usually lasts for many hours. Figures 1.18 and 8.2 (pages 57 and 419) illustrate the TS plastic behaviors of viscosity and temperature with time; Chapter 12 gives information on TS cycle times, etc.

Exothermic reactions

The embedment of large devices can cause an excessively exothermic reaction (generally excess heat). The exothermic reaction occurs internally, so heavy devices or devices with a large cross section (a large mass) should be made of material that prevents overheating, subsequent resin degradation, and/or damage to the embedded device.

A runaway exothermic reaction may cause excessive stresses and even cracking throughout the casting because it creates temperature variations which lead to localized shrinkage. The high temperature caused by the exothermic reaction may volatilize components such as the curing agents and modifiers, causing bubble entrapment. Temperature-sensitive solid-state devices can be damaged, and wire insulation may be degraded, causing electrical short circuits, etc. Exothermic reactions can be monitored by plotting temperature versus time in which the values of gel time, peak temperature, and time to peak temperatures are measured.

Molds

Molds and cases for casting, potting, etc., are usually made of metal, plastic, or wood. Other materials include plaster of paris, ceramic, and glass. Metal molds may be machined from aluminum or steel; they may also be constructed of welded sheet metal (Chapter 15). Molds need provision for part removal after cure. They are often in two parts for those requiring a closed mold that may be bolted or clamped together during curing and disassembled for part removal. Open molds are also used with dip encapsulation and similar processes.

If TP molds or cases are used, the casting process must operate below the temperature that would melt the TP. Detail molds for decorating cast or potted parts that would look like wood grain are often made from flexible silicone rubber; its flexibility allows easy removal of the part without damaging the mold when the part is peeled off.

Since most of the TS resin systems tend to act as strong adhesives, mold release is necessary for cured part removal. Typical mold releases include carnauba wax, zinc stearate powder, and silicone sprays; they are similar to the agents used in processing TS reinforced plastics (Chapter 12). With silicone, the surface of the cured plastic makes painting, or any other material attachment impossible. Surface-cleaning systems can be used but it is more practical not to use silicone if secondary operations are required (Chapter 17). The required amount of release agent depends on the cavity surface. Very little is required with a smooth surface, more is used on rough surfaces. Too much mold release can cause rough or porous surfaces. For optimum cosmetic results, the mold cavity should be carefully prepared on nonporous material using the smallest amount of release agent.

MATERIALS

Different TS and TP materials are used, but predominately TS. In summary, materials include plastics (epoxies, polyesters, polyurethanes, allylics, etc.), fillers and extenders (finely ground solids such as silica, quartz, calcium carbonate, metal, iron oxide, aluminum oxide, mica, and glass), and flexibilizers that reduce the hardness and rigidity of epoxy plastics. Also used are elastomers such as flexible polyurethanes, polysulfides, flexible silicones, and room temperature vulcanizing (RTV) rubbers. TS foams are used where voids in the protective plastic are not objectionable (Chapter 9). Foams are lightweight, they give good mechanical protection, and they improve thermal insulation.

ADVANTAGES AND DISADVANTAGES

In general, all the processes ensure the mechanical, physical, electrical, and/or environmental integrity of any devices or parts included in the plastic. They also lend themselves to very low volume or prototype production because of their simplicity and low cost.

The principal disadvantages are the relatively high cost for high-volume production (as compared to other methods reviewed in Chapter 7), the high material waste, and the frequent clutter or messiness. Sometimes a line is set up for long production runs, such as with ornaments, where the insert shape and other factors result in the lowest cost. Another disadvantage, but not always a problem, is shrinkage on curing. Shrinkage is progressive; it starts where the heat is first generated by the exothermic reaction and it ends where the last area or volume section cures. Nonuniform shrinkage puts physical stress on inserts and could damage delicate devices. Shrinkage can also cause warpage, sink marks, and loss of close dimensional tolerances. Change in the plastic and/or its components may eliminate shrinkage or bring it under control.

COSTING

There are wide variations in processing techniques; often they are labor intensive, with or without expensive processing equipment. Costings vary from part to part. Equipment costs may be very low, but they start to increase if special items are required. Examples include disposable containers, static mixing heads, waste removal, solvents and solvent cleaners. Automatic dispensing systems and sophisticated vacuum potters often require lengthy and expensive setups.

TROUBLESHOOTING

This chapter covers an extensive range of problems and their remedies; collectively they should provide an adequate troubleshooting guide. Among the factors reviewed are the development of stresses on curing, the difference in thermal expansion between the plastic and the insert, the crucial importance of viscosity for an embedment plastic system, and the creation of adhesions and voids.

7

Coatings

INTRODUCTION

Plastic coatings have widespread industrial and commercial applications. They are applied by direct contact of the liquid coating with the substrate or by deposition using an atomization process. Direct methods include brushing, roller coating, dipping, flow coating, and electrodeposition. Deposition methods include conventional spray, airless spray, hot spray, and electrostatic spray.

The term *plastic coating* encompasses conventional paints, varnishes, enamels, lacquers, water-emulsion and solution finishes, nonaqueous dispersions (organosols), plastisols, and power coatings. However, the term *paint* is often used nonspecifically to cover all these categories as though it were synonymous with coating; the terms are often used interchageably. Paint coatings consume by far the largest quantity of coating material, but the other coating processes are important and useful. All these surface coatings represent a large segment of the chemical industry.

Film formation

A solid film can usefully be defined as a material that does not exhibit detectable flow under the conditions of observation during the interval of observation. Thus, a film can be defined as a solid under a set of conditions by stating the minimum viscosity at which flow is not observable in a specified time interval. A film is called dry or solid when it develops a viscosity high enough to pass a certain test. For example, it is reported that a film is dry-to-touch if its viscosity is higher than 1cP.

Film coating can involve chemical reaction, polymerization, or crosslinking. Some films merely involve coalescence of plastic particles. The various mechanisms involved in the formation of plastic coatings are as follows: (1) coating formed by chemical reaction, polymerization or cross-linking of epoxy, TS polyester, polyurethane, phenolic, urea, silicone, etc. (Chapter 1); (2) dispersions of a plastic in a vehicle; after removal of the vehicle by evaporation or bake, the plastic coalesces to form a film of plastisol, organosol, water-based or latex paint, fluorocarbons, etc.; (3) plastic dissolved in a solvent followed by solvent evaporation to leave a plastic film of vinyl lacquer, acrylic lacquer, alkyd, chlorinated rubber, cellulose lacquer, etc.; (4) pigments in an oil that polymerizes in the presence of oxygen and drying agents of alkyd, enamels, varnishes, etc.; (5) coatings formed by dipping in a hot melt of plastic such as polyethylene or acrylic; and (6) coatings formed by using a powdered plastic and melting the powder to form a coating using many different thermoplastics.

Surface coatings are usually composed of viscous liquids; they have three basic components: a film-forming substance or combination of substances called the binder, a pigment or combination of pigments, and a volatile liquid. The combination of binder and volatile liquid is usually called the vehicle. It may be a solution or a dispersion of fine binder particles in a nonsolvent. No pigments are included if a clear, transparent coating is required. The composition of the volatile liquid provides enough viscosity for packaging and application, but the liquid itself rarely becomes part of the coating. When the coating is applied to the surface, the volatile liquid evaporates, leaving the nonvolatile binder–pigment combination as a residual film; it may or may not require chemical conversion to an insoluble condition. Small amounts of additives are often included to improve application, pigment settling, drying, and film properties. Most binders are either high molecular weight, nonreactive plastics or low to medium molecular weight, reactive plastics capable of being further polymerized via chain-extension or cross-linking reactions to high molecular weight films (Chapter 1).

Most coatings are manufactured and applied as liquids; they are converted to solid films once they are on the substrate. Powder coatings are applied as a solid powder, converted to a liquid on the substrate, then formed into a solid film.

Coating films are viscoelastic, so their mechanical properties depend on the temperature and the rate of stress application (Fig. 1.31, page 82). Their behavior approaches the elastic mode with increased tensile strength to failure (breakage) or decreased elongation to failure, and with a more nearly constant modulus as a function of stress when the temperature is decreased or when the rate of application of stress is increased [3]. The shifts can be especially large if results are compared above and below the glass-transition temperature, T_g (Chapter 1). Below T_g the coatings have an elastic response and are therefore brittle; they break if the relatively low elongation to failure is exceeded. Above T_g the viscous component of the deformation response is larger; the films are softer (lower modulus) and less likely to break during forming. Caution is required in considering the relationship of T_g to formability because some materials, such as acrylic and especially PC, are ductile at temperatures far below T_g. Above T_g the modulus is primarily controlled by the density of the plastic cross-linkages.

Thermoplastics

Serviceable thermoplastic films must have a minimum level of strength, depending on the end use of the product. Film strength depends on many variables, but a critical factor is molecular weight (Chapter 1). Molecular weight (MW) varies with the chemical composition of the plastic and the mechanical properties required in the application. As an example, in solvent evaporation from solutions of TPs for spray applications, consideration has to be given to its evaporation behavior. With a methyl ethyl ketone (MEK) solvent and vinyl copolymer that has a relatively high vapor pressure under application conditions, MEK evaporates quickly.

With this type of system, a large fraction of the solvent evaporates in the time interval between the coating leaving the orifice of the spray gun and its deposition on the surface being coated. As the solvent evaporates, the viscosity increases and, soon after application, the coating reaches the dry-to-touch state and does not block. However, if the film is formed at 77 °F (25 °C), the dry film contains several percent of retained solvent.

In the first stages of solvent evaporation from such a film, the rate of evaporation depends on the vapor pressure at the temperatures encountered during evaporation, the surface area-to-volume ratio of the film, and the rate of airflow over the surface; it is essentially independent of the presence of plastic. The rate of solvent diffusion through the film depends not only on the temperature and the T_g of the film, but also on the solvent structure and any solvent–plastic interactions. The coating thickness is another parameter involved in solvent loss and film formation.

TP-based coatings have a low solids content because their relatively high MWs require large amounts of solvent to reduce the viscosity to levels low enough for application. Air pollution regulations limiting the emission of volatile organic compounds (VOCs) and the increasing costs of solvents have led to the increasing replacement of these coatings with lower-solvent or solventless coatings. But large-scale production means that solvent coating systems become economically beneficial when used with a solvent recovery system.

Latex or **paint emulsion** has low solvent emission as well as other advantages. A latex is a dispersion of high-MW plastic in water. The dispersion is stabilized by charge repulsion and entropic repulsion (also called steric repulsion and osmotic repulsion). Because the latex plastic is not in solution, the rate of water loss is almost independent of composition until the evaporation gets close to its end. When a dry film is prepared from a latex, the forces that stabilize the dispersion of latex particles must be overcome and the particles must coalesce into a continuous film. The rate of coalescence is controlled by the free volume available, which in turn depends mainly on T_g. The viscosity of the coalesced film also depends on the free volume.

With latex or paint emulsion, coating material is made of two dispersions: (1) dry powders with colorants, fillers, and extenders, and (2) plastic dispersions. These emulsion paints have the binder in a water-dispersed form. Principal types are styrene-butadiene, polyvinyl acetate and acrylic plastics. Percentage composition by volume is usually 25–30% dry ingredients, 40% latex, and 20% water plus stabilizer. Their unique properties are ease of application, absence of disagreeable odor, and nonflammability; they are used both indoors and out.

Thermosets

A problem in TS systems is the relationship between storage stability of the coating before application and the time and temperature required to cure the film after application. The processing of TSs is different than TPs, as explained in Chapter 1. It is desirable to store a coating for many months or even years without significant increase in viscosity, caused by cross-linking reaction during storage. However, after application, the cross-linking (cure) should proceed in a short time at as low a temperature as possible.

Since reaction rates depend on the concentration of the polymer's functional groups [9], the storage life can be increased by using more dilute systems; adding more solvent increases storage life. When the solvent evaporates after application, the reaction rate increases initially. Although it is advantageous to reduce solvent concentration as much as possible, the problem of storage stability has to be considered for systems with a higher solids content.

Much less solvent is required when formulating a TS coating from a low-MW plastic that can be further polymerized to a higher MW after application to the substrate and evaporation of the solvent. The average functionality must be more than 2 per molecule in order to ensure the MW of the final cured film is high enough for good coating properties.

The mechanical properties of the final film depend on T_g for the cross-linked polymer and the degree of cross-linking, i.e., the cross-link density. The cross-link density is affected by the average functionality, the equivalent weight of the system, and the completeness of the reaction (complete cure of the TS).

Film formation and curing

After a coating is applied, solvent evaporation and rheological factors contribute to the setting or solidification of the coating film (Chapter 1). Solvent initially evaporates from the surface of the film at about the same rate as it would in the absence of a binder. As the film solidifies, evaporation slows down because the diffusion rate to the surface is usually slower than the evaporation rate. Lacquer films do not cure by chemical reaction to achieve the required hardness and toughness. They just dry by solvent evaporation and depend on the high MW of these TP materials to provide the required performance. Latex paints behave in a similar manner.

Different methods can be used to accelerate curing or solidification, e.g., an electron-beam radiation system. This requires a power supply and an electon-beam accelerator. The part to be coated is sprayed electrostatically and passed behind the accelerator in an inert atmosphere. The curing rate is almost instantaneous (Chapter 17).

PROCESSING METHODS

Many different methods are used to apply plastic coatings to substrates of all sizes and types, ranging from the simple to the complex. They are generally composed of one or more plastics, a mixture of solvents (except with powder coatings), commonly one or more pigments, and frequently several additives. Coatings can be classified as TPs or TSs. Coating methods are categorized in different ways by the different industries that require them. A listing of methods is given in Table 7.1 [68]. The next few sections describe some of the more popular methods, based on quantity consumed and/or capabilities.

Extrusion coating, one of the principal methods, has already been reviewed in Chapter 3. Coating via calenders, another important method, is reviewed in Chapter 5.

PAINT COATINGS

A paint is a suspension of pigment in a liquid that dries or cures to form a solid film. Traditional paints contain a vehicle, a solvent, and a pigment. Some are applied by spraying or dipping. Other systems involve heating parts and spraying them with a dry plastic powder which coalesces on the hot part to form a film. The differences among the various coating systems are the mechanism of film formation and the type of plastic being applied. Many important details are involved in surface preparation and in application techniques.

Both solvent-borne and aqueous paints are used. Paints are usually classified on the basis of the binder (vehicle) used. The most often used are

(1) acrylics (aqueous acrylic emulsions, solvent-borne enamels, melamine, and other modified acrylic emulsions); (2) polyurethanes (aqueous and solvent-borne); (3) alkyds and modifications; (4) epoxies and modifications; (5) polyesters; (6) vinyls and modifications (latex or solvent-borne); (7) nitrocelluloses (solvent-borne); and (8) polyamides (solvent-borne).

The distinction between paints and **enamels** is not straightforward. However, enamels generally contain higher-MW binders and are formulated with lower solids concentration. They are also formulated at lower pigment/binder ratios to create a superior gloss.

Lacquers differ from paints and enamels because they are compounded with TPs, soluble plastics of much higher MW and low chemical reactivity. Film formation occurs by solvent evaporation. Conventional lacquers are normally solvent-borne. Dispersions of plastics in water, **latexes**, or organic vinyl liquids, **organosols**, yield soluble films of TP; they also qualify as lacquers. **Plastisols** are dispersions of finely divided vinyl in plasticizers that are nonsolvents at room temperature, but good solvents at high temperatures. They are stable under normal storage conditions and can be coalesced into films at elevated temperatures (Chapter 14).

Some plastic products that require painting may need special considerations because of their surface condition. Table 7.2 is a guide for painting plastics. Some plastics may be sensitive to certain solvents, so take care to understand the situation.

COATING FILMS AND FABRICS

Plastic coating is big business; the substrates may be films such as plastics, aluminum foils, and papers, or fabrics, woven and nonwoven. The coating material provides many properties required to make the substrates more useful in commercial and industrial applications. Considerations in selecting the plastic coating include chemical environment, mechanical properties, processing characteristics, and costs. Coating fabrics include PVC (principally used for the coating of fabrics), PUR, and various elastomers.

Films

Films are coated to extend the utility of the substrate by improving existing properties or adding new and unique properties. The coatings can provide heat sealability, impermeability (moisture, water, vapor, perfumes, and other gases), heat and UV barriers, modified optical or electrical properties, altered coefficients of friction, and a tendency toward blocking.

Coatings are different from **laminations** of two or more films. Laminates vary in construction: plastic film to aluminum foil, two or more

Table 7.1 Examples of coating methods

Coating method	Base material[a]	Coating composition[b]	Usual coating speed ($m\,min^{-1}$)	Viscosity range, ($mPa\,s$)	Wet-coating thickness range (μm)
Air knife	B,D	R,T,X	15–600	1–500	25–60
Brush	B,C,E,F,G	R,S,X,Z	30–120	100–2000	50–200
Calender	A,B,D,E	U,V,W	5–90		100–500
Cast-coating	A,B,D	Q,R,S,T,V,Y	3–60	1000–5000	50–500
Curtain	A,B,C,D,E,F	R,S,V,X,Z	20–400	100–20000	25–250
Dip	A,B,D,E,F,G	R,S,V,X,Y,Z	15–200	100–1000	25–250
Extrusion	A,B,D,E	T,U,V,W	20–900	30000–50000	12–50
Blade	A,B	R,S,T,V,X,Y,Z	300–600	5000–10000	12–25
Floating knife	A,B,D	R,S,T,V,X,Y,Z	3–30	500–5000	50–250
Gravure	A,B,D,E	R,S,T,U,V,Y,Z	2–450	100–1000	12–50
Kiss roll	A,B,C,D,E,F	R,S,V,X,Z	30–300	100–2000	25–125
Knife-over-blanket	A,B,D	R,S,T,V,X,Y,Z	3–30	500–5000	50–250
Knife-over-roll	A,B,C,D,E	R,S,T,U,V,X,Y,Z	3–60	1000–10000	50–500
Offset gravure	B,D	R,S,T,Z	30–600	50–500	12–25
Reverse roll	A,B,C,D,E,F	R,S,T,U,V,X,Y,Z	30–300	50–20000	50–500

Reverse-smoothing roll	A,B	R,T,X	15–300	1000–5000	25–75
Rod	B,D	R,S,T,V,X,Y,Z	3–150	50–500	25–125
Sprays					
Airless spray	A,B,C,D,E,F,G	S,T,V,X,Y,Z	3–90	c	2–250
Air spray	A,B,C,D,E,F,G	S,T,V,X,Y,Z	3–90	c	2–250
Electrostatic	A,B,C,D,E,F,G	S,T,V,X,Y,Z	3–90	c	2–250
Squeeze roll	A,B,C,D,E,F	R,S,T,U,V,X,Y	30–700	100–5000	25–125
In situ polymerization	A,B,C,D,E,F,G	Y,Z	undetermined	liquid or vapor	6–2.5
Powdered resin	A,B,C,E,F,G	Q	3–60		25–250[c]
Electrostatic spray		Q			20–75[c]
Fluidized bed	E,G	Q			200–2000[c]

[a] Key: A = woven and nonwoven textiles; B = paper and paperboard; C = plywood and pressed fiberboards; D = plastic films and cellophane; E = metal sheet, strip, or foil; F = irregular flat items; G = irregularly shaped [9].

[b] Key: Q = powdered resin compositions; R = aqueous latexes, emulsions, dispersions; S = organic lacquer solutions and dispersions; T = plastisol and organosol formulations; U = natural and synthetic rubber compositions; V = hot-melt compositions; W = thermoplastic masses; X = oleoresinous compositions; Y = reacting formulations, e.g., epoxy and polyester; Z = plastic monomers [9].

[c] Dry thickness.

Table 7.2 Guide to painting plastics[a,b]

Plastic	Urethane	Epoxy	Polyester	Acrylic lacquer	Acrylic enamel	Acrylic waterborne
ABS	R	R	NR	R	R	R
Acrylic	NR	NR	NR	R	R	R
PVC	NR	NR	NR	R	R	NR
Styrene	R	R	NR	R	R	R
PPO/PPE	R	R	R	R	R	R
Polycarbonate	R	R	R	R	R	R
Nylon	R	R	R	NR	NR	NR
Polypropylene	R	R	R	NR	NR	NR
Polyethylene	R	R	R	NR	NR	NR
Polyester	R	R	R	NR	NR	NR
RIM	R	NR	NR	NR	R	R

[a] Primers may be required in order to obtain adequate paint adhesion.
[b] R = recommended; NR = not recommended.

plastic films combined, plastic film to paper to plastic film, paper to plastic film to paper, and so on. With plastic films, the coatings are usually thinner than the base film. Coatings are generally 0.05–0.2 mil (1.3–5.1 μm) thick. In laminations most films are at least 0.25 mil (6.4 μm) thick, and more commonly 0.5–2 mil (13–51 μm) thick.

Fabrics

Different desirable properties of a fabric can be supplemented by plastic coating. The fabric provides at least tensile and shear strengths with elongation control. Coatings can protect the fabric, reduce porosity, provide decorative effects, and so on. Coated fabrics are designed for specific applications. The three major considerations are the physical environment, the chemical environment (water, acid, solvents, and so on), and cost.

Fibers used in the fabric patterns include nylon, cotton, PP, TP polyester, glass, and others. Mono- or multifilament fibers or staple fibers (chopped, etc.) are used. The properties of the substrate are determined by the fiber properties and fabric construction [3]. Estimates suggest that 70 wt% are fabric coated or filled woven, 20% are nonwoven, and 10% are knitted. To obtain proper adhesion and melt flow with woven, nonwoven, and knitted fabrics usually requires removal of any existing sizing or lubricant used to prepare them for weaving, so that appropriate impregnation occurs.

Impregnation is the process of throughly soaking, filling of voids and interstices of the substrate (as well as wood and paper) with the plastic coating. The porous materials generally serve as reinforcements for the plastic after the coating treatment.

Processing is dictated by the properties of the substrate and the coating. The viscosity of the coating must permit flow around the yarn or fiber surface. In extrusion and calendering, the coating is fluidized by pressure and heat. In other processes, viscosity can be reduced by solution or dispersion.

Wall coverings, upholstery, and apparel are examples of decorative coated fabrics. Inks are applied with one or more gravure printers to correct the color or to add a pattern. Relief patterns are obtained by applying heat and pressure with embossing rolls.

Leatherlike materials

Leather substitutes are designed to initiate the appearance of leather with its surface grain. This is accomplished by coating substances that are capable of forming a uniform film. This requirement was first met by plasticized PVC during the 1940s. When plasticized PVC (solid or foam) is coated onto a substrate, it produces a leatherlike material called vinyl-coated fabric. It exhibits high density, very low water-vapor permeability, cold touch, poor flex endurance, and poor plasticizer migration. But it has good scratch resistance and colorability as well as being inexpensive.

Polyurethane (PUR) coated fabrics, developed in the 1960s, were an improvement. PUR is coated on woven or knitted fabrics. With a T_g below 32 °F (0 °C), PUR is very flexible at room temterature without a plasticizer (Chapter 1). Also very important, its molecular structure allows water-vapor permeability. In addition, the solvents normally used for PUR permit coagulation by a nonsolvent with formation of a porous structure. The result is increased flexibility and water-vapor permeability. Ordinary PUR-coated fabrics are produced by drying a cast PUR solution to form a film that is laminated onto the substrate. Significant improvement in appearance, feel, and grain is accomplished by using a brushed fabric as the substrate. It is laminated with a cast PUR film. Alternatively, an organic solvent solution of PUR is applied to a brushed, woven fabric immersed in a nonsolvent bath for coagulation.

Poromerics, also called synthetic leather, were developed during the 1960s as an improvement over leatherlike coated fabrics, whose applications were limited by the properties of the knitted or woven substrate. Poromerics use a nonwoven fabric impregnated with plastic, thereby creating a substrate resembling leather. Fine fiber construction provides the desired softness. Prepared with PUR, the poromeric coating layer corresponds to the grain of the leather.

Equipment

Different methods are used. Each has its performance advantages and cost benefits. Coating equipment is used to apply a surface coating, a laminating adhesive, and any compounds for saturation/impregnation of a fabric. The equipment has three basic components: the coating head, a dryer or other coating solidification unit, and web-handling hardware (drives, winders, edge guides, controls, etc.). It can generally coat various substrates in roll or sheet form. Coatings can be applied directly to the substrate or transferred to the substrate from another surface, such as a roll. Transfer from another surface is used when the substrate is sensitive to the coating material, when it may be damaged by exposure to oven temperatures, for special secondary operations such as applying pressure-sensitive labels, and so on.

During its application the coating must be sufficiently fluid to be spread into a uniformly thin layer across a web. Coatings can be applied as solutions in organic solvents, as aqueous solutions or emulsions, or as molten or softened solids. Solutions and emulsions require drying to obtain solid coatings. Hot melts are solidified by cooling. Some coatings may be applied as reactive liquids (Chapter 1) then polymerized by IR or heat.

Heat and mass transfer take place simultaneously during the drying process. The heat is transferred by convection in an air dryer, by radiation in infra-red dryers, or by conduction in contact-drum dryers. The drying equipment usually has a means to remove and recirculate the vapor with heat-exchange equipment to conserve energy.

The coating head accomplishes two functions. It applies the coating to the substrate, distributing it in metered amounts uniformly over the surface. Most coaters fall into the following categories: roll, knife, blade, or bar. There are also extrusion or slot-orifice coaters. Roll coaters, the most widely used, are subdivided by their construction into direct, reverse, gravure, and calender.

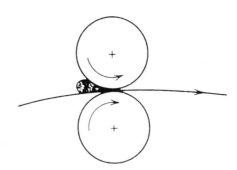

Figure 7.1 Direct roll coater.

Figure 7.1 shows the direct roll coater. The coating is metered by a gap between accurately machined metering and applicator rolls. Sheets are fed through the nip formed by the applicator and the carrier rolls. Kiss roll coaters (Fig. 7.2) apply the coating to the web from a pan. This type of coating is not accurate. Reverse roll coaters allow wide ranges of materials and viscosities to be used. Solvent solutions, aqueous solutions, and hot melts are coated on this type of machine. Reverse roll coaters are expensive because the rolls must be accurately machined.

In the direct gravure arrangement (Fig. 7.3) the engraved roll contacts the web directly. In offset gravure, the coating is first transferred to a rubber-covered roller or an apron then on to the web. Offset gravure allows the use of higher nip pressures and is therefore more suitable for coating substrates or web with rough surfaces. It also allows the coating to level itself on the roll surface before it transfers to the web. An important action in a gravure coater is the engraved roll. The roll is wetted with the coating, excess is removed by a doctor knife, and the coating remaining in the engraved cells below the roll surface is transferred to the web at the backup roll nip of the gravure roll.

Knife and bar coaters are metering devices which remove excess coating and allow only a predetermined amount to pass through. A knife-over-roll coater consists of a straight-edged knife placed against a roll (Fig.

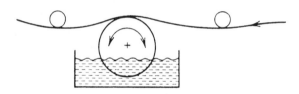

Figure 7.2 Kiss roll coater.

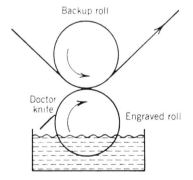

Figure 7.3 Direct gravure coater.

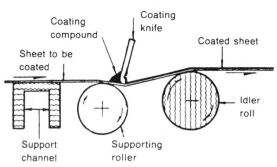

Figure 7.4 Knife coating.

7.4). The coating weight is adjusted by setting a gap between the roll and the knife. An excess of coating is delivered to the bank before the knife and the desired amount is metered by the gap. The steel roll and the knife are both accurately machined. Flexible-blade coating is the dominant coating process along with air-knife coating for applying pigment types over paper and paperboard. Pigment coatings give a smooth and printable surface. Blade-coating processes are suitable for high-speed applications such as required in paper converting.

The slot-orifice and curtain coaters are variations of the extrusion coating process (Chapter 3). Slot-orifice coaters are equipped with an orifice of an adjustable width through which a coating is extruded. They are often used for hot melts and to apply waterproof coatings over paper. They are also used to apply aqueous coatings. Slot-orifice coaters have a lighter construction than extruders.

A curtain coater consists of a slotted head through which a liquid curtain is allowed to fall onto the moving web. The excess coating is collected in a pan underneath and returned to the coater head. Curtain coaters are especially suitable for coating sheets (plastic, corrugated paperboard, etc.) and may be used with hot melts.

SPRAY COATINGS

Overview

Plastic-spray coating is especially attractive when parts are already assembled and have irregularly shaped and curved surfaces. The material applied is usually in the form of a paint. Many different types of spray equipment are in use. They can be classified by their method of atomization (airless, air, rotary, electrostatic, etc.) and by their deposition assist (electrostatic or nonelectrostatic, flame spray, etc.). Spraying techniques may fall into several of these categories. They range from simple systems with one manual applicator to highly complex, computer-controlled, automatic systems. They can incorporate hundreds of spray units. Automatic systems may have their applicators mounted on fixed stands, on reciprocating or rotating machines, on robots, and so on.

Airless atomization

Airless atomization uses pressure to force the coating material (paint, etc.) through a small nozzle at a sufficiently high velocity, or to impinge a material stream against the part, causing a sudden change in flow direction. In either case, the fluid stream disintegrates into small particles. The degree of atomization is rather poor, the particle size is large, and the fluid must have a sufficiently low viscosity.

Nevertheless, this spraying method is more energy-efficient than air atomization. It requires less paint and the transfer efficiency is much better. The paint viscosity may be lowered simply by heating. Airless spraying can be improved by providing a low-pressure airstream to assist atomization and direct the paint to the target. Although the transfer efficiency is slightly decreased, the degree of atomization is substantially improved.

Electrostatic spray coatings

The thickness with electrostatic spray coating is in the range 3–30 mil (80–800 μm). A dry powder of coating plastic is withdrawn from a reservoir in an airstream and electrostatically charged in a high-voltage corona field within the spray gun. The charged particles are attached to the electrically grounded part being sprayed and the coating adheres to it by electrostatic attraction. Metal parts are naturally able to be grounded, but plastics require a conductive coating surface.

After spraying, the coated substrate is placed in a heating oven. The coating is fused to form a continuous film. If the powder is sprayed on a preheated part, the powder melts and fuses directly on the hot surface; further heating may be required to completely fuse or cure the coating, but this depends on the type of coating powder and the substrate material. When the powder builds to a certain thickness, usually about 8 mil (200 μm), a dielectric barrier is created and more powder is rejected.

Flame spray coatings

Flame spray coating consists of blowing a powder through a flame which partially melts the powder and fuses it as it contacts the substrate. The part's surface is preheated with the flame usually to about 400 °F (204 °C) when using PE. The usual approach is to coat only a few square meters at

Table 7.3 Transfer efficiency for spray methods

Spray method	Transfer efficiency (%)
Air atomization	35
Air atomization with electrostatic charge	55
Airless atomization	50
Airless atomization with electrostatic charge	70
Disk-and-bell atomization with electrostatic charge	90

a time, so the temperature can be controlled. The flame is then adjusted. When coating is completed, the powder is shut off and the coating is postheated with the flame. Flame spraying is particularly useful for coating products with surface areas too large for heating in an oven. Disadvantages are the problems associated with an open flame and the need for skilled operators to apply the coating.

Spray-paint transfer efficiency

This is a measure of the percentage of paint that reaches the target. Table 7.3 provides a guide for various methods. This table takes into account the sensitivity to part geometry. It does not take into account factors such as large, flat parts versus small, irregularly shaped parts; the values are only guides.

POWDER COATINGS

Powder coating is a solventless system; it does not depend on a sacrificial medium such as a solvent, but is based on the performance constituents of solid TP or TS materials. It can be a homogeneous blend of the plastic with fillers and additives in the form of dry, fine particles of a compound similar to flour.

Advantages of powder coating include minimum air pollution and water contamination, increased performance with coating, and consequent cost savings. It is basically a chemical coating, so it has many of the same problems as solution painting. If not properly formulated, the coating may sag, particularly for thick coatings, show poor performance when not completely cured, show imperfections such as craters and pinholes, and have poor hiding with low film thickness. Various methods are used to apply powder coatings.

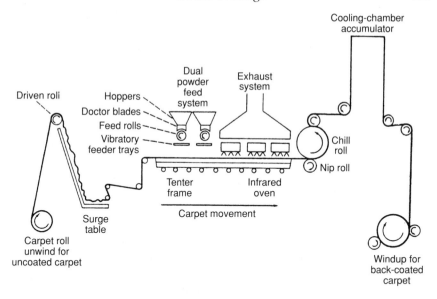

Figure 7.5 Powder coating on a carpet-coating line.

Flood coatings

Simplest of all the dry coatings, flooding is best for regularly shaped products where minor variations in coating thickness are not objectionable. The nonfluidized bed, a form of flood coating, dips a heated product into a container of the powder. A coating is formed then withdrawn for possible postcuring. The amount of coating depends on factors such as type of coating material, amount of preheat on product, and time in the powder. The insides of hollow products can be coated by filling the preheated part with powder and dumping out the excess.

An offshoot of this process is the powder coating of textiles, plastics, papers, etc. A large-volume application is the fusible interlining of drapery, upholstery, and carpet fabrics (Fig. 7.5). Other important markets are coating metal pipes and tanks, etc.

Probably the single biggest advantage with powder is the nearly 100% utilization of material with proper recovery and recycling.

Fluidized-bed coatings

This method, originally called sinter coating, uses a two-chambered container (one on top of the other) separated by a porous medium that retains plastic powder in the top container but allows free passage of gas (usually compressed air) through the powder. The porous flat plate can be sintered

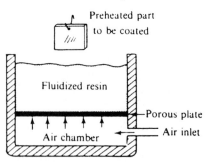

Figure 7.6 Fluidized-bed coating system: the preheated part will be dipped into the powdered fluid.

metal, steel plate with many minute holes, or simple acoustic celing tile. Air is forced from the bottom chamber through the porous medium into the powder (Fig. 7.6). The rising gas separates and suspends the powder, causing its volume to increase to many times the height of the powder at rest. When it reaches a steady state, the pressure drop of the rising gas will no longer support the weight of the powder. At this point, the gas being vented through the top of the bed forms an equilibrium with the gas being supplied. The top container has its top side opened.

Even though the gas–powder mix is uniform, most systems also use a mechanism to vibrate the bed; this improves distribution, especially in shallow beds, and assists powder flow during the dipping of parts to ensure uniform coating.

The powder is dry and free of agglomeration, so the gas can surround individual particles completely. The gas-suspended powder acts very much like a liquid, a **fluidized bed**. Powdered particles range in size from less than 20 μm to about 200 μm. Particles larger than 200 μm are difficult to suspend. Particles smaller than 20 μm may create excessive dusting and release of particles from the top of the bed. Powders may also contain fluidizing aids to keep them flowing freely.

The part to be coated (usually metal) is heated to a temperature above the melt temperature of the powder and is then immersed into the fluidized bed. As the powder contacts the hot substrate, the particles adhere, melt, and flow together to form a continuously conforming coating. The part is removed from the bed when the desired coating thickness is obtained, or when its maximum thickness is reached (this depends on the heat-transfer characteristic of powder and the heat retention in the substrate). With thermoplastics, the coating cools and solidifies. Thermosets usually require more time at elevated temperature to complete the cure, once they have been removed from the bed. Many TPs can be used, particularly if they have low melting points. Higher-melting plastics must have a sufficiently low melt viscosity, so the particles can flow and fuse together to form a coating with good integrity. Most common TPs are nylon, PVC, PMMA, PE, PP, PUR, silicone, EVA, PS, etc. TSs are limited largely to epoxy, polyester, and epoxy/polyester hybrids. TSs such as phenolics or DAPs give off by-products during cure. These volatiles can create voids in the coating, damaging its integrity and function.

The stability of the fluidized-bed process derives from the stability of the bed itself. The mixing of the gas keeps the powder well distributed; the dipping process does not selectively remove particles in a manner that would change the character of the bed. Fluidized beds have been known to function for years with little attention other than replenishment of the powder supply.

Disadvantages include (1) high product cost, (2) requirement that substrates withstand heating, (3) thick coatings, (4) dust as a health hazard, and (5) for large parts, a large fluidized bed and much powder.

Certain part configurations are not easily coated by this process. Parts with small, deep holes tend to plug off at the opening rather than coat uniformly. Tight U-shapes or parts that present large heat sinks in one area also coat unevenly. In general, the heat configuration of the product significantly influences the evenness of the coating.

Electrostatic coatings

This process uses the principle that oppositely charged particles attract, the principle used in spray painting. Electrostatic spraying can produce coating thicknesses of 1–2 mil 25–50 μm. A continuous process, it is suited to automated production lines.

Plastic powder is first fluidized in a bed to separate and suspend the particles. It is then transferred through a hose by air to a specially designed spray gun. Powder can also be introduced into the transfer hose using a reservoir system. As it passes through the gun, direct contact with the gun and ionized air applies an electrostatic charge to the particles of powder. The contact area may be a sleeve that extends the length of the gun or merely small pins that extend into the passageway of the powder. The powder continues through the gun and exits past a specially designed tip that forms the spray pattern of the powder.

The part to be coated is electrically grounded, attracting the charged particles. This not only reduces overspray, but also produces a more even coating. Parts to be coated may be preheated, thereby forming the coating immediately, or they may be coated cold. The electrostatic charge will hold the particles in place until heat is applied, although care must be taken not to jar the particles loose. Once heat is introduced, the particles melt and flow together, forming a continuous protective coating.

Low melt viscosity is crucial for any TP or TS to be applied; it enables the individual particles to flow together and form a continuous coating. The selection of the plastic depends on the end use of the coated part and its environmental conditions. The particle size of sprayed powders is smaller than for powders used in a fluidized bed. The average particle size is 30–60 μm; the largest, 100–120 μm. A smaller particle size is dictated by the mechanics of spraying and the need to apply a thin coating. The advantages and disadvantages of powder coating are similar to those of fluidized-bed coatings, but there are some restrictions on electrostatic spraying of powdered plastics. Airborne dust from sprayed powder is a much greater hazard to worker health, and dust control systems are mandatory. An enclosed spray booth improves worker safety and assists powder recovery. Airborne dust also constitutes a significant explosion hazard. When sufficient quantities are airborne, any finely divided, organically based powder can explode if an ignition point is provided. Proper preventive measures must be followed strictly.

Thickness is limited due to the natural insulating properties of plastics. Coatings 50–75 μm thick can be applied electrostatically to cold objects; coatings up to 250 μm thick to hot objects. There are also some limitations on the configuration of the part. Depressions or recesses may not be deeper than their width. A deep recess attracts the particles to the sides at the top of its opening, producing a thicker coating near the opening and little or no coverage at the bottom.

Electrostatic fluidized coatings

The electrostatic fluidized bed is a combination of the electrostatic spray and the fluidized beds. A current of air passing upward through the bed fluidizes and suspends the powder, which is charged by electrodes in the permeable membrane. The applied voltage determines the density of the cloud when a ground is passed over the bed. A cold and grounded object is passed over the bed, not immersed in it. The electrostatic principles that apply to powder spray are pertinent here, too. Coating thicknesses can range from the low values of electrostatic spraying to the high values of the fluidized bed.

VACUUM COATINGS

This technology and use is important in different industries, particularly in advanced applications such as microelectronics. Coatings for these hi-tech parts are employed to maintain certain desirable surface properties that differ from those of bulk material for protection of the underlying product. Good coatings can be deposited using vacuum techniques, either by physical or chemical vapor deposition. Physical methods include evaporation and sputtering; chemical methods include polymerization through pyrolysis and glow discharge (plasma). The starting materials may be solids or gases, and the type of energy may be plasma or thermal. Sputtering employs a solid target and a plasma energy source.

MICROENCAPSULATION COATINGS

Microencapsulation is where particulate matter is individually coated for protection against environmental influences. In the broadest sense, it provides a means of packaging, separating, and storing materials on a microscopic scale for later release under controlled conditions. Minute particles or droplets of almost any material can be encased by an impervious capsule wall and thus isolated from reactive, corrosive, or otherwise hostile surroundings. The contents of a capsule can be made available by mechanical rupture of the capsule wall, its disintegration by electrical or chemical means, or by leaching action carried out in an appropriate liquid environment. In most applications the particles are very small, so the term *microencapsulation* is appropriate. The term *encapsulation* refers to the process in which a larger part or assembly is coated, usually by dipping in a highly viscous or thixotropic medium. Both terms have been used loosely. However, microencapsulation involves capsules that start at a few micrometers in diameter and go up to 4 mm or larger. There is no definite industry rule to say when encapsulation ceases to be microencapsulation, but the general size is 200 μm.

ELECTRODEPOSITION COATINGS

The process is also called electrocoating, electroplating, and e/coat. It is the precipitation of a material at an electrode by the passage of an electric current through a solution or suspension of material, including paint, latex, epoxy, and elastomers. The electrode is the shape of the part.

An important advantage of this process is that very complex products can be coated with accurate thickness control. It offers very high corrosion protection on metals, low cost, and compliance with environmental regulations. It is used for the coating of articles of various sizes, including steel building trusses, car bodies, furniture, appliances, toys, and nuts and bolts. The success of this process is due to the use of water-dispersible, electrodepositable macroions as film formers. Use of water as a carrier virtually eliminates fire hazard and environmental pollution, and reduces the cost of control equipment.

The low viscosity of the paint bath, similar to water, facilitates agitation and pumping and allows fast entry and drainage of workpieces. Freshly deposited coats are composed of 97% nonvolatile substances and therefore allow immediate gentle handling; there is no tendency to sag or wash off during cure. A second coat, usually a color coat, of waterborne or solvent-borne spray paint can be applied directly over the uncured electrocoat. Approximately 95% of the applied paint is utilized because the liquid paint, which adheres or fills cavities of freshly coated pieces, is rinsed back into the coating tank. Overall savings, accounting for materials, labor capital investment, energy, etc., are about 20–50% when compared with spray, electrostatic spray, or dip-coat painting.

FLOC COATINGS

Floc spraying or flocking is a method of coating by spraying finely dispersed powders or fibers. With the usual hot-flocking techniques, materials are dispersed in air and sprayed or blown onto a preheated substrate, where they melt and form a coating. In a variation of this process, small parts are preheated and dropped into a bed of powder or fibers kept in a mobile state of vibration. In this method the parts are completely coated with an unfused layer on the surface.

In order for flocculation to occur, chemicals called flocculants or coagulants are introduced to the solid suspension. Most of these chemicals are water-soluble polymers.

Plastics are used to rapidly expand the scale of flocculants in a wide variety of industrial applications. More than half of all plastic flocculants, including ultrahigh-MW flocculants, are copolymers in which the principal monomer is acrylamide.

PINHOLE-FREE THIN COATINGS

This is a coating process capable of producing pinhole-free coatings of outstanding conformity and thickness uniformity by the unique chemistry of the xylylene monomer, a substrate is exposed to a controlled atmosphere of pure gaseous monomer, p-xylylene (PX). The coating process is best described as a vapor deposition polymerization (VDP). The monomer itself is thermally stable but kinetically unstable. Although it is stable as a gas at low pressure, on condensation it spontaneously polymerizes to produce a coating of a high molecular weight, linear poly(p-xylylene) (PPX).

The p-xylylene polymers (PPXs) formed by the Gorham process are generically known as the parylenes. The terms Parylene N, Parylene C, or Parylene D refer specifically to coatings produced from the Union Carbide

Corporation dimers. The polymerization process takes place in two stages that must be physically separate but temporally adjacent.

The parylene process has certain similarities with vacuum metallizing. The principal distinction is that truly conformal parylene coatings are deposited even on complex, three-dimensional substrates, including on sharp points and into hidden or recessed areas. Vacuum metallizing, on the other hand, is a line-of-sight coating technology. Areas of the substrate that cannot be 'seen' by the evaporation source are 'shadowed' and remain uncoated.

Freestanding films can be produced of Parylene. These **ultrathin (35–3000 nm)** films, called pellicles, are used as beam splitters in optical instruments, windows for nuclear radiation measuring devices, dielectric supports for planar capacitors, and for extremely fast responding, low-mass thermistors and thermocouples.

Applying Parylene requires special, though not complex or bulky, equipment: a vaporizer, a pyrolysis unit, and a deposition chamber. The objects to be coated are placed in the deposition chamber, where the vapor coats them with a polymer. A condensation coating like this does not run off or sag as in conventional coating methods, neither is it 'line-of-sight technology,' as in vacuum metallizing. In condensation coating, the vapor evenly coats edges, points, and internal areas. Although the vapor is all-pervasive, holes can still be coated without bridging. Masking can easily prevent chosen areas from being coated. The objects to be coated can also remain at or near room temperature, thus preventing possible thermal damage. The quantitative nature of this reaction allows the coating thickness to be accurately and simply controlled by manipulating the polymer composition charged to the vaporizer.

These thermoplastics are generally insoluble up to 150 °C (302 °F). At 270 °C (518 °F) they will dissolve in chlorinated biphenyls, but the solution gels upon cooling below 160 °C (320 °F). Their weather resistance is poor. Embrittlement is the primary consequence of their exposure to UV radiation.

The first significant commercial application of Parylenes was as a dielectric film in high-performance, precision electrical capacitors, followed by use in circuit boards and electronic module coatings. These coatings are to protect units from airborne contaminants, moisture, salt spray, and corrosive vapors while maintaining excellent insulator protection. The coatings are also extensively used in the protection of hybrid circuits. Such coatings do not affect part dimensions, shapes, or magnetic properties.

DIP COATINGS

Dip coating involves the use of PVC plastisol (Chapter 14). The process consists of preheating a metal object – the product to be coated. This heated part is dipped into the plastisol. After a specified time, it is removed from the PVC compound; any surplus is allowed to drain back into the tank. The coated product is cured by passage through an oven. The thickness of the coating is controlled by varying the viscosity of the plastisol and the temperature of the product before dipping.

8

Compression molding and transfer molding

INTRODUCTION

Compression molding (CM) and transfer molding (TM) are the two main methods used to produce molded parts from thermoset (TS) resins. Compression molding (CM) was the major method of processing plastics during the first half of this century because of the development of a phenolic resin (TS) in 1909 and its extensive use at that time. By the 1940s this situation began to change with the development and use of thermoplastics (TPs) in extrusion and injection molding (IM) processes. CM originally processed about 70 wt% of all plastics, but by the 1950s its share of total production was below 25 wt% and now, at the time of writing, that figure is about 3 wt%. This change does not mean that CM is not a viable process; it just does not provide the much lower cost to performance of TPs, particularly at high production rates. In the early 1900s resins were almost entirely TS (95 wt%); that proportion had fallen to about 40 wt% by the mid-1940s, and is now about 15 wt%.

During this century, TSs experienced an extremely low total growth rate, whereas TPs expanded at an unbelievably high rate. Regardless of the present situation, CM and TM are still important, particularly in the production of certain low-cost parts as well as heat-resistant and dimensionally precise parts. CM and TM are classified as high-pressure processes.

The basic process of CM and TM is as follows. A thermosetting molding compound is exposed to sufficient heat, approximately 300 °F (149 °C), to soften or plasticize it. The fluid plastic is held at the molding temperature under pressures as high as 2000–4000 psi (13.8–27.6 MPa) for a sufficient length of time for the material to undergo polymerization or cross-linking, which renders it hard and rigid. The part is then removed from the mold

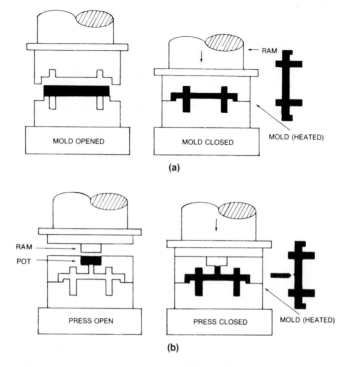

Figure 8.1 (a) Compression molding and (b) transfer molding.

cavity (Fig. 8.1). But some TSs, such as polyesters, require lower pressures of only 50 psi (340 kPa or even just contact (zero pressure).

CM is the most common method of molding TSs. Material is compressed into the desired shape using a press containing a two-part closed mold, and is cured with heat and pressure. This process is rarely used with TPs.

MATERIALS

Different TS resins are used, such as phenolics, TS polyesters, DAPs, epoxies, ureas, melamines, and silicones, all with their own parameters and performance properties based on the additives used (Tables 8.1 and 8.2). Note there are TS and TP polyesters; TSs are used primarily in CM and TPs in injection molding and blow molding. In this chapter the emphasis is on TSs, which have completely different processing characteristics and properties. This resin is used in the very popular CM materials known as bulk molding compounds (BMCs); in Europe they are called dough molding compounds (DMCs). They are formulated from different percentages of TS polyesters filled with glass fibers of lengths up to $\frac{1}{2}$ in.

Table 8.1 Common thermosetting molding compounds and their principal applications

Material	Advantage	Applications
Phenol-formaldehyde	Low cost	
General-purpose	Durable	Small housings
Electrical grade	High dielectric strength	Circuit breakers
Heat resistant	Low heat distortion	Stove knobs
Impact resistant	Strong	Appliance handles, legs
Urea formaldehyde	Color stable	Kitchen appliances
Melamine formaldehyde	Hard surface	Plastic dinnerware
Alkyd	Arc resistant	Electrical switchgear
Polyester	Arc resistant	Electrical switchgear
Diallyl phthalate	High dielectric strength	Multipin connectors
Epoxy	Soft flowing	Encapsulating electronic components
Silicone	Withstands high temperature	Encapsulating high-power electronic components

(13 mm) and fillers. The BMCs flow easily and provide high strength. Chapter 12 covers BMC and SMC materials. Still popular as CM molding materials are the TS vinyls, used for phonograph records. TP vinyls are cross-linked to turn them into TS vinyls (Chapter 1).

Very soft-flowing TS materials are required for molding around very delicate inserts. Vast quantities of electronic components, such as resistors, capacitors, diodes, transistors, integrated circuits, etc., are encapsulated with such soft-flowing thermoset compounds, principally epoxies, by transfer molding. Silicone molding compounds are used occasionally, where higher environmental temperatures are required of the encapsulated part, up to 500 °F (260 °C) or more. TS polyester compounds, which are somewhat less expensive than epoxies or silicones, may also be selected.

With most TS resins there exists a wide range of flow characteristics, cure times, and ultimate mechanical, chemical, and electrical properties. These molding compounds are mixtures of constituents, usually of different size and shape; so the compounds themselves present the greatest number of variables that must be understood and properly applied. Knowledge of these resins is principally gained through molding experience.

From a processing point of view, the viscosity–time curve is often the most critical characteristic. Most compounds are granular at room temperature; when exposed to higher curing heats, the granules melt and

Table 8.2 Examples of various reinforcing fibers and fillers used with thermoset resins

Thermosets	Alumina	Calcium carbonate	Carbon black	Clay	Cotton flock	Glass bubbles	Glass fibers	Graphite	Mica	Quartz	Talc	Wood flour
Alkyds	X	X	X	X			X		X		X	X
Diallyl phthalate		X		X		X	X				X	
Epoxy	X	X	X	X		X	X	X				
Phenolic	X	X	X	X	X	X	X	X	X	X	X	X
Polyester	X	X	X	X		X	X	X				X
Melamine					X		X				X	
Urea				X	X				X			X
Silicone	X			X			X	X		X		
Urethane	X	X	X	X		X	X	X		X		

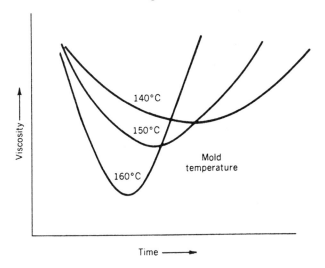

Figure 8.2 Shortening cycle time: plots of mold temperature versus flow time.

become fluid. Under continued heat, cross-linking occurs, and the plastic compound solidifies. As the material goes from solid to fluid to solid (Figs 8.2 and 1.18 on page 57), its viscosity changes accordingly. At some optimum heat for a given part in a specific mold, the viscosity of the fluid plastic will generally be at a minimum for the period of time needed to ensure filling of the cavities at an acceptable pressure. If the heat is too low or too high, the filling of the cavities may be problematic, even if higher pressures are used. Proper pressure is ideally applied when the lowest viscosity occurs, producing good surface finish and relatively stress-free parts.

The majority of TS compounds are heated to about 300–400 °F (149–204 °C) for optimum cure. Higher heats could degrade their performance or could cause them to solidify rapidly, particularly in TM, where material could solidfy before the cavity completely filled. Lower heats extend the cycle time. The molds are heated by electricity, steam, or hot circulating heat-transfer fluids.

PROCESSING CHARACTERISTICS

Compared to other processes, particularly injection molding (IM), for shaping plastics, CM and TM are fairly labor-intensive, even if they are semiautomated, but they require lower capital investment. Molding cycles for CM and TM are generally longer than for IM. If the material used in TM is preheated or preplasticized before it is placed in the transfer pot, TM cycles may be comparable to IM.

PREHEATING

As compounds generally have good heat insulation, preheating is often used to reduce the molding cycle; it can aid in providing even heat through the material and can cause a more rapid rise in heat than occurs in the mold. Preheating may be accomplished by a warm surface plate, infra-red lamps, a hot-air oven, or a screw/barrel preheater, but the best and quickest method is high-frequency (dielectric) heating. Preheating is usually carried out at 150–300 °F (66–149 °C), followed by quick transfer to a mold cavity. The actual heat depends on the material, the heater capability, and the speed of transfer. Circular preforms are used with dielectric heaters so they can be rotated to obtain uniform heating. Pills of compressed compound are used to produce preforms to reduce the bulk factor, facilitate handling, and control the uniformity of charges for mold loading. Preforms can be of the shape desired in the mold cavity.

MOLD HEATING

When molding thermoset compounds by compression or transfer methods, the mold is maintained at a constant temperature set for optimum polymerization of the material at each cycle. Such temperatures range from 300–400 °F (149–204 °C). For optimum molding operations, temperature must be uniform across the surface of the mold and in the cavity areas, ideally to ±2 °F (±1.1 °C). Electric heating is presently the most common technique, utilizing multiple electric heating cartridges inserted in the top and bottom halves of the mold, positioned to supply heat to all cavity areas. On larger molds, temperature controllers and sensing elements are often used in several zones.

An earlier method for mold heating channeled saturated steam around the cavities. Steam has the advantage of rapid temperature recovery because of its tendency to condense in the steam channels when any lower temperature occurs, rapidly releasing the heat of condensation. But steam is hard to contain in a mold and in the vapor lines leading to and from the mold. It requires excessively high pressures when higher mold temperatures are required.

By contrast, circulating hydraulic oil can easily bring mold temperatures to 400 °F (204 °C) or higher. Self-contained oil heating and cooling systems are available that make this type of mold heating practical, although the cost of fluid-heated molds is generally higher than the cost of electric cartridge-heated molds. Some molding compounds and molding cycles may require one temperature for cure but a lower temperature for optimum ejection. Such molds have provisions for both heating and chilling, and require longer molding cycles because of the time needed for heat transfer. They may have electric heaters for heating and water channels for cooling, or may utilize the circulating hydraulic oil system with pro-grammed heating and chilling.

OUTGASSING

Many TSs, particularly phenolics, ureas, and melamines, emit a gas dur-ing curing that must be allowed to escape; major exceptions are the TS polyesters (Chapter 1).

Some of the gases can escape through clearances at the two mold halves and around ejection pins. To ensure they do not become entrapped and weaken the part or cause surface blemishes, during the start of the cure cycle the mold is slightly opened about 1–5 mil (25–130 μm) to allow the gases to escape. This action is called breathing or bumping. If the cure cycle permits, it is best to repeat this bumping cycle.

The length of time for single or multiple bumping depends on the material used and the size and configuration of the part. This dwell time with release of gases can actually reduce the total duration of the cure cycle. Outgassing also can occur after parts are molded. With certain materials this condition can last for months or years. If parts are to be metallized or plated, outgassing tendencies can actually rupture the plated film.

SHRINK FIXTURES

After a molded part is removed from the cavity, it is still hot and the material is not fully rigid. Any internal stresses in the material may there-fore cause the shape of the part to change while it is cooling. Where close tolerances are required, and especially where parts have thin sections, dimensional accuracy is achieved by placing the hot, molded part on a fixture near the press that will hold it until it has cooled. The fixture may be a spindle placed in a molded hole so the hole dimension remains fixed; it may simply be a clamping mechanism to hold the modled parts flat against a cold, flat surface; or it may be any other restraining arrangement to enable the part to cool in the desired shape. Once cool, the part can be removed from the fixture with no further shrinkage or tolerance change.

POSTCURING

Molded thermosets are frequently postcured by baking at a material sup-plier's recommended times and heats, to enhance mechanical properties, thermal properties, and dimensional stability as well as to eliminate outgassing. Baking also improves creep resistance and reduction in stresses. This postcuring is also used with certain TPs after IM or extrusion to improve their performance.

The postcuring heat is usually below the actual molding heat. Postcuring should be done with a multistage heat cycle such as is used for glass fiber and/or mineral reinforced phenolic compounds: (1) for parts $\frac{1}{8}$ in. (3 mm) or less in thickness, times are 2 h at 280 °F (138 °C), 4 h at 330 °F (166 °C), and 4 h at 375 °F (191 °C); and (2) for parts exceeding $\frac{1}{8}$ in. (3 mm) thickness, the time periods are doubled for each $\frac{1}{16}$ in. (1.5 mm) of thickness. The use of longer times is more effective than simply increasing the heat.

The reinforcement system of the compound will largely dictate the heating cycles. Parts molded from compounds using organic reinforcements are postcured at lower heats than those using glass and mineral reinforcements. Parts of uneven thickness will exhibit uneven shrinkage. This shrinkage effect is included in the mold design.

COMPRESSION MOLDING

CM machines can be discussed in three major categories: platen frame, driving means, and controls (Fig. 8.1(a)). Two or more platens can be used so that molds are placed between platens. Most of these CM (and TM) machines have the platens moving vertically to simplify locating and feeding the molds. As in other press operations (IM, blow molding, etc.), the platens and their supporting frames must have sufficient strength to minimize deflections. Beware of 'lightweight' machines; there may not be enough steel to contain the molds, preventing them from bending, twisting, etc.

The driving means to move platens and apply clamping pressures is usually a double-acting hydraulic ram; that is, it uses high-pressure oil to bring the platens together and to separate them. Electric pumps provide 2000–3000 psi (13.8–20.7 MPa) oil pressure. The driving means consists of an oil reservoir, valving, and so on, resembling an IM machine. Most CM is performed in machines that only provide relatively constant, high clamping pressure. For better operation, the processor should consider variable-pressure loading, similar to in IM. It significantly improves the melt flow during curing. Machine controls, as with IM, include manual, semiautomatic, and sometimes fully automatic.

Although most presses are hydraulically operated, some are pneumatically operated. Platens range from 6 in. (15 cm) square to 8 ft (2.4 m) square or more, exert clamping pressures of 6–10 000 tons. Virtually all CM presses are of vertical design; platen moving up and down. Most lower-tonnage machines have the top platen moving (upward-acting); others have the lower platen moving (downward-acting). Some presses are built with a shuttle-clamp arrangement that moves the mold out of the clamp section to facilitate setup (inserts, etc.) and part removal. The top platen on some machines can be tilted, a book-type operating press. The different types of press provide different means to handle different mold shapes and types; they improve access for mold cleaning and maintenance, etc. The clamp table can have two mold halves, increasing productivity by allowing two operators to work on opposite sides of the press.

TRANSFER MOLDING

In TM the mold halves are brought together under pressure, as in CM. The charge of molding compound is then put into a pot, and is driven from the pot through runners and gates into the mold cavities using a plunger. Its basic construction, driving system, and controls are similar to those of CM except for the additional action required by the transfer pot (Fig. 8.1(b)).

The process differs from CM in that the plastic is heated to a point of plasticity in the pot before it reaches and is forced into the closed mold. This procedure facilitates the molding of intricate parts with small holes,

Figure 8.3 Front view of a 64-cavity mold (with integrated circuits) showing layout, top-transfer pot, and plunger used in a transfer molding system. (Courtesy Hull Corp.)

Table 8.3 Comparison of compression and transfer molding

Characteristic	Compression	Transfer
Loading the mold	1. Powder or preforms 2. Mold open at time of loading 3. Material positioned for optimum flow	1. Mold closed at time of loading 2. Preforms RF heated and placed in plunger well
Material temperature before molding	1. Cold powder or preforms 2. Preforms RF heated to 220–280 °F	Preforms RF heated to 220–280 °F
Molding temperature	1. One step closures: 350–450 °F 2. Others: 290–390 °F.	290–360 °F
Pressure via clamp	1. 2000–10 000 psi (3000 optimum on part) 2. Add 700 psi for each inch of part depth	1. Plunger ram at 6000–10 000 psi 2. Clamping ram having minimum tonnage of 75% of load applied by plunger ram on mold
Pressure in cavity	Equal to clamp pressure	Very low to maximum of 1000 psi
Breathing the mold	Frequently used to eliminate gas and reduce cure time	1. Neither practical nor necessary 2. Accomplished by proper venting
Cure time (time pressure is being applied on mold)	30–300 s but will vary with mass of material, thickness of part, and preheating	45–90 s but will vary with part geometry
Size of pieces moldable	Limited only by press capacity	About 1 lb maximum
Use of inserts	Limited because inserts may be lifted out of position or deformed by closing	Unlimited but complicated; inserts readily accommodated
Tolerances on finished products	1. Fair to good: depends on mold construction and direction of molding 2. Flash = poorest, positive = best, semipositive = intermediate	Good: close tolerances are easier to hold
Shrinkage	Least	1. Greater than compression. 2. Shrinkage across line of flow is less than with line of flow

numerous metal inserts, and so on. Less force is used, and less melt action occurs in the cavity (Fig. 8.3 and Table 8.3).

Automation of the transfer/plunger concept is accomplished by the addition of a hopper-fed screw plasticator, in a system called screw transfer, which can replace the preform and preheat operations and automatically load the pot in CM (and is used to load a CM mold cavity). Conventional screw injection molding (IM) is another so-called transfer system. As explained in Chapter 2, for TSs the processor uses a screw compression ratio of 1, the barrel heat is kept relatively low (below the curing/hardening heat), and the mold is at a higher heat to permit final cure.

Insert molding

Any TM inserts must be adequately supported in the cavity when the mold is closed, so they are not displaced by the influx of molding material. Inserts are pushed into holes or slots in the cavity that satisfactorily support them. Inserts may be flexible or supported by weak extensions, such as soft lead wires on a resistor (which are not rigid enough to prevent the part from shifting in the cavity during the molding operation). To make certain the inserts are positioned correctly, retractable positioning pins may be incorporated in a mold. They are the ejector pins and are actuated so that, during the initial introduction of material into the mold cavity, the pins project into the cavity to restrain any shifting of the inserts.

Once the cavity is essentially filled with the molding material, but prior to polymerization, the pins are automatically retracted until their ends are flush with the surface of the cavity. Because there is still residual fluid material in the transfer pot and the transfer plunger is still under pressure, further flow of material into the cavity fills the voids left by retraction of the positioning pins. Inasmuch as the insert is well surrounded by molding compound at this stage, there is no further displacement. The molded part is cured with the insert in the desired position. Retractable positioning pins can be designed into a mold without adding appreciably to the cost. They are moved by small hydraulic cylinders mounted on the outside of the mold and actuated by a timer.

MOLDS

In the process of loading material into a compression mold, the charge of material to be molded is somewhat smaller in length and width but is thicker than the final part and is loaded near the geometric center (Fig. 8.4). The closing of the mold spreads the preform to fill out the cavity. Instead of preforms, loose granules can be used. With sheet molding compound (SMC), described in Chapter 12, the material is cut to a prede-

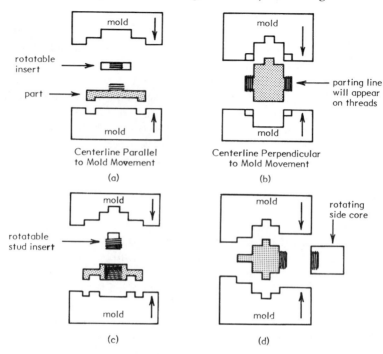

Centerline Parallel
to Mold Movement

(a)

Centerline Perpendicular
to Mold Movement

(b)

(c)

(d)

Figure 8.4 Compression molding threads in various positions.

termined size and weight then placed on the lower half of an open mold in a specific arrangement called the charge pattern.

Placing material in an open mold rather than injecting it into a closed mold, as in TM or IM, helps to preserve the integrity of SMC reinforcing fibers. For this reason, SMC parts usually have better strength when made in an open mold. The pattern used to mold a part can have a critical effect on both quality and performance.

A key feature of compression molds for SMC is a telescopic shear edge, located around the periphery of the mold cavity (Fig. 8.5). Its function is to seal off the mold when closed and to vent air and gases from the mold cavity. The shear edge also allows the cavity half of the mold to slide over the core half. Because of this feature, full molding pressure can be applied to each part and the plastic retained within the cavity. The closed position of the mold can vary, depending on the charge volume and the pressure on the part. This variance in turn causes an equal variation in the thickness of the molded SMC part; so it is very important to control the SMC charge size (thickness, length, and width), weight, and volume. One should use the same approach with other materials, such as granule and bulk molding compound (BMC); see Chapter 12.

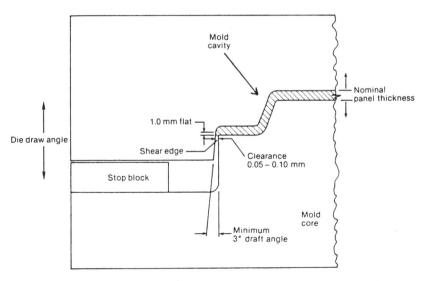

Figure 8.5 Closed mold using a shear edge around the cavity periphery.

Pressure

The curing pressures required for TSs depend on the formulation of the compound. With CM, the cavity depth can have a significant effect on pressure, as well as on the benefits of using preheated rather than cold material delivered into the cavity (Table 8.4).

Cavity plating

TS molds are usually hard chrome-plated on areas that are exposed to the molding compound during the molding process, such as the plunger, pot, sprue bushing, runners, gates, actual cavity, and any mold area subject to the clamp force. Plating may be done after the mold has been through initial trial runs and approved for production. With some compounds, the mold surface is coated with material that will not permit plating (poor adhesion would occur). One should consult the material supplier about this potential problem, particularly when using additives in the compound such as silicones and fluorocarbons. Once cavity surfaces have been treated, it is difficult or impossible to remove an unwanted coating. This problem also affects secondary operations such as decorating and painting.

An allowance is provided in the cavity for tight part tolerance requirements and plating. As plating highlights the mold surface, precise cavity polishing is required before plating. Plating will not cover a poorly polished surface.

Table 8.4 Examples of pressure (psi) required to compression mold based on cavity depth using phenolic molding compounds

Depth of molding (in.)	Conventional phenolic		Low-pressure phenolic	
	Dielectric preheat	*Not preheated*	*Dielectric preheat*	*Not preheated*
$0-\frac{3}{4}$	1000–2000	3000	350	1000
$\frac{3}{4}-1\frac{1}{2}$	1250–2500	3700	450	1250
2	1500–3000	4400	550	1500
3	1750–3500	5100	650	1750
4	2000–4000	5800	750	2000
5	2250–4500	[a]	850	[b]
6	2500–5000	[a]	950	[b]
7	2750–5500	[a]	1050	[b]
8	3000–6000	[a]	1150	[b]
9	3250–6500	[a]	1250	[b]
10	3500–7000	[a]	1350	[b]
12	4000–8000	[a]	1450	[b]
14	4500–9000	[a]	1550	[b]
16	5000–10 000	[a]	1650	[b]

[a] Add 700 psi for each additional inch of depth. Preheating is desirable beyond 4 in. depth and essential beyond 12 in. depth.
[b] Add 250 psi for each additional inch of depth. Preheating is desirable beyond 4 in. depth and essential beyond 12 in. depth.

Some of the reasons for plating are (1) to provide an excellent part finish; (2) to enhance compound flow, producing more uniform and denser parts; (3) to lengthen mold life by improving wear resistance, particularly when coarse materials are included in the compound, such as glass and metals; and (4) to provide resistance to mold staining, permitting longer production runs with less mold cleaning time. The cost of hard chrome plating is minor when compared to the savings in the molded part.

Venting

All closed molds, as in CM, TM, IM, blow molding, etc., require a means by which air and volatiles (for certain TSs and TPs) are evacuated from the cavities. Although bumping may be employed, the usual technique is to incorporate 'openings,' usually located at the mold parting line. Their size depends on the plastic compound's viscosity, but they are usually 0.25 in. (6 mm) wide and 1–3 mil (25–76 μm) deep, located where the cavity will be filled last. The location of the vent opening also depends on the heating

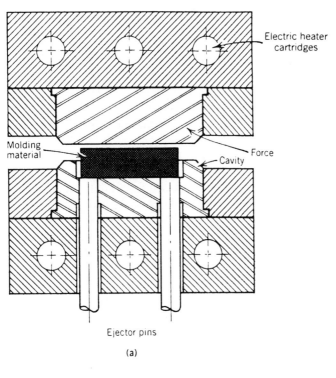

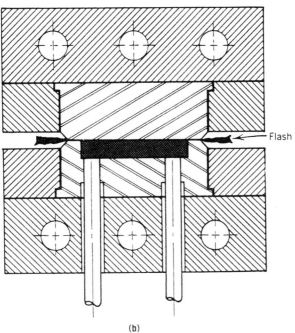

Figure 8.6 Operation of a flash mold: (a) open, preform in the cavity and (b) mold closed with flash.

pattern of the mold, particularly if the heat flow pattern was not logically planned. CAD programs are available to provide the proper heating pattern. With excessive heat in one section of the mold, the viscosity of the compound could be low enough to require a vent opening in that area. Knockout pins often provide a means for venting, and they may require recessed sections, such as 'flats' ground on the OD, that will allow venting.

A special feature is vacuum venting of the cavities. This is not normally required for CM, where venting and/or bumping is adequate, but may be needed in TM, where the mold is closed before introducing the material. Vacuum-vented molds are comprised of vents in the cavities leading to a vacuum chamber (via a resin trap). In addition to the vent area itself, a vacuum seal consisting of a high-temperature 0-ring, usually silicone rubber, is placed around the cavities at the parting line or around the entire parting surface of the mold. Adding vacuum to TM and when required to CM, can significantly reduce the number of defects caused by trapped air in the molding compound. It also draws moisture and other contaminants out of the material, rasing the quality of molded parts.

Flash

During the thermal curing cycle, the low viscosity of TSs tends to cause flash, particularly in CM (Fig. 8.6). There is no flash with TPs, as the viscosity is not as low, and the molds are closed. The costly removal of flash from the parting line and holes, as well as slower molding cycles, difficulty in molding side holes or sections, and problems in molding flash-free metal inserts, have all reduced the application of CM in favor of closed-mold TM and IM.

Flash removal is most often accomplished with tumbling machines. As their name implies, the parts are tumbled against each other to break off the flash. The simplest tumblers are merely wire baskets driven by a small electric motor and pulley belt. Tumblers can also be very elaborate, involving not only the tumbling operation, but often 'blasting' with granulated peach or apricot pits, walnut shells, plastic pellets, and so on, to provide additional flash removal. This type of action can also improve the surface polish and/or toughness for certain molding compounds. A steam jet is sometimes used to minimize the accumulation of static charges. Tumblers can be of the batch type or provide continuous movement of parts to more accurately control the time of flash removal on each part.

MOLD CONSTRUCTION

The information presented on molds in Chapter 2 applies to CM and TM. In TM the transfer pot and plunger should have a diametrical clearance of about 1 mil (25 μm) per inch of diameter. With one or two small, half-round, O-ring type grooves cut circumferentially around the plunger, considerable plunger wear can be tolerated without leakage of material. The grooves fill with material and remain filled during any production run. The pot should be hardened, polished, and plated; the plunger should be well polished and hardened.

Molds are generally produced from various grades of tool steel. Most are polished or chrome-plated to improve the material flow and the surface of the molded part. Mold temperatures are typically 300–400 °F (149–204 °C), but can go as high as 1200 °F (650 °C). They are heated by electric strip heaters, electrical cartridges, steam, hot oil, hot water, and gas flames; electric heaters are the most common.

Runners: cold and hot

The runner size depends on the material being processed and whether it is a TS cold or hot runner. As reviewed in Chapter 2 and Fig. 2.32 (pages 172–3), with TPs a hot runner solidifies with the injection-molded part. If a cold runner is used (with TP), only the molded part solidifies; there is no runner scrap.

With TS closed molds (TM, IM, etc.), the complete mold can be at the maximum heat, which cures and solidifies both the runner and the molded parts. With TSs this runner is called a hot runner. If the mold is designed so the maximum heat surrounds only the cavity and the lower heat (similar to the lower pot heat) surrounds the runner, that is called a cold runner. In this case, material in the runner remains a fluid, so the next shot into the mold cavity receives the runner material before material flows from the pot during transfer molding.

Unfortunately, solidified TS runners, flash, and defective parts cannot be recycled as they are with TPs. Some operations can use this material as a filler after it has been granulated. One should consider using 'cold runners' with TSs. Materials must be of a type that will permit their use; for example, the time cycle should permit material to remain in the fluid state. As different hardeners, accelerators, and so on, can be used, it is practical to use cold runners. They have limited use in TM but are used more widely in IM.

SHRINKAGE

Shrinkage with TPs can be rather substantial, but with TSs it is lower, principally because of mixing with inorganic mineral additives that are not affected by heat changes. The degree and direction of shrinkage will depend on the molded part density, part configuration, ratio of resin to additives/reinforcements, and/or orientation of fibers (Chapters 1, 2, and 12).

THERMOPLASTICS

Even though CM and TM are used principally with TSs, these processes are sometimes used to mold conventional TPs for short runs with low-cost molds. They are also used with TPs that are difficult to melt, have too short a heat melt range for IM or other processes, and require high pressures. A typical example is ultrahigh molecular weight polyethylene (UHMWPE).

DESIGN ANALYSIS

With experience, logic, and trial-and-error runs, molds can be designed to maximize the efficiency of material flow and heat control. The result is that parts are molded to meet performance requirements at the lowest cost. Heat flow/control in molds is generally rather inefficiently evaluated, so there is excess or unnecessary heat, and part performance may not be maximized. The cycle time could be affected.

The flow and heating analyses that have been developed into CAD systems for IM (Chapter 2) are also being applied in CM and TM. They facilitate the construction of molds, by locating cavities, sizing runners, and heater lines (electric, steam, etc.) as well as other devices to meet cost and performance targets.

COSTING

Either CM or TM may prove to be an ideal process for low-volume production and for development or prototype work because the equipment required is less expensive than injection molders. Parts up to 10–12 in. (254–305 mm) in any dimension are suited, but larger parts warrant consideration of other processes as well. For very large parts, machinery and mold costs increase rapidly. Extremely tiny parts, on the other hand, can often be made in multicavity molds at relatively low cost per part. Molds of several hundred cavities are not uncommon in some applications. Using the semiautomatic process, parts requiring metal or other molded-in inserts are simple to produce as compared to molding in inserts in fully automatic IM.

The low cost of many of the TSs, principally phenolics, permits relatively inexpensive molded parts. The more exotic materials, such as the heavily glass-filled materials or the extremely soft-flowing materials, are more costly, making them less competitive. When applications call for maximum impact strength and dimensional stability, optimum materials are the fiber-reinforced phenolics or polyesters processed by semiautomatic compression molding. TPs can often withstand greater distortion without failure, but they lack acceptable dimensional stability under stress in many applications. The molding of fiber-reinforced materials with the IM process leads to high impact strength; however, compression molding causes less fiber breakage and produces more random fiber distribution, leading to even greater impact strength.

Because TSs flow at relatively low viscosity, there is often a flash line or gate scar on the molded parts that must be removed for cosmetic purposes. Such finishing of the molded parts may be accomplished simply by tumbling the parts against each other in a rotating barrel or passing them through an air-blast deflasher. The deflasher blows a grit material that is softer than the molded plastic and can be directed against the flash and gate areas for flash removal without damage to the molded part. Occasionally, with heavily reinforced plastics, such flash must be removed with a trimming die or by manual filing or breaking.

Finishing costs are often higher with CM than with TM or IM, principally because TM and IM have a fully closed mold before the material is introduced into the cavity. Other finishing costs may include tapping threads in molded holes or possibly drilling and tapping where the holes are not molded in. Inserts are occasionally pressed into molded-in holes after molding as a secondary operation. Inserting such parts after automatic molding may prove more economical than using molded-in inserts with semiautomatic molding.

Though more than a century old, compression and transfer molding continue to maintain their niche in the market. The exceptional economy and performance of thermoplastics, coupled with the variety of other processes, have shrunk the market for thermosets, but new applications continue to evolve. The dental and medical markets have seen a proliferation of CM orthodontic retainers and pacemaker castings. This is because the tools to produce the same parts would be eight times more expensive.

TROUBLESHOOTING

An adequate troubleshooting guide may be synthesized from the problems and remedies discussed throughout the chapter.

9

Foam molding

BASIC PROCESS

For about a century foamed plastics, whether TPs or TSs, have been a large and special category within the plastics industry. They are known by different names such as cellular plastics, expanded plastic foams, structural plastic foams, and plastic foams. A plastic foam material is a plastic whose apparent density is decreased by the presence of numerous cells throughout its mass; it is a two-phase gas–solid system in which the solid is continuous and composed of plastic material.

Foams are available with open-celled construction, closed or interconnecting construction, or in combination. Their densities range from $1.6\,\mathrm{kg\,m^{-3}}$ to over $960\,\mathrm{kg\,m^{-3}}$ ($0.1\,\mathrm{lb\,ft^{-3}}$ to over $60\,\mathrm{lb\,ft^{-3}}$). They can be rigid, semirigid, or flexible, colored or plain, and the foam can be virtually any TP or TS. They offer an extensive range of insulating properties, rigidity, compressive strength, cushioning and loading, structural characteristics, and others properties [3]. Their performance depends to a great extent on the type of base plastic, the type of blowing system, and the method of processing. Each plastic can include fillers or reinforcements to provide certain desirable properties.

There are many ways in which foams can be processed and used: slabs, blocks, boards, sheets, molded shapes, sprayed coatings, extruded profiles, and foamed-in-place within an existing cavity, where the liquid material is poured and allowed to foam. Conventional equipment can be used: extruders, injection molding machines (see relevant chapters).

The foaming methods vary widely. One is to whip air into suspension or a solution of the plastic, which is then hardened by heat curing. A second is to dissolve a gas in a mix, then expand it when the pressure is reduced. Another is to heat a mixture until one of its liquid components volatilizes. Similarly, water produced in an exothermic chemical reaction can be volatilized within the mass by the heat of reaction. A different technique uses a chemical reaction to produce carbon dioxide gas within a solid mass. A related way is for a gas such as nitrogen to be liberated within a mass by thermal decomposition of a chemical blowing agent. Other techniques disperse small solid particles, tiny beads of resin, or even glass microballoons within a plastic mix or syntactic foam.

The most common method disperses a gaseous phase throughout a fluid plastic phase then preserves the resulting combination; it is called the dispersion process. The expansion process consists of the following actions: (1) creation of small discontinuities or cells in a plastic fluid phase, (2) growth of these cells to a desired volume, and (3) stabilization of the resultant cellular structure by physical or chemical means. The gas phase is usually distributed in voids or pockets called cells. They can be foamed open-cell but usually they are foamed closed-cell.

FOAM DEFINITIONS

Effective communication requires agreed definitions. At times there can be more than one definition in order to meet the different requirements of different organizations, industries, etc. [9]. Various definitions are used within the foam industry; here is a guide to some of them.

Foamed closed-cell

Individual cells are noninterconnecting. The cells have no access to the surrounding air or fluids; cells are not communicating.

Foamed open-cell

Individual cells are generally interconnected; pores are accessible to surrounding air or fluids. This construction is called reticulated or web-like.

The open-cell or reticulated foams are made by removing the cell membrane from conventional closed-cell foam leaving a 3D skeletal structure which consists of the interconnected framing edges of the individual cell faces. These strut-like members are formed during the course of the foaming reaction, where adjacent cell sides meet, and are generally appreciably thicker than the thin membranes attached to them. The removal of the membranes can be accomplished by a variety of methods, all of which depend upon the cells being interconnected by an opening in at least one of the membranes connected to its neighbors; thus open-cell foam. Polyurethanes are the most important class of reticulated foams.

Foamed-in-place or in situ

Deposition of foams when the foaming machine must be brought to the work, as opposed to bringing the work to the foaming machine. Foam mixed in a container and poured into a mold or into cavity brickwork, where it rises to fill the cavity.

Foamed airflow rise

The airflow parallel to foam rise is the airflow value obtained when the air enters and leaves the mounted specimen parallel to the foam rise.

Foamed casting

A simple nonmechanical version of reaction injection molding or liquid injection molding. Liquid components of the plastic with suitable additives are mixed and poured into an open mold. Polymerization and foaming take place in the mold cavity, which could include a matching mold cavity to enclose the foaming action. Molds are generally heated or oven-cured, sometimes both.

Foamed cold-cure

Used to produce highly resilient, flexible foams.

Foamed collapsed

Inadvertent densification of foam or cellular plastic during fabrication; it is caused by the breakdown of cell structure.

Foamed cream time

The length of time between pouring mixed foam (usually PUR) and the instant it turns creamy or starts to foam. Also called foamed rise time.

Foamed extruded film

Foamed film uses extruders. See Foamed sheet stock.

Foamed flame resistance coating

A plastic coating formulated with an intumescent action to protect wood, etc., from intense heat of flames by decomposing into a foam barrier.

Foamed frothing

This process is similar to the process used for making dessert topping (lemon meringue, etc.). A gas is dispersed in a fluid which has surface properties to produce a foam of transient stability. The foam is then permanently stabilized by chemical reaction. The fluid may be a homogeneous material – a solution of a heterogeneous material.

Foamed metal

A foamed or cellular metallic structure, usually aluminum or zinc alloys, made by incorporating titanium or zirconium hydride in the base metal. Hydrogen evolves during heat processing. It is principally used in absorption of shock impact without elastic rebound. Fiber-reinforced light-metal composite foams can be used in structural applications such as furniture or automobiles; during the early 1940s it was used for de-icing the leading edges of aircraft wings.

Foamed orientation

Plastic foams are produced by expanding tiny gas bubbles in a semimolten plastic, stretching the interfacial walls between the gas bubbles to orient them in a single-step multiaxial stretch. At the corner where three bubbles intersect, the interfacial walls between them may be subjected to some type of three-dimensional orientation, particularly when incorporating short fibers in the mix.

Foamed polyethylene ionization

The process of foaming PE by exposing it to ionizing radiation which evolves hydrogen from the PE, causing it to foam.

Foamed pouring

A method of foam processing where the complete foam mix is poured into an open mold. Although usually molded into a large bun, other shapes can be formed.

Foamed pour-in-place

See under Foamed-in-place or *in situ*.

Foamed preform

Preformed material such as polystyrene is commonly extruded in the form of large 'logs' several meters across and in densities usually ranging from 1–2 lb ft^{-3} (16–32 kg m^{-3}). They are cut into boards, planks, and blocks for installation in buildings. Closed-cell foams, with the low vapor permeability of PS, provide integral vapor barriers; they are used in basements, roof slabs, insulating plaster base, etc.

Foamed polyurethane mixing head

The mechanism in which polyol and isocyanate streams are combined by impingement mixing, such as used in reaction injection molding.

Foamed polyurethane molding one-shot

A system in which the isocyanate, polyol, catalyst, and other additives are directly mixed and a foam is produced immediately; this is in contrast to a prepolymer.

The formation of a polymer from a precursor polymer by chain extension and/or cross-linking reactions. The term is especially used for the formation of PURs, a two-stage process. In the first stage a polyester or polyol is reacted with a diisocyanate to form an isocyanate-terminated prepolymer. It is chain extended and/or cross-linked in the second stage. Cross-linking occurs by reaction with a diol, amine, water, or further isocyanate groups for PUR elastomer formation, or with water, catalysts, and other ingredients for polyether-based flexible PUR foam. The liquid prepolymer is supplied to fabricators (the first-stage mix) with a second premixed blend of additional polyol, catalyst, blowing agent, and so forth. Foaming occurs when the two components are subsequently mixed.

The polymer process was once the most widely used because of the low reactivity of polyether polyols. But since the discovery of more active catalysts, it has been overtaken by the one-shot process. Rigid PUR foams, especially those using TDI, are also produced by the prepolymer process. A disadvantage, apart from having an extra processing step, is that the isocyanate-terminated prepolymers often have a limited stability, due to reaction with atmospheric water. And the prepolymers may have high viscosities, making them difficult to mix during the second stage.

Foamed polyurethane reticulate

Very low density PUR foams characterized by a three-dimensional skeletal structure of strands with few or no membranes between the strands. They contain up to 97% or more of void space. Conversion is usually made by treating an open-cell foam with a dilute aqueous sodium hydroxide solution under controlled conditions; this dissolves the thin membranes but leaves the strands substantially unaffected. Ultrasonic vibration is sometimes used to assist the solution process. Their uses include air filters, air cleaners, and acoustic panels.

Foamed rise time

In plastic foam molding, particularly with PUR, the time between the pouring of the PUR mix and the start of foaming. Also called foamed cream time.

Foamed rubber

Made from a liquid starting material, there are two types of foamed rubber: (1) latex foamed rubber, produced by mechanically whipping air into a rubber latex, gelling it, then vulcanizing it; and (2) foamed polyurethane rubber, produced from a liquid monomer mix.

Foamed rubber, microcellular

A cellular or foamed rubber with small cells produced by molding a mixture of a rubber with a blowing agent and high-impact polystyrene plastic. Expansion occurs only after removal from the mold.

Foamed sandwich structure

A foamed core material surfaced on both sides with relatively thin, dense, high-strength or high-stiffness facing materials [3].

Foamed security system

A novel system based on polyurethane foam to secure valuable articles in vehicles, etc. Once activated, this Instant ARMY (instantaneous antirubbery mass foam-yielding system) will deliver a mass of rapidly expanding PUR foam to engulf the valuables being carried (armored cars, etc.). It uses a foamed polyurethane molding prepolymer. The expanding foam (the foam mass is 150 times the volume of the liquids) jams the doors of a security vehicle as well as its contents within 8 s. It was designed by automatic equipment specialists in a joint venture between two Italian companies, Pa.je.t Srl and Apco Italia Snc.). During the early 1940s some low-flying airplanes had PUR prepolymer mixes that activated when the pilot was in trouble but unable to parachute because of low altitude. They foamed in place and, on impact, a semiflexible foam was targeted to protect the pilot.

Foam, self-skinning

Foam that cures to produces a tough outer skin over a foamed core. Also called integral skin foam.

Foamed sheet stock

Usually refers to expandable sheet stock produced by calendering in which a decomposable blowing agent is milled into a plasticized polyvinyl chloride below the decomposition temperature of the blowing agent. Subsequent heating decomposes and expands the blowing agent. The expanded plastic has to be stabilized by cooling. Its uses include egg cartons, meat trays, and fast-food containers.

Foamed silicone

Processes include the use of fluid silicone plastic made by mixing with a catalyst and blowing agent, pouring the mixture into molds, and curing at room temperature for about 10 h; at elevated temperatures much less time is required. Silicone foam sponge is made by mixing unvulcanized silicone rubber with a blowing agent and heating at the vulcanization temperature.

Foamed spray

Very fast reacting polyurethane or epoxy foams are fed in liquid streams to the spray gun and sprayed on the surface. The liquid foams on contact. Other plastic foams are used such as quick-acting mixes of TS polyester.

PROCESSES

Almost all fabrication processes are used to produce foamed products. There are various combinations of plastics and blowing agents. The blowing agent expands the plastic, initiating cells that grow to produce the final foam. As gas is produced, an equilibrium is established between material in the gas phase and the material dissolved in the solid state. The gas dissolved in the solid state migrates from the solution into the gas phase. The cells formed are initially under higher than ambient pressure because they must counteract the effects of surface tension. The pressure due to surface tension depends on the reciprocal of the cell radius, so the pressure within the cell is reduced as the cell grows. Small cells tend to disappear and large cells tend to get larger. This is because the gas migrates through the substrate (plastic) or the cell walls break.

After forming cells, the foam has to be stable; the gas must not diffuse out of the cell too quickly, thereby causing collapse or excessive shrinkage. The stability of the foam depends on the solubility and diffusivity of the gas in the matrix. The many processes make for many methods of cell initiation, cell growth, and cell stabilization. And these methods lead to a convenient process classification.

The growth of the cell depends on the pressure difference across the cell membrane: between the inside of the cell and the surrounding medium. Such pressure differences may be generated by lowering the external pressure (decompression) or by increasing the internal pressure in the cells (pressure generation). Other methods of generating the cellular structure are by dispersing gas (or solid) in the fluid state and stabilizing this foamed state, or by sintering plastic particles together in a structure that contains a gas phase. Foamable compositions in which the pressure is increased within the cells with respect to the surroundings have generally been called expandable types. Both physical blowing agents (PBAs) and chemical blowing agents (CBAs) are used. There is no single name for the group produced by the decomposition processes; the principal processes are extrusion, injection molding, and compression molding. Both PBAs, and CBAs are also used.

Based on production output, the most important processes are (1) extrusion (using PUR, PS, PE, PVC, CA, etc.), (2) expandable (PS, PE, PVC, PUR, phenolic, epoxy, PF, EP, SI, etc.), (3) spray (PUR, EP, UP, etc.), (4) froth (PUR, PVC, UF, EP, etc.), (5) injection molding (PE, PP, PS, PVC, etc.), (6) compression (PE, PVC, UP, etc.), (7) sintering (PS, PE, PTFE), and (8) leaching (PE, PVC, CA). The processes are identified by different names, with some overlap. They include bead molding, calender foaming, expandable plastic foam, expandable PS, expandable sheet stock, expandable PVC, extruded foam, injection-molded foam (low-, high-, and counterpressure types), mechanical foaming, reaction injection molding, reticulated foam, spray foam, steam foam molding, structural foam, syntatic foam, and the following, all starting with the word *foamed* (e.g., foamed blow molding): blow molding, casting, extruded film, frothing, gas counterpressure, injection molding, liquid, reservoir molding, rubber, PE, PS, PUR, PVC, preform, sandwich structure, sheet stock, silicone, spray, steam molding, in-place, pouring, short-shot molding, and mechanical.

BLOWING AGENTS

For the production of foamed or cellular plastics, depending on the basic material and process, different blowing agents (also called foaming agents) are used to produce gas and thus to generate cells or gas pockets in the plastics. They can produce rigid or flexible types and may be divided into two broad groups: physical blowing agents (PBAs) and chemical blowing agents (CBAs). PBAs are represented by compressed gases and volatile liquids (Table 9.1). The compressed gases most often used are nitrogen and **carbon dioxide**. These gases are injected into a plastic melt under pressure (higher than the melt pressure) and form a

Table 9.1 Properties of physical blowing agents

| Blowing agent | Molecular weight | Boiling point (°C) | Blowing efficiency[a] | |
			At boiling point	At 100 °C
Pentanes				
n-Pentane	72.15	36.1	216	261
2,2-Dimethylpropane	72.15	9.5	196	260
1-Pentene	70.15	30.0	227	280
Hexanes				
n-Hexane	86.17	68.7	212	232
2-Methylpentane	86.17	60.2	207	232
3-Methylpentane	86.17	63.3	211	234
2,2-Dimethylbutane	86.17	49.7	204	229
Cyclohexane	84.17	80.8	266	281
Heptanes				
n-Heptane	100.20	98.4	206	207
2,2-Dimethylpentane	100.20	79.2	193	204
2,4-Dimethylpentane	100.20	80.6	193	204
3-Ethylpentane	100.20	93.4	204	212
1-Heptene	98.20	93.2	212	216
Toluene	92.13	110.6	294	286
Trichloromethane	119.39	61.2	342	382
Tetrachloromethane	153.84	76.7	296	316
Trichlorofluoromethane	137.38	23.8	261	329
Methanol	32.04	64.6	679	752
2-Propanol	60.09	82.3	378	397
Isopropyl ether	102.16	67.5	198	217
Methyl ethyl ketone	72.10	79.6	324	344

[a] Data computed according to the formula

$$\frac{22\,400}{\text{molecular weight}} \times (\text{density at } 25\,^\circ\text{C}) \times \frac{(273+t)}{273}$$

where t = boiling point (°C) for the penultimate column and 100 °C for the last column. Vapors are treated as ideal gases.

cellular structure when the melt is released to atmospheric pressure or low pressure, as in a mold cavity with a short shot.

The volatile liquids are usually aliphatic hydrocarbons, which may be halogenated, and include materials such as carbon dioxide, pentane, hexane, methylene chloride, etc. Chlorofluorocarbons were formerly used but they have now been phased out. The liquids act as a gas source by vaporizing during the process. Regardless of their physical form, they rely solely on pressure for controlling gas development in a foaming process.

CBAs, generally solid materials, are of two types: inorganic and organic. Inorganics include sodium bicarbonate, by far the most popular, and carbonates such as zinc or sodium. These materials have low gas yields and, compared with organic CBAs, the cell structure they create is not as uniform. Organics are mainly solid materials designed to evolve gas within a defined temperature range, usually called the decomposition temperature range. This is their most important characteristic and allows control over gas development through both pressure and temperature. This increased control produces a finer and more uniform cell structure as well as better surface quality in the foamed plastic. There are over a dozen different types available that decompose at temperatures from 104 °C (220 °F) to 371 °C (700 °F), possibly higher. Many of these CBAs can be made to decompose below their decomposition temperature through the use of activators.

Typical activators are zinc oxide, various vinyl heat stabilizers and lubricants, acids, bases, and peroxides. The temperature reduction depends upon the chosen activator, its concentration, and the particle size of the blowing agent.

Besides selecting the proper CBA to meet temperature requirements, other factors to be considered relate to the plastic and the process. They include the types of gases evolved, the solid decomposition residues, the degree of self-nucleation, any discoloring or staining characteristics, and regulatory approval. The relevant authority in the United States is the Food and Drug Administration (FDA). CBAs could affect certain plastics going through certain processes; they could become yellow, and their lifespan could be shortened. CBAs should possess the following desirable qualities: (1) long-term storage stability; (2) gas release over a controlled time and temperature range; (3) low toxicity, odor, and color of the blowing agent and its decomposition products; (4) no deleterious effects on the stability and processing characteristics of the plastic; (5) the cells they produce have uniform size; (6) the foams they produce are stable; gas is not lost from the cell, causing it to collapse; and (7) acceptable cost–performance relation and availability.

Examples of blowing agents for certain plastics are (1) physical blowing agents that evaporate at the processing temperature; (2) hydrocarbons such as pentane for the production of expandable polystyrene; (3) chemical blowing agents which split off, e.g., nitrogen at the temperature of processing; these are used in the foaming of PE and PVC; and (4) blowing agents which, through the reaction of the raw material components, release a gas (usually carbon dioxide); PUR soft foam is produced in this manner.

The overwhelming majority of foams are themoplastics, but thermosets are also foamed with CBAs, although some of them do create problems. A popular TS foam is polyester. This plastic is an unsaturated polyester dissolved in styrene. Thermal decomposition of the blowing agent cannot be applied in this system because the heat of polymerization is not high enough to induce decomposition. But chemical reactions simultaneously produce gas and free radicals; they typically involve oxidation and reduction of a hydrazine derivative and a peroxide. The reactions are catalyzed by metals, which can be used repeatedly.

Direct injection of nitrogen requires special equipment which will allow it to be injected into a hot melt, such as an extruder or injection barrel, whereas a CBA does not require this type of equipment. The most common means of using CBAs is by drum tumbling with the resin. A wetting agent such as white mineral oil is commonly used to ensure good adhesion to the resin pellets; this is particularly important if the mix is to be air-conveyed to the hopper. Another mixing method is to use one of the many hopper metering and blending units available, which eliminates the labor required in drum tumbling and avoid a potentially poor mix. A third popular method uses a liquid dispersion of CBA compatible with the resin being foamed. An especially popular approach involves putting the CBA in a resin concentrate. A variety of resins are used and provide excellent dispersion of the CBA.

FOAMED POLYSTYRENE SHEETS

PS foamed sheets are produced using extruders (Chapter 3). Their densities have a wide range such as 3–12 lb ft^{-3} (0.05–0.19 g cm^{-3}) with thicknesses of 15–150 mil (0.38–3.8 mm). Use of these materials can be in flat sheet form (building insulation, packaging, etc.). They are used to go through other processes; principally thermoforming (Chapter 10). Product applications include disposable packages such as meat and produce trays and egg cartons, and as containers and trays for carryout meals and disposable dinnerware. They are also used for drinking cups, bottle labels, and cushion-type packaging.

These sheets are laminated with other materials (plastic films or sheets, aluminum foil, wood, etc.). Laminating improves performance for industrial use and in commercial applications such as decorative, insulated wall panels.

Sheets are usually produced by the gas-injection systems used in extruders (Chapter 3). The process can form biaxially oriented sheets. The basic fabricating processes are (1) an extruder with a large-diameter single screw with a large L/D (length/diameter) ratio, (2) twin-screw extruders, (3) the Winsted system using a single-screw extruder with a patented cooling system, and (4) two extruders in tandem. Of the four, the last process is most successful and widely used, having two extruders in tandem. One extruder is in line with the other extruder, so the first extruder feeds its melt to the second. The first extruder is used for melting, mixing, gas injection (usually nitogen) and feeding the extrudate to the secondary extruder. The second extruder is used for cooling and additional mixing of the extrudate before it exits a circular or annular die, where the foaming action is initiated. Most tandem systems use a $4^{1}/_{2}$ in. (11.4 cm) screw in the primary extruder. The secondary extruder uses a 6 in. (15.2 cm) screw that provides an output of 500–1200 lb h^{-1} (227–544 kg h^{-1}). To meet higher output rates, the trend has been to use a 6 in. (15.2 cm) primary screw and an 8 in. (20.3 cm) secondary screw with an output of 900–1800 lb h^{-1} (408–816 kg h^{-1}).

The secondary extruder is set up to ensure a uniform temperature in the melt before it exits the die, thus a uniform cell structure will be produced in the sheet at all times. After exiting the die, the extrudate has the form of a circular tube and is immediately drawn over an accurately designed expansion-cooling sizing mandrel that provides a controlled three-dimensional expansion of the extrudate and the start of the gas-blowing action. During this blowup, an accurately controlled air-cooling system is used around the mandrel. The exiting melt needs to have a uniform temperature, otherwise uniformity in the sheet becomes uncontrollable. The melt is cut at the mandrel so it forms itself into a flat sheet before entering a series of rolls. Finally it is wound on the last roll. The plastic foam continues to cool as it is rolled.

The mixture to produce foam sheet is plastic, nucleator, and blowing agent. The plastic is generally a high-heat, general-purpose polystyrene (GPPS). Nucleators, such as talc or a citric acid–sodium bicarbonate mixture, are added to provide foaming sites to obtain the desired cell size and uniformity. The blowing agent is usually injected in liquid form under a pressure of $\geq$1000 psi (7 MPa) or at least higher than the melt pressure existing in the barrel, usually about 500 psi (3.5 MPa). Injection is into the primary extruder and normally uses a positive-displacement volumetric pump. The gas (liquid converts to gas with heat) enters the screw/barrel plasticator about two-thirds of the way up the barrel from the feed opening, just before the metering zone of the extruder (Chapter 3). The amount and type of blowing agent controls the density of the foam produced. To have the sheets handled or worked, such as thermoforming, they are aged at least 3–5 days on the rolls. This aging action allows the gas pressure of the cell to reach equilibrium [3, 63].

EXPANDABLE POLYSTYRENES

Expandable polystyrene (EPS) illustrates the use of resin concentrates that include the blowing agents. Resin beads containing the blowing agent are supplied to the molder as accurately formed solid spheres. The beads may be about 0.1–0.2 mm in diameter and they contain a blowing agent, usually pentane. Producing EPS or foamed PS involves an unusual processing technique that originated in Germany during 1951. This process is also called bead molding or steam molding, but the most popular name is expandable polystyrene (EPS). Other plastics are foamed by the same or similar methods, but EPS is produced in the largest quantities.

The process involves two major steps. The first consists of preexpansion of the virgin beads by heat (usually steam, but also hot air, radiant heat, or hot water). Steam is extensively used because it is the most practical, the most economical and has other advantages. The preexpansion step brings the beads almost to the required density within the molded part, then they are stored for 6–12 h to allow them to reach an equilibrium. Figure 9.1 shows preexpansion within the mold (bottom view) to compare expansion before and after heating; the preexpander was as described above.

The next step conveys these beads, usually by air through a transport tube, to a two-cavity mold. Final expansion occurs in the mold, usually with steam heat, either by having live steam go through perforations in

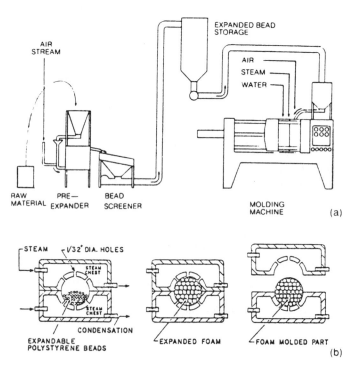

Figure 9.1 Expandable polystyrene process: (a) schematic of a basic EPS processing line and (b) action in the mold during expansion.

the mold itself or by steam probes that are withdrawn as the beads are expanding. During expansion the beads melt together, adhering to each other and forming a relatively smooth skin, filling the cavity or cavities. Multiple cavities can be used with small parts. After the heat cycle comes the cooling cycle. Because EPS is an excellent thermal insulator, it takes a relatively long time to remove the heat before demolding; if the heat were to remain, the part would distort. Cooling is usually by water spray over the mold. To facilitate removal, particularly for complex shapes mold-release agents are used. The final density is about 0.7–$10\,lb\,ft^{-3}$ (11–$160\,kg\,m^{-3}$) or in normal molding the density of the part will closely approximate the bulk density of the unheated beads.

EPS molds have double walls; the inner wall is the actual shape to be formed. It is perforated with vents to allow steam to penetrate the foam; the hot gases that develop leave the part through these vents. Thus, the double wall allows for encasing the steam that is delivered to the mold and in turn flows throw the vents.

Before removal from the mold, parts are stabilized by creating a vacuum and spraying water on the inner mold wall, causing diffusion of gasses from the many cells as well as a reduction in temperature. Since the product is a foam, it takes longer to cool than solid plastics.

Different preexpansion equipment is available; each type has advantages and disadvantages. Different controls exist because they are needed. The type of expander and controls will depend on the production quantity. There are continuous, single-stage, multistage, and discontinuous preexpansion systems. As an example of performance, consider the advantages of the continuous system over the other types: lower unit cost, higher throughput, easier maintenance, and greater reliability. Its principal disadvantage is the time it takes to change between materials of different bulk density. This disadvantage is avoided with the discontinuous unit.

EPS molding generates pressures of less than $30\,psi$ ($2\,kPA$) in most molding applications. This low pressure allows the use of inexpensive molds, perhaps made of aluminum. To process the other expandable plastic foams (EPFs), such as PE, PP, and PMMA, the equipment for EPS can be used with only slight modifications.

EXPANDABLE POLYETHYLENES

Expandable polyethylene (EPE) is a low-density, semirigid, closed-cell, weather-stable, PE homopolymer. It is easier to compress than EPS but less compliant than flexible PUR. EPE foam uses a similar molding process to EPS foam. The material is supplied by the manufacturer in a cross-linked expanded form ready for molding. Conventional EPS molding presses can process EPE with the addition of a filling device, provision for higher molding pressure, and postmold oven curing. The expanded particles do not contain a blowing agent and can be stored for long periods at room temperature. Their density range is 1.8–7.5 lb ft^{-3} (29–120 kg m^3). The most commonly used density is 1.8 lb ft^{-3} (29 kg m^{-3}).

EXPANDABLE POLYETHYLENE COPOLYMERS

Moldable expanded polyethylene copolymer (EPC) is a combination of approximately 50% polyethylene resin and 50% polystyrene resin. Combining the properties of both resins widens the selection of resilient materials for packaging engineers and designers. EPC is a low-density, semirigid, closed-cell material that requires refrigerated storage blow 40 °F (4.4 °C) in its raw granular form and has a shelf life of at least one month. The material expansion and conveying of the sensitive prepuff requires special handling and molding within a short period of time. The molding process and equipment are similar to EPS, but with slower molding cycles. Unlike EPE, EPC contains its own blowing agent.

EPC is a material that falls between EPS and EPE in performance, but exceeds both materials in toughness. The tensile and puncture resistance of EPC are superior to all of the moldable resilient foams available. It has good multiple-impact performance characteristics with better memory than EPS, but not as good as EPE. The cushion performance of EPC parallels EPE but at higher levels (7–12 g) even after repeated drops. EPC is especially good for reusable material-handling trays and packaging applications which require a nonabrasive, solvent-resistant, impact-absorbing material with a superior toughness that elongates, compresses, and flexes without material fatigue. EPS molders are capable of molding this material with minor adjustments to their existing equipment.

EXPANDABLE SANS

Styrene-acrylonnitrile (SAN) copolymer is a moldable, lightweight, semi-rigid, closed-cell, highly resilient styrene copolymer resin. The resin relies on a blowing agent to preexpand and fuse the material during its molding phase. The raw materials can be stored without refrigeration and they have a shelf life of 6 months, provided some simple precautions are observed. Their processing is much like EPS, but with extended cycles due to the higher level of blowing agent. EPS equipment and standard EPS mold designs can be used without any major changes or adaptations. Postmold oven curing is not necessary. A low density of more than 40 times expansion can be attained during the preexpanding phase, as low as 1.0 lb ft^{-3} (16 kg m^{-3}) on the first pass through the expander and 0.8 lb ft^{-3} (13 kg m^{-3}) on the second pass.

FOAMED POLYURETHANES

Liquid (pour-in-place), froth, and spray foaming techniques are some of the different methods used with PUR foaming (Fig. 9.2). When injected, it is commonly called reaction injection molding (Chapter 11).

When the PUR liquid ingredients are mixed, gases are produced which cause the mass to expand as it stiffens and hardens. The reaction is complete in a few minutes. In a sandwich structure the liquid mix is poured between the cover sheets and foams between them, bonding directly to the sheets without an adhesive. (EPS needs an adhesive on the sheets if bonding is required.) As foamed-in-place materials expand, they can exert appreciable pressure, so the sandwich sheets have to be held in a rigid frame to prevent bulging until the reaction is completed. To overcome this pressure, the liquid mix may be allowed to form a froth of almost its ultimate volume prior to pouring. The result is little or no pressure and a rigid foam. Flexible PUR foam, such as that used in upholstery, is made by continuous deposition on a belt before being cut into

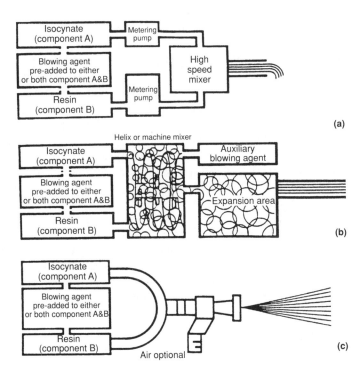

Figure 9.2 Producing PUR foams: (a) liquid process, (b) froth process, and (c) spray method.

blocks or sheets of desired shape and size. Other foam materials may be handled in somewhat similar ways, and may be prefoamed or foamed-in-place.

Water is the component-forming blowing gas in the formulation for PUR soft, flexible foam; it forms carbon dioxide with isocyanates. The evolution of heat and the change in temperature caused by this gas-generating reaction mean that other blowing agents (such as dichloromethane) have to be used, along with coolants, in order to produce low-density foam of less than $20 \, kg \, m^{-3}$. Flexible PUR foams have open cells and are used in the furniture, automotive, packaging, insulation, mechanical (sealants, sponges, etc.), and other industries.

A rigid, foamed cross-linked PUR, usually with closed cells, is formed by the reaction of a diisocyanate (sometimes containing components of a higher functionality) and often methane diisocyanate (MDI) or polymeric MDI with a polyester or more usually with a polyether polyol. Foaming may result from the incorporation of water, which reacts with isocyanate groups to form carbon dioxide, but is usually the result of using other blowing agents, sometimes in combination with water. They are more rigid than flexible foams because they contain more cross-links. This is accomplished by the use of polyols, usually polyoxypropylene glycols of low molecular weight (about 500 units), which are highly branched by mixing of higher-functionality comonomers (such as sorbitol or pentaerythritol). Prepolymer processes and quasi-prepolymer processes are used with TDI (toluene diisocyanate) to reduce the toxic hazard of this material. One-shot processes are used with MDI and polymer MDI with polyethers; normally a tertiary amine and/or organotin catalyst system is used, together with a silicone surfactant, as with flexible foams. And catalysts are normally used; this is to obtain the right balance of reaction rates so the gas bubbles are trapped in the liquid plastic melt as the viscosity increases during polymerization and cross-linking. A surfactant, usually a silicone-polyether block copolymer, is also often present to control cell morphology. The major application for these foams is thermal insulation because they have exceptionally low thermal conductivity.

A variation of the PUR foam-in-place method is foam frothing. The liquid mixture of PUR chemicals is dispensed as a partially expanded froth. Frothing is achieved by using a mixture of blowing agents in the basic mix to give a two-step blowing action. The first expands the mix into a froth and cools, delaying the second (and final) expansion for about 30 s.

PUR dispensers

Controllable and safely operating dispensers are used to process PUR. They are designed to meet the precision mixing actions required when using these highly reactive chemicals. Ease of handling and safe pouring are required. When dispensing into open molds or large cavities, pour-in-place requires reduced-output, splatter-free mixing heads for the best results. The reaction of the mixed liquids takes place in the mold or cavity, which can be left open or closed after the shot. Open-pour is used with rigid foam systems for various molding operations and continuous panel production. Numerous flexible-foam products are also produced using the open-pour method such as foam bunstock.

With froth pour-in-place, additional blowing agent is used in the system to create a substance more like shaving cream from the mixed liquid before it actually reacts and becomes rigid foam. Common uses for froth are in large open-pour panel production, where the fixtures or jigs may be loosely constructed and the thicker liquid has no tendency to run or leak from the tool. A benefit of the process is that the material flow is slow and uniform, preventing voids in the product.

When spray foaming, instead of mixing the reactive components in a mixing head and pouring or injecting the mixture, spray systems employ a high-pressure, impingement-mix spray gun. The materials exit the gun in an atomized state and are applied to various substrates in either round or fan spray patterns.

There are several types of machinery for dispensing PUR foam. Machinery requirements and the method of dispensing are most often determined by the product specification. Polyurethanes are dispensed and applied with spray machinery, high-pressure RIM machinery, or low-pressure meter-mix-dispense machinery. Each has advantages.

FOAMED POLYVINYL CHLORIDES

Although vinyl foams can be produced by many methods, including mechanical frothing, decompression techniques, and leaching out of soluble additives, the most widely used procedure is chemical blowing. Between 1 and 2wt% of a blowing agent such as azobisformamide is incorporated in PVC compound or dispersion, remaining inert until it is decomposed by processing heat to release gas. These compounds are processed by conventional methods such as calendering, extrusion, injection molding, compression molding, slush casting, and rotational molding. In most of these processes either rigid or flexible foams can be made, depending upon the amount and type of plasticizer used. Vinyl foams are widely used in clothing, flooring, footware, furniture, packaging, etc.

FOAMED INJECTION MOLDING

Low pressure

Low-pressure or short-shot conventional foam processing methods are the most commonly used (compared to other IM methods) because they

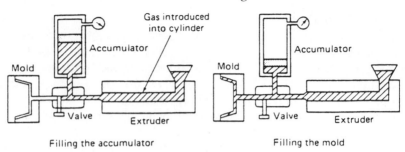

Filling the accumulator   Filling the mold

Figure 9.3 Low-pressure foam injection molding.

are easy, simple, and best suited to economical production, particularly of large, complex, three-dimensional parts. A controlled melt mixture (plastic and blowing agent) is injected into a mold cavity, creating a low cavity pressure, usually 200–500 psi (1.4–3.5 MPa). Along with about 0.5 wt% of CBA, this mixture can be injected directly from the barrel of a conventional injection molding machine (with limited modifications) or, as shown in Fig. 9.3, via an accumulator. The mixture only partially fills the mold (short shot), and the gas bubbles, having been at higher pressure, expand immediately and fill the cavity. As the cells collapse against the mold surface, a relatively solid skin of melt is formed over the rigid foam core.

Skin thickness is controlled by the amount of melt injected, the mold temperature, the type and amount of blowing agent, the temperature and pressure of the melt, and the capabilities of the molding machine, particularly its speed of injection.

Low pressure with coinjection

This technique involves the usual separate injection of two compatible plastics that are coinjected using two injection plasticators. A solid plastic is injected from one plasticator to form a solid, smooth skin against the surfaces of the mold cavity. Simultaneously a second material, a measured short shot containing a blowing agent, is injected to form the foamed core. This approach can also take a relatively full-core shot and have the mold open (as in coining) after the skin solidifies, having the melted core expand with mold-opening action.

Low pressure with surface finish

In low-pressure surface-finish (SF) molding, not using coinjection or coining, the volume of the molding cavity is always larger than the volume of the plastic in the unfoamed state. The low pressure allows microbubbles to nucleate and grow. Foam expansion occurs during filling, and growing bubbles are carried to the mold surface, creating unacceptable surface irregularities and imperfections called splay or swirl pattern. The irregularities can be seen and felt; the surface roughness can be as much as 1000 μin. (25 μm). Parts needing smooth, finished surfaces require secondary operations, usually sanding, filling, and painting. There are techniques to improve surface appearance during fabrication. The principal process variables are melt and mold temperatures, injection rate, the nature or type of blowing agent and its concentration. Cyclic heating and cooling of the mold surface and direct injection of blowing agent into the melt as it is being injected into the mold are two of the methods used.

Gas counterpressure

This method uses a sealed mold pressurized to 400–500 psi (2.8–3.5 MPa) with an inert gas, sufficient pressure to suppress foaming as the plastic mix enters the mold cavity. After the measured shot is injected, the mold pressure is released, allowing the instantaneous foaming to form the core between the already formed solid skins. The mold action is similar to coining. Another technique is gas injection molding, used to develop similar foamed structures (Chapter 2). Once the plastic at the mold surface has solidified, the gas pressure is released to permit the remaining melt mix to foam, creating the part's core.

High pressure

High-pressure molding using expandable molds, so-called structural foam molding, is an offshoot of conventional injection molding. The heated melt mix (with blow agent) is injected into the mold, creating a cavity pressure higher than the blowing-agent gas pressure but usually much higher; this is to ensure no loss in pressure during injection. Pressure for certain machines could be 5000–20 000 psi (34.5–138 MPa). The mold is entirely filled so the pressure prevents any foaming from occurring while the skin portion starts solidifying against the mold surfaces. As soon as the skin surface hardens to a desired thickness, the cavity mold pressure is reduced to allow the remaining melt to foam between the skins. Depending on the type of equipment and the size, as well as on the part configuration, these two provisions are made either by withdrawing cores or by special press motions that partially open the mold halves (such as the compression molds used in coining to provide two-dimensional action; three-dimensional mold actions have been used). The degree of foam density, wall thickness, and surface finish depends on the foam mixture (constitutents and amounts); the machine controls depend on the mold action.

STRUCTURAL FOAMS

So-called structural foam is also called integral skin foaming or reaction injection molding (RIM) and can overlap in its performance as well as use with the significantly larger market of the more conventional foamed plastics. Up until the 1980s in the United States, the RIM and structural foam processes were kept separate. Combining them in the marketplace was to aid in market penetration. From the beginning of the 1930s to the end of the 1960s, liquid injection molding (LIM) was the popular name for what later became RIM and structural foam.

Structural foam is characterized as a plastic structure with foam cores of nearly uniform density and integral skins that are almost solid, usually with 20–40 wt% reduction in density. It can mold small parts but its greatest use is for large parts. Equipment is available for all sizes of shot up to at least 750 tons clamping, processing 2000 lb (900 kg) of plastics per hour with a 150 lb (68 kg) shot capacity.

The structural foam industry defines its output as a plastic product with integral skins, a cellular core, and enough strength-to-weight ratio to be classified as structural. When they are used in load-bearing applications, structural foams have a bulk density of typically 50–90% of the value for the unfoamed plastic. Around 90% of structural foams are made from different thermoplastics, principally PS, PE, PVC, and ABS. Polyurethane is the primary thermoset. Unfilled and unreinforced plastics represent about 70% of products. Three-quarters of all processing is modified low-pressure injection molding. Extrusion and RIM account for about 10% each.

FOAMED RESERVOIR MOLDING

Also known as elastic reservoir molding, this process creates a sandwich of plastic-impregnated, open-celled, flexible polyurethane foam between the face layers of fibrous reinforcements. When this plastic composite is placed in a heated mold and squeezed, the foam is compressed, forcing the plastic and air outward and into the reinforcement. The elastic foam exerts sufficient pressure to force the plastic-impregnated reinforcement into contact with the mold surface (Fig. 9.4).

SYNTACTIC FOAMS

In syntactic foams, instead of using a blowing agent to form bubbles in the mass, preformed hollow spheres of plastic (usually phenolic), glass, or ceramic are embedded in a matrix of an unblown plastic. They use conventional plastic such as epoxies, TS polyesters, and phenolics. The diameter of the tiny hollow spheres is 10 mil in (300 μm) or less. The foam

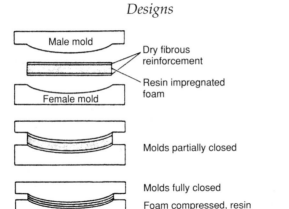

Figure 9.4 Foamed reservoir molding.

systems are rigid and generally have a specific gravity of less than 0.1; they are very strong in compression and flexure.

Syntactic foams may also be made in combination with conventional gas-filled plastic foams. A mixture of syntactic microspheres and conventional plastic can be formulated into a moldable mass then shaped or pressed into cavities and molds similar to molding plaster of paris, sand, and clay. When finished and cured, the properties of the hardened mass, usually a TS plastic, can be tailored by different formulations to create products with different properties. Synthetic wood is produced by a mixture of TS polyester and hollow glass spheres.

DESIGNS

Foamed plastics, like their solid counterparts, can be used for an almost unlimited range of products, and most industries use them [3]. Figure 9.5 shows an architectural example that was an R&D project in 1966. Dome-shaped buildings in Lafayette, Indiana, were built using polystyrene foamed boards produced by spiral generation, a technique patented by Dow Chemical. Boards were heated on site so they could be bonded to each other in layers (Fig. 9.5(a)), forming a continuous pattern to produce the dome. The self-supporting domes required no internal or external support during or after manufacture, they provided their own insulation, and it was easy to cut out sections to create the openings for doors, windows, and connecting halls from dome to dome. Figure 9.5(b) is a model of the medical clinic and Fig. 9.5(c) a cross section of the central medical dome, which interconnected with smaller domes and formed connecting rooms. Their outsides were covered with steel wire mesh and concrete then waterproofed. The inside walls were covered with appropriate plaster and paneling.

COSTING

The cost of foamed products is determined by the usual constituents, such as raw materials, as well as the machine or process costs, including labor, secondary operations, operating overhead, and equipment amortization. Although each processor will develop an individual pricing philosophy, an important aspect that differs with each type of process involves handling the foam components and, very important, determining the amount of scrap. The scrap factor can be very little but is sometimes as high as 10 wt%. Another variable is the cost of setting up the fabricating process; certain processes are very labor intensive. Table 9.2 includes cost comparisons for some of them.

TROUBLESHOOTING

This chapter has included information concerning problems and remedies. Table 9.3 provides a guide for plastic foam films. Since chemicals

(a)

Figure 9.5 Dome-shaped buildings using PS foamed boards.

(b)

(c)

Figure 9.5 *Continued*

play an important part in most of the foam processes, to help in eliminating their problems the following factors should be considered: (1) long-term storage stability under fairly ordinary conditions, (2) gas release over a controlled time and temperature range, (3) low toxicity, odor, and color of the blowing agent and its decomposition products, (4) no deleterious

Table 9.2 Comparison of structural foam with five other processes

Foam vs. sheet metal	Foam vs. die casting	Foam vs. sheet molding compound	Foam vs. hand layup fiberglass	Foam vs. injection molding
		Advangtages		
1. Fabrication economy: less assembly time; tighter dimensional tolerances; increased product integrity; less final product-inspection time	1. Much lower tooling costs	1. Uniform physical properties throughout the part	1. More consistent part reproduction	1. Flexibility for functional engineering
2. Fewer parts required for assembly	2. Longer tool life, lower maintenance	2. Warping and sink marks reduced or eliminated	2. Lower labor	2. Better low- to medium-volume economics
3. Dent resistance	3. No trim dies required	3. No resin-rich areas to cause configuration problems	3. Simplified assembly	3. Lower tooling costs
4. Elimination of oil canning	4. Lighter weight	4. Higher impact resistance	4. Better dimensional stability	4. Better large-part capability
5. Greater design freedom	5. Higher impact resistance	5. Greater inherent structural capabilities	5. More design freedom	5. Better sound damping
6. Better sound damping	6. Better sound damping	6. Lower shipping costs	6. More uniform physical properties	6. Lower internal stresses
7. Reduced damage from shipping	7. Better strength-to-weight ratio	7. Large parts more economical	7. Better sound damping	7. Sink marks reduced or eliminated
8. Reduced tooling costs for complex configurations	8. Better impact resistance	8. Lower tooling costs		8. Inherent structural strength
		9. Better sound damping		*Many process similarities exist*

Limitations

1. Smaller variety of finishes available, such as chrome or baked enamel 2. No RFI and grounding capabilities 3. Harder to retrofit to frame or skins 4. Thicker wall 5. Higher tool costs than with brakeforming	1. No heat-sink capabilities 2. No RFI and grounding capabilities 3. Fewer available finishes for cosmetic appearance 4. Higher finishing costs 5. Thicker walls 6. Possible internal voids	1. Increased finishing costs (surface swirl) 2. Heat distortion 3. Thicker wall 4. Lower physical properties 5. Possible internal voids	1. More prone to heat distortion 2. Poorer economies of part size vs. quantity 3. Thicker walls 4. Higher tooling costs	1. Poorer surface finish 2. Application of cosmetic detail for appearance parts 3. Longer cycle time 4. Thicker walls 5. Poorer high-volume economics 6. Less equipment available for various shot sizes

Potential savings from structural foam (%)

>50[a]	15–30	<30	>50	15–20[b]

[a] Even with limited quantities.
[b] Depending on unit volume and part size.

Table 9.3 Troubleshooting guide for plastic foam film

Problem	Probable cause(s)	Correction
Random, poor cell structure	Low melt pressure	Increase screw speed
		Reduce die-lip temperature
		Decrease gauge of screen packs
		Use resin of lower melt index
		Reduce die gap
	Hangup in die	Reduce land length
		Clean die
	Stagnant low-pressure areas in head	Increase screw speed
	Irregular cells in spider area or opposite die-ring feed	Use bottom-fed spiral die
		Increase head pressure
Poor skin formation	Too much blowing agent	Reduce blowing-agent level
	Linear skin speed too low	Increase screw speed
	Loss of melt pressure in die land	Reduce land length
		Increase L/D ratio
		Increase screw speed
	Die-block temperature too low	Increase temperature
Pinholes in film or bubble burst on surface	Too much blowing agent	Reduce blowing-agent level
	Die temperature too high	Reduce die temperature
	Resin melt index too high	Decrease resin melt index
		Reduce processing temperature
	Blowing agent decomposing too soon	Reduce processing temperature
		Increase screw speed
		Reduce blowing-agent level level
	Poor flow within polymer skin	Improve flow in head and die
Cells collapsing	Resin melt index too high	Decrease melt index
		Reduce processing temperature
	Cooling too fast	Reduce cooling rate

effects on the stability and processing characteristics of the plastics, (5) the ability to produce cells of uniform size, (6) the ability to produce a stable foam, i.e., the gas must not be lost from the cell and cause it to collapse, and (7) good cost–performance relation and availability.

10

Forming

INTRODUCTION

Formed or shaped plastics provide a great variety of marketable products, in a wide size range from drinking cups to large products (Figs 10.1 and 10.2). Different techniques are used, but thermoforming is the most productive and the most diversified. Other techniques are similar to thermoforming but normally use less heat and are more limited in their choice of plastic; these processes include cold forming, stamping or compression forming, flow molding, rubber pad molding, diaphragm forming, coining, and forging. Formed parts are used in many different applications and production lines (form, fill and seal, etc.). Food, electronic devices, medical products, and other parts use continuous thermoforming operations at the end of high-speed production lines to reduce the handling of products, provide hermetically sealed contents, reduce costs, and so forth.

Thermoforming has many advantages over other methods of thermoplastic fabrication:

1. Parts with a large surface area can be formed with relatively low mold and equipment cost, because of the low pressures required.
2. Very thin-walled parts can be readily formed, which is not feasible by any other method.
3. High-volume thin-walled products, such as drinking cups, can be produced at the lowest cost per capital investment, at production rates of 50 000 to over 200 000 units per hour.
4. Low-volume heavy-gauge products, such as computer housings, are competing favorably with injection molding in price, through lower tooling costs, and in product detail with 690 kPa (100 psi) forming pressures.

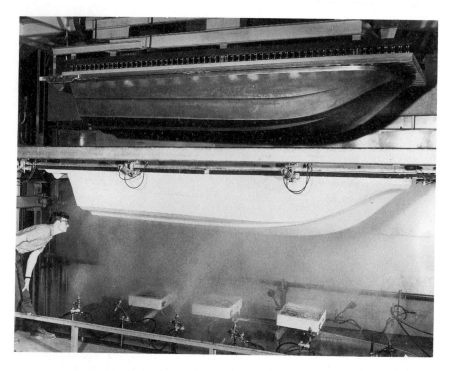

Figure 10.1 Precise-timed cooling for ABS thermoformed hull for a 15 ft ($4^1/_2$ m) runabout.

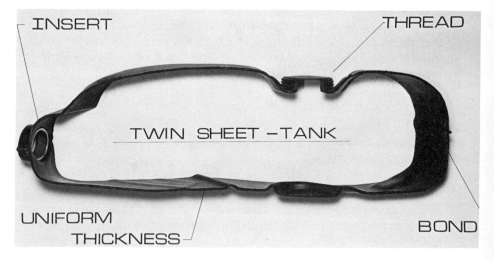

Figure 10.2 Two thermoformed parts, made from coextruded sheets, are bonded together to create a fuel tank. The tank's inside has a nylon surface to resist gasoline; the outside uses PP to provide the required support strength while keeping the cost low.

THERMOFORMING

Thermoforming usually begins with plastic sheet or film: sheet thickness tends to be 10 mil (250 µm) and greater; film thicknesses are normally less than 10 mil (250 µm). Extrusion is the most common method of producing sheet and film for thermoforming (Chapter 3); very small amounts are cast or calendered (Chapter 5).

Thermoforming usually consists of heating extruded thermoplastic (TP) sheet, film, and profile to its softening heat and forcing the hot and flexible material against the contours of a mold by pneumatic means (differentials in air pressure are created by pulling a vacuum between the plastic and the mold, or the pressure of compressed air is used to force the material against the mold), mechanical means (plug, matched mold, etc.), or combinations of pneumatic and mechanical means.

The process involves (1) heating the sheet (film, etc.) in a separate oven then transferring the hot sheet to a forming press, (2) using automatic machinery to combine heating and forming in a single unit, or (3) a continuous operation feeding off a roll of plastic or directly from the exit of an extruder die (postforming). Almost all the materials are TPs. To date very few thermosets (TSs) have been used, as markets have not developed. These TSs can be reinforced or unreinforced (Chapter 12). Almost any TP can be used, but certain types make it easier to achieve deep draws without tearing or excessive thinning in areas such as corners. Ease of forming depends on material characteristics; it is influenced by minimum and maximum thickness, pinholes, the ability of the material to retain heat gradients across the surface and the thickness, the controllability of applied stress, the rate and depth of draw, the mold geometry, the stabilizing of uniaxial or biaxial deformation, and most important, minimizing the thickness variation of the sheet.

Bending, one of the oldest thermoforming techniques, is relatively easy to handle. It is often accompanied by joining (adhesive or welding) or mechanical operations (milling, drilling, polishing). If the sheet is heated only locally in the bending operation, no special forming tools are needed. The width of the heating zone and the thickness of the sheet determine the bending radius. Limitations are related to the softening point of the sheet and the intrinsic rigidity of the heated sheet (sag should be minimized). Transparent plastics (such as PMMA and PC) with thicknesses up to $3^{1}/_{2}$ in. (90 mm) are frequently bent for use in store displays, staircases, partitions in banks, aircraft windows, and so on. With this type of plastic, if restrictions in the bending area are minimized, the thickness at the bend can remain unchanged.

PROCESSING

All thermoforming systems include a means of receiving sheets cut to size, from rolls, or directly from an extrusion line. Sheet thicknesses over

1.5 mm are generally cut to size, and thinner sheets are supplied on rolls. There are no gauge limitations on extrusion lines. The thermoformer contains the mold, may or may not include the trimming means, and provides the pin chain or gripping system that indexes the sheet through the heating, forming, and sometimes the trimming operations. Trimming is not necessarily an integral part of the forming cycle, but few applications can use the formed web without some kind of trimming. Specific variations of basic thermoforming include processing sheets cut to size, or from rolls, or directly from an extrusion line.

The most elementary thermoforming techniques include basic vacuum forming, drape forming, pressure forming, and matched-mold forming. These terms describe the single operation of clamping the heated sheet against the male or female mold and removing air between sheet and mold surface by vacuum, external air pressure, or both. All thermoforming molds contain fine holes for air evacuation. A secondary operation within the forming cycle prestretches the hot sheet in order to control thickness in the final product. Prestretching may be accomplished with a plug or ring in plug-assist or ring-assist forming, or by air in pressure-

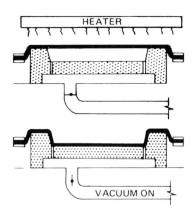

Figure 10.3 Straight forming: vacuum.

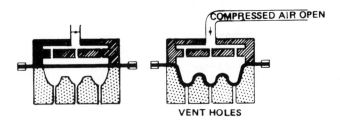

Figure 10.4 Straight forming: pressure.

bubble plug-assist forming, vacuum snapback forming, pressure-bubble vacuum snapback, and air-slip forming. Another popular technique is trapped-sheet contact-heat pressure forming.

The various thermoforming techniques are generally described in terms of the means used to form the sheet, such as bending, vacuum forming, pressure forming, plug-assist forming, and matched-mold forming. The different methods enable the processor to form different product shapes to meet various performance requirements. Most of these techniques are reviewed in Figs 10.3 to 10.12. The range of formable shapes runs from the

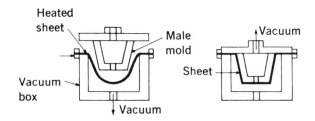

Figure 10.5 Snapback forming.

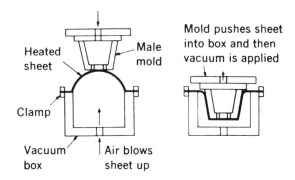

Figure 10.6 Forming with a billow snapback is recommended for parts requiring a uniform, controllable wall thickness.

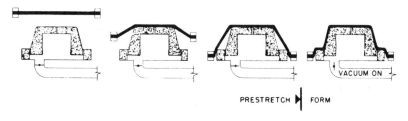

Figure 10.7 Drape form.

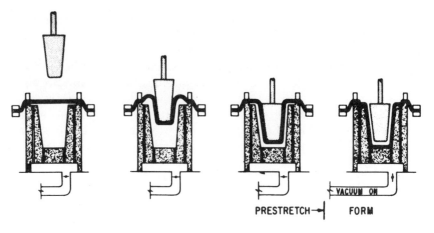

Figure 10.8 Plug-assist forming.

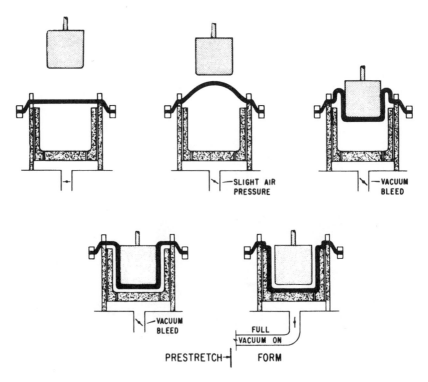

Figure 10.9 Plug-assist, reverse-draw forming.

simple to the very complex, and the shape as well as the surface condition can be accurately controlled outside, inside, or on both sides.

There are two basic forming methods from which all others are derived: drape forming over a positive (male) mold, and forming into a cavity (female) mold. Product configuration, stress and strength requirements,

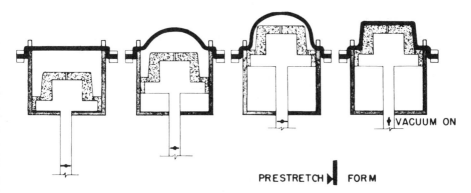

Figure 10.10 Air-slip forming.

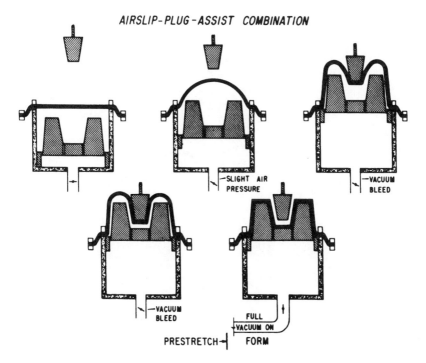

Figure 10.11 Plug-assist, air-slip forming.

and material specifications all play a part in determining the process technique. Forming into a female mold is generally used if the draw is relatively deep, e.g., cups. Female molds generally provide better material distribution and faster cooling than male molds. Male-mold forming is preferred for certain product configurations, particularly if product tolerances on the inside of a part are critical. Male-mold forming produces heavier bottom strength; female-mold forming produces heavier lip or perimeter strength. An advantage of straight forming into a female mold is that parts with vertical sidewalls can be formed and extracted, stress-free, from the molds because of the shrinkage that occurs as the part cools.

In drape forming, when the hot plastic sheet touches the mold as it is being drawn, it chills and starts to solidify. Successful drape forming

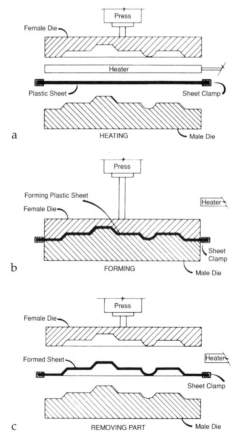

Figure 10.12 Matched-die forming.

needs to consider several variables. One of the most significant is shrinkage. Since these types of plastic material have a high coefficient of thermal expansion and contraction, i.e., about 7–10 times that of steel, care must be taken when designing the mold to provide sufficient draft on the sidewalls so the part can be extracted from the mold. It is not unusual for parts to rupture upon cooling on an improperly designed drape mold. Another potential problem is the part may become so highly stressed during forming and cooling that it loses most of the physical properties the sheet would otherwise provide.

Natural process evolution has combined the two systems to take advantage of the better parts of each method. The plug-assist process, similar to matched-die forming, involves a male mold (or plug) having a volume about 60–90% of the cavity. By controlling the geometry and size of the plug and its rate and depth of penetration, material distribution can be improved for a broad range of products. The plus-assist technique is used to manufacture cups, containers, and other deep-draw products.

Many thermoforming techniques have been developed to obtain better material distribution and broaden the applicability of the process. Some of the more popular methods are illustrated in Figs 10.3 to 10.12. Most of these techniques can employ vacuum, pressure, or a combination to apply the force necessary to shape the heat-softened plastic sheet.

Forming definitions

Air-assist

Methods in which airflow or air pressure is employed to preform the sheet partially before the final pulldown onto the mold using vacuum.

Air-slip

A variation of snapback thermoforming in which the male mold is enclosed in a box so that, when the mold moves forward toward the hot plastic, air is trapped between the mold and the plastic sheet, creating a cushion. As the mold advances, the plastic is kept away from it by the air cushion until the full travel of the mold is reached. A vacuum is then applied to remove the air cushion and form the part against the plug.

Billow

Heated sheet is clamped over a billow chamber. Air pressure in the chamber is increased, causing the sheet to billow upward against a descending male mold.

Bubble

Sheet is clamped into a frame suspended above a mold, heated, blown into a blister shape by air, then molded to shape by a descending plug applied to the blister, forcing it downward into the mold.

Clamshell

A variation of blow molding and thermoforming in which two preheated sheets are clamped between halves of a split mold (like the two-part mold used to form the final blow-molded part). The two sheets are drawn into the mold cavity by a vacuum and kept separate by injecting air between them. An end contact surface could include an integral hinge.

Cold

A process of changing the shape, primarily TP sheet or billet in the solid phase, through plastic (permanent) deformation with the use of pressure dies (Fig. 10.13). The term implies that deformation occurs with the plastic at room temperature. However, its range has been widened to include forming at higher temperatures, or warm forming, but much below the plastic melt temperature and lower than those for thermoforming. Within this process there are special methods, such as **solid-phase pressure forming** (SPPF), which use cold and warm forming as well as thermoforming.

Cold plastic forming is very similar to cold metal forming. The main differences are the time dependence of TP deformation and springback, or recovery. Formed TPs exhibit molecular orientation along the principal strain directions, thus increasing performance (Chapters 1 and 3).

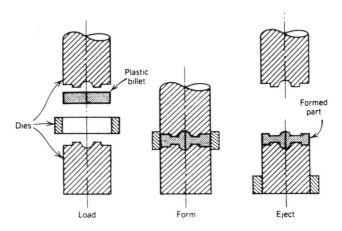

Figure 10.13 Cold forming.

Comoform cold molding

An extension of the cold molding process using a thermoformed plastic skin to impart excellent surface and other characteristics (weather resistance, etc.) to a cold-molded reinforced plastic or laminate.

Drape

Sheet is clamped into a frame, heated, and draped over the mold. A vacuum can be used to pull the sheet into conformity with the mold.

Drape assist frame or drape forming

A frame, made of anything from thin wires to thick bars, is shaped to the peripheries of the depressed areas of the mold and suspended above the sheet to be formed. During forming, the assist frame drops down, drawing the sheet tightly into the mold, thereby preventing webbing between high areas of the mold and permitting closer spacing in multiple molds.

Forging

A production method whereby TP stock, usually heated, is shaped to a desired form by compression forces (impression molding) or by sharp hammer-like blows. Virtually all ductile materials may be forged and preheating may not always be required. When a material is forged below the melt temperature, it is cold forged. Cold forging of plastic is generally called cold forming or solid-phase pressure forming. When a material is forged above the melt temperature, it is hot forged.

Form and spray

A technique to strengthen thermoformed sheets by applying a sprayup of reinforced plastics to one side.

Form, fill, and seal

Form, fill, and seal (FFS) pouch, extensively used in packaging, involves inline thermoforming of film or sheet, inserting the product being packaged, and sealing the package using heat, adhesive, etc. As shown in Fig. 10.14, three types of pouch are used. Each has many variations.

Form, fill, and seal versus preform

Current FFS technology generates a variety of concerns over its visual clarity. Preforms are typically thick-walled and rigid; this is to protect

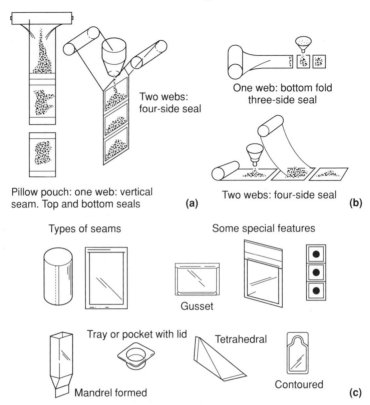

Pillow pouch: one web: vertical
seam. Top and bottom seals **(a)**

Two webs:
four-side seal

One web: bottom fold
three-side seal

Two webs: four-side seal **(b)**

Types of seams

Some special features

Gusset

Tray or pocket with lid

Tetrahedral

Mandrel formed

Contoured **(c)**

Figure 10.14 Form, fill, and seal processes: (a) vertically formed, (b) horizontally formed, and (c) shape formed.

expensive, heavy, and/or sharp devices, to accommodate intricate shapes and compartments, and to prevent corner thinning. But thickness is often accompanied by haze. FFAs are usually less thick and rigid than preforms. Although thinness improves clarity, it limits the depth of draw. Hence, FFS is well suited for packaging inexpensive, high-volume products, such as syringes and gloves, which do not require much physical (as opposed to microbiological) protection. When a deeper draw is required as well as clarity, coextrusions may be an option. There are combinations that are flexible yet tough enough for deeper draws. Nylon is an example.

Form, fill, and seal with zipper online

The zipper is usually put in the film by a secondary operation, which adds cost and tends to be wasteful (up to 30% becomes scrap). Solutions to this problem were developed during 1990 (by Bodolay Pratt Division of Package Machinery Co., Stafford Springs CT and Klockner-Bartelt Co., Sarasota FL). Both systems essentially combine the pouch web and zipper strip from a supply roll and pass them through a strip-seal station, where the zipper is heat-sealed to the web. The system has two zipper sealing stations, plus an ultrasonic spot-seal station for packages that are hermetically sealed or gas-flushed. The speed of the line is identical to conventional FFS lines.

Plug

Also called plug-assist, a process in which a plug, male mold, or male stretching device is used to partially preform the part before forming is completed using vacuum and/or pressure.

Plug-and-ring

A plug functioning as a male mold is forced into a heated plastic sheet held in place by a clamping ring.

Prebillow

Prestretching of the heated plastic sheet by differential air pressure prior to thermoforming.

Preprinting

Printing of a distorted pattern on a plastic sheet which is then thermoformed to the desired shape, bringing the printed pattern into the proper undistorted shape.

Pressure

Application of air pressure onto the sheet to force it into the cavity to form the part rather than using vacuum to draw the sheet against the mold cavity.

Prestretched

Stretching of heated sheet either by mechanical means or by differential air pressure prior to the final shaping by differential air pressure.

Sandwich heating

The usual method of heating the sheet, prior to forming, which consists of heating both sides of the sheet simultaneously.

Snapback

A variation of vacuum forming. Heated sheet is pulled to a concave form by the vacuum box underneath; it is then snapped upward against a male plug by vacuum through the plug. The process can extend deep drawing.

Stretch

Heated sheet is stretched over a male mold then drawn into shape by vacuum and/or pressure.

Twin-sheet

The twin-sheet process produces hollow parts from cut-sheet or roll-feed machinery. With a typical web-fed, twin-sheet system, two rolls of plastic materials are simultaneously fed, one above the other. The webs are transported through the oven on separate sheet-conveyor chains and heated to a formable temperature. At the forming station, a specially designed blow pin enters the space between the two sheets before the mold closes. Air pressure is introduced between the sheets through the blow pin and a vacuum is simultaneously applied to each mold half. Twin-sheet forming is done by a slightly different method on specially designed rotary thermoformers.

MATERIALS

The following TPs are thermoformed in large volumes: high-impact and high-heat PS, HDPE, PP, PVC, ABS, CPET, and PMMA. Other polymers of lesser usage are transparent styrene-butadiene block copolymers, acrylic multipolymer, polycarbonate, cellulosics, and ethylene-propylene thermoplastic vulcanizates (TPE). Coextruded structures of up to seven layers include barriers of EVAL, Saran, or nylon, with polyolefins, and/or styreneics for functional properties and decorative aesthetics at reasonable costs.

   All TPs that can be processed into films or sheets can be thermoformed, provided the heated area to be formed does not exceed the hot-strength capabilities of the material to support itself. Among them is foam sheet such as PS foam sheet (Chapter 9).

Some plastic sheets stretch as much as 600%, others as little as 15%. This behavior directly influences what shapes can be formed and their quality. Those with a putty-like appearance respond to very small pressures; others, which tend to be stiff, require heavier operating equipment. The pressure response is somewhat related to the ability to be stretched while hot, but the correspondence is not exact.

The most useful formable TPs do not have sharp melting points (Chapter 1). Their softening with increasing heat is gradual. Each material has its own range of heat, wide or narrow, within which it can be effectively formed. Thus, one plastic may have a forming heat of 275–400 °F (135–204 °C), whereas another may become soft enough for forming at 350 °F (177 °C) but melt at 400 °F (204 °C). And a plastic may stretch well at a given heat but tear easily if heated a few degrees higher or cooled a few degrees. This single property is one of the most important of all the factors involved in forming.

Films (<10 mil, <250 µm) of formable resins exhibit different behavior depending on the plastic: PS is unstable with heat and requires extra cooling; PVC and PVDC are excellent, with no restrictions; nylon is difficult; PCTFE is sensitive to heat and pressure fluctuations; HDPE is difficult without a support film; and PP has a very narrow heat range. In fact, PP is extremely unstable within the conventional range of forming heat, so it is processed by other techniques. Conventional PP has the major deficiency of lacking a rubbery plateau at the forming heat; it just sags and falls apart. A process was developed to form PP just below its softening point, avoiding any sag. Known as solid-phase pressure forming (SPPF), it forces PP into the desired shape by mechanical plugs and pressure. In turn researchers changed PP to overcome its deficiencies. They developed a proprietary catalyst and reactor technology to extrude thermoformable sheet and film. Their material has a rubbery plateau region and a high dynamic modulus, so it is processable in conventional thermoforming machines.

Similar changes have also been made to plastics that were difficult or impossible to form but which had properties desirable in a formed product. PET is an example involving large production quantities. To make it formable, researchers produced crystallized PET (CPET).

Other important materials are coextruded sheets and films (Chapter 3). These multilayer extruded materials provide synergism between physical properties and chemical resistance. They include barrier layers of ethylene-vinyl alcohol (EVOH) copolymers and others, including those required for aseptically packaged food products with a long shelf life at room temperature. Crystallized polyethylene terephthalate (CPET) has been used in dual-ovenable thermoformed trays for packaging frozen foods. These trays can be heated in microwave and conventional ovens; the formed parts are not affected.

Recycling

With most forming (not including bending) there can be up to 50% scrap trim. This material could be wasted, but it is actually recycled and blended with virgin materials. Individual sheet or film stock formed into round shapes could have 50% or more scrap. With square forms, there could be up to 25% scrap.

Quality products

The wall-thickness distribution is a decisive quality criterion for thermoformed products. Online registration and control of the wall thickness occurs during thermoforming. Pertinent measuring systems, the control strategy, and practical testing procedures are used.

The different shapes and forming techniques mean that different requirements need to be set up. Fastness, rigidity, surface properties (gloss, structure, etc.), diffusion properties, shaping precision, and thermoforming stability are largely material-specific demands. However, they are considerably affected by the choice of process parameters.

Some component properties are directly influenced by the quality of wall thickness and its distribution, i.e., for most processes, poor control of thickness tolerances will lead to poor repeatability of formed shapes. The wall thickness indirectly affects the shaping precision and the thermoforming stability. Thus, the cooling behavior and the local relaxation of molecular orientation as well as the shrinkage of the material differ, depending on the thickness of the product. This is why the wall thickness and its distribution are important quality characteristics that have to be set. Important considerations are the defined tolerances and the limits on distribution variation of sheet and film to be processed, especially if the formed product must meet performance requirements and/or cut costs by using the minimum wall thickness to reduce material consumption.

In plug-assist forming of small parts, the wall thickness depends primarily on the molding plug and partly on the temperature. But when the parts have large surface areas, the surface-temperature distribution has a considerable influence on the wall-thickness distribution. This behavior is very important since it provides a parameter that can be drastically modified from one forming cycle to the next [53].

OTHER FORMING METHODS

Certain TPs, when formed, require handling that is normally unavailable with conventional thermoforming machines, so other processes have evolved. Most of these methods tend to reduce the required heat or even eliminate it entirely. One popular technique is high-pressure forming, which is like conventional compression molding (Chapter 8).

The techniques that are used modify conventional metalworking tools. They can be classified as (1) cold forming (performed at room temperature with unheated tools), (2) solid-phase forming (plastic is heated below the melting point and formed), and (3) compression molding of reinforced/composite sheets (using heat). Other methods are classified as forging (including closed-die forming, open-die forming, and cold pressing), stamping, rubber pad or diaphragm forming, fluid forming, coining, spinning, explosive forming, scrapless forming [3, 9, 63, 67–69], and so on.

Cold forming and solid-phase forming include the use of ABS/PC, PC, conventional PP, and HMWHDPE. By using solid-phase forming, processors can make more efficient use of ultrahigh molecular weight, high-density plastics that are difficult or impossible to process by other methods. Forming by these techniques can normally use existing metalworking equipment with minor modifications. Tooling is inexpensive, and production rates can be high. Flash, trim, or weld lines can be eliminated by using some of these processes.

Thermoplastic composites can be stamped to produce high-performance parts. Fiber reinforcements can be used, including glass, graphite, aramid, and so on, in different patterns (short fibers, woven, etc.; see Chapter 12). Products are molded in quick, high-productivity processes, using less energy than needed to manufacture comparable aluminum and steel parts. Tooling costs decrease because of part consolidation. Stamping involves two very different forming processes: solid-state forming and flow molding (or fast compression molding). Each has its advantages and disadvantages.

Solid-state forming uses a male metal-plug mold that matches a female metal-cavity mold and can be used only with crystalline resins. Below their glass-transition temperatures (T_g) amorphous resins are generally too stiff to be rapidly formed into stable products. Crystalline types can be permanently deformed at temperatures between their T_g and their melting point (Chapter 1). Molecular orientation, the mechanism that allows this to occur, relates to the draw ratio. Draw ratios can vary from 5:1 for PET and nylon to 10:1 for low molecular weight PP.

The major advantage of solid-state forming is that parts can be produced in very fast cycle times, usually 10–20 s. The surface finish of these parts is rather smooth, as the fibers do not surface.

Flow molding is not limited to crystalline types because the resin is melted prior to forming. The forming temperature is usually lower than for IM or extrusion. Plastics need not be trimmed, as the composite is 'compression-molded' to completely fill the mold cavity. Most important, flow molding permits more complex parts to be formed than solid-state forming. The process cycle time is usually about 1 min, which is faster than the time needed for most thermoset composites.

The surface is molten during forming, so the surface finish tends to have a fibrous finish. Fiber separation could occur for extremely complex parts. Braided woven fabrics and continuous-fiber mat reinforcements practically eliminate separation. Discontinuous fiber-reinforced composites, such as those made by the slurry process, can be molded into complex shapes without separation.

Postforming

A popular forming process that has provided both performance and cost savings, principally for long production runs, is applied as the plastic sheet, film, tape, and different profiles (tube, rod, etc.) exit the die of an extruder. Upon leaving the die, and retaining heat, the plastic is continuously postformed. With this type of inline system, the hot plastic has only to be reduced to the desired heat of forming. All it may require is a fixed distance from the die opening. Cooling can be accelerated with blown air, a water spray, a water bath, or combinations thereof. Examples of postforming products are given in Chapter 3 (Figs 3.46 to 3.49 on pages 285-6). This equipment requires precision tooling with very good registration.

EQUIPMENT

Thermoforming machines usually have sheet feeders or web feeders. Sheet-fed machines operate from sheet cut into definite lengths and widths for specific applications (Fig. 10.1, page 462). Web-fed machines use either coil stock or a web which is fed directly from an extruder. The machines range from simple, perhaps homemade, single-stage outfits to multistage operations with computerized process control. With single-stage machines, precut sheets are loaded individually into a clamping frame, moved into a heating chamber, and moved back to their original position, where the forming takes place. Figure 10.15 shows a single-stage, shuttle-type, plug-assist former. A two-stage unit consists of two forming stations with one heating chamber.

Another type of machine uses three or more stages. These rotary or carousel types are usually built on a horizontal circular frame that rotates (Fig. 10.16). The rotary table operates like a merry-go-round, indexing through the various stations. A three-stage machine would have stations for loading and unloading, heating, forming, and cooling; stations would be indexed 120° apart.

To speed up output, inline sheet-fed machines are used. Two parallel continually moving tracks hold and move a clamped sheet through the required stations of heat and forming. All movements are indexed so all

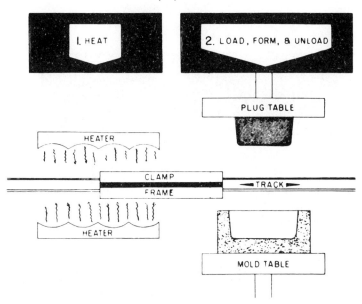

Figure 10.15 Single-stage, shuttle-type thermoformer.

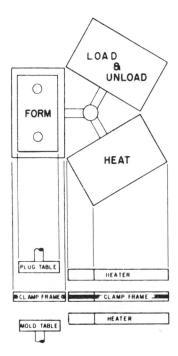

Figure 10.16 Three-stage rotary or carousel-type thermoforming machine.

Figure 10.17 A high-speed continuous thermoformer.

actions are repeatable. To further increase the output, continuous rolls of sheet or sheet material are fed directly from an extruder (Fig. 10.17). A set of continuously conveying chains/tracks indexes the sheet as it moves accurately through the heating, forming, trimming, and packaging stations. Other stations can be included, such as decorating. Multicavity molds are used extensively. They can have sophisticated computer controls to ensure proper operation of all machine and material functions.

Roll-fed thermoforming machinery can produce beverage cups at rates of at least 75 000–100 000 pieces per hour while consuming plastic sheet in excess of 1 ton per hour (910 kg per hour). Production rates of integrated inline extrusion thermoforming have reached 1135 kg of trimmed products per hour.

Sheet consumption for larger and heavier containers can exceed 2 tons per hour (1800 kg per hour), but unit production rates may be lower because fewer mold cavities can be mounted in the machines and more time is required to cool the thicker walls.

Incorrect trimming can damage formed plastics and slow down or stop the output. Tools for trimming include shear disks, steel-rule dies, and saws. The cutting action can be done with the usual punch press, as well as press brakes and other devices. Punch and die clearances should be held to a minimum. The generally accepted rules that are applied to metals are not applicable to plastics. And what is good for one plastic may not work on another (Chapter 15). Plastics have different cutting habits: some tend to be brittle, some rubbery, and so on. Material suppliers and tool manufacturers can provide useful information about trimming.

Inline thermoforming installations, particularly production lines fed directly from an extruder, have to be completely synchronized, else their products may be inferior and their operating costs may rise. If the trimmer operation has to be slowed down, the extruder output has to be reduced. In fact, a slowdown can lead to a shutdown if the extruder cannot operate at the slower speed required. All functions and stations have to be properly interrelated.

Pneumatic controls

Vacuum thermoforming can be related to most of the other forming processes. A vacuum system may use heaters to bring the sheet to its processing temperature and forcing techniques to impart the shape of a mold. The hot, pliable material is moved rapidly to the mold (perhaps by gear drives) and/or moved by an air pressure differential, which holds it in place as it cools. When the proper set temperature is reached, the formed part can be removed without losing its shape.

Two important requirements in this cycle are to sustain the pressure and to maintain uniform heating of the plastic. Faster evacuation generally produces higher-quality parts. It is important to have the correct mold heating so the fast vacuum will produce a part with no internal stress (or very little). While the part is formed, the vacuum gauge should never fall below 20 in. Hg (68 kPa). As a TP cools, this pressure cannot provide sufficient force to form the part and will not hold the plastic tight against the mold (Table 10.1).

A vacuum under 20 in. Hg (68 kPa) is not satisfactory; at least 25 in. Hg (85 kPa) is required. For proper pressure regulation, a vacuum storage or surge tank is necessary to retain a minimal even vacuum. For long forming cycles, a surge tank will permit the use of a smaller vacuum pump than would otherwise be required. To determine the vacuum surge tank size in cubic feet, use the following formula (229):

$$V_0 \times P_0 + V_m \times P_m = V_1 \times P_1$$

Table 10.1 Vacuum pressure measurements

Pressure (psi)		
Gauge[a]	Absolute[b]	Pressure (in. Hg)
0.0	14.7	0.0
−1.0	13.7	2.04
−2.0	12.7	4.07
−4.0	10.7	8.14
−6.0	8.7	12.20
−8.0	6.7	16.30
−9.0	5.7	18.32
−9.9	4.9	20.00
−10.0	4.7	20.36
−11.0	3.7	22.40
−12.0	2.7	24.43
−12.3	2.4	25.00
−13.0	2.7	26.47
−13.7	1.0	27.89
−14.0	0.7	28.50
−14.2	0.5	28.91
−14.3	0.4	29.00
−14.6	0.1	29.73
−14.7	0.0	29.92

[a] Amount of pressure exceeding atmospheric pressure.
[b] Measured with respect to zero (absolute) vacuum; in a vacuum system, absolute pressure (psia) is equal to the negative gauge pressure (psig) subtracted from the atmospheric pressure.

where

V_0 = surge tank volume, including piping to vacuum control valve
V_m = volume of the molding area
$V_1 = V_0 + V_m$
P_0 = absolute pressure in surge tank (0.5 psi, 3.4 kPa)
P_m = initial pressure in the mold (at sea level 14.7 psi, 101 kPa; with prestretched forming use 17.7 psi, 122 kPa)
P_1 = desired atmospheric working pressure

In an example where the volume of the mold and piping is $4 \, ft^3$ $(0.11 \, m^3)$, the vacuum pump can pull about 29 in. Hg (98 kPa), so the surge tank pressure is 0.5 psi (3 kPa). The desired working pressure is 2.42 psi (16.7 kPa) in the tank, and the initial mold pressure is 14.7 psi (101 kPa).

$$V_0 \times 0.5 + 4 \times 14.7 = (V_0 + 4) \times 2.42$$
$$0.5 V_0 + 58.8 = 2.42 V_0 + 9.68$$
$$V_0 = 25.58 \, ft^3 \left(191 \, gal \text{ or } 723 \, dm^3\right)$$

Suppose a lower pressure of 20 in. Hg (68 kPa) is used, which is 4.88 psi (33.6 kPa) in the tank.

$$V_0 \times 0.5 + 4 \times 14.7 = (V_0 + 4) \times 4.88$$
$$0.5 V_0 + 58.8 = 4.88 V_0 + 19.52$$
$$V_0 = 8.97 \, ft^3 \left(671 \, gal \text{ or } 2540 \, dm^3\right)$$

In thermoforming it is sometimes necessary to prestretch (or preblow) the hot sheet before final forming. Compressed air at 3–5 psi (21–34 kPa) is normally used, producing a greater amount of air at atmospheric pressure than for nonprestretched parts. In the above formula add the volume of the prestretched bubble to the volume of the mold and the pressure differential needed for blowing the bubble to the initial atmospheric pressure in the mold.

The objective is to have the vacuum surge tank as close as possible to the forming station and the vacuum control valve. Flexible vacuum hose with connections eliminates elbows, tees, and tubing reducers. All valves must be capable of operation at their fully open position. To capitalize on the rapid vacuum capability of the surge tank, the mold must be able to exploit all the available vacuum pressure. Vacuum holes should be drilled as large as possible, and a maximum number should be used.

Backdrilling of large holes (to 0.125 in., 3.18 mm) on the underside can be used when smaller holes are required on the part side. Male molds can be mounted on a vacuum plate with thin washers or shims, and large vacuum holes can be drilled under the mold. Narrow slots can also be used, and they offer much less resistance than holes when air is evacuated through the mold. Flat areas, segmented sections, or male portions of a mold can be joined with shims, providing long slots.

Temperature controls

Even though TPs have specific processing heats, forming requires thorough, fast, and uniform radiant heat from the surface to the core to the surface. To achieve these sheets, plastics over 0.040 in. (1.02 mm) should use sandwich-type (bottom and top) heater banks. To ensure sufficient heat is used, heaters should have capacities of at least 4–6 kW ft^{-2} (43–65 kW m^{-2}). Various types of radiant heating elements and their performances are shown in Table 10.2. Figure 10.18 shows ABS sheets, 76 in. × 230 in. (193 cm × 584 cm), being conveyed to an IR oven in the back of the console, where the sheets are individually heated and formed into 15 ft (4.6 m) outboard-powered runabouts. The complete automatic process of conveying the sheet, heating, forming, and cooling takes 10 min.

Table 10.2 Types of radiant heating elements[a]

	Efficiency (%)			
Element	When new	After 6 months	Average life (h)	Performance
Ceramic panel	65	55	12 000–15 000	Best buy; heats uniformly and is efficient and capable for profiling heat
Quartz panel trademark	58	50	8000–10 000	Same as ceramic heaters
Coiled Nichrome wire	18–20	8–10	1500	Initially lowest cost; is very inefficient, and heats nonuniformly with use
Tubular rods[b]	45	20	3000	Inexpensive; heats nonuniformly with use and is difficult to screen or mask for profiling heat
Gas-fired infra-red	40–45	25	5000–6000	Lowest cost to operate; has many disadvantages, including wavelength variations and frequent maintenance

[a] Steel clamping frames should be plated with nickel–copper–chrome to reflect heat to sheet edges. After 6 months consider replacing side and back reflectors in order to regain 4–8% efficiency.
[b] Sanding and polishing oxidized tubular heaters can improve their efficiency by 10–15%.

Figure 10.18 Heating sheets in an IR oven; the heated sheets are thermoformed into a boat.

The cycle time is controlled by the heating and cooling rates, which in turn depend on the following factors: the temperature of heaters and the cooling medium, the initial temperature of the sheet, the effective heat-transfer coefficient (Table 1.28, page 86), the sheet thickness, and thermal properties of the sheet material. Different materials absorb radiant heat most efficiently at various wavelengths, which in turn are affected by the temperature of the emitting heater. The most appropriate wavelengths for TPs fall within the infra-red spectrum of 6 μm (400 °F, 204 °C) to 3.2 μm (1200 °F, 649 °C). For example, ABS, PE, and HIPS absorb radiant heat most efficiently when the heating elements emit 3.5–3.3 μm, whereas PC requires 3.4 μm.

Typical material and process heats for a variety of plastics are given in Table 10.3. The normal forming heat should be attained throughout the sheet, and should be measured just before the mold and sheet come together. Shallow-draw projects with fast vacuum and/or pressure forming allow somewhat lower sheet heats and thus a faster cycle. Slightly

Table 10.3 Guide to thermoforming processing temperatures (°F)

Plastic	Mold heat[a]	Lower processing limit[b]	Normal forming heat[c]	Upper limit[d]	Set heat[e]
HDPE	160	260	295	330	180
ABS	180	260	325	380	200
PMMA	190	300	350	380	200
PS	185	260	295	360	200
PC	265	335	375	400	280
PVC	140	210	275	300	160
PSU	320	390	475	575	360

[a] The mold temperature is important in the forming process. High mold heats provide high-quality parts with high impact strength, low internal stress, and good detail, material distribution, and optics (clarity and lack of distortion). However, thin-gauge materials can frequently be thermoformed on molds at lower heat, such as 35–90 °F, as the additional stresses produced are not pronounced in the thin gauges and do not interfere with product performance.

[b] The lower processing limit represents the lowest heat at which the sheet can be formed without undue stresses. This means that the sheet should touch every corner of the mold prior to reaching this lower limit; otherwise problems develop such as stresses and strains that can cause warpage, brittleness, or other physical changes in the part.

[c] The normal forming temperature is the heat at which the sheet should be formed under normal operation. This temeprature should be reached throughout the sheet. Shallow draws with fast vacuum and/or pressure forming will allow somewhat lower sheet heat and thus a faster cycle. Higher heats are required for deep draws, prestretching, detailed mold decorations, etc.

[d] The upper limit is the heat point where the sheet begins to degrade or becomes too fluid and pliable to form. These temperatures can normally be exceeded only with an impairment of the plastic's physical properties (higher heats obtain for IM and extrusion).

[e] The set temperature is the heat at which the part may be removed from the mold without warpage. Parts can sometimes be removed at higher heats if postcooling fixtures are used.

higher heats may be required for deep draws, prestretching, and highly detailed molds.

When extrusion and thermoforming are separate operations, the heat energy supplied for extrusion is completely lost by chilling the sheet. Reheating for thermoforming requires additional heat energy. The inline process offers the advantage of using a high percentage of the energy contained in the sheet to condition it to the forming heat. Savings of about 30–40% can actually be obtained. The inline process provides a more even heat distribution, and weight distributions can be reduced without changing physical properties. At equal output rates, an inline process needs only half the floor space of separate operations.

The time required to cool the heat-softened plastic below its heat-deflection temperature while it is in contact with the mold is often the key to determining the overall forming cycles. Cooling is accomplished by conductive heat loss to the mold and convective heat loss to the surrounding air [3]. The cooling rate depends upon the tooling because, in all methods except matched mold, the plastic is in contact with the mold on one side only. The opposite side is cooled convectively by forced air and ambient air. Water sprays are sometimes used but often pose as many problems, e.g., water spotting. Pressure forming helps to minimize cooling time because the higher air pressure keeps the sheet in more intimate contact with the mold surface.

Molds

Molds can range from hardwood for short runs to filled and unfilled high-temperature polyester (TS) and epoxy resins, cast solid urethane, sprayed metal, cast aluminum, cast porous aluminum, and machined steels. The most common material is cast aluminum, which provides a good combination of durability, light weight, thermal conductivity, ease of manufacture, and cost.

In tooling design, a male primary mold will allow a deeper draw than a female mold because the plastic can be draped or prestretched over the male mold. However, when a male plug-assist is used to prestretch the sheet for a primary female mold, the advantage is nullified. In general, female molds provide easier release, are less likely to get scratched or damaged, produce thicker and stronger rims in containers, can use smaller sheet blanks, and provide the sharpest definition on the outside of the part. Female molds usually have the disadvantage of producing parts with thin bottoms; however, good plug-assist design and operation can largely eliminate this problem. Male molds are generally lower in cost.

Molds used with vacuum or pressure techniques require holes, channels, slits, ducts, and so on, to evacuate the air or to build up the pressure. To avoid visible marks on the surface of thermoformed parts, holes should be kept as small as possible, such as 10–25 mil (250–635 µm). Careful placement of the holes will be helpful in providing fast, efficient airflow during forming. Logic and experience provide guidelines for the placement of openings.

In cast-resin molds, vacuum holes can be provided by including greased wires in the casting for later removal. Cast porous aluminum molds (also used in blow molding) should be considered for greater detail, such as graining, stitching, and relief work.

Undercuts can be included by the use of split molds. Some molds use a removable section that pulls out of the mold after forming (or can be left inside to provide a threaded insert, etc.). In the design of all molds one should consider a draft angle of at least 2–3° per side for the female molds and 5–7° for the male molds (the larger the better). A straight-sided angle in the direction of the draw makes the parts difficult to release. This is

Table 10.4 Shrinkage guide for thermoformed plastics

Plastics	Shrinkage (%)
LDPE	1.6–3.0
HDPE	3.0–3.5
ABS	0.3–0.8
PMMA	0.2–0.8
SAN	0.5–0.6
PC	0.5–0.8
PS	0.3–0.5
PP	1.5–2.2
PVC, rigid	0.4–0.5
PVC, flexible	0.8–2.5

especially true with male molds, where the natural shrink is toward the mold. With advanced forming techniques, such as collapsible molds, parts with zero degrees of draft or even negative drafts can be successfully formed.

Various sheet materials have different mold shrinkage factors, ranging from almost no shrinkage up to as much as $3\frac{1}{2}\%$. Typical basic shrinkage values are given in Table 10.4. Shrinkage can be changed significantly when additives or fillers are used in the resin blends; it can go from zero to practically any preengineered value. However, the percentage of shrinkage is not as important as the consistency of the factor. Molds can be designed to allow for the shrinkage. Careful pretesting is required for precision parts.

Cooling conditions also affect the rate of shrinkage. Restraining the part, either before or after release, will tend to limit the total shrinkage. The mold heat, cooling speed, and cooling fixtures should remain constant to ensure the uniformity of final part shapes. About 70–80% of the dimensional change due to shrinkage occurs as the sheet cools from its forming heat to its set heat. Stabilization to the final dimension can take several hours, or even longer. Most of the change may be due to plastic relaxation once forming stresses are removed.

DESIGNS

Designers should consider the nature of thermoforming [3] which uses flat panels instead of the solid, enclosed, boxlike, cylindrical, rodlike, or structural shapes of other processes. They should be aware of and observe the material's depth-of-draw limitations, which can vary depending on the type of TP, the thickness tolerance of the sheet, and the degree of pinhole freedom it enjoys. For straight vacuum forming into a female mold, the depth/width ratio should generally be ≤0.5. For drape forming over a male mold, this ratio should be ≤1. For parts to be used with the plug-assist, slip-ring, or one of the reverse-draw methods, the ratio can exceed 1 and may even reach 2 under normal circumstances. However, shallow drafts are generally formed more readily than deep drafts and they produce more uniform wall thicknesses.

Undercuts and reentrant shapes are possible in many designs. They require movable or collapsible mold members, but with small undercuts they can often be sprung from a female mold while the formed part is still warm. This type of action works best when the plastic has some flexibility, as do the TPEs, or is very thin. Guidelines for the maximum amounts of undercutting that can be stripped from a mold are as follows: 0.04 in. (1.02 mm) for acrylics, PCs, and other rigid plastics; 0.060 in. (1.52 mm) for PEs, ABSs, and PAs; and 0.100 in. (2.54 mm) for flexible plastics such as the PVCs.

As reviewed on p. 253, coextruded films and sheets are used to gain product performances. The coextrusion can also be sectionalized to gain product advantages (Figs 10.19 and 10.20).

When female tooling is split to permit the removal of parts with under-cuts, a parting line of the split halves becomes visible on the formed part. If this is objectionable, the designer can sometimes incorporate the parting line in the decoration of the part or at some natural line on the part.

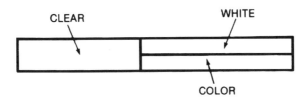

Figure 10.19 A coextruded sheet for producing a three-color thermoformed container with an integrally hinged lid.

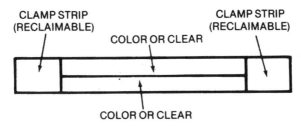

Figure 10.20 Addition of a single-plastic clamping strip at each side of a coextruded sheet permits scrap reclaim of the thermoformed trim waste.

Sharp corners should never be specified, since they hamper the flow of material into the mold's corners. This results in excessive thinning of the materials and causes concentrations of stress. A minimum radius of twice the stock thickness is recommended. It is also more desirable from several standpoints to have large, flowing curves in a thermoformed part than to have squared corners or rectangular shapes. The best parts have smooth, natural curves and drawn sections that are spherical or nearly spherical. Their walls will be more uniform, they will be more rigid, their surfaces will have a lower tendency to show tool marks, and their tooling and molds will be lower in cost. Notches or square holes should be avoided when punching formed parts. Round holes are preferred to oval ones for minimizing stress buildup.

Some draft is required in sidewalls to facilitate the easy removal of the part from the mold. Female molds require less draft since parts tend to pull away from mold walls as they shrink during cooling. With female tooling, for most plastics the draft on each sidewall should be at least $\frac{1}{4}°$. For male tooling, it should be 1° (Fig. 10.21).

Metal inserts are usually not feasible, because thin walls are not sufficiently strong to hold inserts, particularly if thermal expansion and contraction take place. Figure 10.22 shows a method of holding metal fittings. It may be desirable to increase the stiffness of thermoformed parts. Many

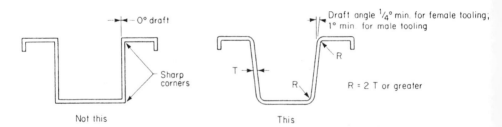

Figure 10.21 The draft required in sidewalls to facilitate easy removal of a thermoformed part from a mold.

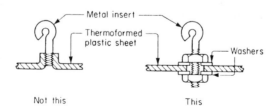

Figure 10.22 A recommended method for holding metal fittings in thermoformed parts.

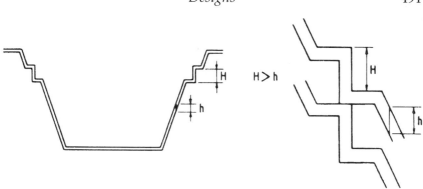

Figure 10.23 A boss design can stack thermoformed parts without jamming.

such parts are panel shaped and made of thin walls, so they may lack rigidity. It corrugations are used it is preferable to have them in two directions, or an embossed pattern can add to their rigidity [3]. With short-run production it may be more economical just to use thicker sheet plastic to gain stiffness. If the function of the part permits, use curved, dished, or domed surfaces to gain stiffness (Fig. 10.23).

When thermoformed parts are stacked, without controlled spacing, they will jam together, which could create sufficient stress to cause them to split. To avoid jamming and control the space between parts, a stacking boss or shoulder system can be used (Fig. 10.23). Within this stacking area the plastic must be sufficiently rigid to prevent the deflection of bosses that would cause jamming. The height of the bosses is generally greater than their vertical cross sections at the point of least taper; otherwise the tapered walls will interfere before the stacking sections can engage. Other designs can also be used to eliminate jamming.

Tolerances

Thermoformed parts lack the dimensional accuracy of injection- and compression-molded parts. With its low pressure, thermoforming reduces the degree to which the sheet being formed is forced to conform to the mold. Sheet variations, mainly in their thickness and degree of existing pinholes, affect the final accuracy of the part. This is particularly true because tooling is generally one-sided. The objective should be to use a sheet with tight thickness controls which is pinhole-free, rather than just to determine its weight (some fabricators 'buy' by the lower-cost method, where weight is the controlling factor).

Parts are dimensionally affected by the difference between their forming temperature and their product-use temperature. Thus, a plastic's coefficient of thermal expansion and contraction has a significant effect. The chosen tooling is generally inexpensive. High-precision tooling is usually not produced.

The pressure, time, and temperature variations that can exist will affect the final part dimensions. Of these factors, evenness in heating the sheet before forming is usually the most important. An allowance must also be made for postforming shrinkage (Chapter 12). Molds should be designed oversize so that, when shrinkage is complete, the part dimensions will be correct to within the design tolerances.

Accuracy is much more prevalent with really precise tooling, especially with matched male and female molds and careful control of temperature, time, and pressure. The dimensional tolerances with the more conventional single-mold system are generally ±0.6% (±0.35% for close tolerances) with female molds, ±0.5% (±0.3% close) with male molds under 3 ft (0.9 m), ±0.8% (±0.4% close) with male molds over 3 ft (0.9 m), and ±30% (±10% close) for wall thicknesses.

COSTING

The central element of any thermoforming system is the tooling, comprising the mold and trimming means. The design of the product will determine the thermoforming technique to be used. Cost and volume will influence the size of the forming machinery, the number of mold cavities, and the rate of production.

The selection of a thermoplastic material for a thermoformed product first requires that it be extrudable into sheet form. Then its cost, availability, and manufacturing continuity are considered. The performance requirements of the product dictate the selection of the lowest-cost plastic that offers adequate formability; tensile, elongation, and impact strength; chemical resistance, low or high temperature resistance, and other properties that may apply; clarity; dielectric strength; moisture, vapor, and oxygen permeability; and recyclability.

If the properties of expensive engineering plastics are required, they may be achievable through multilayer extrusion or lamination or by alloying to achieve such properties at lower cost with a structure composed mainly of commodity plastics. Other alternatives to lower the cost of the product material include density reduction by introducing foaming agents, and incorporating inexpensive fillers to extend the primary resin.

TROUBLESHOOTING

This chapter has presented different problems and solutions along with some guidelines. Like other processes, forming is subject to many variables that influence appearance, performance, and cost. All the variables

Table 10.5 Troubleshooting guide for thermoforming

Problem	Cause(s)	Solution(s)
Blisters or bubbles	Overheating	Lower the heater temperature Increase distance of heater from sheet Attach masks or baffles
	Wrong sheet type or formulation	Obtain correct formulation
	Poor storage conditions	Do not remove material from moistureproof wrap until ready to use
Blush or change in color intensity	Insufficient heating Mold is too cool Assist is too cool Sheet cools before it is completely formed Too deep a draw Poor mold design	Lengthen heating cycle Warm the mold Warm the assist Speed the drape action Add vacuum holes Use heavier-gauge sheet Use mold of proper design
Sticking to the mold	Rough or improperly designed mold	Make mold smoother Increase the taper of male plugs Use mechanical release assists Use air pressure to blow piece from mold Use mold-release agents
Incompletely or improperly formed pieces	Sheet is cold Insufficient vacuum Vacuum holes are plugged up	Lengthen the heating cycle Bring heater closer to sheet Check vacuum system Clean, relocate, or add vacuum holes
Warped or distorted pieces	Poor mold design Sheet removed while too hot	Redesign mold using proper tapers and ribs Increase cooling cycle Use water-cooled molds
Webbing or bridging	Insufficient vacuum Sheet is overheated Long parallel molds with extrusion direction parallel Poor mold layout or design	Check vacuum system Add more vacuum holes Shorten heating cycle Increase heater distance from the sheet Move sheet 90° in relation to mold Use mechanical drape or plug assists

Table 10.5 *Continued*

Problem	Cause(s)	Solution(s)
	Sharp corners on deep draw	Increase radius
Bad surface markings	Markoff (due to trapped air)	Slow draping action
		Add more vacuum holes
	Markoff (due to accumulation of plasticizer on mold)	Use a temperature-controlled mold
		Have mold as far away from the sheet as possible during the heating cycle
		Shorten the heating cycle (if too long)
		Wipe the mold
	Mold is cool	Warm the mold
		Bring the heater closer to the sheet
	Mold is too hot	Provide cooling for mold
	Improper mold composition	Avoid phenolic molds with clear transparent sheet
	Mold surface too highly polished	Remove high surface gloss from mold
	Mold surface too rough	Smooth surface
Excessive post shrinkage	Sheet removed from mold while still hot	Increase cooling time
Pinholing or rupturing	Vacuum holes too large	Partially plug up holes with wood or solder or completely plug and redrill
	Uneven heating	Attach baffles to the top clamping frame

are controllable, and logical steps can be taken to manage them, as reviewed in Chapters 2 through 4.

Major influences are sheet thickness, plastic viscosity, and melt index (Chapter 1), regrind (Chapter 16), sheet orientation (Chapter 1), draw ratio, forming temperature and pressure, and surface blemishes, blisters, blushing, scratch marks, and so on. A guide to troubleshooting the thermoforming process is given in Table 10.5.

11

Reaction injection molding

INTRODUCTION

The RIM process involves high-pressure impingment mixing of two or more very reactive liquid components, and injection of the mixture into a closed mold at low pressure. Large and thick parts can be molded using fast cycles with relatively low-cost materials based on the performances they provide. Its low energy requirements with relatively low investment costs make RIM attractive.

Polymerization of a monomer mixture in the mold allows for the custom formulation of material properties and kinetics to suit a particular application (Chapter 1). For reinforced RIM (RRIM), also called structural RIM (SRIM), the materials along with milled glass and short glass fibers are added to one or both components, so the reinforcement is in the mix entering the mold. Woven and nonwoven fabrics can be tailored to fit the mold; the liquid components are metered to fill the mold. Chapter 12 reviews the different types of fibers that can be used (glass, carbon, graphite, etc.) and methods to enclose these reinforcements in the cavity so they are kept in their proper orientation.

Different materials can be used, such as nylon, TS polyester, and epoxy, but TS polyurethane (PUR) is predominant (Fig. 11.1) because almost no other plastic can offer as great a range of properties. PUR has a modulus of elasticity in bending of 29 000–203 000 psi (200–1400 MPa) and heat resistance in the range 122–392 °F (90–200 °C); higher values are for chopped glass-fiber SRIM.

RIM and SRIM products are created by a volatile chemical reaction. Compounds containing active hydrogens – alcohols in the form of polyols – react with isocynanates in an exothermic reaction to form a PUR; this process produces the plastic or polymer by starting with the monomer (Fig. 1.3, page 4). Polyols and isocyanates are the basic materials used. The isocyanates may be diphenylmethane-4,4-dioscyanate (MDI) or toluene

SCHEMATIC RIM PROCESS

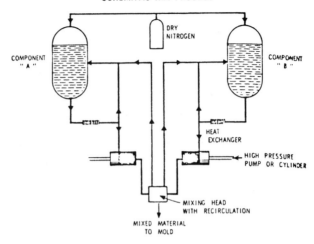

Figure 11.1 PUR-RIM schematic.

diisocyanate (TDI); MDI is the more common. Polyols may be polyethers
but are more commonly polyesters. Additives such as catalysts, sur-
factanta, and blowing agents are also incorporated. Their purpose is to
propagate the reaction and form a finished product possessing the desired
properties (Chapter 1).

The two major classifications for RIM parts are (1) high-density, high-
modulus, flexible elastomers and (2) lower-density structural foams.
Automotive trim and fascia are usually elastomers. Furniture and equipment housings are frequently molded as structural foams (especially
when texture and/or sound deadening are included in the product
specifications).

About 85% of the processed PUR is elastomeric. The rest is rigid, usually structural foam, which has a solid skin encasing a foamed core. PURs
can be used with physical blowing agents such as halocarbons. Foaming
is an integral part of the RIM process, even for solid products, because it
compensates for the shrinkage that occurs during polymerization. That is
why most elastomeric products are also foamed.

PROCESSING

The RIM equipment consists of metering and mixing units combined with
a mold-and-clamp facility. The unique feature of RIM equipment is the
mixhead, where the two or more components are dynamically mixed.
Unlike injection molding, the clamping press does not have to be close to

the material source. The components can be transferred safely across the floor of the processing plant. A metering unit can accommodate as many as five mixheads or molding stations because the lapsed time for the metering shot is only a small fraction of the overall molding cycle.

Inside the mixhead, the two streams of PUR chemicals collide with each other violently and under high pressure, generally 2000–3000 psi (13.8–20.7 MPa); this is called high-pressure impingement. Once the streams collide, the flow is very turbulent and the reaction begins. The stream exits the mixhead and is deposited into the mold. After the pour, a piston inside the mixhead scrapes the walls of the chambers completely clean so that no reacted foam is left inside the mixhead. The straight-through mixhead with its straight chamber has largely been superseded by the more popular L-shaped mixhead with its bent chamber. The straight-through mixhead is designed so the chemicals collide and exit along a straight path through the chamber.

Processors usually prefer the L-shaped mixhead because there is laminar flow when the mix exits the head, and an aftermixing action can be built into the mixing head instead of into the mold (where it occurs for straight-through mixheads).

Important features of a mixhead are size and recirculation capabilities. Smaller mixheads are lighter and more easily clamped for SRIM processing. Recirculation ensures cleaning and temperature conditioning at the start of the pour. If the temperature is not properly controlled, the viscosity of the mix will change, reducing throughput, lowering efficiency, and impairing the quality of the products, perhaps even damaging them.

A metering unit measures the chemicals and delivers the required amount to the mixhead. Whether the processor plans to pour into a bucket, a prototype mold, or a production-scale mold clamp, a metering unit is needed to deliver the chemicals. An important feature in a metering unit is the ability to work at high pressures; this avoids having to flush the mixhead after each shot. Another feature is modularity. Initial setup may be accomplished on a small, economical scale, but it is important to have the ability to increase the parameters of the existing equipment when production increases. Flow transducers, automatic day-tank chemical-level control, automatic filling, agitators, and other accessories can be added as the need arises.

Electronics and closed-loop controllers exist for pump-type metering units. Although the cheaper systems can process quality PUR foam, they may not be able to upgrade as requirements increase. For example, the use of smaller tanks may limit the shot-size capability.

Other features to review are clamping, nucleation, turntables, programmable logic controllers (PLCs), and mold release. Clamps come in a variety of shapes and sizes; most are custom-built. A clamp should have a smooth action through its entire operational sequence. Any error in movements can damage an expensive mold, and improper sequencing can lead to poor production quality. With nucleation the system injects air into the chemicals. Nucleators are used to improve the flowability of chemicals or to improve the cell structure of the final product. Improved flowability makes the melt flow more laminar and increases the throughput. Improved cell structure means there is a highly consistent, very fine pattern of cells all through the molded product.

Turntables are used as a production solution when the cure time of the chemicals is much longer than the duration of the molding cycle. By shuttling the mold, the metering unit can be used to greatest effect, optimizing the time interval between shots. PLCs are used to control the operating sequences of metering units and to communicate with a personal computer that interfaces between the operator and the machine. Software programs are available so that users can monitor and control the complete foaming process. The software generates a graphic illustration of process parameters such as pressures, temperatures, mixture levels, mixture ratios, and output rates. Software is also available for preventive maintenance and troubleshooting.

It is important to provide easy release of parts from the mold. This requires special mold treatments and careful physical removal, both of which increase costs. PURs can also include an internal mold-release (IMR) agent that improves part performance and process economics. But the release agent may have to be inside the mold when producing complex parts with thin walls.

Epoxy molds can be used with RIM for small parts and small production quantities, but a steel mold is still preferred for high surface quality. To make RIM more profitable than injection molding, molds must be designed to exploit their technical advantages, such as proper gating, controlled venting, rheological mold layout, mold rigidity, separating edge definitions, proper heat profiling, and automatic separation of gates (Chapter 2). Gates should be designed to maximize the flow performance of the melt and to conserve material. Most design requirements are developed through experience or trial and error; it also helps to have a knowledgeable instructor.

ADVANTAGES AND DISADVANTAGES

RIM is the logical process to consider for molding large and/or thick parts. It is less competitive for small parts. Capital requirements for RIM processing equipment are rather low when compared with injection molding equipment that would be necessary to mold parts of similar size. Energy requirements are much lower. The injection of liquid materials requires very little energy; however, the liquids are pressurized to 2000–3000 psi (13.8–20.7 MPa) to provide the energy required for impingement

Table 11.1 Production of parts with large surface areas: comparison of RIM and injection molding using unreinforced and reinforced plastics

	PUR-RIM	*Injection molding*
Plastic temperature (°C)	40–60	200–300
Plastic viscosity (Pas)	0.5–1.5	100–1000
Injection pressure (bar)	100–200	700–800
Injection time (s)	0.5–1.5	5–8
Mold cavity pressure (bar)	10–30	300–700
Gates	1	2–10
Clamping force (t)	80–400	2500–10 000
Mold temperature (°C)	50–70	50–80
Time in mold (s)	20–30	30–80
Annealing	30 min at 120 °C	rarely
Wall/thickness	1/0.8	1/0.3
Part thickness, typical maximum (cm)	10	1
Shrinkage (%)		
Unreinforced	1.30–1.60	0.75–2.00
Glass reinforced		
Parallel to fiber	0.25	0.20
Vertical to fiber	1.20	0.40
Inserts	easy	costly
Sink marks around metal inserts	almost none	distinct
Mold prototype (months)	3–5 (epoxy)	9–12 (steel)
Mold alterations	cost-effective	costly

mixing. Molding pressures seldom reach 100 psi (0.7 MPa). In comparison, injection molding pressures are measured in many tons.

COSTING

Elements to consider in making a cost study include (1) normal raw material inventory, (2) normal finished part inventory, (3) cost of raw materials, (4) capital cost of equipment which includes auxiliary and handling equipment, (5) equipment reliability as it relates to output rates, (6) cycle time versus production schedule, and (7) availability of spare parts. The cost analyses and comparisons given throughout this chapter can be used as guides. Table 11.1 compares RIM with injection molding.

TROUBLESHOOTING

This chapter includes many ideas for troubleshooting, including the use of computer software.

12

Reinforced plastics

INTRODUCTION

When plastics processing is being discussed, reinforced plastics (RPs) and composites represent many different techniques. This chapter includes the usual processes (IM, CM, etc.) as well as their specialized techniques. Their details are reviewed in other chapters but will be summarized here, too. Some of the different RP processes are shown in Figs 12.1 and 12.2.

RP and composites are combinations of plastic materials and reinforcing materials, usually in fiber form (chopped fibers, porous mats, woven fabrics, continuous fibers, etc.). Both thermoset (TS) and thermoplastic (TP) resins are used. The modern RP industry began in 1940 with glass fiber reinforced, unsaturated polyesters (TSs) and low-pressure or contact-pressure curing systems. By 1944 the design, fabrication, preliminary static and dynamic tests, and flight tests were successfully completed on the AT-6 military airplane. Its primary and secondary structures were completely fabricated of glass fiber–polyester materials by hand layup, now called open molding (Fig. 12.3). Since then RPs for commercial and military aircraft have made significant technical strides, they have found a wider range of applications, and they have cut operating costs by producing fuel savings; an example is the Boeing 777 (Fig. 1.45, on page 111) [3–8].

Today about 60 wt% of the plastics industry uses different forms of glass fiber–polyester composites. This chapter is confined to TS polyester resins. Although they are used elsewhere, the TP polyester resins are also used in RPs. Overall, at least 90 wt% of all RPs (polyesters, epoxies, etc.) use glass fibers, and 50 wt% are glass fiber TP polyesters (8).

A processor can produce products whose mechanical properties in any direction are both predictable and controllable [3, 8]. This is done by carefully selecting the resin and the reinforcement in terms of composition and orientation, followed by the appropriate process. All types of shape

TYPE OF
REINFORCEMENT

METHOD OF
PROCESSING

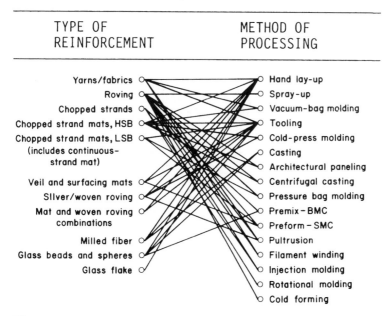

Figure 12.1 Reinforcements and processes used for reinforced plastics.

can be produced: flat and complex shapes; solid and tubular rods/pipes; molded shapes, housings, and complex configurations; structural shapes (angles, channels, box and I-beams, etc.); and so on. RPs can produce some of the strongest materials in the world, as summarized in Fig. 1.15, on page 50.

REINFORCEMENTS

The molder has a variety of alternatives to choose from, regarding the kind, form, and amount of reinforcement to use (Figs 12.1 to 12.4, Tables 12.1 and 1.27 on page 76). With the many different types and forms (organics, inorganics, fibers, flakes, etc.), practically any performance requirement can be met, molded into any shape. Shapes range from extremely small to extremely large, from simple to extremely complex.

Directional properties

The reinforcement type and form (woven, braided, chopped, etc.) depend on performance requirements and the method of processing the RP. Fibers can be oriented in many different patterns to provide the directional properties desired (Fig. 12.5). Depending on their packing arrangement, different reinforcement-to-resin ratios are obtained. In its simplest presentation, using glass fiber with epoxy resin, if the fibers were packed as close as possible (like stacked pipe) the glass would occupy 90.6% of the

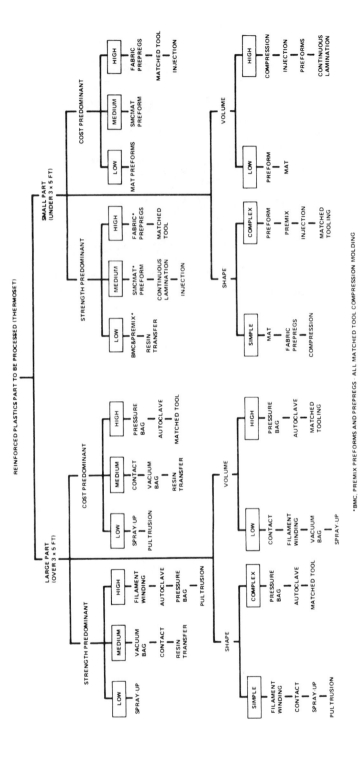

Figure 12.2 Detailed guide to selecting reinforced thermoset plastics.

Figure 12.3 Final flight tests of the Air Force AT-6 (all RP, hand layup); later, during 1953, BT-15 airplanes were fabricated in Long Island.

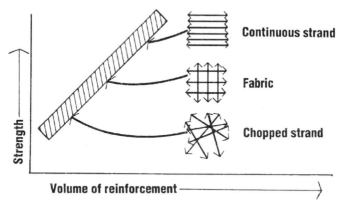

Figure 12.4 Strength–volume relationship for reinforcements used in composites.

volume. With a 'square' packing (fibers directly on top of and alongside each other) the glass volume would be 78.5%. Glass fibers and most other reinforcements require special treatment to ensure maximum performance, such as selecting materials compatible with the resins used, and protecting individual filaments during handling and/or processing.

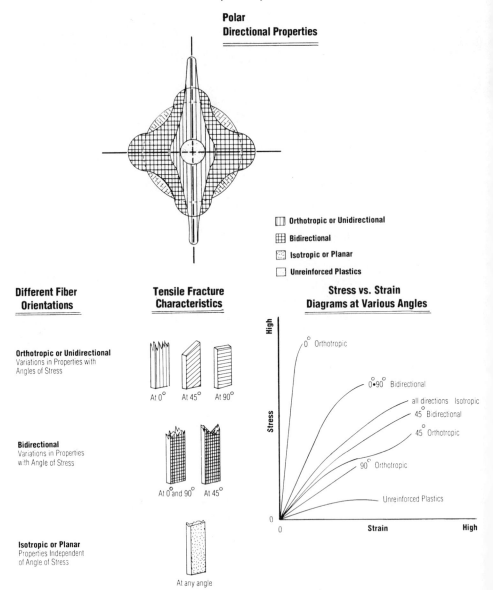

Figure 12.5 Performance of RP and composites with different orientations of their fiber reinforcements.

Table 12.1 Examples of fiber reinforcements used in reinforced plastics[a]

Type of fiber reinforcement	Specific gravity	Density (lb in.$^{-3}$)	Tensile strength/ 10^3 (psi)	Specific strength/ 10^6 (in.)	Tensile elastic modulus/ 10^6 (psi)	Specific elastic modulus/ 10^8 (in.)
Glass						
E monofilament	2.54	0.092	500	5.43	10.5	1.14
12-end roving	2.54	0.092	372	4.04	10.5	1.14
S monofilament	2.48	0.090	665	7.39	12.4	1.38
12-end roving	2.48	0.090	550	6.17	12.4	1.38
Boron (tungsten substrate) 4 mil or 5.6 mil	2.63	0.095	450	4.74	58	6.11
Graphite						
High strength	1.80	0.065	400	6.15	38	5.85
High modulus	1.94	0.070	300	4.29	55[a]	7.86
Intermediate	1.74	0.063	360	5.71	27	4.29
Organic						
Aramid	1.44	0.052	400	7.69	18	3.46

[a] The principal reinforcement, with respect to quantity, is glass fiber. Many other types are used (cotton, rayon, polyester/TP, nylon, aluminum, etc.). Of very limited use because of cost and processing difficulty are 'whiskers' (single crystals of alumina, silicon carbide, copper, or others), which have superior mechanical properties.

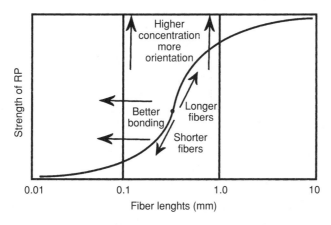

Figure 12.6 Effect of fiber length on RP strength.

The acceptance and use of nonwoven fabrics as reinforcements has led to the development of major products. These reinforcements include felts and paper structures, which usually contain a binder that retains these structures and is compatible with the resin matrix. Combinations of different chopped fibers (glass and aramid, etc.) are also used, including long filaments and woven fabrics. The combinations provide unique properties and they usually permit the molding of different shapes that would not be possible otherwise. The longer fibers are best for optimizing mechanical properties. With short, chopped fibers, the fiber length can range from extremely short 1 mil (250 μm) to at least 0.5 in. (13 mm), and on up to 2 in. (50 mm). The length used usually depends on processing and performance requirements (Fig. 12.6). To obtain the best mechanical performance with fibers in a properly molded part, it is only necessary for them to have an aspect ratio (length/diameter) of about 10.

The use of conventional industrial cutting methods on these reinforcements is not usually the best procedure and may not even work (what cuts glass will not cut aramid, etc.). Woven and other constructions can present a variety of problems. Methods applicable to each type are available, and it is best to contact the material supplier for recommendations.

MATERIALS

Nearly all resins (TS and TP) are used in RPs, but a few predominate (Tables 12.2 to 12.5), with TS polyesters being the major type. Typical polyesters are summarized in Table 12.4. The polyester RPs are used in all processes, but their principal uses are in the low-pressure methods (sprayup, hand layup, pressure bag molding, casting, pultrusion, rotational molding, filament winding, and compression molding).

RPs can be processed in different ways. The individual components (reinforcement and resin) can be put together by the processor. Although TP resins generally need no extra material, the TS resins usually require different additives and fillers (Chapter 1).

Most of the TPs used in RPs are injection molded with compounds prepared by material suppliers. It is estimated that more than half of TSs are prepared by the processor. There are compounds available from material suppliers, as well as in some processing plants, that are ready to be processed; the most popular are sheet molding compounds (SMCs) and bulk molding compounds (BMCs).

Sheet molding compound

A sheet molding compound (SMC) is a reinforced plastic compound in sheet form (Fig. 12.7). Most SMCs combine glass fiber with a polyester

(TS) resin. Any combination of reinforcement and resin can be produced. The reinforcements can have long continuous fibers or any size of chopped fibers laid out in a different orientation from the resin (Figs 12.8 and 12.9). The different orientation makes it feasible to use SMCs on flat and complex-shaped molds. These SMCs will contain various additives and fillers to provide a variety of processing and performance properties.

SMCs are made to meet the shelf life required. These B-staged compounds are usually used in a few weeks or months; some have a shelf life of 6 months. Follow the supplier's recommendations and keep them at a low temperature, else a curing action will occur (Chapter 1).

TPs are also used in sheet form with different reinforcements and resins (Table 12.2). They are called stampable sheets rather than SMCs. These compounds provide unique properties as well as quick and easy processing.

Bulk molding compound

A bulk molding compound (BMC) is a molding compound that is not produced in sheet form. It consists of the mixture used in an SMC, except that it contains only short fibers.

Recycle scrap

TPs can be granulated and recycled, whereas TSs cannot be remelted but could be used as fillers. However, recycling of TPs can degrade performance (Chapters 1–4). When RPs are granulated, the length of the fibers is reduced. When they are reprocessed with virgin materials or alone, their processability and performance definitely change. So it is important to determine whether the change will affect the performance of the final part; if it will, a limit to the amount of regrind mix should be determined.

Void content

Voids are generally the result of the entrapment of air during the construction of a layup, particularly with the use of hand layup and very low pressure processing methods. It is possible to have void contents of 1–3%. Depending on the application, voids can cause a reduction in part performance, particularly in certain environments and after lengthy outdoor exposure. If voids are undesirable, procedures can be used to reduce or eliminate them, such as applying a vacuum during the process. Another preventive method is to squeeze out air during layup by a roller or a spatula.

Table 12.2 Guide to mechanical, thermal, and processing properties of glass fiber reinforced plastics[a]

Material		Glass fiber (wt%)	Specific gravity
			D792
Glass fiber reinforced thermosets (RTS)	Polyester SMC, compression	30.0	1.85
	Polyester SMC, compression	20.0	1.78
	Polyester SMC, compression	50.0	2.00
	Polyester BMC, compression	22.0	1.82
	Polyester BMC, injection	22.0	1.82
	Epoxy filament wound	80.0	2.08
	Polyester, pultruded	55.0	1.69
	Polyurethane, milled fibers (RRIM)	13.0	1.07
	Polyurethane, flaked glass (RRIM)	23.0	1.17
	Polyester spraying/layup	30.0	1.37
	Polyester, woven roving, layup	50.0	1.64
Glass fiber reinforced thermoplastics (RTP)	Acetal resin	25.0	1.61
	Nylon-6,6	30.0	1.48
	Polycarbonate	10.0	1.26
	Polypropylene	20.0	1.04
	Polyphenylene sulfide	40.0	1.64
	Acrylonitrile-butadiene-styrene terpolymer (ABS)	20.0	1.22
	Polyphenylene oxide (PPO)	20.0	1.21
	Styrene-acrylonitrile copolymer (SAN)	20.0	1.22
	Polybutylene terephthalate	30.0	1.52
	Polyethylene terephthalate	30.0	1.56
Unreinforced thermoplastics (TP)	Acetal resin		1.41
	Nylon-6,6		1.13
	Polycarbonate		1.20
	Polypropylene		0.89
	Polyphenylene sulfide		1.30
	Acrylonitrile-butadiene-styrene terpolymer (ABS)		1.03
	Polyphenylene oxide (PPO)		1.10
	Styrene-acrylonitrile (SAN)		1.05
	Polybutylene terephthalate		1.31
	Polyethylene terephthalate		1.34
Metals	ASTM A-606 HSLA steel, cold rolled		7.75
	SAE 1008 low-carbon steel, cold rolled		7.86
	AISI 304 stainless steel		8.03
	TA 2036 aluminum wrought		2.74
	ASTM B85 aluminum, die cast		2.82
	ASTM AZ91B magnesium, die		1.83
	ASTM AG40A zinc, die cast		6.59

[a] See the appendix for English to metric conversions ASTM test methods used.

Thermal coefficient of expansion	Heat deflection at 1.8 MPa (°C)	Thermal conductivity ($Wm^{-1}K^{-1}$)	Specific heat ($Jkg^{-1}K^{-1}$)	Tensile strength (MPa)	Tensile modulus (GPa)
D696	D648	C177		D638	D638
	200+		1.26	83	11.7
	200+		1.26	36.5	11.7
9.4	200+		1.26	158	15.7
6.6	260	8.37	1.26	41.3	12.1
6.6	260	8.37	1.26	33.5	10.5
2.0	200+	1.77	0.96	552	27.6
5.0		6.92	1.17	207	17.2
78.0	29			19.3	
53.1				30.4	
12.0	200+	2.60	1.30	86.2	6.9
4.0	200+			255	15.5
4.7	161			128	8.6
1.8	254	2.60	1.26	159	8.3
1.8	141	7.97	1.21	83	5.2
2.4	132	14.5		45	3.7
1.1	266	3.47	1.05	152	14.1
2.1	99	2.42		76	6.2
2.0	143	6.57	0.84–1.67	100	6.3
2.1	102	4.84		100	8.6
1.4	213	12.1	0.46	131	8.3
1.7	216	11.2		145	9.0
4.7	110	2.80	1.46	81	2.6
4.5	75	2.94	1.26	79	2.8
3.7	132	2.34	1.26	66	2.3
3.8	46–60	2.10	1.88	34	0.7
	135	2.89		66	3.3
3.2	93–104	1.61		41	2.1
68.0	100	1.59	0.84–1.67	54	2.6
36.0	104	1.21	1.38	66	2.8
4.5	50–85	1.76–2.89		57	1.9
	38–41	1.51	1.42	59	2.8
6.8		43.3	0.46	448	207
6.7		60.6	0.42	331	207
9.6		16.3	0.50	552	193
13.9		159	0.88	338	70
11.6		91.8		331	71
14.0		72.5	1.05	228	448
15.2		113	0.42	283	75

Table 12.2 *Continued*

Material		Elongation (%)	Flexural modulus (GPa)
		D638	D790
Glass fiber reinforced thermosets (RTS)	Polyester SMC, compression	<1.0	11.0
	Polyester SMC, compression	0.4	9.7
	Polyester SMC, compression	1.7	13.8
	Polyester BMC, compression	0.5	10.9
	Polyester BMC, injection	0.5	9.9
	Epoxy filament wound	1.6	34.5
	Polyester, pultruded		11.0
	Polyurethane, milled fibers (RRIM)	140.0	0.26–0.37
	Polyurethane, flaked glass (RRIM)	38.9	1.0
	Polyester spraying/layup	1.3	5.2
	Polyester, woven roving, layup	1.6	15.5
Glass fiber reinforced thermoplastics (RTP)	Acetal resin	3.0	7.6
	Nylon-6,6	1.9	5.5
	Polycarbonate	9.0	4.1
	Polypropylene	3.0	3.6
	Polyphenylene sulfide	3.0	13.1
	Acrylonitrile-butadiene-styrene terpolymer (ABS)	2.0	6.0
	Polyphenylene oxide (PPO)	5.0	5.2
	Styrene-acrylonitrile copolymer (SAN)	1.8	7.6
	Polybutylene terephthalate	4.0	8.1
	Polyethylene terephthalate	6.6	8.6
Unreinforced thermoplastics (TP)	Acetal resin	30.0	2.7
	Nylon-6,6	60.0	2.9
	Polycarbonate	110.0	2.3
	Polypropylene	200.0	0.9–1.4
	Polyphenylene sulfide	1.0	3.8
	Acrylonitrile-butadiene-styrene terpolymer (ABS)	5.0	2.4–2.8
	Polyphenylene oxide (PPO)	50.0	2.3–2.8
	Styrene-acrylonitrile (SAN)	0.5	3.8
	Polybutylene terephthalate	50.0	2.3–2.8
	Polyethylene terephthalate	50.0	2.4–3.1
Metals	ASTM A-606 HSLA steel, cold rolled	22.0	
	SAE 1008 low-carbon steel, cold rolled	37.0	
	AISI 304 stainless steel	40.0	
	TA 2036 aluminum wrought	23.0	
	ASTM B85 aluminum, die cast	2.5	
	ASTM AZ91B magnesium, die	3.0	
	ASTM AG40A zinc, die cast	10.0	

Compressive strength (MPa)	Impact strength Izod at 22 °C (Jm^{-1})	Hardness	Water absorption in 24h (%)	Mold shrinkage (%)
D695	D256	D785	D570	D955
166	854	Barcol 68	0.25	
159	438	Barcol 68	0.10	0.002
221	1036	Barcol 68	0.50	
138	227	Barcol 68	0.20	0.001
	154	Barcol 68	0.20	0.004
310	2400	M98	0.50	0.008
207	1335	Barcol 50	0.75	
		Shore D65–75		
	112			
152	690–800	Barcol 50	1.30	
186	1760	Barcol 50	0.50	
117	96	M79	0.29	0.004
183	117	M95	0.50	0.002
97	107	M80	0.14	0.005
172	59	R103	0.05	0.003
145	80	R123	0.01	0.002
97	64	R107	0.30	0.002
121	96	R107	0.24	0.003
121	59	R122	0.06	0.002
124	96	R118	0.06	0.003
172	96	R120	0.05	0.003
90	32	R119	1.3–1.9	0.005
103	43	R120, M83	1.0–1.3	0.008
86	854	M70	0.15	0.005–0.007
24	50–1000	R50–96	0.03	0.020
110	<27	R123	<0.02	0.007
69	160–320	R107–115	0.20–0.45	0.004–0.009
83	270	R115	0.07	0.005–0.007
97	16–24	M80–85	0.20–0.35	
59	43	M68–78	0.08–0.09	0.015–0.020
76	13–35	M94–101	0.1–0.2	0.02–0.025
448		B80		
331		B34–52		
552		B88		
338		R80		
331		Brinell 85		
227		Brinell 85		
283		Brinell 82		

Table 12.3 Example of properties and processes for the major thermoset resins used in composites

Thermosets	Properties	Processes
Polyesters	Simplest, most versatile, economical, and most widely used family of resins; good electrical properties, good chemical resistance, especially to acids	Compression molding, filament winding, hand layup, mat molding, pressure bag molding, continuous pultrusion, injection molding, sprayup, centrifugal casting, cold molding, encapsulation
Epoxies	Excellent mechanical properties, dimensional stability, chemical resistance (especially to alkalies), low water absorption, self-extinguishing (when halogenated), low shrinkage, good abrasion resistance, excellent adhesion properties	Compression molding, filament winding, hand layup, continuous pultrusion, encapsulation, centrifugal casting
Phenolic resins	Good acid resistance, good electrical properties (except arc resistance), high heat resistance	Comprssion molding, continuous lamination
Silicones	Highest heat resistance, low water absorption, excellent dielectric properties, high arc resistance	Compression molding, injection molding encapsulation
Melamines	Good heat resistance, high impact strength	Compression molding
Diallyl o-phthalate	Good electrical insulation, low water absorption	Compression molding

The following equation can be used to estimate void content:

$$\text{Percent voids} = 100 - 100a\left(\frac{d}{c} + \frac{e}{b} + \frac{f}{g}\right)$$

where

a = specific gravity of product
b = specific gravity of fiberglass = (2.55)
c = specific gravity of cured resin = (1.18–1.24)
d = resin content, by weight

Table 12.4 Characteristics of glass fiber reinforced polyesters (TS)

Polyester type	Characteristic	Typical uses
General-purpose	Rigid moldings	Trays, boats, tanks, boxes, luggage, seating
Flexible resins and semirigid resins	Tough, good impact resistance, high flexural strength, low flexural modulus	Vibration damping; machine covers and guards, safety helmets, electronic part encapsulation, gel coats, patching compounds, auto bodies, boats
Light-stable and weather-resistant	Resistant to weather and ultraviolet degradation	Structural panels, skylighting, glazing
Chemical-resistant	Highest chemical resistance of polyester group; excellent acid resistance, fair in alkalies	Corrosion-resistant applications such as pipe, tanks, ducts, fume stacks
Flame-resistant	Self-extinguishing, rigid	Building panels (interior), electrical components, fuel tanks
High heat distortion	Service up to 500°F, rigid	Aircraft parts
Hot strength	Fast rate of cure (hot), moldings easily removed from die	Containers, trays, housings
Low exotherm	Void-free thick laminates, low heat generated during cure	Encapsulating electronic components, electrical premix parts; switchgear
Extended pot life	Void-free uniform, long flow time in mold before gel	Large complex moldings
Air dry	Cures tack-free at room temperature	Pools, boats, tanks
Thixotropic	Resists flow or drainage when applied to vertical surfaces	Boats, pools, tank linings

e = glass content, by weight
f = filler content, by weight
g = specific gravity of filler

Table 12.5 Trade-offs in thermoplastic composites

| Desired modification | How achieved | Sacrifice (from base resin)[a] | | Comments |
		Amorphous	Crystalline	
Increased tensile strength	Glass fibers Carbon fibers Fibrous minerals	Ductility, cost Ductility, cost NA[b]	Ductility, cost Ductility, cost Ductility	Glass fibers are the most cost effective way of gaining tensile strength. Carbon fibers are more expensive; fibrous minerals are least expensive but only slightly reinforcing. Reinforcement makes brittle resins tougher and embrittles tough resins. Fibrous minerals are not commonly used in amorphous resins.
Increased flexural modulus	Glass fibers Carbon fibers Rigid minerals	Ductility, cost Ductility, cost Ductility	Ductility, cost Ductility, cost Ductility	Any additive more rigid than the base resin produces a more rigid composite. Particulate fillers severely degrade impact strength.
Flame resistance	FR additive	Ductility, tensile strength, cost	Ductility, tensile strength, cost	FR additives interfere with the mechanical integrity of the polymer and often require reinforcement to salvage strength. They also narrow the molding latitude of the base resin. Some can cause mold corrosion.
Increased heat-deflection temperature (HDT)	Glass fibers Carbon fibers Fibrous minerals	Ductility, cost Ductility, cost NA	Ductility, cost Ductility, cost Ductility	When reinforced, crystalline polymers yield much greater increases in HDT than do amorphous resins. As with tensile strength, fibrous minerals increase HDT only slightly. Fillers do not increase HDT.

Property	Additive			Comments
Warpage resistance	5–10% glass fibers	NA[b]	Ductility, cost, tensile strength	Amorphous polymers are inherently nonwarping molding resins. Only occasionally are fillers such as milled glass or glass beads added to amorphous materials because they reduce shrinkage anisotropically. Addition of fibers tends to balance the difference between in-flow and cross-flow shrinkage usually found in crystalline polymers. When a particulate is used to reduce and balance shrinkage, some fiber is needed to offset degradation.
	5–10% carbon fibers		Ductility, cost, tensile strength	
	Particulate fillers	Cost		
Reduced mold shrinkage (increased mold-to-size capability)	Glass fibers	Ductility, cost	Ductility, cost	Reinforcement reduces shrinkage far more than fillers do. Fillers help balance shrinkage, however, because they replace shrinking polymers. The sharp shrinkage reduction in reinforced crystalline resins can often lead to warpage. The best 'mold-to-size' composites are reinforced amorphous composites.
	Carbon fibers	Ductility, cost	Ductility, cost	
	Fillers	Tensile strength, ductility, cost	Tensile strength, ductility, cost	
Reduced coefficient of friction	PTFE, Silicone, MoS_2, Graphite	Cost	Cost	These fillers are soft and do not dramatically affect mechanical properties. PTFE loadings commonly range 5–20%; the others are usually 5% or less. Higher loadings can cause mechanical degradation.
Reduced wear	Glass fibers	—	—	The subject of plastic wear is extremely complex and should be discussed with a composite supplier.
	Carbon fibers	—	—	
	Lubricating additives	—	—	
Electrical conductivity	Carbon fibers	Ductility, cost	Ductility, cost	Resistivities of 1–100000 Ω cm can be achieved and are proportional to cost. Various carbon fibers and powders are available with wide variations in conductivity yields in composites.
	Carbon powders	Tensile strength, ductility, cost	Tensile strength, ductility, cost	

[a] See Chapter 1.
[b] Not applicable.

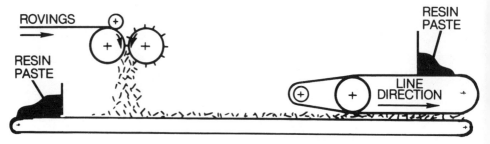

Figure 12.7 A method for manufacturing sheet molding compound (SMC).

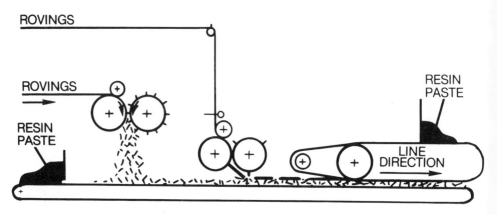

Figure 12.8 Production of SMCs incorporating long, high-performance fiber reinforcements.

However, the equation for estimating void condent is not exact because it is assumed that the resin system has the same density with reinforcement as it does in an unreinforced casting. This can sometimes overstate the void content. Besides air, volatiles can also be entrapped within resins that release them during processing.

PROCESSES

Choosing the optimum process encompasses a broad spectrum of possibilities. Sometimes only one process can be used, but generally there are options. Influencing the selection are quantity, size, thickness, tolerances, type of material, and performance requirements (Fig. 12.10 and Table 12.2). Resins with fillers and/or reinforcements are generally far more stable in meeting tight tolerances. (In fact, reinforced or unreinforced thermosets are dimensionally more stable than other resins.)

RP parts are fabricated by processes using pressures that range from contact (or no pressure), through moderate 50–100 psi (3.4–6.8 kPa), on up

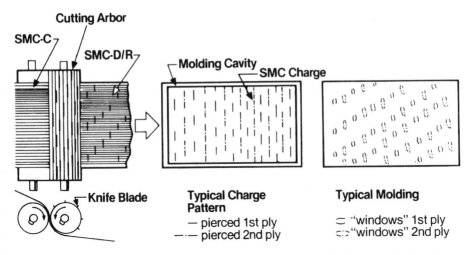

Figure 12.9 An online or off-line production process used when required to cut directional SMC for a specific mold contour; this significantly reduces or even eliminates unwanted wrinkles during layup and processing.

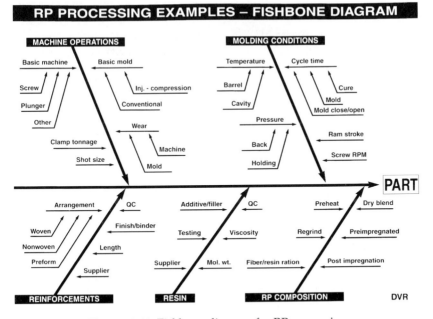

Figure 12.10 Fishbone diagram for RP processing.

to thousands of psi (several megapascals). Temperatures can range from room temperature to the usual 250–600 °F (120–315 °C) and beyond, particularly for certain high-performace TPs. The time cycles can range from seconds to minutes, hours, or even days. Parts cover a wide size range, from small parts to those as large as boat hulls, 80 ft (25 m) or longer (Figs 12.11 and 12.12). The actual process conditions of pressure, heat, and time depend on the material to be processed and on whether a long cure time is required to form the part into a shape (hand layup, etc.). Chapter 1 describes the A-, B- and C-stages of TSs.

Information on processing requirements for materials is reviewed throughout this book. Figures 1.18 and 8.2 (pages 13 and 419) show that the moment to process flowing TS plastics occurs when the viscosity of the melt is at its lowest. Different plastics can be used, with shorter or longer times at this low, level of viscosity. When working with new materials, the processor should obtain relevant details from the material supplier.

Processing may involve equipment that is simple to operate, or it may require extensive specialized equipment. Among the most common processes are contact molding methods (hand layup, sprayup, vacuum bag, pressure bag, autoclave, etc.), matched mold methods (compression molding, transfer molding, resin transfer molding, injection molding, compression–injection molding, stamping, etc.), and other methods (filament winding, cold-press molding, pultrusion, continuous laminating, centrifugal casting, encapsulation, rotational molding, reaction injection molding, etc.). Tables 12.6 through 12.8 summarize some of these processes.

Open mold

This is the oldest and in many ways the simplest and most versatile process. Previously it was called hand layup. However, it is slow and very labor-intensive. It consists of the hand tailoring and placing of layers of (usually glass fiber) mat, fabric, or both on a one-piece mold and simultaneously saturating the layers with a liquid TS resin (usually polyester). The assemblage is then cured with or without heat, and commonly without pressure. An alternative may be to use preimpregnated, B-stage, partially cured, dry material (such as SMC), but in this case heat is applied with the probability of applying low pressure.

Fabrication begins with a pattern from which a mold is made. The mold may be of any low-cost material, including wood, hard plaster or hydrostone, concrete, a metal such as aluminum or steel, and glass fiber reinforced polyester or epoxy. If only a few parts are to be made, a single mold will suffice; otherwise multiple molds may be required. If the volume is large enough and speed is important, heating elements such as lines for steam or other fluids, or electrical heating units, may be

Figure 12.11 Glass fiber–polyester RP water filtration tank: diameter 20 ft (6.1 m), height 32 ft (9.6 m); built and shipped in one piece (by water barge).

Figure 12.12 Glass fiber–polyester 4000 gallon (15.2 m³) tanks in a highly corrosive environment (salt water).

incorporated. Automated equipment may also be installed (Fig. 12.13). The mold may be male (plug) or female (cavity), depending upon which side of the formed part is to have the accurate configuration (the other side will be rough).

Before the actual layup, the mold must be sealed if it is porous, such as wood, and coated with a mold-release agent to prevent sticking of the molded part. Waxes, silicones, thin films, and other agents are used. Layup consists of tailoring the sheet materials to fit, and placing them in layers on the mold, saturating the layers by brush, spray, or any other suitable means, and working them with a serrated roller to consolidate the layers, reducing or eliminating voids and porosity.

To provide resistance to weathering, erosion, or chemical attack, a resin-rich surface coat (gel coat) is frequently added. This surface layer is applied first then allowed to stiffen into a tough layer (not cured) before additional layers are applied. It is usually reinforced with a surface veil using C-glass (rather than the usual E-glass) if chemical resistance is

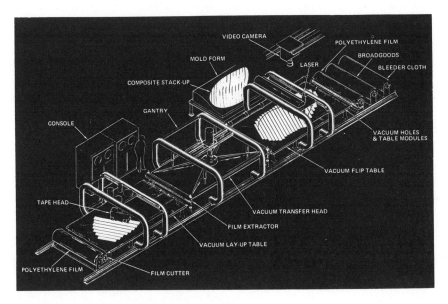

Figure 12.13 Automated, integrated reinforced plastics layup process using TS-preimpregnated reinforced sheet.

required, or a synthetic fiber veil may be used for resistance to weather, particularly sunlight. The resin may be a special formulation that includes TP to improve the surface appearance. Subsequent layers of mat, fabric, or combinations are then applied.

Inserts, strengthening ribs (of wood, metal, or fiberglass shapes), and other devices can be incorporated. They are placed in the layup.

Sprayup

An air spray gun includes a roller cutter that chops fiberglass rovings to a controlled length before they are blown in a random pattern onto a surface of the mold at the same time as a spray of catalyzed resin (Fig. 12.14). The chopped fibers are coated with resin as they exit the gun's nozzle. The resulting, rather fluffy, mass is consolidated with serrated rollers to squeeze out air and reduce or eliminate voids. As in hand layup, the first layer of a gel coat may be applied over the mold, followed by successive passes of the sprayed-on composite before a final gel coat is applied. Inserts or other items can be included during spraying, if required. Thixotropic agents (Chapter 8) may be employed in the resin.

Like hand layup, sprayup usually produces little material waste. One can tailor the formation; the charge follows the contours of three-dimensional shapes very easily. There is no practical limit to size, local

Table 12.6 Process comparison of various RP manufacturing techniques

| | Resin transfer molding | Open molding | | Cold-press molding | Compression molding | |
		Sprayup	Hand layup		Mat/preform	Sheet molding compound
Mold construction	FRP,[a] spray metal, cast aluminum; gasket seal, air vents, self-sealing injection port		FRP	FRP, spray metal, cast aluminum, pinch (land)	Metal, shear edge	High-grade steel; shear edge
Pressure	Pressure feed pumping equipment required; mold halves clamped (methods range from clamp frame to pressure pod)		None	Low-pressure press, capable of 50 psi (345 kPa) (hydraulic or pneumatic mechanical); resin-dispensing equipment not required but recommended	Hydraulic press, normal range of 100–500 psi (0.7–3.4 MPa)	Hydraulic; as high as 2000 psi (14 MPa)
Pressure						

Cure system	Room temperature			Heated; normal range of 225–325°F (107–163°C)	Heated; normal range of 275–350°F (135–177°C)
Resin compounding equipment	High-shear type	Not needed		High-shear type	
Reinforcement	Continuous-strand mat, preform, woven roving	Continuous roving	Chopped-strand mat, woven roving cloth	Continuous-strand mat, perform, woven roving	Continuous roving (specific orientations for higher strength)
Part trim equipment	Yes			With optimum shear edges, minor trimming only	
Generally expected mold life (parts)	3000	1000	3000	≥150000	≥150000

[a] FRP = fiberglass-reinforced plastics.

Table 12.7 Comparison of resin transfer molding, SMC compression, and injection molding

	RTM	SMC compression	Injection
Process operation			
Production requirement, annual units per press	5000–10 000	50 000	50 000
Capital investment	Moderate	High	High
Labor cost	High	Moderate	Moderate
Skill requirements	Considerable	Very low	Lowest
Finishing	Trim flash, etc.	Very little	Very little
Product			
Complexity	Very complex	Moderate	Greatest
Size	Very large parts	Big flat parts	Moderate
Tolerance	Good	Very good	Very good
Surface appearance	Gel-coated	Very good	Very good
Voids/wrinkles	Occasional	Rarely	Least
Reproducibility	Skill-dependent	Very good	Excellent
Cores/inserts	Possible	Very difficult	Possible
Material usage			
Raw material cost	Lowest	Highest	High
Handling/applying	Skill-dependent	Easy	Automatic
Waste	Up to 3%	Very low	Sprues, runners
Scrap	Skill-dependent	Cuts reusable	Low
Reinforcement flexibility	Yes	No	No
Mold			
Initial cost	Moderate	Very high	Very high
Cycle life	3000–4000 parts	Years	Years
Preparation	In factory	Special mold-making shops	
Maintenance	In factory	Special machine shops	

reinforcement is readily provided by building up the thickness or incorporating reinforcements, and their costs are about the same. The production speed is usually less with sprayup, but its thickness control is less efficient, and its strength is likely to be lower and more variable than with hand layup. Compared with hand layup, sprayup is more dependent on the skill and care of the operator. To help overcome or improve on these negative factors, sprayup can be automated, a practical approach in long production runs.

Table 12.8 Guide to compatibility of materials and processes

	Thermosets					Thermoplastics									
	Polyester	*Polyester SMC*	*Polyester BMC*	*Epoxy*	*Polyurethane*	*Acetal*	*Nylon-6*	*Nylon-6,6*	*Polycarbonate*	*Polypropylene*	*Polyphenylene sulfide*	*ABS*	*Polyphenylene oxide*	*Polystyrene*	*Polyester*
Injection molding	X			X	X	X	X	X	X	X	X	X	X	X	X
Hand layup	X				X										
Sprayup	X				X										
Compression molding	X	X	X	X	X						X				
Preform molding	X				X										
Filament winding	X				X										
Pultrusion	X				X										
Resin transfer molding	X				X										
Reinforced reaction injection molding	X				X	X			X						

Vacuum bag

A molder part made by hand layup or sprayup is allowed to cure without the application of external pressure. For many applications, this approach is sufficient, but it rarely achieves maximum consolidation. There is some porosity; fibers may not fit closely into internal corners with sharp radii but tend to spring back; and resin-rich or resin-starved areas may occur because of drainage, even with thixotropic agents. With moderate pressure these defects can be overcome, with an improvement in mechanical properties and better quality control of parts.

One way to apply such moderate pressure is to enclose the 'wet liquid-resin' composite, mold in a flexible membrane or bag, and draw a vacuum inside the enclosure. Atmospheric pressure on the outside presses the bag or membrane uniformly against the wet composite. Pressures are commonly 10–14 psi (69–97 kPa). Chapter 10 explains vacuum pressure. Withdrawal of the air inside the bag not only causes external pressure, but tends to draw air bubbles out of the wet material, thus reducing porosity. Hand working over the bag with rollers, when vacuum is applied, helps to consolidate the structure.

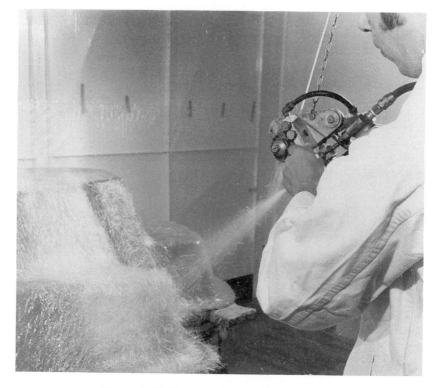

Figure 12.14 Sprayup process for composites.

Flexible materials for a bag or membrane include silicones, neoprene, natural rubber, PVC, PVOH, and others that are not affected by the resin. The bag or membrane needs to be carefully arranged with the vacuum hoses to avoid local dams that entrap air instead of allowing it to escape. Bleeder mats or porous sheets should be placed around the edges of the molded part and/or over the mat assemblage under the bag. Consider using a bleeder with a vacuum line at any place on the part that permits a rough surface at the point of contact but improves the removal of trapped air. As in layup and sprayup, curing can be accelerated by heating the mold.

Pressure bag

If more pressure is required than offered by a bag system, a second envelope can be placed around the whole assemblage and air pressure admitted between the inner bag and outer envelope. This method is also called the vacuum–pressure bag process.

Seamless containers, tanks, pipes, and other products can be made using this process. A preform is used that is made by a sprayup, a mat, or a combination of materials (using a perforated screen to produce the preform). Only enough resin is included in the preform to hold the fibers in place. It usually amounts to about 0.5 wt% and is compatible with the matrix resin.

An inflatable elastic pressure bag is positioned within the preform, and the assembly is put into a closed mold. (The mold could be a drum, etc.) Resin is injected into the preform, and the pressure bag is inflated to about 50 psi (345 kPa). Heat is applied, and the part is cured within the mold. When curing is complete, the bag is deflated and pulled through an opening at the end of the mold before the part is removed.

Autoclave

Still higher pressures can be obtained by placing the vacuum assemblage in an autoclave (Fig. 12.15). Air or steam pressures of 100–200 psi (690–1380 kPa) are commonly achieved. If still higher pressures are required and the danger of extremely high air pressures is to be avoided, a hydroclave may be used, employing water pressures as high as 1000 psi

Figure 12.15 Autoclave process for composites.

(6900 kPa). The bag must be well sealed to prevent infiltration of high-pressure air, steam, or water into the molded part. These processes may or may not employ an initial vacuum.

Foam reservoir

This process, also known as elastic reservoir molding, consists of making a sandwich of resin-impregnated, open-celled, flexible polyurethane foam between the face layers of fibrous reinforcements. When this composite is placed in a mold and squeezed, the foam is compressed, forcing the resin outward and into the reinforcement. The elastic foam exerts sufficient pressure to force the resin-impregnated reinforcement into contact with the mold surface.

Resin transfer

RTM is a closed-mold, low-pressure process (Figs 12.16 and 12.17) in which a dry reinforcement preform is preplaced and impregnated with a

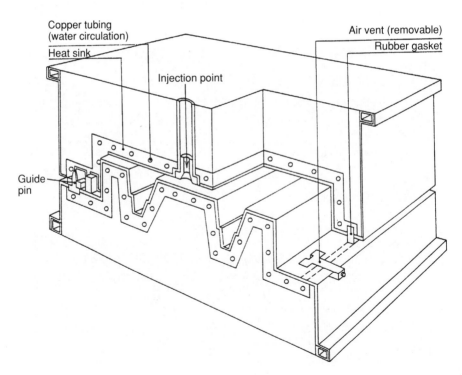

Figure 12.16 RTM with plastic liquid entering from top of mold.

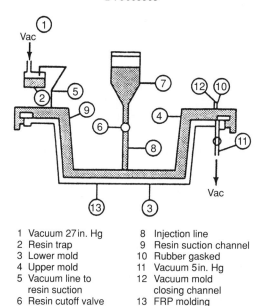

Figure 12.17 Cross section of RTM mold.

1	Vacuum 27 in. Hg	8	Injection line
2	Resin trap	9	Resin suction channel
3	Lower mold	10	Rubber gasked
4	Upper mold	11	Vacuum 5 in. Hg
5	Vacuum line to	12	Vacuum mold
	resin suction		closing channel
6	Resin cutoff valve	13	FRP molding
7	Resin supply reservoir		

liquid resin (usually polyesters, although epoxies and phenolics may be used) in an injection or transfer process, through an opening in the center of a mold (similar to the setup of Fig. 8.1(b) on page 416). The preform is placed in the mold, and the mold is closed. A two-component resin system (including catalyst, hardener, etc.) is then mixed in a static mixer (Chapter 3) and metered into the mold through a system of runners. The air inside the closed mold cavity is displaced by the advancing resin front, and escapes through vents located at the high points or the last areas of the mold to fill (as in injection molding, Chapter 2). When the mold has filled, the vents and the resin inlets are closed. The resin within the mold cures, and the part can be removed. During the 1950s this basic concept involved the use of a vacuum; it was called the Marco process [8]. There was a dam around the outside opening of the two-part mold that contained the preform. This dam was filled with the mixed resin, and in the center of the mold, there was an opening which drew a vacuum. Thus resin could be drawn through the reinforcement, producing a cured part subjected to a maximum pressure of up to 14 psi (97 kPa). This type of vacuum at the vents or parting line is sometimes used with RTM to aid resin flow. Reverse pressure (through center opening) was added later and eventually only 'center' pressure was used (the start of RTM).

Advantages of RTM are that the molded part has two finished surfaces, and the overall process may emit a lower level of styrene vapor if the chosen polyester resin contains styrene. Molded parts may range from the small to the extremely large. Unlike a compression or TP stamping mold, an RTM mold is completely closed to defined stops before the final part formation or curing. This gives a more reproducible part thickness and tends to minimize trimming and deflashing of the final part.

Use of a reinforcement preform allows the preplacement of a variety of reinforcements in precise locations. The preforms remain in position during mold closure and resin injection. If large amounts of random reinforcements are used, consideration must be given to minimizing the washing or movement of the fibers due to resin flow near the resin-inlet gates. A low injection pressure is another characteristic. Simple parts with a low proportion (10–20 vol%) of reinforcement will fill rapidly at pressures of 10–20 psi (70–140 kPa). More complex parts, with 30–50 vol% reinforcement, may require resin injection pressures in the range 100–200 psi (700–1400 kPa) for rapid mold filling.

If low pressure is a requirement, as with large panels, a low injection pressure can generally be maintained and the fill time extended. When low pressure and a fast fill-time are required, the preform construction must be carefully tailored to promote a rapid low-pressure fill without fiber movement. For a high volume, the cycle time will vary, depending on the complexity of the part and the degree of part integration achieved. For a simple component, a 1 min cycle is commercially achievable at a production rate of 1.2 million parts per year. For complex parts, a cycle time of 6 min or longer may be needed.

Reaction injection

RIM is similar to RTM. In the reinforced RIM (RRIM) process a dry reinforcement preform is placed in a closed mold. Next a reactive resin system is mixed under high pressure in a specially designed mixhead. Upon mixing, the reacting liquid flows at low pressure through a runner system to fill the mold cavity, impregnating the reinforcement in the process. Once the mold cavity is filled, the resin quickly completes its reaction. The complete cycle time required to produce a molded part can be as little as 1 min (Chapter 11).

The advantages of RRIM are similar to the advantages of RTM. However, RRIM uses simpler preforms and a lower reinforcement content than RTM. Current resin systems for RRIM will build up viscosity rapidly, producing a higher average viscosity during mold filling. This action follows the initial filling with a low-viscosity resin.

Compression and transfer

Compression and transfer molding processes are high-volume, high-pressure methods suitable for molding simple or complex parts (Chapter 8). Compression molding uses a combination of rather rigid glass fibers with 'soft' plastics, so it can be difficult to obtain smooth surfaces when coating molded parts. Ripples, sink marks, and other imperfections can occur, particularly with a high fiber content. To produce a smooth surface, various methods are used, such as blending TP resins with the base TS resin (polyester), using special coating veils, and so on. There is also a system where a surfacing resin (usually polyurethane) is injected in a compression mold just before the final cure. At the appropriate time in the cycle, the mold is 'cracked' open a few thousandths of an inch, and within a second the resin is injected (Fig. 12.18). The mold closes, and the cure cycle is completed. The coating provides a smooth surface, as well as excellent adhesion to the composite.

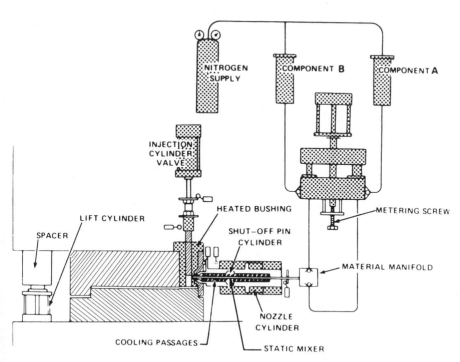

Figure 12.18 A complete system to coat compression-molded parts.

Cold press

This process is an economical press molding method for manufacturing an intermediate number of parts, such as 200 to 2000. It uses low-pressure, room temperature curing resins, and inexpensive molds. Cold press is similar to compression molding except that the resin curing action occurs via its own exothermic heat of reaction (after resins are mixed with catalyst, etc.). Pressures are moderate, usually 20–50 psi (1.4–3.4 kPa). Thus molds can be made of relatively inexpensive metals, plaster, or reinforced plastics. The edges need not be trimmed. Ribs, bosses, and other fairly complex shapes are not easily produced. Two good mold surfaces are obtained.

Stamping

In the stamping process, a reinforced thermoplastic sheet material is precut to required sizes. The precut sheet is preheated in an oven; the heat depends on the TP used, such as PP or nylon, where the heat can range upward from 520 °F (270°C) or 600 °F (315 °C). Dielectric methods provide rapid heating and, most important, the heat is uniformly distributed through the thickness and across the sheet. After heating, the sheet is quickly formed into the desired shape in cooler matched-metal dies using

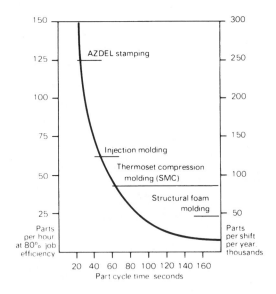

Figure 12.19 Stamping versus other processes: production cycle time compared to production rates based on single-cavity molds.

conventional stamping presses or SMC-type compression presses. Stamping is a highly productive process, capable of forming complex shapes with the retention of the fiber orientation in selected locations. The process can be adapted to a wide variety of configurations, ranging from small components to large box-shaped housings and from flat panels to thick, heavily ribbed parts (Fig. 12.19).

Reinforcements of different types and layouts are used. To make a stampable sheet, two layers of reinforcing sheet are typically laminated to extruded sheets of thermoplastic (Chapter 3). The resultant sheet is homogeneous and has uniform mechanical properties.

Pultrusion

In contrast to extrusion, which pushes, pultrusion pulls a combination of liquid resin and continuous fibers continuously through a heated die of the required profile shape (Fig. 12.20). Shapes include structural I-beams, L-channels, tubes, angles, rods, and sheets; the resins most commonly used are polyesters with fillers. Other resins such as epoxies and urethanes are used where their properties are needed. Longitudinal fibers are generally continuous rovings. Glass fiber (mat or woven) is added for cross-ply properties.

There are six key elements to a pultrusion process, three of which precede the use of the pultrusion machine. The line starts with a reinforcement-handling system (called a creel, as used in a textile weaving), a resin impregnation station, and the material-forming area. The machine consists of components designed to heat, continuously pull, and cut the profiles to a desired length. With machines producing profiles, line speeds may be 1–15 ft min^{-1} (4.6 m min^{-1}) with large to small cross sections.

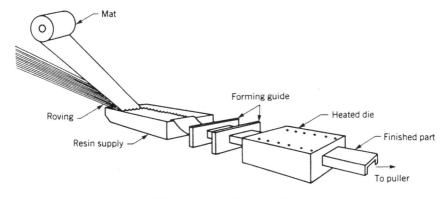

Figure 12.20 Pultrusion line.

The process starts when reinforcements are pulled from the creels and through a resin bath, where they are impregnated with formulated or mixed resin. In some operations, previously resin-impregnated reinforced tapes replace the creel and bath stations. The resin-impregnated fibers are usually preformed to the shape of the profile to be produced. This composite then enters a heated steel die, precision-machined to the final shape of the part. Wheel pullers, clamps, or other devices continuously pull, and when the profile exits the die/mold, it is cured. The profile finishes its cooling in ambient air, water, and/or forced air as it is continuously pulled. The product emerges from the puller mechanism to be cut to the desired length by an automatic device such as a flying cutoff saw.

Control devices must be used to ensure that proper resin impregnation occurs and is held within the required limits. Simple devices, such as doctor rolls or squeeze rolls, are usually sufficient. It is important to control the resin viscosity. The most difficult part to set up is the shape of the opening in the die/mold. Experience and/or trial and error are required. Processing techniques are also used where resin impregnation occurs in the die; takeoff from extrusion coating is covered in Chapter 3.

Continuous laminating

Flat or corrugated, translucent or opaque, panels can be made in presses, but they are more commonly made by continuous laminating. A layer of liquid polyester is deposited on a moving belt (covered with a film) and passed under a chopper that cuts fiberglass rovings into lengths, commonly 1.5–2 in. (37.5–50 mm) and deposits them as a random mat on the resin. The fibers and resin are compacted and covered with an upper film, such as cellophane or PE, and passed through squeeze rolls. The composite envelope may be passed through corrugators or remain flat. Next it passes through an oven to cure. Subsequently, the cover films are stripped, and the continuous sheet is cut into the desired lengths.

Centrifugal casting

Cylindrical shapes such as pipe, tubing, and tanks can be made by placing chopped-strand mat and/or directional chopped rovings against the inner wall of a hollow mold (such as a cylinder, etc.). The mold is heated and rotated, and liquid resin is applied against the fiber. The resin can be delivered through a pipe that moves in the center of the mold, horizontally forward and backward at a controlled rate. Centrifugal force distributes and compacts the resin and fibers against the wall. Hot air may be blown through the mold to accelerate the cure. When the part is cured, the mold is stopped and the part removed.

Encapsulation

This extremely simple process lends itself to high-volume production and automation at a very low cost. Inserts of any size, shape, and quantity can be encapsulated. Little material is wasted, if any. Milled fibers or short, chopped glass strands are combined with slightly exothermic polyesters, epoxies, silicones, and other materials, then mixed and poured into open molds. Molds can be made in simple forms or complicated shapes, using many different materials (wood, plaster of paris, concrete, Kirksite, aluminum, etc.). Depending on the mix, cures can be achieved at room temperature or heat. The cure cycle can be controlled so that an unblemished part is produced. If required, a vacuum is used with the composite during mixing and/or during curing to eliminate air bubbles or voids (Chapter 6).

Filament winding

In filament winding (FW), continuous filaments are wound onto a mandrel after passing through a resin bath; preimpregnated (prepreg) filaments or tapes need not pass through the resin bath. The shape of the mandrel is the internal shape of the finished part. The configuration of the winding depends upon the relative speed of rotation of the mandrel and the rate of travel of the reinforcement-dispensing mechanism. The three most common types are helical winding, in which the filaments are at a

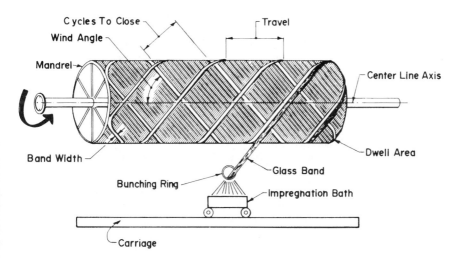

Figure 12.21 Helical filament winding.

Figure 12.22 This giganic filament-winding machine is 22 ft (6.7 m) high, 60 ft (18.3 m) wide, 125 ft (38.1 m) long and has a 100 ton metal mandrel. The 150 000 gallon (569 m^3) tank measures 21 ft (6.4 m) long by 156 in. (3.96 m) in diameter, contains 156 million miles (251 million kilometers) of glass fiber.

significant angle with the axis of the mandrel (Fig. 12.21); circumferential winding, in which the filaments are wound like thread on a spool; and polar winding, in which the filaments are nearly parallel to the axis of the mandrel, passing over its ends on each pass [3, 8, 12, 22, 40, 58, 76].

Different configurations can be employed on successive passes, and the orientation of the filaments tailored to the stresses set up in the part (Figs 12.22 to 12.24 [3]). For example, pipe allows continuous helical winding on a segmental mandrel, on an extruded mandrel, or on release film placed on a stationary mandrel. Other filament winders include braiding machines, loop wrappers, small to very large storage-tank machines, rectangular box-frame machines, and many special fiber-placement machines with several degrees of freedom for intricate shapes.

Curing can be performed by ovens, autoclaves, vacuum bags, heating lamps, and so on. Room temperature curing systems also are used. After being wrapped around the mandrel and before curing, the very high glass content, preimpregnated and oriented composite is sometimes removed from the mandrel, usually by cutting in the axial direction. This sheet can be cut to required sizes and placed in a heated two-part compression mold

Figure 12.23 Glasshopper: the first railcar using RP by Cargill Inc./Southern Pacific and ACF Industries. It had two major advantages: a lower tare weight (8000 lb (3632 kg) lighter than steel cars) and corrosion resistance (principally against the contents to be carried, such as fertilizers) [3].

that provides the shape and the curing stage, in a procedure similar to the stamping of TP sheets. The compression mold can be open but is usually closed. In a matched two-part closed mold, the cutting of trim edges can be used to provide a finished part (Fig. 8.5, page 427). To take advantage of the FW orientation capability, TP resins may be used instead of the usual TS resin. In turn, the TP sheet is used in the conventional stamping procedure (see above).

The FW mandrel must be strong enough to withstand rather high accumulated tension loads due to the filament winding, and must be stiff enough not to sag between end supports. At the same time, it must be possible to remove the mandrel from the finished part after curing, which may require the use of an intricate collapsible mandrel.

As filaments are continuous and tightly packed, they permit a high ratio of filament to resin. This capability often creates products having the highest strength-to-weight ratio obtainable in any structures.

Even though most filament windings use glass filaments, all types can be used (Table 12.1, page 502). Precautions must be observed if superior properties are to be achieved. Glass fibers are strong, but as glass they are subject to a severe loss in strength with surface abrasion. They must be

Type of winding	Considerations	Machinery required
Hoop or circumferential	High winding angle. Complete coverage of mandrel each pass of carriage. Reversal of carriage can be made at any time without affecting pattern.	Simple equipment. Even a lathe will suffice
Helix with wide ribbon	Complete coverage of mandrel each pass of carriage. Reversal of carriage can be made at any time without affecting pattern.	Simple equipment with provision for wide selection of accurate ratios of carriage to mandrel speeds. Powerful machine and many spools of fiber required for large mandrel.
Helix with narrow ribbon and medium or high angle	Multiple passes of carriage necessary to cover mandrel. Programmed relationship between carriage motion and mandrel rotation necessary. Reversal of carriage must be timed precisely with mandrel rotation. Dwell at each end of carriage stroke may be necessary to correctly position fibers and prevent slippage.	Precise helical winding machine required. Ratio of carriage motion to mandrel rotation must be adjustable in very small increments. Relationship of carriage to mandrel positions must be held in selected program without error through carriage reversals and dwells. Relationship between carriage position and mandrel rotation must be progressive so that pattern will progress.
Helix with low winding angle	Fibers positioned around end of mandrel close to support shaft. Characteristics of 'helix with narrow ribbon' apply. Fibers tend to go stack and loop on reversal of carriage. Fibers tend to group from ribbon into rope during carriage reversal. Mandrel turns so slowly that extremely long delay occurs at each end of carriage stroke and speed up of mandrel at each end of carriage stroke is highly desirable to shorten winding time.	Similar machinery required as for 'helix with narrow ribbon.' Take-up device for slack fibers is necessary if cross-feed on carriage is not used. Cross-feed on carriage is required for very low winding angles. Programmed rotating eye can be used to keep ribbon in flat band at carriage reversal. Mandrel speed-up device must be programmed exactly with carriage motion or pattern will be lost. Polar wrap machine can be used for narrow ribbons with winding angle below about 15° without take-up device or mandrel speed-up being required.
Zero or longitudinal	Mandrel must remain motionless during pass of carriage and then rotate a precise amount near 180° while carriage dwells. Fibers must be held close to support shaft during mandrel motion or fibers will slip.	Precise mandrel indexing required. Simple two-position cross feed on carriage sufficient. Vertical mandrel machine and pressure follower for ribbon sometimes required to preserve ribbon integrity.
Polar wrap	Low angle wrap. Fibers may be placed at different distances from centers at each end when geodesic (non-slipping) path does not have to be followed.	Polar wrap machine with swinging fiber delivery arm desirable for high-speed winding. Helical machine with programmed cross feed will wind polar wraps more slowly.

Figure 12.24 Different filament-winding patterns can be used to meet different performance requirements.

Cone

General considerations same as for helical winding except that carriage motion is not uniform.

Programmed non-linear carriage motion required. Other machine requirements same as for helical winding.

Simple spherical
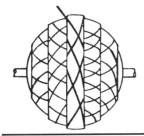
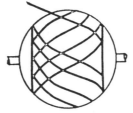

Planer windings at a particular angle result in a heavy build up of fibers at ends of wrap. For more uniform strength, successive windings at higher angles are required.

Sine wave motion of carriage is required for carriage with no cross-feed. At low angles of wind, cross-feed is necessary because carriage travel becomes excessive. Polar wrap machine may be used if range of axis inclination is large enough.

Simple ovaloid
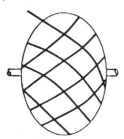

Similar to simple spherical winding but with different carriage or cross-feed motion.

Helical machine with programmed carriage or cross-feed. Polar wrap machine can be used where geodesic (non-slipping) path is in a plane.

True spherical
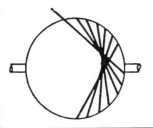

Path of fibers programmed to give uniform wall thickness and strength to all areas on sphere.

Special machine best approach. Otherwise complex programming of all motions of helical machine required.

Miscellaneous
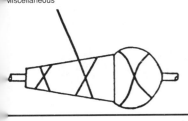

For successful filament winding. It must be possible to hand-wind with no sideways slipping of fibers on mandrel surface.

Machine to reproduce motions of hand winding. Programmed motions in several axes may be required.

Figure 12.24 *Continued*

carefully handled and processed to avoid such deterioration. FW and other layups may include abrasion-resistant plastic fibers or film; film is more usual. This construction permits parts to operate in severe load environments, such as vibration and twisting, and it eliminates or significantly reduces glass-to-glass abrasion when the fiber-to-resin ratio is high. Other types of fibers should be studied to determine whether fiber damage could occur when the part is in service. Certain fibers, with or without resins, might be brittle, and other problems could develop. The designer of the part should have knowledge of potential problems. If problems do develop, they can be overcome by taking the appropriate steps during processing. If unwanted porosity occurs, liners (gel coatings, elastomeric materials, etc.) can be included during FW.

Injection

As explained in Chapter 2, most injection-molded (IM) parts are made from TP, and some of the TP uses milled glass fibers to improve part performance. Other fibers have seen limited use to date. TS compounds usually include reinforcements (Fig. 12.25). Details on IM and factors that influence machine and mold performance, such as wear and abrasion, are reviewed in Chapter 2 [4, 8].

Figure 12.25 Automobile front panel injection-molded from short glass fiber–TS polyester resin compound.

TS reinforced molding compounds processed in IM require water- or oil-cooled barrels, rather than the electrically heated barrels used for TP. The heat of the TS during screw plastication has to be kept lower than its curing heat. The mold heat is higher than the resin heat, causing the TS finally to react and solidify. If there is excess heat in the barrel, the resin cross-links and the machine stops operating; the required cleanup may mean lengthy downtime. A liquid heat control is more effective at regulating melt heat. And the compression ratio of the screw used to process the TS is nominally 1, which helps to keep the heat low and, more important, under control.

IM machines are available that can process BMC using stuffer mechanisms in the hopper (as in certain extrusion operations; see Chapter 3). Depending on the 'stiffness' of the BMC, these machines use a plunger rather than a screw.

Coining

Chapter 2 gives details on the coining process. With reinforced plastic this process (also called injection–compression molding) provides a means of controlling fiber orientation, and so on.

DESIGNS

RPs differ from many other materials because they combine two essentially different materials – fibers and a resin – into a single composite. In this way they are somewhat analogous to reinforced concrete, which combines concrete and steel. But RPs generally have their fibers much more evenly distributed throughout their mass and the ratio of fibers to resin is much higher.

In designing RPs it is necessary to take into account the combined actions of the fiber and the resin as well as the process (Tables 12.9 to 12.12). The combination can sometimes be considered homogeneous but these cases are rare. Thus, it is necessary to allow for the fact that two widely dissimilar materials have been combined into a single unit. This section sets out the basic elements of design theory and begins by making certain fundamental assumptions. The most important assumption is that the two materials act together and that the stretching, compression, and twisting of the fibers and resin under load is the same, i.e. the strains in the fiber and resin are equal. This implies that a good bond exists between the resin and the fiber to prevent slippage between them and wrinkling of the fiber.

Table 12.9 Design recommendations for choosing an RP process

	Contact molding, sprayup	Pressure bag	Filament winding	Continuous pultrusion	Premix/molding compound	Matched-die molding with preform or mat
Minimum inside radius, in.	$\frac{1}{4}$	$\frac{1}{2}$	$\frac{1}{8}$	NA[a]	$\frac{1}{32}$	$\frac{1}{8}$
Molded-in holes	Large	Large	NR[b]	NA	Yes	Yes
Trimmed-in mold	No	No	Yes	Yes	Yes	Yes
Built-in cores	Yes	Yes	Yes	NA	Yes	Yes
Undercuts	Yes	Yes	No	No	Yes	No
Minimum practical thickness, in. (mm)	0.060 (1.5)	0.060 (1.5)	0.010 (0.25)	0.037 (0.94)	0.060 (1.5)	0.030 (0.76)
Maximum practical thickness, in. (mm)	0.50 (13)	1 (25.4)	3 (76.2)	1 (25.4)	1 (25.4)	0.25 (6.4)
Normal thickness variation, in. (mm)	±0.020 (±0.51)	±0.020 (±0.51)	±0.010 (±0.25)	±0.005 (±0.1)	±0.002 (±0.05)	±0.008 (±0.2)
Maximum buildup of thickness	As desired	As desired	As desired	NA	As desired	2:1 max.
Corrugated sections	Yes	Yes	Circumferential only	In longitudinal direction	Yes	Yes

	Mold size	Bag size	Lathe bed length and swing	Pull capacity	Press capacity	Press dimensions
Metal inserts	Yes	Yes	Yes	No	Yes	Yes
Surfacing mat	Yes	Yes	Yes	Yes	No	Yes
Limiting size factor	Mold size	Bag size	Lathe bed length and swing	Pull capacity	Press capacity	Press dimensions
Metal edge stiffeners	Yes	NR	Yes	No	Yes	Yes
Bosses	Yes	NR	No	No	Yes	Yes
Fins	Yes	Yes	No	NR	Yes	NR
Molded-in labels	Yes	Yes	Yes	Yes	No	Yes
Raised numbers	Yes	Yes	No	No	Yes	Yes
Gel coat surface	Yes	Yes	Yes	No	No	Yes
Shape limitations	None	Flexibility of the bag	Surface of revolution	Constant cross section	Moldable	Moldable
Translucency	Yes	Yes	Yes	Yes	No	Yes
Finished surfaces	1	1	1	2	2	2
Strength orientation	Random	Orientation of ply	Depends on wind	Directional	Random	Random
Typical glass content (wt%)	30–45	45–60	50–75	30–60	25	30

[a] NA = not applicable.
[b] NR = not recommended.

Table 12.10 Design guide for processes versus product requirements[a] (Courtesy Owens Corning Fiberglas Corp.)

		Compression molding			Injection molding (TPs)	Cold-press molding	Sprayup and hand layup
		Sheet molding compound	Bulk molding compound	Preform molding			
Minimum inside radius, in. (mm)		$\frac{1}{16}$ (1.59)	$\frac{1}{16}$ (1.59)	$\frac{1}{8}$ (3.18)	$\frac{1}{16}$ (1.59)	$\frac{1}{4}$ (6.35)	$\frac{1}{4}$ (6.35)
Molded-in holes		Yes	Yes	Yes	Yes	No	Large
Trimmed in mold		Yes	Yes	Yes	No	Yes	No
Core pull and slides		Yes	Yes	No	Yes	No	No
Undercuts		Yes	Yes	No	Yes	No	Yes
Minimum recommended draft		$\frac{1}{4}$ in. to 6 in. (6.35–152 mm) depth, 1–3° $\geq$6 in. (152 mm) depth, 3° or as required				2–3°	0°
Minimum practical thickness, in. (mm)		0.050 (1.3)	0.060 (1.5)	0.030 (0.76)	0.35 (0.89)	0.080 (2.0)	0.060 (1.5)
Maximum practical thickness, in. (mm)		1 (25.4)	1 (25.4)	0.250 (6.35)	0.500 (12.7)	0.500 (12.7)	No limit
Normal thickness variation, in. (mm)		±0.005 (±0.1)	±0.005 (±0.1)	±0.008 (±0.2)	±0.005 (±0.1)	±0.010 (±0.25)	±0.020 (±0.51)
Maximum thickness buildup, heavy buildup and increased cycle		AR[b]	AR	2:1 max.	AR	2:1 max.	AR
Corrugated sections		Yes	Yes	Yes	Yes	Yes	Yes
Metal inserts		Yes	Yes	NR[c]	Yes	No	Yes
Bosses		Yes	Yes	Yes	Yes	NR	Yes
Ribs		AR	Yes	NR	Yes	NR	Yes
Molded-in labels		Yes	Yes	Yes	No	Yes	Yes
Raised numbers		Yes	Yes	Yes	Yes	Yes	Yes
Finished surfaces (reproduces mold surface)		2	2	2	2	2	1

[a] Parallel or perpendicular to ram action and with slides in a tooling or split mold.
[b] AR = as required.
[c] NR = not recommended.

Table 12.11 Example of wall-thickness ranges and tolerances for RP parts

| Molding method | Thickness range[a] | | Maximum practicable buildup within individual part | Normal thickness tolerance, mm (in.) |
	Minimum, mm (in.)	Maximum, mm (in.)		
Hand layup	1.5 (0.060)	30 (1.2)	No limit; use cores	±0.5 (0.020)
Sprayup	1.5 (0.060)	13 (0.5)	No limit; use many cores	±0.5 (0.020)
Vacuum bag molding	1.5 (0.060)	6.3 (0.25)	No limit; over 3 cores possible	±0.25 (0.010)
Cold-press molding	1.5 (0.060)	6.3 (0.25)	3–13 mm ($1/8$–$1/2$ in.)	±0.5 (0.020)
Casting, electrical	3 (0.125)	115 (4.5)	3–115 mm ($1/8$–$4\,1/2$ in.)	±0.4 (0.015)
Casting, marble	10 (0.375)	25 (1)	10–13 mm; 19–25 mm ($3/8$–$1/2$ in; $3/4$–1 in.)	+0.8 (0.031)
EMC molding	1.5 (0.060)	25 (1)	Min. to max. possible	±0.13 (0.005)
Matched-die molding; SMC	1.5 (0.060)	25 (1)	Min. to max. possible	±0.13 (0.005)
Pressure bag molding	3 (0.125)	6.3 (0.25)	2:1 variation possible	±0.25 (0.010)
Centrifugal casting	2.5 (0.100)	45% of diameter	5% of diameter	±0.4 mm for 150 mm diameter (0.015 in. for 6 in. diameter); ±0.8 mm for 750 mm diameter (0.030 in. for 30 in. diameter)
Filament winding	1.5 (0.060)	25 (1)	Pipe, none; tanks, 3:1 around ports	Pipe, ±5%; tanks, ±1.5 mm (0.060 in.)
Pultrusion	1.5 (0.060)	40 (1.6)	None	1.5 mm, ±0.025 mm ($1/16$ in. ±0.001 in.); 40 mm, ±0.5 mm ($1\,1/2$ in. ±0.020 in.)
Continuous laminating	0.5 (0.020)	6.3 (0.25)	None	±10 wt%
Injection molding	0.9 (0.035)	13 (0.5)	Min. to max. possible	±0.13 (0.005)
Rotational molding	1.3 (0.050)	13 (0.5)	2:1 variation possible	±5%
Cold stamping	1.5 (0.060)	6.3–13 (0.25–0.50)	3:1 possible as required	±6.5 wt% for flat ±6.0 wt% parts

[a] Thickness may be varied within parts, but it may cause prolonged cure times, slower production rates, and warpage. If possible, the thickness should be held uniform throughout a part.

Table 12.12 Recommended dimensional tolerances for RP parts[a]

Dimension, mm (in.)	Class A (fine tolerance), mm (in.)	Class B (normal tolerance), mm (in.)	Class C (coarse tolerance), mm (in.)
0–25 (0–1)	±0.12 (±0.005)	±0.25 (±0.010)	±0.4 (±0.016)
25–100 (1–4)	±0.2 (±0.008)	±0.4 (±0.016)	±0.5 (±0.020)
100–200 (4–8)	±0.25 (±0.010)	±0.5 (±0.020)	±0.8 (±0.030)
200–400 (8–16)	±0.4 (±0.016)	±0.8 (±0.030)	±1.3 (±0.050)
400–800 (16–32)	±0.8 (±0.030)	±1.3 (±0.050)	±2.0 (±0.080)
800–1600 (32–64)	±1.3 (±0.050)	±2.5 (±0.100)	±3.8 (±0.150)
1600–3200 (64–128)	±2.5 (±0.100)	±5.0 (±0.200)	±7.0 (±0.280)

[a]Class A tolerances apply to parts compression molded with precision matched-metal molds. BMC, SMC, and preform are included.
Class B tolerances apply to parts press molded with somewhat less precise metal molds. Cold-press molding, casting, centrifugal casting, rotational molding, and cold stamping can apply to this classification when molding is done with a high degree of care BMC, SMC, and preform compression molding can apply to this classification if extra care is not used.
Class C applies to hand layup, sprayup, vacuum bag, and other methods using molds made of RP/C material. It applies to parts that would be covered by Class B when they are not molded with a high degree of care.

The second major assumption is that the material is elastic, i.e., the strains are directly proportional to the stresses applied, and when the load is removed the deformation will disappear. In engineering terms the material is assumed to obey Hooke's law [3]. This assumption is probably a close approximation of the material's actual behavior in direct stress below its proportional limit, particularly in tension, if the fibers are stiff and elastic in the Hookean sense and carry essentially all the stress. This assumption is probably less valid in shear, where the resin carries a substantial portion of the stress. The resin may then undergo plastic flow, leading to creep or relaxation of the stresses, especially when the stresses are high.

More or less implicit in the theory of materials of this type is the assumption that all the fibers are straight and unstressed or that the initial stresses in the individual fibers are essentially equal. In practice this is quite unlikely to be true. It is expected, therefore, that as the load is increased some fibers will reach their breaking points before others. As they fail, their loads will be transferred to the unbroken fibers, so failure will be caused by the successive breaking of fibers rather than the simultaneous breaking of all of them. The effect is to reduce the material's overall strength and reduce its allowable working stresses accordingly,

(a)

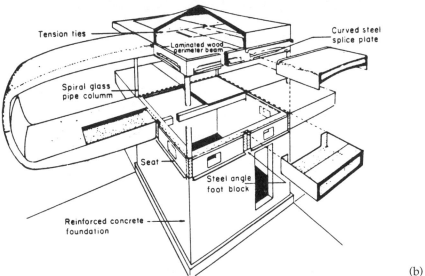

Tension ties

Laminated wood
perimeter beam

Curved steel
splice plate

Spiral glass
pipe columm

Seat

Steel angle
foot block

Reinforced concrete
foundation

(b)

Figure 12.26 Built almost exclusively from RP, this 1957 house had four cantilevered 'wings' extending 16 ft (4.9 m) from a central foundation core to support living quarters. To provide sufficient stiffness, both the U-shaped floor, 8 ft (2.4 m) wide by 16 ft (4.8 m) long, and the roofing sections were monocoque box girders that included sandwich structures (RP facings with kraft paper honeycomb core): (a) partial view of house and (b) schematic of construction [3].

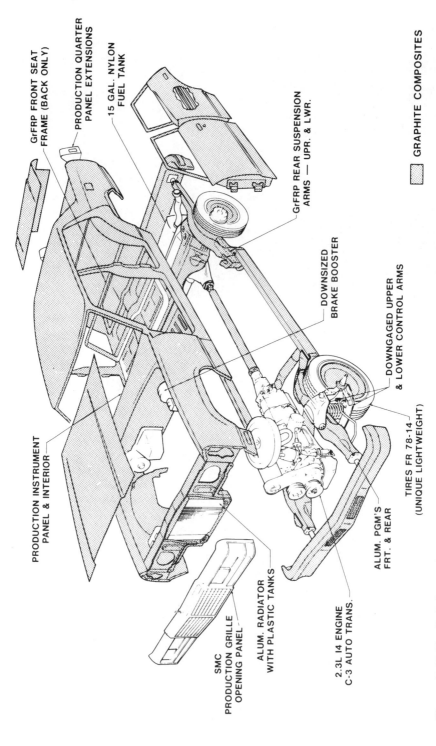

Figure 12.27 Ford's lightweight concept vehicle of the 1960s made extensive use of high-performance RP and composites employing graphite fibers.

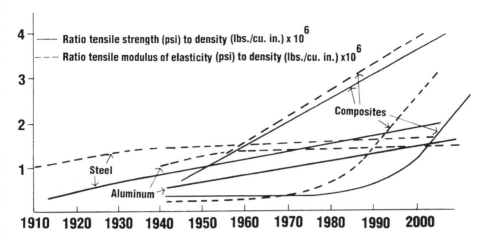

Figure 12.28 Growth pattern for the structural properties of RP and composites compared to steel and aluminum.

but the design theory is otherwise largely unaffected, as long as the behaviour is approximately elastic. The development of higher working stresses is thus largely a question of devising fabrication techniques like filament winding to make fibers work together to obtain their maximum strength [3, 8].

Many different structural and nonstructural products have been molded since the 1940s with future properties of RPs continuing to increase (Figs 12.26 to 12.28).

COSTING

RP may be simple or complex. Simple processes tend to have higher labor costs and lower equipment costs, whereas complex processes tend to have lower labor costs and higher equipment costs. Factors such as quantity, size, shape, and the product cost have a direct influence on the choice of process. Tables 12.8 to 12.13 provide some guides.

TROUBLESHOOTING

Although RP processes share a common set of troubleshooting procedures, each process has its own variations; taken together they should provide an adequate troubleshooting guide. Troubleshooting is discussed throughout the book and those discussions can provide approaches and ideas for setting up guides. Besides those given in Table 12.14, here are some general suggestions:

Table 12.13 Property comparison and design guidelines of resin transfer molding versus sprayup, hand layup, mat/preform, and SMC molding

Design parameter	Resin transfer molding	Sprayup	Hand layup	Mat/preform	SMC
Minimum inside radius, in. (mm)	$\frac{1}{4}$ (6.35)	$\frac{1}{4}$ (6.35)	$\frac{1}{4}$ (6.35)	$\frac{1}{4}$ (6.35)	$\frac{1}{16}$ (1.59)
Molded-in holes	No	Large	Large	Yes	Yes
In-mold trimming	No	No	No	Yes	Yes
Core pull and slides	Difficult	Difficult	Difficult	No	Yes
Undercuts	Difficult	Difficult	Difficult	No	Yes
Minimum recommended draft	2–3°	0°	0°	$\frac{1}{4}$ in. to 6 in. depth, 1–3° >6 in. depth, 3° or as required	
Minimum practical thickness, in. (mm)	0.080 (2.0)	0.060 (1.5)	0.060 (1.5)	0.030 (0.76)	0.050 (1.3)
Maximum practical thickness, in. (mm)	0.500 (12.7)	No limit	No limit	0.500 (12.7)	1 (25.4)
Normal thickness variation, in. (mm)	±0.010 (±0.25)	±0.020 (±0.50)	±0.020 (±0.50)	±0.008 (±0.2)	±0.005 (±0.1)
Maximum thickness buildup, heavy buildup (ratio)	2:1	Any	Any	2:1	Any
Corrugated sections	Yes	Yes	Yes	Yes	Yes
Metal inserts	Yes	Yes	Yes	Yes	Yes
Bosses	Difficult	Yes	Yes	Difficult	Yes
Ribs	Difficult	No	No	Yes	Yes
Hat section	Yes	Yes	Yes	Difficult	No
Raised numbers	Yes	Yes	Yes	Yes	Yes
Finished surfaces	2	1	1	2	2

1. Check mixing and/or pumping equipment. Adjust the resin mix or pumping equipment to achieve the proper time period.
2. Determine whether the mold requires proper preparation to achieve part release, using wax, PVA, and so on.
3. Where appropriate, check whether the reinforcement is properly located so as not to interfere with mold stops, seals, and/or bleed ports.

Table 12.14 Troubleshooting RP processes

Problem	Possible cause	Solution
Nonfills	Air entrapment	Additional air vents and/or vacuum required
	Gel and/or resin time too short	Adjust resin mix to lengthen time cycle
Excessive thickness variation	Improper clamping and/or layup	Check weight and layup and/or check clamping mechanisms such as alignment of platens, etc.
Blistering	Demolded too soon	Extend molding cycle
	Improper catalytic action	Check resin mix for accurate catalyst content and dispersion
Extended curing cycle	Improper catalytic action	Check equipment, if used, for proper catalyst metering Remix resin and contents; agitate mix to provide even dispersion

4. Where appropriate, ensure the clamping frame does not interfere with mold closing.
5. Where heating and/or cooling systems are used, ensure they are operating properly. Verify that recording instruments are working properly.

13

Rotational molding

INTRODUCTION

Rotational molding (RM), also called rotomolding or rotational casting, is a process for producing hollow, seamless items of all sizes and shapes. The molded products range from domestic tanks to industrial containers, from small squeeze-bulbs to storage vessels for highly corrosive materials. Commercial and industrial containers for packaging and materials handling may be covers and housings, water-softening tanks and tote boxes; baby cribs, balls, doll parts; display figures; sporting equipment such as golf carts, surfboards, footballs, juggling pins, helmets; playground equipment and games; housings for vacuum cleaners, scrubbers, and lawnmowers; traffic barricades; display cases; boat hulls (Fig. 13.1); and so on.

Rotational molding can produce quite uniform wall thicknesses even when the part has a deep draw of the parting line or small radii. It is also useful if other techniques make it impossible to obtain a uniform wall thickness when molding a flat surface, especially if the surface has a large area. The liquid or powdered plastic used in this process flows freely into corners or other deep draws upon the mold being rotated and is fused/ melted by heat passing through the mold's wall.

This process is particularly cost-effective for small production runs and large part sizes; the molds are not subjected to pressure during molding, so they can be made relatively inexpensively out of thin sheet metal. The molds may also be made from lightweight cast aluminum and electroformed nickel, both of them light in weight and low in cost. Large rotational machines can be built economically because they use inexpensive gas-fired or hot-air ovens with the lightweight mold-rotating equipment.

Large parts range from up to 22000 gal (83 m³) in size, with a wall thickness of 1.5 in. (38 mm). One tank this size used 2.4 tons (5300 lb) of

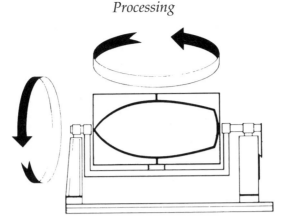

Figure 13.1 A mold and its rotating mechanisms.

XLPE or cross-linked polyethylene (Chapter 1). It was molded by filling the mold three times: the first charge contained about 1.5 tons (3300 lb), the second 0.45 ton (1000 lb), and the third was also 0.45 ton [3, 67–70, 92].

PROCESSING

In most operations the mold cavities are filled with a certain amount of liquid or powder (charging the molds); the mold halves are clamped or bolted together; the charged and closed mold is then placed in a heating oven and the equipment biaxially rotates the mold during the heating cycle. During heating, the plastic material melts, fuses, and densifies into the shapes of the internal cavities by directional centrifugal forces. After heating is completed, the molds are moved into a cooling chamber where they continue to rotate and are slowly cooled by air from a high-velocity fan and/or by a fine spray of water. After removal from the cooling chamber, the molds are opened and the solidified products are removed.

Although RM machines can be built very inexpensively, they are labor-intensive. Both rotations have to be programmed, either at the same rate or at different rates, depending on the product shape. The temperature and duration of the heating cycle also need to be controlled. Most machines that are being built have horizontal rotating arms with closed, recirculating, high-velocity, hot-air ovens, with total automation of the complete process. Many of these machines are also computer programmed to obtain consistent product quality. The most common combine recirculating hot-air, gas-fired ovens with molds made of cast aluminum or fabricated sheet metal. Fast operation is achieved by having four positions: load, heat, cool, and unload. They use four arms, each holding a mold; thus, each position is constantly in use.

More conventional RM uses three-position machines, where one arm is used for unloading and loading, followed by heat and cool (Fig. 13.2). Figure 13.3 shows the feeding inlet to form hollow products inside a closed mold while the mold is heated and rotated about two axes. This system allows different plastics to be molded in multilayers; it is called corotational molding.

Different designs are used to meet different processing requirements: carousel, shuttle, clamshell, rock-and-roll machines, and so on. These designs are similar to those in Chapters 2 to 4, whereby multiple molds can be used to speed up or even simplify production.

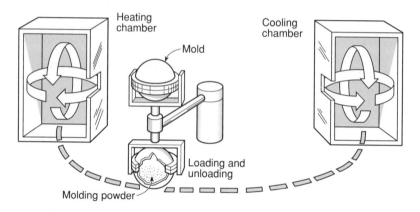

Figure 13.2 Rotational molding actions.

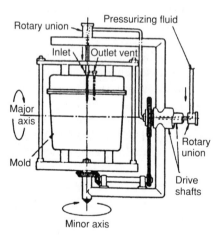

Figure 13.3 Corotational molding: this mold allows multiplastic layers to be created.

Shuttle machines are principally used for rotational molding of large products such as tanks. A frame for holding one mold is mounted on a movable table. The table is on a tract which allows the mold and the table to move into and out of the oven. After the heating period is complete, the mold is moved into an open cooling station. A duplicate table with a mold moves into the heating oven, usually from the opposite side of the oven. As one mold is being cooled, the other mold is in the heating stage, and so on.

The clamshell machines have only one arm. The same location provides mold loading, heating, cooling, and unloading. It uses an enclosed oven that also serves as the cooling station.

The rock-and-roll technique is popular for molding long products such as canoes and kayaks. It rotates on one axis and tilts to provide action in the other axial direction. Most of the motion is in the long direction of the product, which relates to the main rotating motion. The open flame and slush molding machines are the oldest type of equipment used. They rotate only on one axis, fabricating open-ended containers, rubber raingear, shoes, etc. The material moves back and forth (slushing action) during the heating cycle.

MOLDS

Cast aluminum molds are most frequently used, especially for small to medium-size products. Cast aluminum has better heat transfer than steel and is very cost-effective when several molds for the same part are required. Sheet metal molds are normally used for larger parts. They are easy to produce since sections can be welded together. Since the molds are not subjected to pressure during molding, they are not built to take the high loads required in molds for injection, compression, and other pressure operating molds (Chapters 2, 4, 8).

Two-piece molds are usually used but molds in three or more pieces are sometimes required to remove the finished products. Molds can be as simple as a sphere, and molds can be complex with undercuts, ribs, and/ or tapers. Design considerations include heat transfer, mounting techniques, parting lines, clamping mechanisms, mold releases, venting and material stability in storage and during the RM process.

Mold-release agents are usually required because the plastic melt may adhere to the surface of the mold cavity, particularly if the cavity has a very complex shape with contours, ribs, etc. Many molds must have very little or no draft, so they require a mold-release agent. There are mold-release agents that can be baked or applied to the cavity by wiping. By coating with fluorocarbon, the need for mold release could be eliminated. With conventional RM, after the initial mold-release agent is applied, several hundred parts can usually be molded before a stripdown of the mold cavity is required and another baked-on coating is applied. During this time, some touch-up of the mold may be required. Many molders sandblast their mold cavity to remove any buildup of the release agent with the plastic before reapplying the mold release. With the types of plastic that are generally used, melt flow of the plastic during processing will produce a smooth product surface; the sandblasting does not affect the smoothness of the surface. Like other molding operations, a textured cavity can provide a textured product surface. Most texturing of cavities is by chemical etching; it is important to use the appropriate mold material to create a particular texture. An effective release is needed at the parting line to aid in demolding.

Most rotational molds require a venting system to remove the gas that develops during the heat cycle. Venting of molds is also used to maintain atmospheric pressure inside the closed mold during the entire cycle of heating and cooling. The vent will reduce flash and prevent mold distortion as well as lowering the clamping pressure needed to keep the mold closed. It will prevent blowouts caused by pressure and permit use of thinner molds. The vent can be a thin-walled tube with an internal liner of PTFE. The opening where it enters the mold is located where it will not harm the performance or appearance of the molded product. The vent is filled with glass wool to keep the material charge from entering the vent during rotation and particularly during the heating cycle. The outside end of the vent has to be protected so that no water will enter during cooling. A simple method is to put a cap on it; the cap can be automatically controlled to close just before the melt enters the cooling chamber.

MATERIALS

Most RM resins are in powder form with a particle size of 35 mesh (74–2000 μm). The other form is liquid. Some high-flow resins, such as nylon, have been used in small pellet form. About 85 wt% of molding applications use polyethylenes, particularly LDPE, LLDPE, HDPE and cross-linked grades of PE (XLPE and other grades). Ethylene vinyl acetate and adhesive PEs are also used in specialized applications as are PVC, PC, TP polyester, nylon, and PP.

Besides one plastic material, multilayers of different materials are used to obtain improved performance and/or lower costs. Information on the plastic compatibility and performance of multilayers is reviewed in Chapter 2 (coinjection), Chapter 3 (coextrusion), and Chapter 4 (coextrusion and coinjection). In rotational molding the multilayer can be identified as corotation molding. As an example, corotational tanks have been made with a UV-stable exterior and a chemically resistant interior. Use is also made of recycled materials as an interior layer and/or one surface where its properties and performance are met.

DESIGNS

The range of designs is almost unlimited. The Association of Rotational Molders (ARM) recently published a design manual for rotational molding. This was the first publication to address the design guidelines required in the rotational molding industry. ARM represents the rotomolding industry internationally. It includes molders, resin and equipment manufacturers, design organizations, and professional consultants.

ARM is located at 435 North Michigan Avenue, Suite 1717, Chicago IL 60611-4067, USA; telephone 312-644-0828. ARM is also the major information-disseminating organization in the rotational molding industry. It has compiled a comprehensive library with the industry's first design manual. Another important organization on this subject is the Rotational Molding Development Center (RMDS) at the University of Akron. It was founded in 1986 to provide for the industry's future research needs. This information has been presented since they are relatively new and not many are familiar with them.

COSTING

Advantages exist with rotational molding since costs for molds and equipment are lower than those of most other processing methods. Channels for cooling water and resistance to a clamping force are not needed. Different products and colors may be molded on the same machine and in the same cycle. Quick mold changes are possible when several short production runs are required. Large, hollow parts are conveniently molded. Trimming can be eliminated because very little flash is produced. The molded parts are relatively stress-free. Corner sections are thicker than with other processes, which provides additional strength when required. Undercuts, molded-in inserts, intricate contours, and double-wall construction are routinely included.

TROUBLESHOOTING

The important points concerning the behavior of the process during rotational molding have been reviewed in this chapter.

14

Vinyl dispersions

BASIC PROCESS

Vinyl dispersions are fluid suspensions of special, fine particle size, polyvinyl chloride (PVC) plastics in plasticizing liquids. About 10 wt% of all PVC manufactured is processed into these plastisols. When the PVC is heated, fusion or mutual solubilization of the plastic and the plasticizer takes place. The dispersion turns into a homogeneous hot melt. When the melt is cooled, it becomes a tough vinyl product.

Mold preheating saturates the mold with enough heat to gel the plastisol. Because vinyl is a good insulator, it transmits heat only slowly. The thickness of the finished product is determined by the amount of heat transmitted to the plastisol. Preheating temperatures tend to be high for thin molds and lower for thick molds. Finished part thickness is also influenced by the length of time the mold is in the plastisol. The mold should typically be withdrawn from the plastisol slowly and smoothly, otherwise lines will form on the part. With excessive draining, there will be runs and streaks on the product. The mold should ideally contain sufficient residual heat and should be withdrawn slowly enough that the fluid plastisol runs off and the remainder gels immediately without running or dripping. When this cannot be accomplished, it may be possible to rack the parts so that runs drip from one corner; then, by inverting the part, the last drip can flow back.

Many different processing methods are used, all with the essential element of heat. These processes permit products to be made that would otherwise require costly and heavy melt-processing equipment. Different types of dispensing equipment are used to meet different flow rates and delivery amounts. No pressure or mixing is necessary. This means that mold costs are very low and the overall processing equipment costs are low. They are very versatile materials in that almost any additive can be incorporated for special effects as long as it is soluble in the plasticizer or can be ground to a powder sufficiently fine to be suspended in the plastisol.

The term *plastisol* is used to describe a vinyl dispersion that contains no volatile thinners or diluents. Plastisols often contain stabilizers, fillers and pigments along with the essential dispersion resin and the liquid plasticizer. All ingredients exhibit very low volatility under the processing and use conditions. Plastisols can be made into thick fused sections with no concern for solvent or water blistering, as with solution or latex systems, so they are described as being 100% solids materials.

It is sometimes convenient to extend the liquid phase of a dispersion with organic volatiles, which are removed during fusion. The term *organosol* applies to these dispersions.

Vinyl plastisols do have some characteristics that tend to be disadvantages. Dispersion grades of vinyl resin, required to form the suspension, cost a little more than the more common general-purpose resins. Process times are slow, usually 4–20 min in cycle time. Processing time can be minimized by using several low-cost molds when production rate permits. Another factor could be that the plastisol resins contain wetting agents and soaps from the original polymerization reaction. These additives can cause clarity problems and can limit electrical resistance.

PROCESSING PLASTISOLS

This type of plastic is a stable dispersion of fine particles (about 1 μm) of an emulsion material, mainly PVC in a plasticizer, which is a viscous fluid. These flexible vinyl compounds are compounded to meet different requirements. They may contain 25–500 parts plasticizer (possibly more) for every 100 parts vinyl resin. This corresponds to a Shore A hardness of about 98 down to an 8 Durometer [3]. Even harder compounds can be compounded in which part of the plasticizer is replaced by a reactive monomer that is cured when vinyl is fused. This rather large range in hardnesses has enabled plastisols to be used in widely varying applications.

A metallic mold is heated at about 350 °F (177 °C) for a few minutes, in turn heating the plastisol. The plasticizer is absorbed into the particles and solvates them so they fuse together to produce a homogeneous mass. The fusion process is called gelation. Plastisols and the related organosols, plastigels, and rigisols provide a liquid form of PVC to which special processing techniques may be applied. They are often more convenient for producing useful products than conventional melt processing methods reviewed in the other chapters of this book. The plastisol techniques are open moldings, closed moldings, dip moldings, dip coatings, slush moldings, and others that will be reviewed.

During processing, the plastisol is heated slowly. The first change that occurs is a slight lowering of viscosity. At a temperature of 120–200 °F (49–93 °C), the viscosity increases rapidly; this is called the gel point or gel range of a specific plastisol. Different compounded plastisols have different gel points. At the gel point, the resin absorbs the plasticizer. However, in a very soft compound, the resin dissolves into the plasticizer. Because each resin particle remains a separate particle, the resultant gel has no useful physical properties. But on further heating to 350 °F (117 °C), the plasticized resin partially melts and flows into the plasticizer; this occurs at the fusion point or over the fusion range. On cooling, the material comprises the tough rubber compound known as a flexible vinyl.

The viscosity of the plastisol changes as the temperature is raised, starting at a low viscosity, increasing over the gel range, and peaking at the onset of fusion. The viscosity goes down during fusion. Satisfactory processing of vinyl plastisols requires an understanding of gelation and fusion, their mechanisms and their effects on molding and cooling.

At 350 °F the plastisol possesses practically 95% of its ultimate physical properties. Because the vinyl is a relatively good insulator, it takes time for the heat to penetrate completely. It is often advantageous to set the oven at 375–400 °F (191–204 °C) to shorten the fusion time. Care must be taken at the higher temperatures to avoid exceeding the heat stability of the compound and causing it to degrade.

Plastisols are essentially 100% solid materials; when they are fused there is no significant shrinkage or increase in density. A small amount of shrinkage occurs because the vinyl compound usually shrinks more on cooling than occurs with the mold material. This shrinkage is usually 1–6%, with the average at about 3%. The higher shrinkage values are with the softer plastisols.

The rheological behavior of the materials is important during processing. As reviewed in Chapter 1, their viscoelastic behaviors are complex. Plastisols may be shear thinning or shear thickening, depending mostly on PVC particle size, size distribution, and shape, but also on plasticizer type and other additives used.

PROCESSING ORGANOSOLS

Organosols are similar to plastisols in which part of the plasticizer is replaced with a solvent. These less expensive vinyl dispersions were very popular in the past, before legislative action limited the amounts of solvent discharged into the atmosphere. They are still used but now they must have safe ventilation systems (an added cost for the processor).

Organosols are suspensions of finely divided plastic in a volatile organic liquid. PVC is most frequently used. Plastic does not dissolve appreciably in the organic liquid at room temperature, but it does dissolve at elevated temperature. The liquid evaporates at an elevated temperature; the remaining residue upon cooling is a homogeneous plastic mass. Plasticizers may be dissolved in the volatile liquid. Organosol production is more cost-effective with a solvent recovery system.

OPEN MOLDINGS

This very simple process consumes the greatest amount of vinyl plastisols. A measured amount of plastisol is poured into an open mold cavity. The mold and plastic are heated to gel and fuse the plastisol. The mold is then cooled and the vinyl (solidified) part is stripped from the mold. Inserts can be placed in the liquid plastic before it is fused; inserts can also be inserted in the mold before pouring. Two or more colors can be placed in different parts of the mold.

This process is used to produce all kinds of commercial and industrial products. Examples include automotive air filters (big business), table cloths, coin mats, truck flaps, and many other similar flat or relatively flat products. Special applications with special compounding include a major market for automotive oil filters where additives are included in the compound; the additive causes the plastisol to bond to the filter media and to the metal endcap. The plastisol then becomes both an adhesive and an end seal for the oil filter.

CLOSED MOLDINGS

This process is an offshoot of the open molding process except it is closed like a two-part mold. A measured amount of plastisol is poured or pumped into the closed mold cavity, similar to injection molding (Chapter 2) except that no pressure is required. The mold is heated to fuse the plastisol then cooled. Later the mold is opened and the part stripped out. This technique may provide for accurate thickness control, filling very complex surface configurations and so on, by applying a very slight pressure of about 5 psi (34.5 kPa).

DIP MOLDINGS

Dip molding and dip coating are almost identical. With dip molding the plastisol is stripped off the mandrel or mold; in dip coating, the vinyl and mandrel or mold becomes part of the finished product. Dip molding involves (1) preheating the metal mandrel or mold; (2) dipping the mandrel or mold in a tank of plastisol for the required period of time; (3) removing the coated mandrel or mold, allowing any excess plastisol to drain off; (4) returning the mandrel or mold to the oven and heating until the plastisol and the mold reach 350 °F (177 °C); (5) removing from the oven and cooling the part to 130–140 °F (54–60 °C); and (6) stripping the plastic off the mandrel or mold while it is still soft enough to stretch and pull over undercuts but cool enough not to be distorted by stretching. Cooling can be done by hanging in cool air, by water spray, or by actual dipping into water. With water dipping, take care to avoid leaving water-marks on the plastic part.

DIP COATINGS

Processing for dip coating is the same as dip molding except that the mandrel or mold is part of the finished product. Vinyl plastisols do not usually adhere to metals or other mold materials. A primer adhesive is used if the coating requires adhesion. Primer adhesives are usually lacquers that may be dipped, sprayed, or brushed on the metal part before it is preheated. These lacquers are usually solvent-based or water-based adhesives [9].

Dip molding products include automotive bumper guards and gear-shift boots (accordion and/or straight boots), slip-on grips, medical gloves and certain instruments, and electrical devices (transformer and car-battery leads, bus-bar insulation tubes, electronic controls, etc.). It is also used for large-quantity products such as coating tool handles to insulate against heat and cold, insulating field coils for car starters, cushioning kitchen tools, electrically insulating all kinds of devices that perform in hot or cold environments, and providing protective coverings for sharp tools, etc.

SLUSH MOLDINGS

This process was developed many decades ago for manufacturing hollow products from vinyl plastisols; low molecular weight polyethylene is sometimes used in place of the vinyl. Slush molding is the reverse of dip molding and an offshoot of open molding.

Slush molding is carried out in the following steps: (1) preheat the mold; (2) fill the mold cavity with a measured amount of plastisol and hold until gel occurs; it is dispersed evenly over the inner cavity surface by back-and-forth motion of the mold, side-to-side motion, and/or rotation, usually around the vertical axis only; (3) drain excess plastisol out of the mold; (4) heat to fuse the plastisol; (5) cool the mold and plastic part; and (6) strip the part from the mold and trim if required.

As in dip molding, the mold is preheated sufficiently to gel the required thickness of plastisol. The mold is filled and held for several seconds before it is inverted and drained. Another technique is to fill a cold mold then heat it to gel a skin on the mold cavity. This action can improve the reproducibility of the mold texture; any special texture, a grain or engraving on the cavity wall, will be reproduced on the cured product (with preheated or cold mold). Take care when using a cold-mold approach because the plastisol can gel on the air or outside surface of the cavity, producing poor drainage and forming lumps in the plastisol. The lumps can cause the part to have uneven thickness and/or performance. The lumps have to be screened out of the remaining plastisol or they will redeposit on the following slush moldings, causing further problems.

The mold may be split or one-piece. The finished part is removed either by splitting the mold or, in the case of a one-piece mold, by collapsing the part with a vacuum.

Automatic machines were applied to slush molding many decades ago, producing products such as fishing hip-boats, which compete against local products molded in Europe and Asia, particularly Japan. Automatic systems fill molds with plastisol carried by conveyor belts through an oven as it is being 'slushed' (the mold is put into a control motion pattern). The plastisol can gel repeatedly to a thickness of 0.06 in. (15.2 mm). The excess plastisol is poured out of the mold and automatically returned to the main tank for reprocessing. The molds proceed to another oven where curing is completed.

Slush molding has been extensively used for making dolls, balls, and flexible toys of all sorts; other products include automobile gearshift boots, armrests, headrests and road-safety cones.

ROTATIONAL CASTINGS

Rotational casting of vinyl plastisols is similar to the rotational molding of powder resins (Chapter 13); a measured amount of plastisol is placed in the mold and heated while the mold is being rotated. Molded products include volleyballs, basketballs, doll heads and bodies, and various automobile parts.

SPRAY COATINGS

These liquid plastisols can be sprayed onto molds or parts. The viscosity of these special compounds is nonflowing after it is sprayed. Thicknesses of up to 50 mil (1.3 mm) can be obtained in a single pass on a vertical panel, etc. The sprayed parts are heated and cooled, then stripped off the mold or left on as a coating by procedures already reviewed in this chapter. Many small to very large tanks are lined by spraying plastisols or organosols.

With liquid organosols, it is possible to spray or cast a harder film coating than is feasible with plastisols. This is because the organosol's solvent can produce films of thickness 10 mil (0.25 mm) or less. Such films are used to replace paint films in special applications where the chemical resistance of vinyl is required.

CONTINUOUS COATINGS

Plastisols can be spread-coated on different substrates (paper, aluminum foil, plastic sheet or film, etc.). Application is by doctor blade, direct roll, or reverse roll operations (Chapter 7). Examples include plastisols being roll coated on adhesively primed metal for house sidings and cloth fabrics saturated/impregnated and coated in the manufacture of conveyor belts. Foamed vinyl fabrics are made by coating a thin layer of solid plastisol on embossed release paper, then coating a thicker coating of foam plastisol and finally layering on a cloth scrim. The composite is fused and peeled from release paper (Chapter 9). The wear layer of vinyl flooring is usually a coated clear plastisol, making it a no-wax flooring.

SCREENED INKS

A very popular application is to use plastisols for (silk) screening on T-shirts, sportswear, etc. The ink becomes heat-fused. These are highly pigmented systems with application thicknesses of only a few mils (0.25 mm) are possible.

HOT MELTS

Soft plastisols will melt and flow when they are heated to fusion temperatures. Their hardness is 35 Durometer Shore A or softer. This hot, fluid vinyl can be poured or pumped into molds and cooled. Simulated fishing worms, other baits, display items, and various novelties may be encased.

HEATING SYSTEMS

A popular heating system for processing vinyl dispersions is by using ovens. They may be gas-fired or electric, convection or infrared. The key to their success is their uniform heat. An important aspect is providing sufficient exhaust to vent the smoke produced by the hot plastisol.

Any limitations of these ovens may reduce their economy and efficiency at producing the best-performing products. Air is a rather poor medium for transmitting heat when compared to other techniques. IR heats only what it sees. Other methods of heat transfer may be used as required. They include hot plates for open molding, electric resistance heating for preheating rods and mandrels, and hot baths of molten salt for slush molding. Molds may be corded for slush molding despite their very short cycle times. Cording may be accomplished by attaching coils or electric strip heaters to the mold and using hot oil for heating and cold oil for cooling.

MOLDS

Metallic molds are predominantly used. They can be made of almost any nonporous material that can take the heat of the processing plastisols. The most popular are cast aluminum and electroformed nickel since they provide faster heat transfer than steel. However, fabricated steel molds are used particularly for long production runs and/or reducing or eliminating damage to molds. Certain materials, such as zinc and copper, can cause undesirable reactions with plastisols. Zinc is definitely inappropriate; copper should be used cautiously. Zinc can catalyze early degradation of vinyls; copper is okay when the plastisol is not overfused. If overfusion occurs, the part will be stained with copper chloride. Similarly, brass and bronze should only be used where the process is well controlled so that overfusion does not occur. Glasses and glazed ceramic can be used, but they are fragile and need to be handled carefully. Most medical inspection gloves are hot dipped on ceramic molds.

TROUBLESHOOTING

Plastisols go through gelation and fusion stages that relate to temperature, time, and the type of compound being processed. With this cycle of events, the viscosity of the plastisol changes. An understanding of the mechanisms involved during the processing is important to ensure the products meet performance requirements at the lowest cost. Keeping adequate records will provide a direct understanding of this simple process that may easily become complex if not properly controlled.

15

Other processes

OVERVIEW

There are procedures for processing or fabricating plastics. Many overlap or interrelate because different parts of the industry provide their own nomenclature for a process, such as open molding and hand layup, slush molding and rock-and-roll processing, rotational molding and rotomolding.

The procedures reviewed so far pertain to the major production types or special processing types used by the industry; they provide the basis for processing with other techniques. These other techniques are listed in Table 15.1 along with the processes from earlier chapters [1–9, 12–23, 29, 30, 32–40, 51, 55–65, 67–79, 90–93]

WOOD–PLASTIC IMPREGNATION

There are various methods where plastic fabricating processes are used to improve performance of other materials such as steel castings (to seal pores, etc.), concrete (to reinforce, decorate, etc.), wood and others. The wood–plastic impregnation (WPI) fabricating process is a typical example. WPI is also called impreg, compreg, and compressed wood.

WPI started during the early part of the twentieth century after phenolic was developed in 1909. Dry wood has unique properties; its tensile strength, bending strength, compression strength, impact strength, and hardness per unit weight are actually the highest of all construction materials. Plastic-loaded wood (impregnated) has much better performance and gains hardness, rot resistance, long life, etc., with a slight increase in weight.

Originally most of the plastics used were phenolics. By the 1960s a new class of materials containing one or more double bonds was used to treat wood. They consisted of vinyl monomers that could be polymerized into

Table 15.1 Fabricating processes used with plastics

Autoclave molding	Extruder (different types)
Auxiliary equipment	Extrusion blow molding
Bag molding	Fiber spinning
Blister package	Filament placement
Blow molding (different types)	Filament winding (different types)
Bridge, reinforced plastic	Film casting
Bulk molding compound	Flame spraying
Calendering (different types)	Flocculation
Carded package	Flocking or floc spraying
Casting	Flow molding
Centrifugal casting	Fluidized-bed coating
Centrifugal molding	Foamed casting (different types)
Cladding	Foamed-in-place fabrication
Coating (different types)	Foamed molding (different types)
Coextrusion	Foamed reservoir molding
Coining	Forging
Coinjection	Forming, plastic–metal
Cold forming	Forming, scrapless
Cold heading	Forming, solid-phase pressure
Cold molding	Forming, various
Cold-press molding	Foundry molding
Cold stamping	Fourdiner
Cold working	Fusible-core molding
Compounding	Gas counterpressure molding
Compreg molding	Gas-assisted injection molding
Compression–injection molding	Hand layup molding
Compression molding (different types)	Heat sealing
Contact molding	High-pressure molding
Contact pressure molding	Hot-melt molding
Continuous coating	Hot stamping
Continuous injection molding	Hot working
Continuous laminating	Impregnation molding
Die casting	Impulse sealing
Dip casting	Injection blow molding
Dip coating	Injection–compression molding
Dip forming	Injection molding, powder
Dip molding	Injection molding, various
Doctor blade	Insert molding
Double shot molding	Inverse lamination
Draw working	Investment casting
Electroforming	Isotactic molding/pressure
Electroplating	Jet molding
Electrostatic coating	Jet spinning
Embedding	Joining, various
Embossing	Lagging molding
Encapsulation	Laminated molding
Expandable polystyrene	Leatherlike molding

Table 15.1 *Continued*

Liquid injection molding	Screw-plunger transfer molding
Lost-wax molding	Scorim
Low-pressure molding	Sheet molding compound
Machining	Shell molding
Marco vacuum molding	Shrink-wrapping
Melt roll	Sintering
Metallizing	Skiving
Metal powder molding	Slip forming
Metal spraying	Slot extrusion
Microencapsulation	Slush molding
Multilive feed molding	Spline process
Netting extrusion	Soluble-core molding
Notched-die molding	Solvent bonding
One-shot molding	Solvent casting
Open molding	Solvent molding
Orientation process, various	Spinning
Packaging, various	Spraying, various
Perforating	Sprayup molding
Pinhole-free coating	Spread coating
Plastic–concrete	Squeeze molding
Plunger molding	Stamping
Poromeric molding	Stretch blow molding
Potting	Structural foam molding
Powder molding	Superplastic forming
Preform molding	Syntactic molding
Premolding	Tape-placement wrapped molding
Prepolymer molding	Thermal expansion molding
Prepreg molding	Thermoforming, various
Pressure bag molding	Thixomolding
Pressure fabrication	Transfer molding
Processing, art of	Trickle impregnation
Processing, fundamental	Tubing, heat shrinkable
Pulp molding	Two-color molding
Pultrusion	Ultrasonic process
Ram molding	Vacuum bag molding
Reaction injection molding	Vacuum coating
Reactive processing	Vacuum casting
Reinforced plastic, various	Vacuum hot pressing
Resin transfer molding	Vacuum press
Rock-and-roll processing	Vinyl dispersion
Rolling	Viscous process
Rotary molding	Void, plastic impregnated
Rotational casting	Vulcanization
Rotational molding	Welding
Rotomolding	Wet layup molding
Rotovinyl sheet	Wood–plastic impregnation molding
Salt bath process	

solid plastics by free radicals. Vinyl polymerization was an improvement over condensation polymerization with no residue, such as water, left behind that had to be removed from the final WPC (wood–plastic composite).

Processing starts by first evacuating the air from the wood. Any type of mechanical vacuum pump is adequate that can reduce the pressure to 133 Pa (1 mm Hg) or less. The catalyzed monomer containing cross-linking agents, and possibly dyes, is introduced into the evacuated chamber through a reservoir at atmospheric pressure. The wood must be weighted down so that it does not float in the monomer solution.

After the wood is covered with the monomer solution, air is admitted at atmospheric pressure, or dry nitrogen with the radiation process. Immediately the solution begins to flow into the evacuated wood structure to fill the void spaces. The soaking period, like the evacuation period, depends on the structure of the wood. Maple, birch, and other open-cell woods are filled in about 30 min; other woods require more time. WPI using a radiation process rather than regular polymerization reaction gives increased hardness and abrasion resistance.

PLASTIC–CONCRETE

Concrete is the most widely used construction material, representing perhaps 25 vol% of all materials used worldwide; plastic represents about $2\frac{1}{2}$ vol% (Fig. 18.12, page 636). Low cost, convenience, and comprehensive strength are the principal reasons for concrete's universal acceptance. But concrete is not perfect. Most of its shortcomings are caused by Portland cement, one of its important ingredients and also known to engineers of the Roman Empire. Of importance to the plastics industry are freeze–thaw deterioration, destruction by corrosive chemicals, and poor tensile stregth. Concrete does have some outstanding properties such as increasing strength when submerged in ocean water.

To extend the use of concrete, particularly where its shortcomings exist, plastic has been compounded into the concrete mixes since at least 1940. Acrylic, TS polyester, polystyrene, polyethylene, and other plastics were used in special applications by the military and industry worldwide. They have been used in providing decorative and weather-resistant outer structural paneling of buildings, ornaments, synthetic marble floor, wall tiles, architectural cladding, etc. Special concretes are used in commercial and industrial drainage systems, on-top and underground containers, vaults, ocean pilings, bridge supports, road barriers, etc. [6].

Plastic–concrete, also called polymer–concrete and plastic reinforced concrete, is a composite of concrete with plastic acting as a binder. The hardening or curing action is developed by chemical initiators and promoters, or thermal catalytic and radiation polymerization (Chapter 1).

When plastics replace all or part of the Portland cement, the concrete's performance may significantly improve. The concrete becomes impervious to different liquids, particularly corrosive liquids; its compressive strength has at least a twofold increase. Composites may include reinforcing fibers and whiskers, wetting agents, and coupling agents. Unfortunately, these plastic–concrete composites cost at least four times as much as conventional concrete. Plastic–concrete is not used very much, but its markets are expected to grow.

INJECTION MOLDING NONPLASTICS

Since the injection molding evolution/revolution of the 1940s, a variety of materials have been fabricated this way: ceramics, dynamite, food, and many different metals that are usually processed as very fine powder. The metals include aluminum, steel, copper, zinc and its alloys, magnesium and its alloys, and tungsten carbide. Nonmetals are also used. Nonplastics may be processed alone or with a plastic binder, removable by sintering and other methods. Precision moldings can compete with metal parts such as die castings.

Different techniques are used. The processing methods go by different names such as metal injection molding (MIM), ceramic injection molding (CIM), and powder injection molding (PIM). Perhaps it is interesting to realize how plastic injection molding was originally an offshoot of die casting machinery (but using a much lower melting pot). And ever since there has been a relationship between the two. Markets have gradually developed, especially for electrical and electronic products, mechanical devices (automotive, gardening equipment, etc.), and compounding equipment parts that compete with die castings.

MACHINING

Machining can be identified as a fabricating process or as a secondary operation. Products can be machined from fabricated plastics such as compression-molded blocks, thick or thin extruded sheets, extruded rods or profile shapes, and so on. They are machined to produce production products or prototypes. The secondary operations can involve machining after the plastics are fabricated; this may be more cost-effective and/or required to make the product function. It includes drilling holes, threading, providing tighter tolerances, and so on. The following review is a guide to machining plastics under other conditions.

Each type of plastic has unique properties and machining characteristics, which are far different from those of the metallic or nonmetallic materials familiar to many processors. TPs are relatively resilient compared to metals, and require special cutting procedures. Even within a family of plastics (PE, PC, PPS, etc.), the cutting characteristic will change, depending on the fillers and reinforcements.

Elastic recovery occurs in plastics both during and after machining, so the tool geometry should include sufficient clearance to allow for it. The expansion of compressed material during elastic recovery (Chapter 1) causes increased friction between the recovered cut surface and the cutting surface of the tool. In addition to generating heat, this abrasion affects tool wear. Elastic recovery also explains why, without proper precautions, drilled or tapped holes in plastics often are tapered or become smaller than the diameter of the drills that were used to make them.

As the heat conductivity of plastics is very slow, almost all the cutting heat generated will be absorbed by the cutting tool. The small amount of heat conducted into the plastic cannot be transferred to the core of the shape, so it causes the heat of the surface area to rise significantly. This heat needs to be minimized or removed by a coolant to ensure a proper cut.

Many commodity TP resins have low heats of softening, deformation, and degradation. Gumming, discoloration, poor tolerance control, and poor finish are likely to occur if frictional heat is generated and allowed to build up. Engineered TP resins (such as nylon and TFE-fluoroplastic) have relatively high melting or softening points. Thus their tendency to become gummed, melted, or crazed during machining is not as great as for plastics with lower melting points. Heat buildup is more critical in the plastics with lower melting points (Tables 1.28, page 86 and 2.1, pages 132–4). Thermoset plastics generally have the fewest problems of any plastics during machining. Examples of machined plastic products are shown in Table 15.2.

Table 15.2 Machining methods used for plastic parts

Parts	Methods
Bearing, roller	Turning, diamond-cutting, milling, drilling, shaping
Button	Turning, drilling
Cam	Turning, copy turning
Dial and scale	Engraving, sandblasting
Electronic parts	Diamond-cutting
Gear	Turning, milling, gear shaving, broaching
Liner and brake lining	Cutting off, shaping, planing, milling
Optical parts	Diamond-cutting
Pipe and rod	Cutting off, turning, threading
Plates for ceilings and panels	Cutting off, drilling, tapping
Tape for PTFE	Peeling

Cutting guidelines

The properties of plastics must be considered in specifying the best speeds, feeds, depths of cut, tool materials, tool geometries, and cutting fluids. Machining data are available from machinery handbooks as well as plastic material suppliers and cutting machinery suppliers. Note that some plastics may be cut at higher speeds with no loss of reasonable tool life, but higher speeds usually create thermal problems, especially with the commodity resins.

Guidelines for tool geometry start by reducing frictional drag and heat. It is desirable to have honed or polished surfaces on the tool where it comes in contact with the work. The tool geometries should generate continuous chips. Large rake angles will generally achieve this because they create the necessary force directions. Take care that rake angles are not so large as to cause brittle fracture of workpieces and discontinuous chips.

The drill geometry for plastics is not the same as the drill geometry for metals. Drill bits have wide polished flutes combined with low helix angles, to help eliminate the packing of chips, which causes overheating. The normal 118° point angle is generally modified to 70–120°.

Round saws should be hollow-ground, with burrs from sharpening removed by stoning; hand and jig saws should have enough set to give adequate clearance to the back of the blade. This set should be greater than is usual for cutting steel. It is always better to relieve the feed pressure near the end of a cut to avoid chipping.

The proper rate of feed is important, and because most sawing operations are hand-fed, experience is required to determine the best rate. Attempts to force the feed will cause heating of the blade, gumming of the plastic, loading of the saw teeth, and an excessively rough cut. Chrome plating of the blade reduces friction and tends to give better cuts. Above all, band saws and circular saws must be kept sharp. Circular saws are usually $\frac{1}{32}$ to $\frac{1}{8}$ in. thick; bandsaws are usually $\frac{3}{16}$ to $\frac{1}{2}$ in. wide.

Both TP and TS resins can be sawed using cutoff machines with abrasive wheels. This equipment is used to cut rods, pipes, L-beams, and so on. Clean cuts can be made when appropriate wheels are properly used. If necessary, water is used to prevent overheating.

16
Auxiliary equipment

INTRODUCTION

Many different types of auxiliary equipment and secondary operations can be used to maximize overall processing plant productivity and efficiency. Their proper selection, use, and maintenance are as important as the selection of the processing machines (injection molder, extruder, etc.). The processor must determine what is needed, from upstream to downstream, based on what the equipment has to accomplish, what controls are required, ease of operation and maintenance, safety devices, energy requirements, compatibility with existing equipment, and so on. This chapter provides examples of this selection procedure and its importance in evaluating all the equipment required in a processing line.

TYPES

The equipment can be used to maximize overall processing productivity and efficiency, and reduce fabricated product cost. They can cost more than the base machine. Examples of equipment are shown in Figs 16.1 and 16.2. They include the equipment listed in Table 16.1.

All inline equipment has to be properly interfaced so it operates efficently. Much of the equipment used in the past did not properly interface, so operations were rather less efficient than they could have been. A set of rules have been developed that govern the communication and transfer of data between machines and auxiliary equipment.

New-generation auxiliaries that are continually produced are meaningful contributors to plant productivity via greater reliability of operation and pinpoint equipment control. As an example, fluid chillers and temperature control systems are major beneficiaries of the computer revolution. They become more energy-efficient, reliable, and cost-effective with the application of microprocessor- and computer-compatible controls that

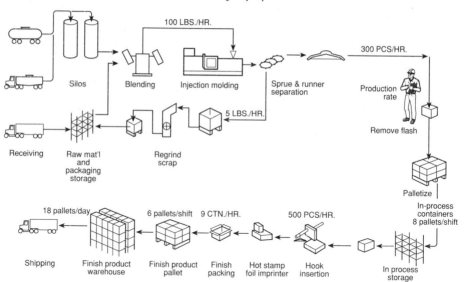

Figure 16.1 This production line starts with upstream auxiliary equipment and continues through an injection molding machine to the downstream equipment.

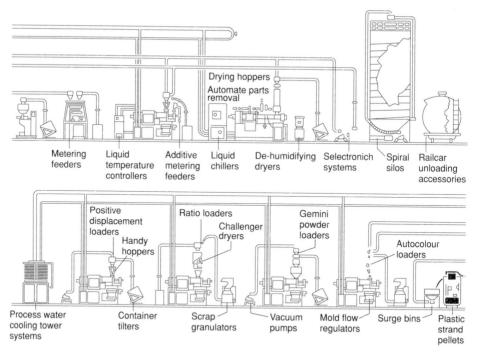

Figure 16.2 Auxiliary equipment next to different processing machines: begin at the top right end and work round to the bottom left.

Table 16.1 Auxiliary equipment used with plastics processes

Adhesive applicator	Process control for individual or all
Bonding	equipment
Cutting	Pulverizing/grinding
Die cutting	Recycling system
Dryer	Robotic handling
Dust-recovery	Router
Freezer/cooler	Saw
Granulator	Screen changer
Heater	Screw/barrel backup
Knitting	Sensor/monitor control
Leak detector	Software
Material handling	Solvent recovery
Metal treating	Testing/instrumentation
Metering/feeding material	Trimming
Mold extractor	Vacuum debulking
Mold heat/chiller control	Vacuum storage
Oven	Water-jet cutting
Pelletizer/dicer	Welding
Printing/marking	Others

can communicate with primary processing equipment. They provide pinpoint control of liquid temperatures. Thus, they have become significant contributors to productivity.

COMMUNICATION PROTOCOL

Within a plastic processing communication protocol cell there are two basic types of auxiliary devices: (1) devices that require minimum configuration or minimum data (chillers, dryers, loaders, and so on), and (2) devices that require large amounts of configuration data or provide large amounts of process data (robots, gauging/sensors, mold/die controllers, cutters, and so on).

A set of rules govern communication or transfer of data between computer hardware and/or software. When related to plastic processing equipment, communication includes reference to exchange of process controls, meeting standards, and following production schedules, activities that permit the interchange of actions, e.g., between molding machines and auxiliary equipment.

The information that is required to monitor and configure a manufacturing operation is distributed among various units of auxiliary equipment. This information is transferred to the central control. Communication interfaces and communication protocols have been developed to allow this information to be exchanged. Successful communication requires a durable interface and a versatile protocol. A communication interface must be mechanically and electrically durable. Mechanical durability is achieved by suitable hardware attachments and cable strain relief. Electrical durability is achieved by suitable transceiver circuitry. Circuitry meeting the requirements of ANSI RS-485 is ideal for half- and full-duplex protocols. It is designed for use where more than one device may talk at a time. Less durable circuitry, such as RS-232 and RS-422, may fail if more than one device talks.

MATERIAL HANDLING

In most processes, for either small or large production runs, the cost of the plastics used compared to the total cost of production in the plant may be at least 60%. The proportion might be only 30%, but it is more likely to exceed 60%; so it is important to handle material with **care** and to eliminate unnecessary production problems and waste. Where small-quantity users or expensive engineering resins are concerned, containers such as bags and gaylords are acceptable; but for large commercial and custom processors, these delivery methods are bulky and costly. Resin storage in this form is also expensive.

Any large-scale resin-handling system has three basic subsystems: unloading, storage, and transfer. For a complete system to work at peak efficiency, processors need to write specifications that fully account for the unique requirements of each subsystem. The least efficient component, no matter how inconsequential it may seem, limits the overall efficiency of the entire system. The guidelines presented here will help one to specify material-handling systems from the time of delivery to the plant on to processing machines that will maximize both efficiency and capacity.

Railcars can be unloaded in many different ways, but some suppliers simply recommend inexpensive, back-to-back, flexible-hose assemblies to unload them. To save time at the loading dock, one should use a stationary-pipe manifold arrangement along the length of the rail siding with connections about every 15 ft (4.5 m). Thus transport is achieved with one short-flex hose from car to car instead of by handling as much as 100 ft (30 ft) of heavy-flex hose. Labor and unloading time will be saved, and the pressure drops inherent in long, multiple-connection, flex-hose runs will be avoided.

The easiest method of unloading uses a vacuum pump/dust collector, which can be located in the silo skirt. The pump induces a vacuum in the line, drawing resin from the railcar into a vacuum loader. When the vacuum loader fills, the pump stops, and the resin dumps into the silo.

This on/off batching effect keeps transfer rates relatively low, typically 6000–7000 lb h^{-1} (2700–3200 kg h^{-1}). If the manifold pickup is far from the silo, the transfer rates will drop.

If unloading requires high speeds or transfer over long distances, one should consider a push/pull system; resin is pulled from the railcar by negative pressure, then pushed by positive pressure to its final location. Some equipment suppliers recommend a one-blower system, but that limits transfer capacities, requires more work-hours, and can actually degrade resin through excessive heat transfer within the airstream. One-blower systems can also lead to line blockages, as they lack purging capabilities.

Another way to unload railcars and transfer resin to storage silos is to use a two-blower system. One blower handles railcar unloading through negative pressure; the other transfers resins to silos using positive pressure. Splitting up the unloading and transfer of resin between two blowers permits peak efficiency in each operation. An improper pressure balance across a single-blower system could greatly reduce transfer rates at a critical point in the system, making the single blower work harder, increasing the heat of the conveying air, and causing resin degradation, particularly for thermally sensitive resins.

A two-blower system times purging to virtually eliminate any line blockages. With this arrangement, the pressure blower continues operating to ensure complete transfer to silos, while the vacuum blower remains out of the transfer loop.

Silos and other containers provide more than just a place to stow away resins until it is ready for use; silos protect resins from environmental damage caused by excess moisture, atmospheric pollutants, and solar radiation. Moreover, silos should require a minimum amount of maintenance; they should not leak, nor should they introduce rust or other contaminants in the resin. They should resist structural damage from environmental corrosion.

There are three types of silo construction: (1) spiral, with silos fabricated from spun aluminum; (2) bolted, with silos made of carbon steel, stainless steel, or aluminum; and (3) welded, also with silos constructed of carbon steel, stainless steel, or aluminum. Silos intended for pellet storage usually have 45° discharge cones at the bottom. Powders and hard-to-flow resins usually have a 60° cone to guarantee that the cone angle is greater than the resin's angle of repose [1].

Each type of silo has advantages and disadvantages. Spirals require the least maintenance. Bolted construction can cost less than spirals, but with erection costs that can be three times higher. Because rubber gaskets are used, bolted silos have a greater tendency to leak. Welded silos have the lowest manufacturing cost of all, but their installation could be problem-

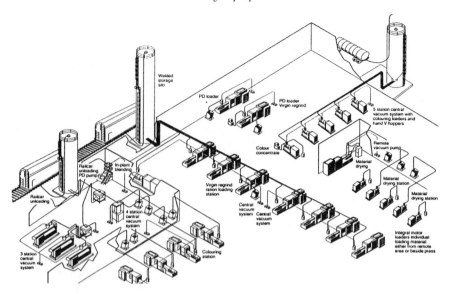

Figure 16.3 Bulk resin handling leads to better production economics.

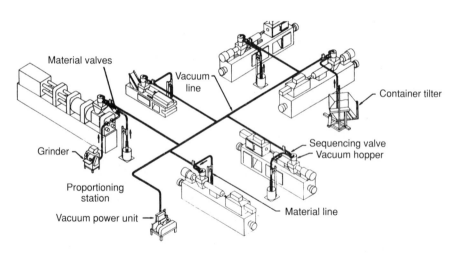

Figure 16.4 Close-up of material handling by different machines.

atic and costly. So it is important to study available storage systems in order to choose the supplier best able to meet individual specifications.

If computer-integrated resin-handling systems are considered, one must compare their operating procedures with one's process requirements. These process requirements describe the flow of resin and product through the system, which determines the system's electronic architecture. Pertinent considerations include batch versus continuous operations, the type and number of conveying lines, resin storage and distribution, quality control means and procedures, inventory control, the type and quantity of process parameter sensors, the type and quantity of controlled devices, modes (automatic, semiautomatic, manual, and/or shutdown modes), process information, process management controls, and centralized versus local operation (Figs 16.3 and 16.4).

Energy conservation

Energy conservation is only one of many factors that should be considered in the selection of an automated materials-conveying system (as well as all equipment used in the processing line). Fortunately, any steps taken to save energy will also save money in most cases. The traditional arguments favoring the silo are savings on resin costs, labor savings through the elimination of handling bags and cartons, the saving of costly warehouse space, and energy savings. For example, if a plant used a large quantity of resins and did not use silos, during the winter months bags or gaylords would be delivered repeatedly through 'open' delivery doors, and warm air would be lost.

With automatic delivery from silos, all resin-handling lines are kept as short as possible. There is no reason for lines to conform to the right angles of the walls; they should follow a straight line from the resin's source to where it has to be delivered. There are graphs from systems suppliers that show the relationship between the length of conveyor lines and power requirements. The graphs also show the horsepower required, based on different factors, such as the length and diameter of the delivery pipe, the position of the pipe, the type of resin being conveyed, the size of the hopper at the machine, and the rate of flow deliverable.

A graph will show, for example, that with an average pellet size of $\frac{1}{8}$ in., 35 lb/ft³ (560 kg m⁻³) bulk density, and a conveying vacuum of 12 in. of mercury (Hg), PE can be moved in different ways. With a 25 hp (19 kW) vacuum hopper unit, a line will convey 18 000 lb h⁻¹ (8200 kg h⁻¹) of PE if it is only 100 ft (30 m) long. Using the same power and a 450 ft (137 m) line, less PE will be moved. Suppliers' data will show the power required to move a material in the fastest way with the least energy consumption. To convert energy consumption to electric-bill charges, one uses the following formula: 1 hp = 0.745 kW, 1 kWh = 3.6 MJ = 3143 Btu.

With a long pipeline, a 25 hp (19 kW) vacuum pump could be used. If the line could be shortened, a 10 hp (7.5 kW) unit would convey the same amount of plastic, resulting in power savings. One metre of vertical height in the line equals 2 m of horizontal distance in its effect on conveying rates. Bends in the pipe add a considerable amount of equivalent footage.

If the lines cannot be kept short or relatively free of bends, then pressure drops will exceed 12 in. Hg (40 kPa), requiring more power. Where the pressure drop is greater than 12 in. Hg (40 kPa), which is the normal operating and limiting pressure for most vacuum systems, it is necessary to use a positive pressure system. All these factors must be considered initially to obtain the best delivery system most economically.

PARTS HANDLING

The logic and approach used in materials handling also applies to the use of handling equipment to move processed parts. Parts-handling equipment (PHE) does not resemble the humanoids of science fiction. Robots are blind, deaf, dumb, and limited to a few preprogrammed motions; but in many production jobs that is all that is needed. They are solutions looking for a problem. Most plants can use some degree of PHE, and it can substantially increase productivity.

Use of PHE can range from simple operations, as reviewed, to rather complex operations with very sophisticated computer controls. Although the concept of automatic operations is very appealing, the ultimate justification for PHE (like material handling, process controls, etc.) must be made on the basis of economics. At times it may provide the solution to

Table 16.2 Example of parts-handling equipment functions

Type	Collect	Remove or pick	Place	Orient	Count/weight	Accumulate
Not integrated with IMM function						
Manual	X	X	X	X	X	X
Box	X	X				X
Conveyor	X	X				X
Unscramble/orient	X	X	X	X	X	X
Integrated with IMM						
Sweep		X				
Extractor	X	X	X			
Cavity separator	X	X	X	X	X	X
Robot/bang-bang	X	X	X	X	X	X
Robot/sophisticated	X	X	X	X	X	X

Table 16.3 Example of parts-handling equipment growth rate

Type	Percentage used with IMM		No. of mold cavities	Part size[a]	Cost/unit dollars
	Current	Future			
Manual	20	12	any	any	Does not include PHE
Box/collector	30	15	any	sm, med	50–500
Conveyor	30	30	any	any	500–3000
Unscramble/orient	10	18	2–24	med	2000–40 000
Sweep	3	5	1–16	sm, med	200–1500
Extractor	4	7	1–24	sm, med	500–5000
Cavity separator	½	2	12–96	sm	5000–50 000
Robot/bang-bang	2	8	4–10	med	5000–25 000
Robot/sophisticated	½	3	1–12	lge	25 000–150 000

[a] sm = small, med = medium, lge = large.

'handling' a part that otherwise would be damaged. Tables 16.2 and 16.3 provide information on PHE for injection molding.

CLEAN PELLETS AND PLANT

When plastics are extruded and pelletized, varying amounts of oversized pellets and strands are produced, along with some fines. When the plastics are dewatered/dried or pneumatically conveyed, more fines, fluff, and streamers may be generated. In many cases, these 'problem products' must be eliminated, significantly reduced, or removed during processing in order to produce good-quality products. The process or product requirement may not be influenced, but with high throughputs, these interferences can present all kinds of problems. Even if acceptable products are produced, costs will be increased by the greater processing time, energy requirements, process controls, and reject rate which result, and/or other problems that may develop.

Fines or dust are created when small particles are torn away from the pellet and hit a rough surface at high speed (rough pipe wall, etc.). With smooth pipe walls, the pellet will slide, generating frictional heat, and could become sticky before it deflects (depending on the melting point). The pellet's surface can actually smear off, leaving a film streak on the surface of the pipe. In time, this film can cover the whole inside of the pipe, and it can tear off in the form of fluff (short threads or hairlike particles) and subsequently streamers (long ribbonlike particles, also called angel hair or snakeskins). In addition to causing dirty products, the streamers can form into balls and create blockages through the rest of the system.

The entire action that produces these contaminants is based on a combination of conditions of heat, weight, and velocity. The first appearance of fluff indicates that an operational threshold has been crossed. As conditions worsen, longer streamers form large quantities of this material.

These fines and other undesirable materials can cause a variety of handling and processing problems; so if they exist, they must be removed. Suppliers have different removal systems such as filters and cyclones.

Some in-plant systems, such as inline filters, are high-maintenance items. As the filter element fills, the pressure increases while the filtering system's efficiency decreases. The filter must be checked and cleaned on a regular schedule to maintain its efficiency.

To remove dust or fines, consider using a compressed-air, backwash-type filter/receiver. Electronically controlled pulsing jets effectively shake fines from filter media automatically, eliminating human error in maintaining the filter system.

The need for cleanliness of the pellets, as well as the plant, is easy to understand. No casual dirt or dust, contaminants, different compounds, metal chips, or other foreign matter can be tolerated. This type of care also protects the processing equipment.

DRYING PLASTICS

All plastics, to some degree, are infuenced by the amount of moisture or water they contain before processing. With minimal amounts in many

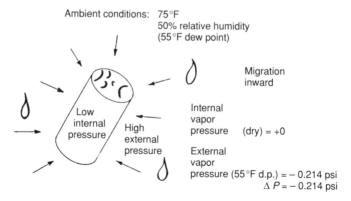

Figure 16.5 Moisture absorption: mechanics.

plastics, mechanical, physical, electrical, aesthetic and other properties may not be affected, or may be of no consequence. However, there are certain plastics that, when compounded with certain additives, could have devasting results.

There are also hygroscopic plastics which require special drying equipment (Chapter 1). It is interesting that moisture is a major problem which has continually influenced degradation of processed plastics. Even those which are not generally affected by moisture can only tolerate a certain amount. (Day–night moisture contamination can be a source of problems if not adequately eliminated; otherwise it has a cumulative effect.)

During the drying process (Fig. 16.5) at ambient temperature and 50% relative humidity, the vapor pressure of water outside a plastic pellet is greater than within. Moisture migrates into the pellet, increasing its moisture content until a state of equilibrium exists inside and outside the pellet.

But conditions are very different inside a drying hopper with a controlled environment. (Fig. 16.6). At a temperature of 177°C (350°F) and −40°C (−40°F) dew point, the vapor pressure of water inside the pellet is much greater than the vapor pressure of water in the surrounding air, so moisture migrates out of the pellet and into the surrounding airstream, where it is carried away to the desiccant bed of the dryer.

Before drying can begin, a wet material must be heated to such a temperature that the vapor pressure of the liquid content exceeds the partial pressure of the corresponding vapor in the surrounding atmosphere. The effect of the atmospheric vapor content on the rate of the dryer as well as the effect of the material temperature is conveniently studied by construction of a psychrometric chart (Fig. 16.7). It plots moisture content,

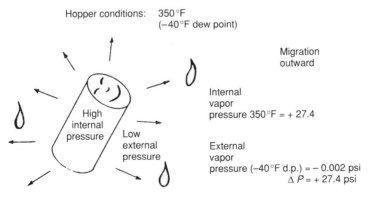

Figure 16.6 Moisture migration: mechanics.

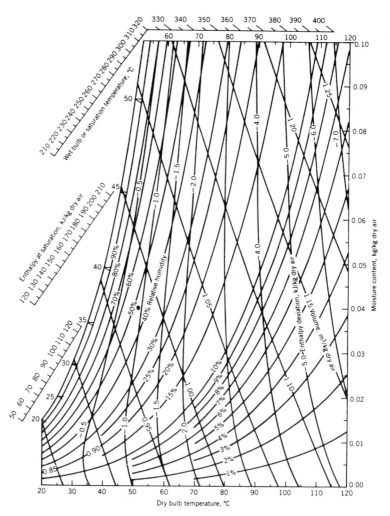

Figure 16.7 Psychrometric chart: air–water vapor at 101.3 kPa (1 atm).

Table 16.4 Drying equipment for coaters

Heat transfer	*Web handling*
Convection dryers	
Parallel airflow	Idler-supported dryers
Impingement air	Conveyor dryers
Airfoil	Catenary dryers
Through dryers	U-type dryers
Infra-red radiation dryers	Arch dryers
Near infra-red (electric)	Tenter frame dryers
Far infra-red (electric or	Floater dryers
gas)	
Conduction dryers	
Hot-roll dryers	

dry-bulb temperature, wet-bulb or saturation temperature, and enthalpy at saturation.

More details about processing plastics are given in Chapter 1. Even though processes may differ in other respects, they often use similar procedures for drying. Specific procedures and equipment for certain processes are given in Table 16.4.

FEEDERS AND BLENDERS

As with other process equipment, suppliers of dryers, blenders, and metering equipment are increasing the feeding accuracy of the equipment by the incorporation of microprocessor-based controllers that can be easily interlinked with the computer facilities of the main plant. Most processors deal with a wide variety of resins and additives, including regrind and color concentrates. As the quality and cost of the product depend directly upon how carefully processors measure and mix components, blending represents a critical stage in material handling (such as meeting tight requirements according to a target; see Fig. 1.36, page 102).

To mix components easily and economically, processors may use blenders mounted on machine hoppers. In this operation, which offers great flexibility, precisely predetermined proportions of each ingredient flow into a mixing chamber, which should discharge a very accurate mix. The key to process success lies in carrying it out **precisely**. To do this, feeders meter by volume or by weight.

In volumetric blending, variable-speed metering augers feed multiple components into the mixing chamber. A microprocessor provides accurate, repeatable programming and closed-loop control over the variable speed of each metering auger. A single master speed control can also increase or decrease throughput while maintaining the desired blend ratios. Operators can calibrate the actual volume easily by occasionally diverting the component to a sampling chute and weighing a sample from each auger. The microprocessor monitors auger speeds at regular intervals to maintain constant feed rates regardless of variations in the metering torque.

Gravimetric blending improves accuracy and requires less operator involvement in calibration, particularly in running processes where great accuracy is needed. Metering by weight eliminates overfeeding of expensive additives. The principle of gravimetric feeding with throughput or metered weight control is well established. Equipment suppliers can assure an accuracy of ±0.25–0.50% for ingredient and blend-ratio setpoints at 2σ (two standard deviations). By comparison, volumetric and quasi-gravimetric blenders that use batch operation usually have accuracy variations of 2–10% or more.

Each hopper is mounted on a separate weigh-load cell specially designed and isolated to minimize the effects of on-machine vibration and electrical interference. This unique load-cell approach updates the controller with weight information so it can adjust the motor speed for any necessary blend corrections. If the bulk density of any component changes, the blend is not influenced, as the unit only works on weights. Controllers are also capable of refilling individual component hoppers automatically with no mixing delays.

Gravimetric metering is particularly advantageous in coextrusion. In many cases it is not possible to continually determine the individual film thickness or thickness profile of the product. This system offers a very simple means of constantly maintaining the average thickness of the individual films and the overall thickness at better than ±0.5%.

This mixing and metering activity is part of the overall conveying process within the plant. Conveying includes not only mechanical and pneumatic transport throughout the plant or within each individual processing line, but also filtration of the exhaust air and safety engineering to prevent dust explosion. With so much mixing required, color changes are becoming increasingly frequent. Table 16.5 explains the advantages of coloring at the hopper throat. The processor should determine what is required and obtain equipment to meet his/her specification. An individual evaluation is the best basis for taking advantage of developments in processing lines.

Motionless mixers, also called static mixers, are used for melt processing and for mixing liquids to produce an almost homogeneous mix. These units are mounted inline on the output end of processing machines, such as extruders (Fig. 3.7, page 222), injection molding machines, and polyurethane mixing units. Their mixing elements, housed within a

Table 16.5 Comparison of central blending versus coloring at the hopper throat

Central blending	At-the-throat coloring
1. If the color blend is not correct (parts too light or more color than necessary in the mix), the mix must be emptied from the hopper and reblended, a costly and time-consuming process. Even then, parts may not have the exact depth of color you desire.	1. Adjustments to color are made with immediate results. If molded parts are light in color, you can make an adjustment and see the results in minutes.
2. Although some parts, as a cost-saving measure, could be molded with less color than others, the mixing of several batches with different let-down ratios is generally considered impractical.	2. Every molded part can be custom blended while in production for optimum color usage and cost savings. Settings are recorded and used for exact repeatability on future runs.
3. Conveying of blended material often produces separation of the color pellets from the natural material due to the much heavier bulk density of color pellets. The result is inconsistent coloring of parts.	3. Material is conveyed without color; separation is not a factor.
4. Normal machine vibration and flow patterns in the hopper cause some separation of the color pellets during residence time in the hopper. Again, the result is inconsistent coloring of parts.	4. Color is metered only as parts are molded. There is no residence time for separation to occur.
5. When a production run is over, any inventory of blended material left in the hopper must be held for use at another time. Storage space, contamination, spillage, mislabeling and tracking of this valuable inventory are all problems.	5. All inventory of blended materials is **eliminated**. Unused color is returned to the container it came from, uncontaminated.
6. When a production run for one color is over, your hopper must be emptied in preparation for the next color. Time is lost. Sometimes material is lost. If hopper and conveying system are not adequately cleaned, color contamination and rejected parts will result.	6. Color changes require no emptying of the main material hopper. Production does not stop. Transition time to the new color is several minutes.
7. If your material must be dried before use, the drying of color blends presents many logistical problems: more driers, careful planning for transition to the next color, cleanout of dryers, etc.	7. You dry only natural material wihout having to concern yourself about color changes.

barrel, are a series of welded plates arranged to divide the melt and redivide it several times.

Motionless mixers provide effective dispersion of colorants and other additives such as plasticizers, antioxidants, flame retardants, stabilizers, and fillers. Their target is to increase the quality of the melt by eliminating any temperature gradients.

GRANULATORS

Most processing plants have to reclaim reprocessable thermoplastic scrap, flash, rejected parts, and so on. If possible, the goal is to eliminate 'scrap' because it has already cost money and time to go through the process; granulating just adds more money and time. Different types of granulators are available from many different suppliers, and selection of a granulator depends on factors such as the type of plastic used, the type of reinforcement, and product thickness and shape (Fig. 16.8). There are some units in which subsystem granulators incrementally reduce the plastic from a large size to the required small size.

An easy 'cutting' unit is required, to granulate with minimum friction, as too much heat will destroy the plastic. General-purpose types have definite limitations. Blending with virgin material definitely influences,

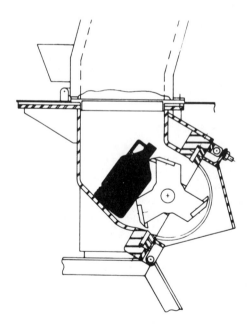

Figure 16.8 Granulator: different types are available; this one granulates blow-molded bottles.

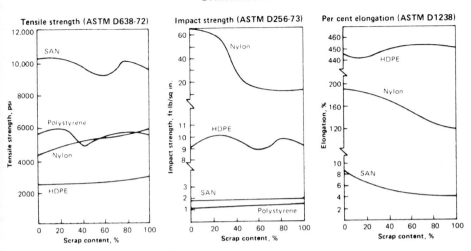

Figure 16.9 How regrind levels affect mechanical properties of certain plastic formulations, 'once through' and blended with virgin plastics.

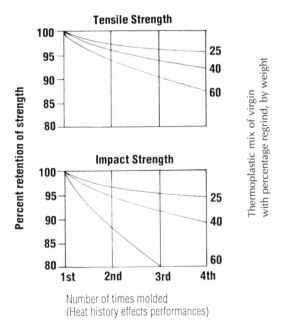

Figure 16.10 Regrind: potential effects on the performance of injection-molded TP mixed with virgin material.

and can significantly change, the melt processing conditions and the performance of the end product. Recycled material is denser and usually has a variable-size regrind that could affect the product's properties, as shown in Figs 16.9 and 16.10.

17

Secondary operations

OVERVIEW

After plastic parts are fabricated in their respective basic functional processing machines, secondary operations may be required to produce the final finished product. These operations can occur inline or off-line during production. Examples include parts handling for finish machining, drilling, reaming, cutting, trimming, postcuring, and annealing.

ANNEALING

Each of the above examples have their respective definitions. An example is annealing as a secondary operation after processing plastics. The plastic is brought to a certain temperature, kept there for a time, then cooled to room temperature. The process is also called physical aging. The primary reasons for annealing include the reduction or removal of residual stresses and strains, dimensional stabilization, reduction or elimination of defects, and improvement of physical properties. In the plastics industry, annealing has been applied primarily to thermoplastic polymers, block copolymers, and amorphous or semicrystalline plastic blends (Chapter 1). Certain materials and/or products after heating, are quenched in liquids such as water, oil, or waxes, exposing them to relatively fast, controllable temperature drops.

Annealing becomes necessary when molded parts are excessively stressed, when maximum dimensional stability and heat resistance are required, or when certain properties must be enhanced. Improper annealing may also cause deterioration in performance.

Theoretically, the most desirable annealing temperatures for amorphous plastics, some polymer blends, and block copolymers is above their glass transition, where the relaxation of stress and orientation is the most rapid (Chapter 1). However, the required temperatures may cause excessive distortion and warping. To anneal as quickly as possible, the plastic is heated to the highest possible temperature at which dimensional changes owing to strain release are within permissible ranges. This temperature can be determined by placing the plastic part in an air oven or liquid bath and gradually raising the temperature by intervals of 3–5 °C until the maximum allowable change in shape or dimension occurs. This distortion temperature is dictated by thermomechanical processing history, geometry, thickness, weight, and size. The annealing temperature should then be fixed about 5 °C lower.

POSTCURING

Postcuring is similar to annealing, particularly for TSs. After fabrication the product receives additional elevated temperature cure, usually without tooling or pressure, to improve final properties and/or complete the final cure. This after-baking following part fabrication can also decrease the volatiles in the part, relieve stresses, and/or improve dimensional stability. With certain plastics, particularly certain TSs, complete cure and ultimate mechanical properties are obtained only by exposure of the plastic to higher temperatures than those of the cure. Property improvements usually include creep resistance and elevated temperatures.

JOINING AND ASSEMBLING

Different methods are used for joining or fastening and assembling plastic products (Fig. 17.1 and Tables 17.1 to 17.5). These techniques can be used with plastic-to-plastic and plastic-to-other materials. It is important to recognize that these techniques have advantages and disadvantages (or limitations), depending on the type material being used. First consideration should be given to utilizing one of the relatively simple means of assembly. Most of the existing applications use techniques of this type. The actual suitability of any of these methods depends greatly on the size and function of the product, and the choice of plastic material. The more complicated techniques are usually more expensive, but they generally lead to the most useful and dependable assemblies.

Adhesives

Adhesives are widely used. There are solvent systems for most TPs, but not TSs. Monomeric or polymerizable cements can be used for most TPs and TSs. Certain plastics with outstanding chemical resistance, such as the polyolefins, preclude the use of many cements; they generally require some form of surface treatment before adhesion, such as flame treatment.

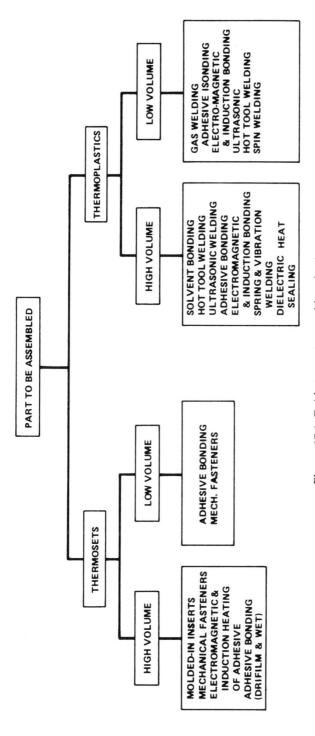

Figure 17.1 Guide to part-assembly selection.

Table 17.1 Reference chart to help in selecting the proper method of fastening thermoplastic materials[a]

Thermoplastics	Mechanical fasteners	Adhesives	Spin and vibration welding	Thermal welding	Ultrasonic welding	Induction welding	Remarks
ABS	G	G	G	G	G	G	Body type adhesive recommended
Acetal	E	P	G	G	G	G	Surface treatment for adhesives
Acrylic	G	G	F–G	G	G	G	Body type adhesive recommended
Nylon	G	P	G	G	G	G	
Polycarbonate	G	G	G	G	G	G	
Polyester TP	G	F	G	G	G	G	
Polyethylene	P	NR	G	G	G–P	G	Surface treatment for adhesives
Polypropylene	P	P	E	G	G–P	G	Surface treatment for adhesives
Polystyrene	F	G	E	G	E–P	G	Impact grades difficult to bond
Polysulfone	G	G	G	E	E	G	
Polyurethane TP	NR	G	NR	NR	NR	G	
PPO, modified	G	G	E	G	G	G	
PVC, rigid	F	G	F	G	F	G	

[a] E = excellent, G = good, F = fair, P = poor, NR = not recommended.

Table 17.2 Reference chart to help in selecting the proper method for fastening thermoset plastic materials[a]

Thermosets	Mechanical fasteners	Adhesives	Spin and vibration welding	Thermal welding	Ultrasonic welding	Induction welding	Remarks
Alkyds	G	G	NR	NR	NR	NR	
DAP	G	G	NR	NR	NR	NR	
Epoxies	G	E	NR	NR	NR	NR	
Melamine	F	G	NR	NR	NR	NR	Material notch sensitive
Phenolics	G	E	NR	NR	NR	NR	
Polyester	G	E	NR	NR	NR	NR	
Polyurethane	G	E	NR	NR	NR	NR	
Silicones	F	G	NR	NR	NR	NR	
Ureas	F	G	NR	NR	NR	NR	Material notch sensitive

[a] E = excellent, G = good, F = fair, P = poor, NR = not recommended.

Table 17.3 Percent tensile strength retention with different welding techniques

	Original tensile strength (psi)[a]	Hot-air welding	Friction welding	Hot-plate welding	Dielectric welding	Solvent welding	Adhesive bonding	Polymerization welding
Thermosetting plastics								
Epoxy	7000–13000	–	10–15	10–15	–	–	50–80	60–100
Melamine	7000–13000	–	–	–	–	–	50–80	60–100
Phenolic	6000–9000	–	–	–	–	–	50–80	60–100
Polyester	6000–13000	–	–	–	–	–	50–80	60–100
Thermoplastics								
Acrylonitrile-butadiene-styrene	2400–9000	50–70	50–70	50–70	50–80	30–60	40–60	–
Acetal	8000–10000	20–30	50–70	20–30	–	–	–	–
Cellulose acetate	2400–8500	60–75	65–80	65–80	–	90–100	50–60	–
Cellulose acetate butyrate	3000–7000	60–75	65–80	65–80	–	90–100	50–80	–
Ethyl cellulose	2000–8000	50–70	50–70	50–70	–	80–90	50–80	–
Methyl methacrylate	8000–11000	30–70	30–50	20–50	–	40–60	40–60	60–90
Nylon	7000–12000	50–70	50–70	50–70	–	–	20–40	–
Polycarbonate	8000–9500	35–50	40–50	40–50	–	40–60	5–15	–
Polyethylene	800–6000	60–80	70–90	60–80	–	–	10–30	–
Polypropylene	3000–6000	60–80	70–90	60–80	–	–	20–40	–
Polystyrene	3500–8000	20–50	30–60	20–50	–	25–50	20–50	–
Polystyrene acrylonitrile	8000–11000	20–60	20–50	20–50	30–50	25–60	20–50	–
Polyvinyl chloride	5000–9000	60–70	50–70	60–70	60–70	50–70	50–70	–
Saran	3000–5000	60–70	50–70	60–70	60–70	50–70	50–70	–

[a] To convert psi to Pas, multiply by 6895.

Table 17.4 Plastic characteristics for different types of ultrasonic welding application[a]

Material	Percent of weld strength[b]	Spot weld	Staking and inserting	Swaging	Welding[c] Near field	Far field
General-purpose plastics						
ABS	95–100+	E	E	G	E	G
Polystyrene						
Unfilled	95–100+	E	E	F	E	E
Structural foam (styrene)	900–100[d]	E	E	F	G	P
Rubber-modified	95–100	E	E	G	E	G–P
Glass-filled (up to 30%)	95–100+	E	E	F	E	E
SAN	95–100+	E	E	F	E	E
Engineering plastics						
ABS	95–100+	E	E	G	E	G
ABS/polycarbonate alloy (Cycoloy 800)	95–100+[e]	E	E	G	E	G
ABS/PVC alloy (Cycovin)	95–100+	E	E	G	G	F
Acetal	65–70[f]	G	E	P	G	G
Acrylics	95–100+[g]	G	E	P	E	G
Acrylic multipolymer (XT-polymer)	95–100	E	E	G	E	G
Acrylic/PVC alloy (Kydex)	95–100+	E	E	G	G	F
ASA	95–100+	E	E	G	E	G
Methylpentene	90–100+	E	E	G	G	F
Modified phenylene oxide (Noryl)	95–100+	E	E	F–P	G	E–G
Nylon	90–100+[e]	E	E	F–P	G	F
Polyesters (thermoplastic)	90–100+	G	G	F	G	F
Phenoxy	90–100	G	E	G	G	G–F
Polyarylsulfone	95–100+	G	E	G	E	G
Polycarbonate	95–100+[e]	E	E	G–F	E	E
Polyimide	80–90	F	G	P	G	F
Polyphenylene oxide	95–100+	E	G	F–P	G	G–F
Polysulfone	95–100+[e]	E	E	F	G	G–F
High-volume, low-cost applications						
Butyrates	90–100	G	G–F	G	P	P
Cellulosics	90–100	G	G–F	G	P	P
Polyethylene	90–100	E	E	G	G–P	F–P

Table 17.4 *Continued*

Material	Percent of weld strength[b]	Spot weld	Staking and inserting	Swaging	Welding[c] Near field	Far field
Polypropylene	90–100	E	E	G	G–P	F–P
Structural foam (polyolefin)	85–100	E	E	F	G	F–P
Vinyls	40–100	G	G–F	G	F–P	F–P

[a] E = excellent, G = good, F = fair, P = poor.
[b] Weld strengths are based on destructive testing. 100 + % results indicate that parent material of plastic part gave way while weld remained intact.
[c] Near-field welding refers to joint $\frac{1}{4}$ in. or less from area of horn contact: far-field welding to joint more than $\frac{1}{4}$ in. from contact area.
[d] High-density foams weld best.
[e] Moisture will inhibit welds.
[f] Requires high energy and long ultrasonic exposure because of low coefficient of friction.
[g] Cast grades are more difficult to weld due to high molecular weight.

Table 17.5 Materials for plastics hardware

Material	Snap-in	Snap-on	Clasp	Drive-pin	Hinges Knuckle and pin	Ball-grip	Integral
ABS	X	X	X		X	X	
Acetal	X	X	X	X	X		
Acrylic			X		X		
Cellulosic		X					
Fluorocarbon	X		X				
Polycarbonate	X	X		X	X	X	
Polyethylene	X	X	X				X
Polyamide	X	X		X	X	X	
Polypropylene		X	X				X
Polystyrene		X	X			X	
Polyurethane				X			
Vinyl	X	X	X		X		

Bonding preparations for the different plastics include sandblasting, plasma etching, water washing, solvent etching, and chemical etching.

Solvent bonds work because they react chemically with the plastic. However, they literally destroy it, so it is important to limit such factors as the length of time and the depth of the plastic soak. The solvent could cause immediate or delayed damage. If a part contains 'excessive' internal strains, the solvent could release the strains and cause cracking, surface defects, and so on. (To evaluate a plastic's reaction with a solvent, after a part is processed, it is immersed in a solvent to determine whether strain patterns exist. The reaction with the solvent can be correlated with processing versus part performance.)

The solvent action described above does not mean that adhesives are harmful; they have been used successfully for over a century. But no matter what action is taken in joining or assembling (Tables 17.1 to 17.3), the processor should determine whether there are limitations to its use.

Certain plastics can be solvent bonded or heat bonded, with or without fillers and reinforcements. One problem that can develop after production starts is to change a plastic's composition while maintaining the same basic plastic. This action may be taken to improve performance or cost. Perhaps the joining action would remain the same, but without checking, there could be a disaster. For example, with a higher percentage of calcium carbonate or glass fibers (additives that are virtually unaffected by solvents or heat), there could be no joining capability. It is therefore important to check the performance of any new material.

Ultrasonic welding

Ultrasonic welding is an economical method for joining small to medium-sized plastic parts of the same or similar plastics. Certain polymers may not weld if they contain specific fillers, in particular glass fibers if they are at a high concentration. This technique is rapid and can be fully automated. Welding occurs when high-frequency (20–40 kHz) vibrational energy is directed to the interface between the two parts, creating localized molecular expansion and causing the plastic to melt. Pressure is maintained between the two parts after vibration stops, and the melted polymer immediately solidifies. The entire welding process normally takes place in less than 2 s. It has high strength, which sometimes approaches the strength of the base material, if the joint design is correct and the equipment is properly set. If it is not properly bonded, poor bonds can be created or nonairtight contact occurs. Table 17.4 shows the types of welds that can be made.

Vibration welding

In vibration welding, two plastic parts are rubbed together in either linear or angular displacement, creating frictional heat that produces a melt at their interface. Different bonding joints can be used to eliminate visible flash at the joints; recesses exist within the bond. The vibration is in the form of high-amplitude, low-frequency, reciprocating motion. A rotary motion is used with circular parts. When the vibration stops, the melt cools and the parts become permanently welded in the alignment that is held. Typical frequencies are 120 and 240 Hz, with amplitudes of 0.10–0.20 in. (2.5–5.0 mm) of linear displacement.

Vibration welding, like ultrasonic welding, produces high-strength joints for materials that can be melted. However, it is much better suited to large parts and irregular joint interfaces. Moisture in materials does not usually have an adverse effect on the weld as it does with ultrasonics.

Spin welding

Spin welding is just a special form of vibration welding. Because it is such a popular technique, it is considered to be a special assembly method.

Radio frequency welding

With this type of process, welding occurs from the heat created by the application of a strong radio frequency (RF) field to the selected joint region on those plastics that are not transparent to RF. The RF is usually applied by a specially formed metal die in the desired shape of the joint, which also applies the clamping pressure needed to complete the weld after the plastic melts. This is a fast process that is sensitive to heat buildup.

This type of welding, usually called heat sealing, is widely used with flexible TP films and sheets such as plasticized PVC and PUR. It can also be used to join film to plastic molded parts.

Heat welding

Many TPs can be heat welded. However, as with other welding techniques, certain fillers or too much of a particular filler, could prevent good bonding. Nevertheless, there are certain fillers that can improve bonding action. Heat welds can also be used with friction or spin welding of TP joints.

Electromagnetic and induction welding

This type of welding uses an RF magnetic field to excite fine, magnetically sensitive particles that are either metallic or ceramic. The particles can be embedded in a preform, filament ribbon, adhesive, coextruded film, molding compound, and other materials. The most common is to include an extra part such as a preform containing the magnetically active particles. The preform is placed at the joint's interface and exposed to an electromagnetic field. Then electromagnetically induced heat is conducted from the particles through the preform and to the part joint as the parts are pressed together.

Table 17.6 Decorating and printing systems

Process	Description	Equipment	Applications	Effect
Conventional spray painting	Paint's sprayed by air or airless gun(s) for functional or decorative coationgs. Especially good for large areas, uneven surfaces or relief designs. Masking used to achieve special effects.	Spray guns, spray booths, mask washers often required; conveying and drying apparatus needed for high production.	Can be used on all materials (some require surface treatment).	Solids, multicolor, overall or partial decoration, special effects such as wood-graining possible.
Electrostatic spray painting	Charged particles are sprayed on electronically conductive parts; process gives high paint utilization; more expensive than conventional spray.	Spray gun, high-voltage power supply; pumps; dryers. Pretreating station for parts (coated or preheated to make conductive).	All plastics can be decorated. Some work, not much, being done on powder coating of plastics.	Generally for one-color, overall coating.
Paint wiping	Paint is applied conventionally, then paint is wiped off. Paint is either totally removed, remaining only in recessed areas, or is partially removed for special effects such as wood-graining.	Standard spray-paint setup with a wipe station following. For low production, wipe can be manual. Very high-speed, automated equipment available.	Can be used for most materials. Products range from medical containers to furniture.	One color per pass; multicolor achieved in multistation units.
Roller coating	Raised surfaces can be painted without masking. Special effects like stripes.	Roller applicator, either manual or automatic. Special paint feed system required for automatic work. Dryers.	Can be used for most materials.	Generally one-color painting, though multicolor possible with side-by-side rollers.

Process	Description	Equipment	Materials	Colors
Screen printing	Ink is applied to part through a finely woven screen. Screen is masked in those areas which won't be painted. Economical means for decorating flat or curved surfaces, especially in relatively short runs.	Screens, fixture, squeegee, conveyorized press setup (for any kind of volume). Dryers. Manual screen printing possible, for very low-volume items.	Most materials. Widely used for bottles; also finds big applications in areas like TV and computer dials.	Single or multiple colors (one station per color).
Hot stamping	Involves transferring coating from a flexible foil to the part by pressure and heat. Impression is made by metal or silicone die. Process is dry.	Rotary or reciprocating hot stamp press. Dies. High-speed equipment handles up to 6000 parts per hour.	Most thermoplastics can be printed; some thermosets. Handles flat, concave or convex surfaces, including round or tubular shapes.	Metallics, wood grains or multicolor, depending on foil. Foil can be specially formulated (e.g., chemical resistance).
Heat transfers	Similar to hot stamp but preprinted coating (with a release paper backing) is applied to part by heat and pressure.	Ranges from relatively simple to highly automated with multiple stations for, say, front and back decoration.	Can handle most thermoplastics. A big application area is bottles. Flat, concave or cylindrical surfaces.	Multicolor or single color; metallics (not as good as hot stamp).
Electroplating	Gives a functional metallic finish (mat or shiny) via electrodeposition process.	Preplate etch and rinse tanks; Koroseal-lined tanks for plating steps; preplating and plating chemicals; automated systems available.	Can handle special plating grades of ABS, PP, polysulfone, filled Noryl, filled polyesters, some nylons.	Very durable metallic finishes.

Table 17.6 *Continued*

Process	Description	Equipment	Applications	Effect
Vacuum metallization	Depositing, in a vacuum, a thin layer of vaporized metal (generally aluminum) on a surface prepared by a base coat.	Metallizer, base- and top-coating equipment (spray, dip or flow), metallizing racks.	Most plastics, especially PS, acrylic, phenolics, PC, unplasticized PVC. Decorative finishes (e.g., on toys), or functional (e.g., as a conductive coating).	Metallic finish, generally silver but can be others (e.g., gold, copper).
Cathode sputtering	Uniform metallic coatings by using electrodes.	Discharge systems to provide close control of metal buildup.	High-temperature materials. Uniform and precise coatings for applications like microminiature circuits.	Metallic finish. Silver and copper generally used. Also gold, platinum, palladium.
Spray metallization	Deposition of a metallic finish by chemical reaction of water-based solutions.	Activator, water-clean and applicator guns; spray booths, top- and base-coating equipment if required.	Most plastics. For decorative items.	Metallic (silver and bronze).
Tamp printing	Special process using a soft transfer pad to pick up image from etched plate and tamping it onto a part.	Metal plate, squeegee to remove excess ink, conical-shaped transfer pad, indexing device to move parts into printing area, dryers, depending on type of operation.	All plastics. Specially recommended for odd-shaped or delicate parts (e.g., drinking cups, dolls' eyes).	Single- or multicolor – one printing station per color.

Method	Description	Equipment	Applications	Colors
In-the-mold decorating	Film or foil inserted in mold is transferred to molten plastics as it enters the mold. Decoration becomes integral part of product.	Automatic or manual feed system for the transfers. Static charge may be required to hold foil in mold.	Most plastics, especially polyolefins and melamines. For parts where decoration must withstand extremely high wear.	Single- or multicolor decoration.
Flexography	Printing of a surface directly from a rubber or other synthetic plate.	Manual, semi- or automatic press, dryers.	Most plastics. Used on such areas as coding pipe and extruded profiles.	Single- or multicolor.
Offset printing	Roll-transfer method of decorating. In most cases less expensive than other multicolor printing methods.	Ranges from low-cost hand presses to very expensive automated units. Drying, destaticizers, feeding devices.	Most plastics. Used in applications like coding pipe.	Multicolor print or decoration.
Valley printing	Uses embossing rollers to print in depressed areas of a product.	Embosser with inking attachment or special package system.	Used largely with PVC, PE for such areas as floor tiles, upholstery.	Generally two-color maximum.
Labeling	From simple paper labels to multicolor decals and new preprinted plastic sleeve labels.	Equipment runs the gamut from hand dispensers to relatively high-speed machines.	Can be used on all plastics. Used mostly for containers and for price marking.	All sorts of colors and types.

Table 17.7 Guide to plastic decorating methods

	Economics	Aesthetics	Product design	Chemistry	Manufacturing	Comments
			Done in the mold			
1. Engraved mold	Unit cost low; labor cost low; investment moderate	Limited	Unrestricted	Not critical Good durability	No extra operations	Best for simple lettering and texture
2. In-mold label	Unit cost high; labor cost high; investment none to moderate	Unlimited	Somewhat restricted	Critical Good durability	Longer molding cycles	Good for thermoplastics and thermosets; automatic loading equipment becoming available
3. Inserted nameplates	Unit cost high; labor cost high; investment moderate	Partially limited	Restricted	Not critical Good durability	Longer molding cycles	Allows three-dimensional as well as special effects
4. Two-shot molding	Unit cost high; labor cost high; investment moderate to high	Limited	Somewhat restricted	Not critical Good durability	Two molding operations	Good where maximum abrasion resistance necessary

			Done after molding			
		Somewhat limited	Unrestricted	Not critical	Hand operation	Allows unusual effects
1. Appliqué	Unit cost high; labor cost high; investment moderate to high	Somewhat limited		Good durability	Hand operation	
2. Electrostatic	Unit cost low to moderate; labor cost low; investment moderate to high	Limited	Somewhat restricted	Critical / Moderate to good durability		Dry process, no tool contact with product
3. Flexoraphic	Unit cost low; labor cost low; investment moderate to high	Somewhat limited	Restricted	Critical / Moderate durability	Automates well	Wet process, tool contacts product; sometimes requires topcoat
4. Hand painting	Unit cost high; labor cost high; investment low	Somewhat limited	Unrestricted	Critical / Good durability	Hand operation	Wet process, tool contacts product

Table 17.7 Continued

	Economics	Aesthetics	Product design	Chemistry	Manufacturing	Comments
5. Heat transfer	Unit cost low to moderate; labor cost low to moderate; investment low to moderate	Unlimited	Somewhat restricted	Critical Good durability	Requires little floor space	Dry process, tool contacts product Multicolor graphics
6. Hot stamping	Unit cost low; labor cost low to moderate; investment low to moderate	Limited	Somewhat restricted	Critical Good durability	Requires little floor space	Dry process, tool contacts product Produces bright metallics
7. Labeling	Unit cost low to moderate; labor cost low to moderate; investment low to high	Unlimited	Somewhat restricted	Less critical Moderate to good durability	Adaptable to many situations	Dry process, no tool contact with product at times Multicolor graphics

Process	Cost	Volume	Restriction	Criticality	Durability	Technology	Process notes
8. Metallizing	Unit cost moderate to high; labor cost moderate to high; investment high	Limited	Somewhat restricted	Critical	Good durability	Requires special technological know-how	Wet and dry process, no tool contact with product; Produces bright metallics
9. Nameplates	Unit cost high; labor cost moderate to high; investment low to moderate	Unlimited	Somewhat restricted	Less critical	Good durability	Adaptable to many situations	Dry process, tool contacts product; Multicolor graphics
10. Offset	Unit cost low; labor cost moderate; investment high	Unlimited	Restricted	Critical	Moderate to good durability	Automates well	Wet process, tool contacts product; Multicolor graphics
11. Offset intaglio	Unit cost low; labor cost moderate; investment moderate	Limited	Unrestricted	Critical	Moderate to good durability	Requires little floor space	Wet process, tool contacts product; New process

Table 17.7 *Continued*

	Economics	Aesthetics	Product design	Chemistry	Manufacturing	Comments
12. Silk screen	Unit cost moderate; labor cost moderate; investment moderate	Somewhat limited	Somewhat restricted	Critical Good durability	Flexible operation	Wet process, tool contacts product
13. Spray	Unit cost moderate; labor cost moderate; investment moderate to high	Limited	Unrestricted	Critical Good durability	Requires much floor space	Wet process, no tool contact with product
14. Wood-graining	Unit cost high; labor cost high; investment moderate to high	Specialized	Specialized	Critical Good durability	Mostly hand operated	Wet process, tool contacts products

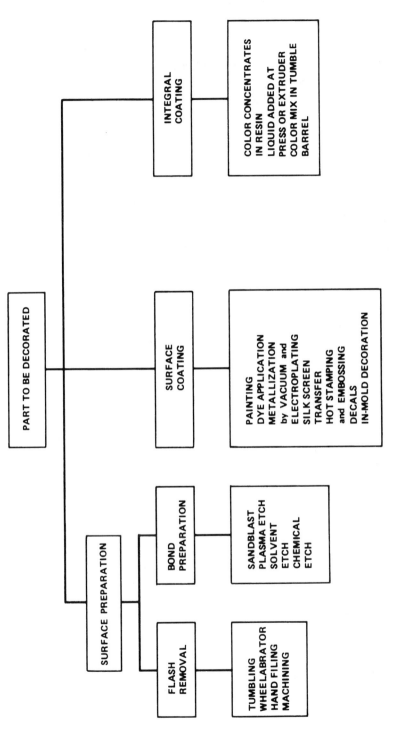

Figure 17.2 Guide to part-decorating selection.

FINISHING AND DECORATING

The finishing of plastics includes different methods of adding either decorative or functional surface effects to a plastic product (Fig. 17.2). Plastics are unique, in that color and decorative effects can be added before, during and/or after processing. Plastic parts of two or more colors are easily processed by coinjection, coextrusion, corotation, etc. (Chapters 2, 3, 13).

Integral coloring includes the use of color concentrates, liquid additives, and color mixing in tumble barrels. Although the older-established methods of painting are still widely used, they are continually being improved by newer and more versatile methods to give the user an almost unlimited choice of decorating designs and effects. The choice of method is governed by several factors, such as the size and shape of the part, the number

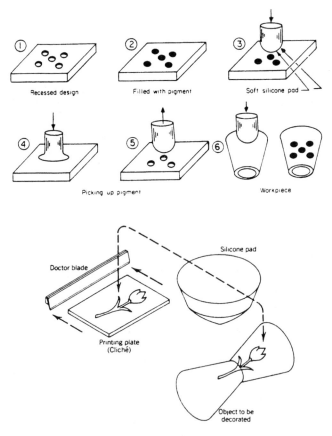

Figure 17.3 Pad printing.

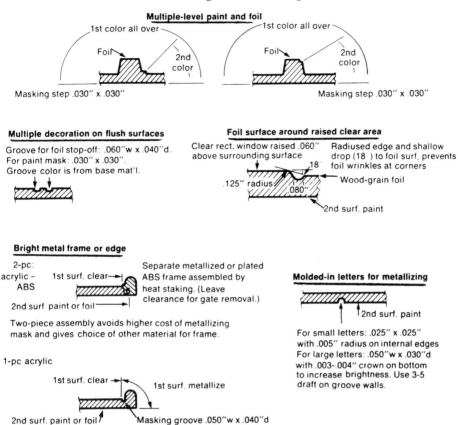

Figure 17.4 Recommended dimensions for masking and stopoff grooves.

of parts to be decorated, and the required variations in color. Many methods are listed here, but the most common are painting, direct screen, pad, and hot stamping. Methods of decorating are included in Tables 17.6 and 17.7 and Figs 17.3 and 17.4.

Surface preparation

Surface decoration requires a clean surface appropriately prepared for the chosen method of decorating. Some of the common causes of failure are contamination from processing lubricants, dust, natural skin greasiness and excess surface plasticizers, surface moisture and humidity conditions, and strain frozen into the processed part. Each plastic (material plus additives, filters, and reinforcements) must be considered separately when selecting coatings, thinners, decals, foils, etc.

Some plastics are prone to solvent attack, particularly certain thermoplastics. It is often necessary to preheat, flame heat, or chemically treat a plastic surface before it can be painted or decorated. This is true of polyolefins. Static electricity on surfaces tends to cause major problems too. Airborne impurities become attracted to the plastic surface and settle on it. Wiping or rubbing with a cloth usually increases the charge and aggravates the problem. There is equipment to remove the static charge. All materials should be decorated in clean, controlled conditions.

18

Summary

INTRODUCTION

Plastics provide the processor with materials that are useful, meet product requirements, produce simple or complex shapes, and are economically beneficial. They can be made to have a long life, to resist corrosive environments, to be degradable, to be recyclable, and to meet practically any performance requirements. They also permit the fabrication of parts whose manufacture is difficult or impossible with other materials (steel, aluminum, glass, wood, etc.).

However, plastics processors must continually update their procedures and/or acquire additional knowledge on how to process plastics. New developments in this field are unlimited. This book has emphasized that it is not difficult to process plastics, and has reviewed the many fabricating processes used to produce many different sizes and shapes of thermoplastic and thermoset commodities and engineering resins, which are used either unreinforced or reinforced (in composites). Process selection depends on product performance requirements, shape, required dimensional tolerances, plastics processing characteristics, production volume, and cost.

Some plastics can be used with many different processes, but others may require a specific process. Process selection can take place before material selection when a range of materials may be available for different processes. Guides are provided throughout this book on methods for selecting an appropriate process. The interplay between processes and materials provides for many useful product performances. Table 18.1 considers extruded film (Chapter 3) where the different materials provide different performances.

Another example is the collapsible squeeze-tube. These tubes are usually identified as airtight, collapsible, lightproof, sterilizable, economical, unbreakable, convenient, and easy to use. They are big business

Table 18.1 Performance of extruded shrink-films

Film type	Advantage	Possible problems
Polyethylene (low density)	Strong heat seals Low-temperature shrink Medium shrink force for broad application Lowest cost	Narrow shrink temperature range Low stiffness Poorer opticals Sealing wire contamination
Polypropylene	Good optical appearance High stiffness High shrink force No heat-sealing fumes Good durability	High shrink temperature High shrink force, not suitable for delicate or fragile products Brittle seals High sealing temperature
Copolymers	Strong heat seals Good optical appearance High shrink force No heat-sealing vapors	High shrink force, not suitable for fragile products Higher shrink temperature Higher heat-seal temperature Lower film slip may give machine problems
Polyvinyl chloride	Lowest shrink temperature range Wide shrink temperature range Excellent optical appearance Controlled stiffness by plasticizer-content control Lowest shrink force for wrapping fragile products	Weakest heat seals Least durable after plasticizer loss Toxic and corrosive gas emission from heat sealing, good ventilation required Durability problem at low temperature Low shrink force inhibits use as a multiple-unit bundling film Low film slip causes machine wrapping difficulties
Multilayer coextrusions	Excellent optical appearance Good machineability Low shrink temperature	In coextruded films, one ply compensates for the deficiencies of the other, so they are superior films with no significant performance shortcomings; the wide variability in layer composition and number of layers makes performance analysis difficult

Table 18.2 Tube product compatibility guide

Products	Metal tubes	Plastic tubes	Laminate tubes
Household/Industrial			
Paint, water-based	X	X	X
Putty	X	X	X
Glue	X		
Shoe polish	X		
Shaving cream	X	X	
Duplicating ink	X	X	X
Detergents	X	X	X
Bicycle rubber solution	X		
Anglers paste	X	X	X
Grease	X	X	X
Artist colors	X		X
Ski wax	X		
Motor oil		X	
Food			
Mayonnaise	X		
Soft cheese	X		X
Mustard	X	X	
Tomato concentrate	X	X	X
Condensed milk	X		
Grated horseradish	X		
Chocolate sauce, etc.		X	X
Meat preparation sauce	X		
Cake garnish	X	X	X
Fish paste	X		
Salad dressing	X	X	X
Bubblegum		X	X
Liver paste	X		
Peanut butter		X	X
Honey		X	
Pharmaceuticals			
Ointment	X	X	X
Paste (cream)	X	X	X
Injectables (Biofill machine)		X	
Laxative	X	X	X
Cosmetics			
Hair dye	X		
Shampoo	X	X	X
Eye shadow		X	X
Hair cream	X	X	X
Hair rinse	X	X	X

Table 18.2 *Continued*

Products	Metal tubes	Plastic tubes	Laminate tubes
Beauty preparations	X	X	X
Suntan cream	X	X	X
Bubble bath		X	X
Deodorant		X	X
Mascara		X	X
Bath and shower gel		X	X
Depilatory cream	X	X	X
Dentifrice			
Paste	X		X
Gel	X		X
Polishes	X		X

worldwide with about 30% made of metal (aluminum), 30% plastic (predominately extruded tube bonded to molded outlet cap of PE or PP) and 40% laminated (paper, aluminum with predominately plastic film, tape, wrap, etc.). Some of the aluminum tubes must be internally coated with a plastic barrier material to protect the aluminum from certain packaged products; exteriors may be coated with plastic to provide special printed decorations.

Traditional tubes were made from impact-extruded lead and zinc; they were first used to hold artist's oil paints. The first tube patent is dated 11 September 1841. Tubes are now extensively used (Table 18.2) and some of their properties are essential to daily life:

Metal tubes are airtight and impermeable with no suckback. They provide superior protection to contents, keeping them fresh and uncontaminated through long periods of intermittent use. Internal (plastic) linings give them added compatibility for products that are highly alkaline or acidic. They are totally seamless.

Plastic tubes are lightweight, leakproof, durable, and unbreakable. Their uses include hand creams and shampoos, cosmetics and toiletries such as cleansing creams, facial masks and tanning lotions, food, pharmaceuticals, household and industrial products, paints, and honey. These tubes have a memory (Chapter 1); they go back to their original shape after squeezing. Plastic tubes maintain their attractiveness throughout the life of the contents and remain smooth after squeezing.

Laminated tubes have the advantages of plastic but with barrier properties that are almost metallic. The tubes have a good and gentle feel, and they maintain their attractiveness throughout the life of their contents. The first major use for laminated tubes was to package toothpaste, but now they are used to package many different products [83].

PLANT CONTROL

It is seldom the case that a processing plant has only one processing machine; if it has more, it is not what happens on the individual machine that determines profitability, but it is the average performance of all machines. With many machines, it can become very difficult to keep track of all the details that go into the plant's overall operation – hundreds or even thousands. It also becomes increasingly difficult for processors, quality control people, maintenance people, and others always to be present when needed; it may become very hard for personnel to make decisions as needed.

Modern central control and management systems have changed this situation. These systems have been called supervisory control, distributed control, CAD/CAM/CAE, and CIM (computer-integrated manufacturing. All these designations refer to a system that can monitor all operating parameters for every machine, every piece of materials-handling equipment, and all other equipment in the plant. The system receives inputs on all parameters and can issue instructions to each machine to ensure efficient and profitable operation.

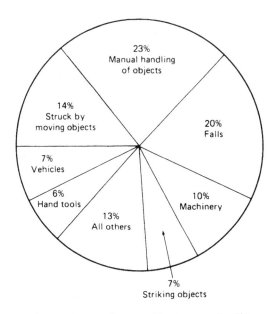

Figure 18.1 General statistics on where accidents occur in all types of manufacturing plants, including plastic plants.

For these systems to operate efficiently, talented people are needed to completely integrate the system. These people must be available and must know what is required for all plant operations, including the processing machines. System control is based on these requirements. In turn, these individuals must establish startup procedures for all plant equipment, making it correspond to the system control. They build limits into the system control, and interface them with control instructions that are best suited to keeping the machines manufacturing parts that meet performance requirements at the lowest processing cost. It is therefore essential that people working with these control systems are properly trained, to understand how best to operate the processing equipment and to pay adequate attention to safety (Fig. 18.1).

As reviewed in Chapter 1, intelligent process control (IPC) is being used with some of the processes. It uses computer simulation to link a process, such as mold filling during injection molding, to real-time machine control. Moldflow IPC, made by the company Australia, begins by requiring the user to perform a mold-filling analysis; it then uses its MF/Optim software to predict an injection speed and pressure profile which will produce good parts.

EQUIPMENT IMPROVEMENTS

New equipment usually offers improvements in processing capabilities. Designers usually plan for equipment to aid in meeting goals of zero

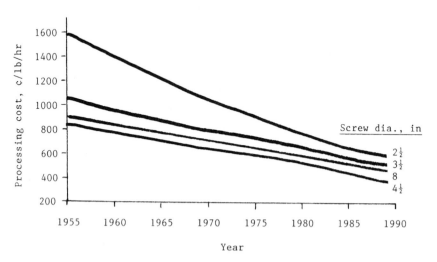

Figure 18.2 Extrusion processing costs are steadily dropping, based on output of different screws ($c\,lb^{-1}h^{-1}$).

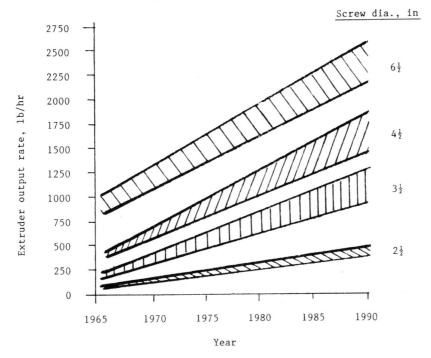

Figure 18.3 Increasing outputs for extruders of various sizes processing the more common resins.

defects, as well as to reduce production costs. Figures 18.2 and 18.3 show the significant improvement in output from single-screw extruders.

Although it is one of the oldest types of machinery for plastics processing, the single-screw extruder has not yet reached the end of its development, as can be seen over the past few years, based on the continuing industrial demand for improvements. Output, once a major focus, remains paramount, but qualities such as wear and versatility are of equal importance. Improvements continue to be made to single-screw extruders, and new developments are also taking place on other types of equipment.

Energy

It is important to evaluate how much energy a machine requires for its operation There are two types of machines: those that require a great deal of energy and those that need less. Molding and extrusion machines are energy-intensive. They require energy to perform different machine operations and also to melt the plastic.

PLASTICS AND ENERGY

Numerous studies have shown that plastics create more energy than they consume in their production. This may sound paradoxical, but a simple energy audit quickly reveals that the energy used in the life cycle of a plastic part (counting raw material, energy for processing, transportation, disposal, and other factors) is generally much less than for competing materials. Plastic products actually save the economy more energy than is used to make them.

COMPUTER AIDS

Computer-integrated manufacturing (CIM) delivers a variety of benefits to plastics fabricators, benefits that have been well documented. And although some processors have not yet taken full advantage of the technology, there is little doubt that the CIM concept is progressing rapidly; in the future it may become a commonplace manufacturing procedure. The ability of CIM to network similar and dissimilar machinery in a processing plant into a smooth-working entity where changes in processing parameters, or even entire job changeovers, can be made swiftly and easily has enormous profit-enhancing potential in a business world where global competition is intensifying year by year.

Many devices are available for process control, such as sensors, actuators, or computers. These devices can be connected with the automation apparatus and integrated into a procedure. Automation improves process efficiency and product quality as well as reducing costs. The applications

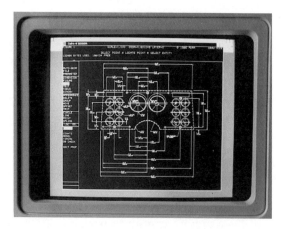

Figure 18.4 Computer readout on mold design.

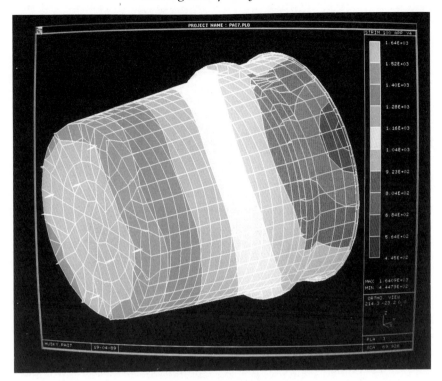

Figure 18.5 Stress and melt-flow analysis of a fabricated product.

range from front-end computers that serve a single control circuit, to process guidance systems that coordinate a complete operation.

With the help of computer technology, one has the capability of managing the enormous number of design, material, and processing applications available when fabricating with plastics, including mold or die designs (Fig. 18.4) and stress analysis of molded products (Fig. 18.5). There are already about 17 000 plastic materials, so to review what is available as well as keeping up to date with the constant proliferation of materials for a specific set of requirements can seem daunting. Software and databases expand capabilities, simplify analyses, speed up evaluations, provide more complete analyses on materials and their processing requirements, predict shrinkage and warpage, and are valuable in other ways, too.

TESTING AND QUALITY CONTROL

Testing yields basic information about a plastic, its properties relative to another material, and its quality with reference to a standard.

Most of all, it is essential for determining the performance of a finished product.

Processors should keep quality under control and demand consistent materials that can be used with a minimum of uncertainty. Plant quality control (QC) is as important to the end result as selecting the best processing conditions with the correct grade of plastic, in terms of both properties and appearance. After the correct plastic has been chosen, its blending, reprocessing, and storage stages of operation need to be frequently or continuously updated. The processor should set up specific measurements of quality to prevent substandard products from reaching the end user.

The properties of plastics are directly dependent on temperature, time, and environmental conditions, and these conditions can be related to raw material, processing, and part performance (Chapter 1). The most important testing is performed on the finished part. And tests on materials or during processing must also be related to part performance.

Unfortunately there is no single set of rules designating which tests are to be conducted in order to manufacture a part repeatedly with zero defects. The tests depend on the required performance. For example, if a part is to operate where any type of failure could be catastrophic to life, it usually requires extensive testing, which may prove costly. Processors are usually aware of some of the available tests. How deeply one gets involved depends on the performance requirements. If weighing the part provides sufficient quality control, then weighing is the method to select. Anything else would be superfluous.

Statistical process control

Statistical process control (SPC) seeks to closely control the manufacturing process and permits the manufacture of tighter-tolerance parts by indicating when a process is starting to drift away from the ideal set point. There are two basic approaches for real-time SPC. The first, done online, involves the rapid dimensional measurement of a part or a nondimensional 'bulk' parameter such as weight, and is the more practical method. Other dimensional measurements of the precision needed for SPC are generally done off-line, the second approach, and produce a response that is too slow. Obtaining the final dimensional stability needed to measure a molded part may also take time; amorphorus resins require at least 30–60 min to cool and stabilize.

There are four common features of SPC: (1) raw material characterization, (2) internal material handling (drying, blending, etc.), (3) machine operation, and (4) implementation of SPC (supported by management, etc.).

Process validation

Process validation (PV) and good manufacturing practices (GMPs) may not at first have received the attention they deserved, but that has begun to change. PV is receiving close attention when existing or planned GMP programs are evaluated or examined by government regulatory agencies and also when subcontracting any plastics to be processed into a product [9].

The Food and Drug Administration (FDA) defines PV as a documented program which provides a high degree of assurance that a specific process will consistently produce a product meeting its predetermined specifications and quality attributes. Elements of validation are product specification, equipment, and process qualifications (installation, performance, and ranging trials), timely revalidation and documentation [1, 5, 7, 9].

PREVENTIVE MAINTENANCE

It is important to set up preventive maintenance procedures on all equipment in the plant. Equipment is built to operate if proper maintenance is used. Processors should make it a habit to perform regular machine checkups and maintenance work. A thoroughly implementad machine maintenance program will reduce downtime and operating costs as well as rejects of poor parts. Periodic machine checkups and regular maintenance should become a habit.

The machine operator or attendant will not always be able to perform all necessary checking and maintenance steps. For example, to maintain such equipment as a drive mechanism or special instruments, the person may need the help of an electrician or an instrument service specialist, who may not be present or available. The goal is to not require this type of service. With a proper maintenance schedule, it may be needed very infrequently or avoided altogether.

Examples of good preventive maintenance include the following activities: (1) establish the frequency of lubrication and what types of oil or grease must be used; (2) check for oil leaks and have a procedure to correct or eliminate them; (3) check heaters, thermocouples, pressure transducers, and so on; (4) set up schedules and procedures to clean machines and molds/dies (barrels, screws, sliding mechanisms, clamps, etc.); (5) check control circuits (electrical, hydraulic, mechanical, etc.); (6) schedule checks of conditions wherever questions of alignment, level, parallelism (mold parts, mold press, die system, etc.), and other similar situations exist; (7) set up a schedule to check safety devices on all equipment; and (8) schedule sessions to repeat instructions on safety equipment procedures to all personnel. Figure 18.1 shows where accidents usually occur.

RECYCLING

Basically, all plastics can be recycled and reused. However, the method and product depend on factors such as the type of plastic part to be recycled, the environment to which the part is exposed, the recycling requirements of the part, and the costs involved. And scrap does not necessarily connote feedstock that is desirable or is usable as the virgin material from which it was generated. Scrap includes material in a processing plant that generates sprues, flash, excess parison material, runners, rejected parts, etc. This type of scrap can be relatively clean and has been reused in most plants since the 1870s, usually by blending with virgin plastics for use in producing the production part (or some other part). Reprocessed plastics may or may not be reformulated by the addition of fillers, reinforcements, plasticizers, stabilizers, pigments, etc. Cost,

Figure 18.6 Recycled PET beverage bottles produced this two-piece suit and tie in 1987.

property performance, and potential metallic contamination of recycled material are factors that have to be included in the use of recycled material with or without virgin plastics.

An example is Goodyear's two-piece suit and matching tie made from recycled two-liter PET stretch injection blow molded bottles in 1978; in 1980 it was donated to the Ripley's Believe It or Not Museum in Wisconsin Dells, Wisconsin (Fig. 18.6). Goodyear developed a recycling process to shred bottles into small flakes that could be processed into a reusable TP polyester. It had the suit made to demonstrate the versatility of recycling PET bottles.

Reusable TPs can be remelted, like virgin TPs, but TSs can rarely be remelted. However, since the 1910s granulated TSs have been used as fillers and reinforcements in TSs and TPs. Chapter 1 reviews these materials and explains how their properties relate to the properties of the virgin plastic; sometimes they are similar, sometimes they are very different.

PLASTICS AND POLLUTION

Although few plastics appear to be naturally biodegradable, whether or not this is a drawback is open to debate. It may even be considered an advantage. Of course, plastics left lying around after use do not disappear from view. Consumer waste such as foam cups, detergent bottles, and discarded film is a visual annoyance. But properly disposed of in a landfill, it becomes a stabilizing factor, presenting no danger of leaching into groundwater or nearby bodies of water. Although plastics contribute only about 1 wt% to the stream of solid waste, their general bulk makes them highly visible and the object of civic concern; recycling has therefore become a major objective of the plastics industry.

Systems have been developed that take all the plastics in the waste stream (plus some paper, aluminum, and other nonplastic junk), and reconstitute this mixture into fence posts, bench slats, and other relatively low-specification products. Other, higher-value projects are under way. The problem of recycling plastics is political and social rather than technological.

Plastics *per se* are not dangerous to human health. Their fire hazards are no different than wood, cloth, or other combustibles involved in a fire, and when treated with fire retardants they can be significantly lower.

TROUBLESHOOTING

Troubleshooting guides have been included throughout this book. In order to find unique, creative solutions to difficult challenges that were not resolved by past tried-and-true techniques, one must get away from the conventional state of mind that is often unimaginative, frustrating, repetitive, and negative. The nature of some problems tends to invite unimaginative suggestions and attempts only to use past approaches.

Problem solving in producing products, as in business and personal problems, generally requires taking a systematic approach, If practical, make rather small changes and allot time to monitor their outcomes. Patience and persistence are required, even though they may be constrained by the amount of time available.

However, when a problem is particularly difficult or only limited time exists, consider a new and imaginative approach with techniques that previously generated creative ideas. First generate as many ideas as possible, including ideas that apparently bear only remote relation to the problem. During the idea-generating phase it is of critical importance to be totally positive; no ideas are bad. Evaluation comes later, so do not attempt to provide creativity and evaluation at the same time; it could be damaging to your creativity. Look for quantity of ideas, not quality. Now all ideas are good; the best will become obvious later.

If possible, relate the problem to another situation and look for a similar solution. This approach can stimulate creative thinking towards other ideas. Try humor; do not be afraid to joke about a problem.

The next step is to evaluate all the ideas. Consider categorizing the list, then add new thoughts, select the best, and try them.

After all this action, if nothing satisfactory occurs, rather than give up, look for that **really creative** solution – it is out there. You may be too close to the problem. Get away from the trees and look at the forest. Climb up one of the trees and look at things from a different perspective.

Use your creative talents and continue to be positive. You have now creatively worked through the frustrations and negativism that problems seem to generate. Your increasingly creative input will generate future opportunities.

Now let us take the thoughts above and improve on them. In doing so, let us avoid saying, 'My mind is made up; do not give me the facts.' Rather, let us believe there is always room for improvement.

PLASTICS BAD AND OTHER MYTHS

Imagine our world denuded of plastics – their cost-effectiveness and their excellent properties – a world where cars would be missing mechanical parts, without interiors or finishes; where supermarkets would suddenly be awash in containerless food and cleaning products. Indeed, it is difficult to identify any mass-produced item of recent decades that is not entirely without plastic.

COST

The production flexibility of the fabrication process is often the single most important economic factor in a plastic product. The component's size, shape, complexity, and required production can be primary determinants. The target is to meet requirements at the lowest cost (Fig. 18.7). As an example, small numbers of large objects tend economically to favor casting, as well as the RP's hand layup or sprayup process, with a minimal tooling cost and maximum freedom for design changes. Many products favor injection and compression molding or long runs in extruders, with their automation capabilities to minimize labor costs. Shape often dictates the process: centrifugal casting or filament winding is used for cylindrical parts, rotational molding for complex hollow shapes or extremely large parts, and pultrusion for constant cross sections requiring extremely high strength and stiffness. These general examples should be considered broadly, since individual processes can be designed for a specific product capability to meet performance requirements at the lowest cost.

Furthermore, major costs may be incurred in postfabrication by operations such as trimming, finishing, joining, and attaching hardware. Selection of any equipment (main or auxiliary) requires close examination of various factors. These factors normally include reliability, performance, ease of use, ease of maintenance, size, safety, and cost. Some processors consider cost to be the most influential factor when choosing equipment to purchase. And cost is definitely important, not just the initial outlay but all the future costs as well. Observing the following practices will help to reduce costs and improve processing and performance, whatever fabrication method is selected: (1) strive for the simplest shape and form; (2) use

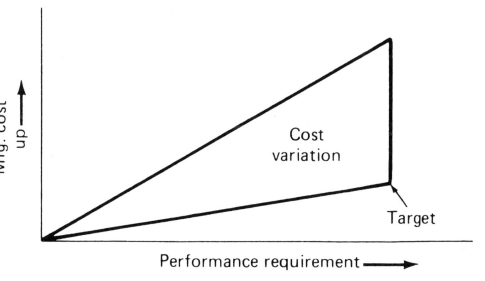

Figure 18.7 Product cost: the influence of manufacturing costs and performance requirements.

the shape of the product to provide stiffness, reducing its required number of stiffening ribs; (3) combine the parts into single moldings or extrusions as much as possible, to minimize assembly time and eliminate designing fasteners, etc.; (4) use a uniform wall thickness wherever possible, and make gradual changes in thickness to reduce stress concentrations; (5) use shape to satisfy functional needs like slots for hoisting, hand grips, and pouring; (6) provide the maximum radii that are consistent with the functional requirements; and (7) keep tolerances as liberal as possible, but once in production aim for tighter tolerances, to save plastic material and probably reduce production costs.

Minimizing costs is generally an overriding goal in any application, whether a process is being selected for a new product application or opportunities are being evaluated for replacing existing materials. The major elements of cost include plastic material, capital equipment, tooling, labor, and inefficiencies such as scrap, repairs, waste, and machine downtime. Each element must be evaluated before determining the most cost-effective process from among the available alternatives. Particularly in

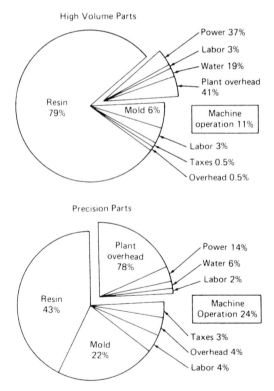

Figure 18.8 Share of cost to injection mold high-volume parts and precision parts.

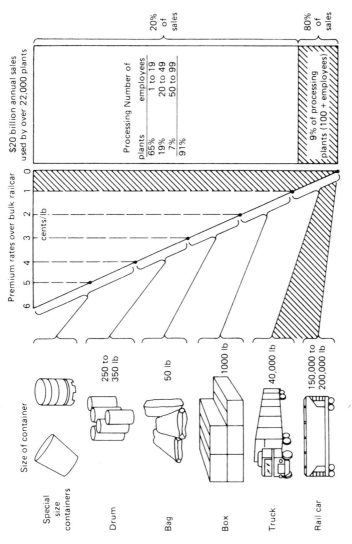

Figure 18.9 Plastic dollar purchases by plant size and size of container.

Table 18.3 Economic comparison of three different processes

Production considerations	Structural foam	Injection molding	Sheet molding compound
Typical minimum number of parts a vendor is likely to quote on for a single setup	250, using multiple nozzle equipment with tools from other sources designed for the same polymer and ganged on the platen	1000–1500	500
Relative tooling cost, single cavity	Lowest; machined aluminum may be viable, depending on quantity required	20% more Hardened-steel tooling	20–25% more Compression molding steel tools
Average cycle times for consistent part reproduction(s)	120–180 For $1/4$ in. nominal wall thickness	40–50	90–180
Is a multiple-cavity tooling approach possible to reduce piece costs?	Yes	Yes; depends on size and configuration, although rapid cycle time may eliminate the need	Not necessarily; secondary operations may be too costly and material flow too difficult

Are secondary operations required except to remove sprue?	No	No	Yes; such as removing material where a 'window' is required (often done within the the molding cycle)
Range of materials that can be molded	Similar to thermoplastic injection molding	Unlimited; cost depends on performance requirements	Limited Cost is higher
Finishing costs for good cosmetic appearance (cents per square foot)	40–60 Depends on surface-swirl conditions	None if integrally colored 10–20 if painted	None if no secondary operations are required, e.g. trimming; else 20–30

high production runs, material cost is usually the largest, ranging from 40 to 80 wt% (Fig. 18.8). For labor-intensive processes, 40–80 wt% would represent labor, the biggest cost factor. Another aspect is the materials cost based on the quantity purchased (Fig. 18.9).

In summary, cost variations may be due to one or more of the following factors:

1. improper performance requirements;
2. improper part design;
3. improper plastic selection;
4. improper hardware selection;
5. improper operation of the complete line;
6. improper setup for testing, quality control, or troubleshooting, whichever is appropriate.

Estimating part cost

Estimating is a critical aspect of custom fabrication, yet it is often practised with very little logic, it is shrouded in mystery and rarely discussed by fabricators; indeed, it is considered among the dullest of topics.

Table 18.4 Mold life and cost[a]

RP/C process	Approximate number of parts from mold (mold life)	Relative mold cost	Type of mold
Hand layup	800–1000	lowest	RP/C
Sprayup	100–200	lowest	RP/C
Vacuum bag molding	100–200	low	RP/C
Cold-press molding	150–200	medium	RP/C
Casting, electrical	3500	high	metal
Casting, marble	300–500+	lowest	RP/C
Compression molding (BMC)	120 000	high	metal
Matched-die molding (SMC)	300 000–400 000	highest	metal
Pressure bag molding	>1 000 000	medium	metal
Centrifugal casting	>1 000 000	low	metal
Filament winding	almost unlimited	lowest	metal
Pultrusion	almost unlimited	minimal	metal
Continuous laminating	almost unlimited	minimal	metal
Injection molding: reinforced thermoplastic (RTP)	300 000–1 000 000	highest	metal
Rotational molding	100 000	low	metal
Cold stamping	1 000 000–3 000 000	highest	metal

[a] RP/C = reinforced plastic/composite.

It is extraordinary if one estimate in ten produces a successful bid. In other words, a 90% failure rate is terrific. No wonder estimating seems like some bizarre sacrificial rite. That does not include all those estimates you just go through the motions of preparing because another company is going through the motions of getting three bids, and you have no chance at all of landing the jobs.

But what more directly represents the heart and soul of your business than estimating? You're pulling together every facet of your operation, distilling it, putting numbers on it, then putting yourself and your company on the line and saying, 'This is what we can do, and this is what we must charge to make a profit.' See Tables 18.2 to 18.5.

There are probably as many estimating techniques as there are estimators. Just who does the estimating varies widely. It could be the company president, the sales manager, the production manager, the treasurer; but it could also be a person or a department devoted to the task.

Much contemporary estimating follows very vague procedures. The number of factors assembled to reach the appropriate figures is some-

Table 18.5 Cost comparison of plastic products and different processes (cost factor × material cost = purchased cost of product)

Process	Cost factor	
	Overall	Average
Blow molding	$1\frac{1}{16}$ to 4	$1\frac{1}{8}$ to 2
Calendering	$1\frac{1}{2}$ to 5	$2\frac{1}{2}$ to $3\frac{1}{2}$
Casting	$1\frac{1}{2}$ to 3	2 to 3
Centrifugal casting	$1\frac{1}{2}$ to 4	2 to 4
Coating	$1\frac{1}{2}$ to 5	2 to 4
Cold-press molding	$1\frac{1}{2}$ to 5	2 to 4
Compression molding	$1\frac{7}{8}$ to 10	$1\frac{1}{2}$ to 4
Encapsulation	2 to 8	3 to 4
Extrusion forming	$1\frac{1}{16}$ to 5	$1\frac{1}{8}$ to 2
Filament winding	5 to 10	6 to 8
Injection molding	$1\frac{1}{8}$ to 3	$1\frac{3}{16}$ to 2
Laminating	2 to 5	3 to 4
Matched-die molding	2 to 5	3 to 4
Pultrusion	2 to 4	2 to $3\frac{1}{2}$
Rotational molding	$1\frac{1}{4}$ to 5	$1\frac{1}{2}$ to 3
Slush molding	$1\frac{1}{2}$ to 4	2 to 3
Thermoforming	2 to 10	3 to 5
Transfer molding	$1\frac{1}{2}$ to 5	$1\frac{3}{4}$ to 3
Wet layup	$1\frac{1}{2}$ to 6	2 to 4

times alarmingly small; many companies do not consider such matters as scrap, colorant, and setup time, to mention some of the more obvious omissions. Some estimates are created by determining part weight, cost of resin, and machine time; scribbling down some numbers; and adding a fudge factor. Some companies do not even have their own standard forms.

Technical cost modeling

The adoption of any technology for producing manufactured products is characterized by a wide range of processing (Fig. 18.10), materials, and economic consequences. Although considerable talent can be brought to bear on the processing and engineering aspects, economic questions remain. Cost problems are particularly acute when the technology that will be employed is not fully understood, as much of cost analysis is based on historical data, past experience, and individual accounting practices.

Historically, technologies have been introduced on the shop floor incrementally, with their economic consequences measured directly. Although incorporating technical changes in the plant to test their viability may have been appropriate in the past, it is economically infeasible to explore today's wide range of alternatives in this fashion. Technical cost modeling (TCM) has been developed as a method for analyzing the economics of alternative manufacturing processes without the prohibitive economic burden of trial-and-error innovation and process optimization.

TCM is an extension of conventional process modeling, with particular emphasis on capturing the cost implications of process variables and economic parameters. By coordinating cost estimates with processing knowledge, critical assumptions (processing rates, energy used, materials consumed, etc.) can be made to interact in a consistent, logical, and accurate framework of economic analysis, producing cost estimates under a wide range of conditions.

For example, TCM can be used to determine the plastic process that is best for production without extensive expenditure of capital and time. Not only can TCM be used to establish direct comparisons between processes, but it can also determine the ultimate performance of a particular process, as well as identifying the limiting process steps and/or parameters.

TCM uses an approach to cost estimating in which each of the elements that contribute to total cost is estimated individually. These individual estimates are derived from basic principles and the manufacturing process. TCM reduces the complex problem of cost analysis to a series of simpler estimating problems, and rather than using intuition, it solves them by bringing to bear genuine processing expertise.

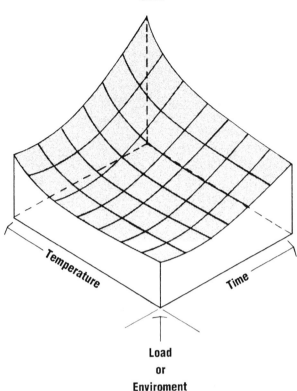

Temperature

Time

**Load
or
Enviroment**

Figure 18.10 Basics for processing with plastics: three-dimensional illustration of the contour plot.

In dividing cost into its contributing elements, the first distinction that can be made is that some cost elements depend upon the annual output of products, whereas others do not. For example, the cost contribution of the plastic is the same regardless of the number of items produced, unless the material price is discounted because of very high volume. On the other hand, the per-piece cost of tooling will vary with changes in production volume. These two elements, variable costs and fixed costs, create a natural division for preparing estimates.

Variable cost elements have values that depend on the number of pieces produced. For most plastics fabrication the variable cost elements are principally material, direct labor, and energy.

Fixed costs are a function of the annual production volume. They are called fixed costs because they typically represent one-time capital investments (building, silo, processing machine, etc.) or annual expenses unaffected by the number of parts produced (building rent, engineering support, administrative personnel, etc.). These costs tend to be distributed over the total number of parts produced in a given period. For plastics fabrication the fixed cost elements are principally the main machines, auxiliary equipment, tooling, buildings, overhead labor, maintenance, and the cost of capital.

To demonstrate the use of such a comparative cost analysis, production of a panel was analyzed according to different processes (Fig. 18.11) and under the following conditions: (1) panels measured 24 in. × 36 in. (61 cm × 91 cm) with the wall thickness dictated by process and part requirements, so the weights of the panels differed; (2) production was at a level of 40 000 items per year; (3) plastics for all panels were of the same type, except that different grades had to be used based on process requirement, so that costs changed; (4) each panel received one coat of paint, except that the structural foam also had a primer coating; and (5) costs were allocated as needed to those processes that required trimming and other secondary operations.

TCM can keep cost data current, based on cost changes from day to day, region to region, and so on. Costs need to be collected on a regular basis

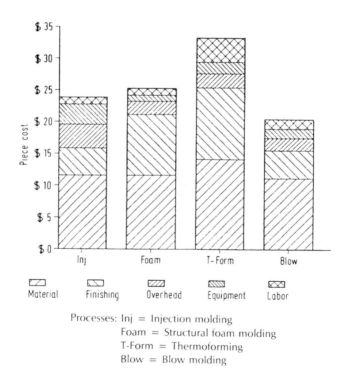

Processes: Inj = Injection molding
Foam = Structural foam molding
T-Form = Thermoforming
Blow = Blow molding

Figure 18.11 Cost comparison of panel production using a technical cost-modeling program, showing blow molding with the lowest piece cost.

and incorporated into the TCM, otherwise the information will become out of date.

PEOPLE

The recipe for productivity includes a list of ingredients: research and development, new technologies, updated equipment, automated systems and modern facilities, to name but a few. However, the one ingredient that ties the recipe together is **people**; none of the other factors has much impact without the right people. Without the appropriate people employed in the fabricating operation, you are not going to be productive, no matter how large your capital expenditure becomes.

Processing and patience

When making process changes, allow enough time to achieve a steady state before collecting data.

PLASTICS GROWTH

Plastics are among the nation's and the world's most widely used materials, having surpassed steel on a volume basis in 1983 (Fig. 18.12). By about the end of the twentieth century, plastics will surpass steel on a weight basis (Fig. 18.13). Plastic materials and products cover the entire spectrum of the world's economy, so that fortunes are not tied to any particular

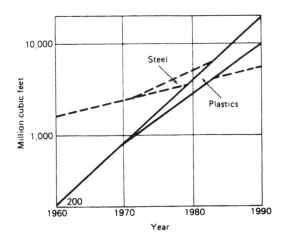

Figure 18.12 World consumption of plastics by volume.

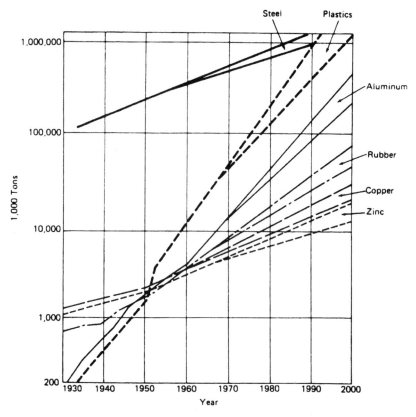

Figure 18.13 World consumption of plastics by weight.

business segment. Plastics manufacturers are in a position to benefit in a wide variety of markets: packaging, building and construction (Fig. 18.14), electronics/electrical devices, furniture, apparel, appliances, agriculture, housewares, luggage, transportation (automotive, aircraft, boat, etc.), medicine and health care, recreation, and so on. But Figs 18.12 and 18.13 represent possibly 10 vol% of all materials used worldwide. They do not show the two major and important categories, **wood** (paper, etc.) and **construction materials** (concrete, stone, glass, etc.). Wood and construction materials each represent perhaps about 45 vol%. Plastics may be about $2\frac{1}{2}$ vol% of the grand total worldwide.

The plastics industry is a highly diversified economic activity which offers abundant opportunities for both employee and entrepreneur. The prospect of growth in plastics is always very bright, even with the ups and downs which typify any business cycle.

Effective exploitation of processing opportunities is the key to success

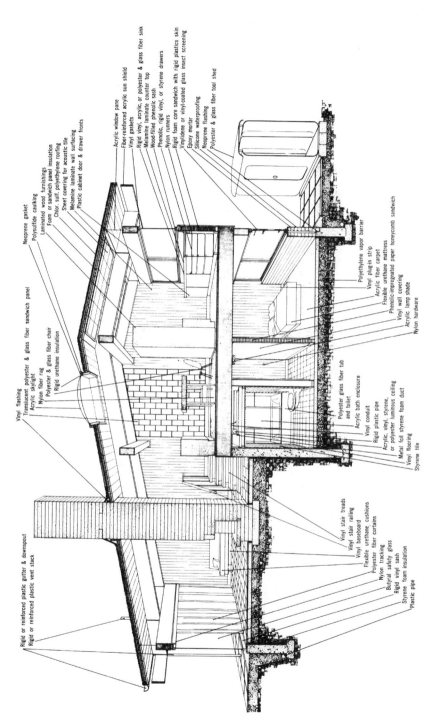

Figure 18.14 The 1940s to 1950s concept of plastics in buildings. (Reproduced with permission from *The House that Plastics Built.*)

Figure 18.15 Location filming: plastic films could be fabricated directly onto farm-land; this tractor is experimenting with two extruders.

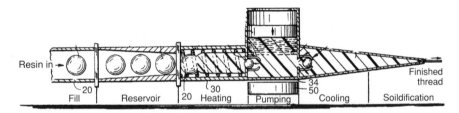

Figure 18.16 A sewing machine that makes its own thread.

(Figs 18.15 and 18.16). And success also hinges on other factors such as proper design, proper selection of materials, and use of the best available processing equipment (Fig. 18.17).

Figure 18.18 serves as a reminder that any activity is a balance between it gains and its losses. Processing equipment provides enormous scope for improving profitability by choosing the best available machines to manufacture the specified product at a competitive price.

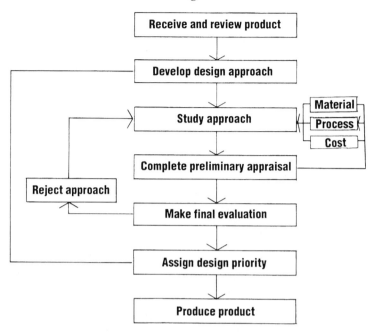

Figure 18.17 New product ideas need to be evaluated in logical steps.

Become aware that for any gain there could be a loss
not originally included in the design performance.

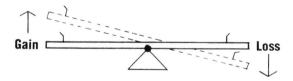

When you gain "something" there will be a loss .....does
that loss influence product performance (for any material:
plastic, wood, steel, glass, etc.).

Figure 18.18 Gains and losses: a judicious balance.

Appendix

METRIC CONVERSION CHARTS

	US to metric			Metric to US	
US	Metric	Multiply by	Metric	US	Multiply by
Density					
lb in.$^{-3}$	kg m^{-3}	27 680	kg m^{-3}	lb in.$^{-3}$	0.000036
lb ft^{-3}	g cm^{-3}	0.0160	g cm^{-3}	lb ft^{-3}	62.43
lb ft^{-3}	kg m^{-3}	16.0185	kg m^{-3}	lb ft^{-3}	0.0624
lb in.$^{-3}$	g cm^{-3}	27.68	g cm^{-3}	lb in.$^{-3}$	0.03613
Temperature					
°F^{-1}	°C^{-1}	1.8	°C^{-1}	°F^{-1}	0.556
°F	°C	(°F − 32)/(1.8)	°C	°F	1.8°C + 32
°F	K	(°F + 459.67)/(1.8)	K	°F	1.8K − 459.67
Pressure					
psi	kPa	6.8948	kPa	psi	0.145
psi	MPa	0.00689	MPa	psi	145
psi	GPa	0.00000689	GPa	psi	145038
psi	bar	0.0689	bar	psi	14.51

Energy and power

Unit	Factor	SI unit	SI unit	Factor	Unit
ft lbf	1.3558	J	J	0.7376	ft lbf
in. lbf	0.113	J	J	8.850	in. lbf
ft lbf in.$^{-1}$	53.4	J m^{-1}	J m^{-1}	0.0187	ft lbf in.$^{-1}$
ft lbf in.$^{-1}$	0.534	J cm^{-1}	J cm^{-1}	1.87	ft lbf in.$^{-1}$
ft lbf in.$^{-2}$	2.103	kJ m^{-2}	kJ m^{-2}	0.4755	ft lbf in.$^{-2}$
kW	1.3596	metric horsepower	metric horsepower	0.7355	kW
US horsepower	0.7457	kW	kW	1.3419	US horsepower
Btu	1055.1	J	J	0.00095	Btu
Btu	0.2931	Wh	Wh	3.412	Btu
Btu. in. h^{-1} ft^{-2} $^\circ$F^{-1}	0.1442	W m^{-1} K^{-1}	W m^{-1} K^{-1}	6.933	Btu. in. h^{-1} ft^{-2} $^\circ$F^{-1}
Btu lb^{-1}	2.326	kJ kg^{-1}	kJ kg^{-1}	0.4299	Btu lb^{-1}
Btu lb^{-1} $^\circ$F^{-1}	4187	J kg^{-1} $^\circ$C^{-1}	J kg^{-1} $^\circ$C^{-1}	0.000239	Btu lb^{-1} $^\circ$F^{-1}
V mil^{-1}	0.0394	MV m^{-1}	MV m^{-1}	25.4	V mil^{-1}

Output

Unit	Factor	SI unit	SI unit	Factor	Unit
lb min^{-1}	7.560	g s^{-1}	g s^{-1}	0.1323	lb min^{-1}
lb h^{-1}	0.4536	kg h^{-1}	kg h^{-1}	2.2046	lb h^{-1}

Velocity

Unit	Factor	SI unit	SI unit	Factor	Unit
in. min^{-1}	0.0423	cm s^{-1}	cm s^{-1}	23.6220	in. min^{-1}
ft s^{-1}	0.3048	m s^{-1}	m s^{-1}	3.2808	ft s^{-1}

Viscosity

Unit	Factor	SI unit	SI unit	Factor	Unit
P	0.1	Pa s	Pa s	10	poise

Source: National Bureau of Standards Metric Information Office.

	US to metric			Metric to US	
US	*Metric*	*Multiply by*	*Metric*	*US*	*Multiply by*
Length					
mil	millimeter	0.0254	millimeter	mil	39.37
inch	millimeter	25.4	millimeter	inch	0.0394
inch	centimeter	2.54	centimeter	inch	0.3937
foot	centimeter	30.48	centimeter	foot	0.0328
foot	meter	0.3048	meter	foot	3.2808
yard	meter	0.9144	meter	yard	1.0936
Area					
$inch^2$	$millimeter^2$	645.16	$millimeter^2$	$inch^2$	0.0016
$inch^2$	$centimeter^2$	6.4516	$centimeter^2$	$inch^2$	0.155
$foot^2$	$centimeter^2$	929.03	$centimeter^2$	$foot^2$	0.0011
$foot^2$	$meter^2$	0.0929	$meter^2$	$foot^2$	10.7639
$yard^2$	$meter^2$	0.8361	$meter^2$	$yard^2$	1.1960

Volume, capacity

inch^3	centimeter^3	16.3871	inch^3	0.061
fluid ounce	centimeter^3	29.5735^3	fluid ounce	0.0338
quart (liquid)	decimeter^3 (liter)	0.9464	quart (liquid)	1.0567
gallon (US)	decimeter^3 (liter)	3.7854	gallon (US)	0.2642
gallon (US)	meter^3	0.0038	gallon (US)	264.17
foot^3	decimeter^3	28.3169	foot^3	0.0353
foot^3	meter^3	0.0283	foot^3	35.3147
yard^3	meter^3	0.7646	yard^3	1.3079
in.^{3}lb^{-1}	m^{3}kg^{-1}	0.000036	in.^{3}lb^{-1}	27680
ft^{3}lb^{-1}	m^{3}kg^{-1}	0.0624	ft^{3}lb^{-1}	16.018

Mass

ounce (avdp.)	gram	28.3495	ounce (avdp.)	0.03527
pound	gram	453.5924	pound	0.0022
pound	kilogram	0.4536	pound	2.2046
pound	metric tonne	0.00045	pound	2204.6
US ton (short)	metric tonne	0.9072	US ton (short)	1.1023

Force

lbf	N	4.448	lbf	0.225

Source: National Institute for Science and Technology, Metric Information Office.

Standard metric symbols				Metric prefixes[b]		
				Numerical value	Term	Symbol
A	ampere	dm^3	liter	10	deka	da
bar	bar	m	meter	10^2	hecto	h
cd	candela	N	newton	10^3	kilo	k
C	celsius[a]	Pa	pascal	10^6	mega	M
g	gram	S	siemens	10^9	giga	G
h	hour	s	second	10^{12}	tera	T
Hz	hertz	t	metric ton	10^{-1}	deci	d
J	joule	V	volt	10^{-2}	centi	c
K	kelvin	W	watt	10^{-3}	milli	m
kg	kilogram			10^{-6}	micro	μ
				10^{-9}	nano	n
				10^{-12}	pico	p

[a] Formerly called centigrade.
[b] These prefixes may be used with all metric units.

MATHEMATICAL SYMBOLS AND ABBREVIATIONS

+	plus (addition)	a', a''	a-prime, a-second
−	minus (subtraction)	a_1, a_2	a-sub one, a-sub two
± ∓	plus or minus, (minus or plus)		
×	times, by (multiplication)	(), [], {}	parentheses, brackets, braces
÷, /	divided by		
:	is to (ratio)	∠ ⊥	angle, perpendicular to
::	equals, as, so is	a^2, a^3	a-square, a-cube
∴	therefore	a^{-1}, a^{-2}	$1/a$, $1/a^2$
=	equals	$\sin^{-1} a$	the angle whose sine is a
~ ≈	approximately equals	π	pi = 3.141592654+
>	greater than	σ, Σ	summation of
<	less than	ε, e	base of hyperbolic, natural or
≧, ≥	greater than or equals		Napierian logs = 2.71828 +
≦, ≤	less than or equals	Δ	difference
≠	not equal to	g	acceleration due to gravity
≐	approaches	E	coefficient of elasticity
∝	varies as	v	velocity
∞	infinity	f	coefficient of friction
‖	parallel to	P	pressure of load
√, ∛	square root, cube root	hp	horsepower
□	square	RPM	revolutions per minute
○	circle		
°	degrees (arc or thermometer)		
′	minutes or feet		
″	seconds or inches		

GREEK ALPHABET

A, α	alpha	H, η	ēta	N, ν	nū	T, τ	tau
B, β	bēta	Θ, θ ϑ	thēta	Ξ, φ	xī	Y, υ	upsīlon
Γ, γ	gamma	I, ι	iōta	O, o	omıcron	Φ, φ	phī
Δ, δ	delta	K, κ	kappa	Π, π	pī	X, χ	chī
E, ε	epsīlon	Λ, λ	lambda	P, ρ	rhō	Ψ, ψ	psī
Z, ζ	zēta	M, μ	mū	Σ, σ, ς	sigma	Ω, ω	ōmega

PROPERTIES OF WATER

Density of water is 62.38 lb ft^{-3} or 0.0361 lb in.$^{-3}$

Specific gravity of water is 1 g cm^{-3}

1 ft^3 of water at 39.1 °F (3.9 °C) equals 0.4335 lb in.$^{-2}$ (ft high)

1 in.^{3} of water weighs 0.576 oz

1 gallon of water contains 231 in.^{3} or 0.13368 ft^3

1 gallon of water weighs 8.3356 lb at 62 °F (17 °C) (air-free, weighed in vacuum)

1 ft^3 of water equals 7.4805 gallons; a cylinder 7 in. diameter and 6 in. high contains 1 gallon of water

The maximum density of water is at 39.1 °F (3.9 °C)

The freezing point of water at sea level is 32 °F (0 °C)

The boiling point of water at sea level is 212 °F (100 °C)

References

1. Rosato, D. V. and Rosato, D. V. (1995) *Injection Molding Handbook*, 2nd edn, Chapman & Hall.
2. Rosato, D. V. and Rosato, D. V. (1989) *Blow Molding Handbook*, Hanser.
3. Rosato, D. V. (1991) *Designing with Plastics and Composites Handbook*, Chapman & Hall.
4. Rosato, D. V. (1996) *Injection Molding Higher Performance Reinforced Plastic Composites*, SPE-ANTEC, May.
5. Rosato, D. V. (1987–1997) *Processing With Plastics*, RISD Course.
6. Rosato, D. V. and Schwartz, R. T. (1968) *Environmental Effects on Polymeric Materials*, Vol. 1: Environment and Vol. 2: Materials, Wiley.
7. Rosato, D. V. (1974–1986) *Fundamentals of Plastics*, University of Lowell Seminars.
8. Rosato, D. V. and Rosato, D. V. (1997) *Designing With Reinforced Plastic Composites*, Hanser.
9. Rosato, D. V. (1993) *Rosato's Plastics Encyclopedia and Dictionary*, Hanser.
10. Rosato, D. V. (1983) Advanced engineering design – short course. *ASME Engineering Conference*.
11. Rosato, D. V. (1983) Designing with plastics. *MD & DI*, July, 26–29.
12. Rosato, D. V. (1964) *Filament Winding*, Wiley.
13. Rosato, D. V., Fallo, W. K. and Rosato, D. V. (1969) *Markets For Plastics*, Van Nostrand Reinhold.
14. Rosato, D. V. (1995) An injection molder's dream comes true, *SPE-IMD News letter*, No. 40.
15. Rosato, Capt. D. V. (1944) All plastic military airplane successfully flight tested. Wright-Patterson Air Force Base.
16. Rosato, D. V. (1963) *Outer Space Parabolic Reflector Energy Converters*, SAMPE, June.
17. Rosato, D. V. (1987) Materials selection, in *Encyclopedia of Polymer Science and Engineering* (eds Mark, Bikales, Overberger, and Menges), Vol. 9, 357–79. Wiley.
18. Rosato, D. V. (1988) Thermosets, in *Encyclopedia of Polymer Science and Engineering* (eds Mark, Bikales, Overberger, and Menges), Vol. 14, 350–91, Wiley.
19. Rosato, D. V. (1980) Polymers, processes and properties of medical plastics – include applications, in *Synthetic Biomedical Polymers* (eds Szycher, M. and Robinson, R. T.), Technomic.

20. Rosato, D. V. (1970) Electrical wire and cable plastic coating – what's ahead? *Wire and Wire Products*, **Mar**, 49–61.
21. Rosato, D. V. (1991) Materials selection, polymeric matrix composites, in *International Encyclopedia of Composites*, Vol. 3 (ed. S. M. Lee), VCH.
22. Rosato, D. V. (1964) Why not use metal wire in filament winding? *The Iron Age*.
23. Rosato, D. V. (1957) Nose cone of first US moon vanguard rocket is made in Manheim-US. *New Era-Lancaster PA Newspaper*, Nov. 30.
24. Rosato, D. V. (1989) What to consider in picking an injection molding machine. *Plastics Today*, Jan.
25. Rosato, D. V. (1985) *Injection Molding Thermosets*, SPI Reinforced Plastics/Composites Annual Meeting Seminar, Jan.
26. Rosato, D. V. (1987) *Extrusion: Technology, Markets, Economics*, University of Lowell Seminars.
27. Rosato, D. V. (1989) Choosing the blow molding machines to meet product demand. *Plastics Today*, Jan.
28. Rosato, D. V. (1986) *Reinforced Plastics/Composites*, University of Lowell Seminars.
29. Rosato, D. V. (1962) Non-woven fibers in reinforced plastics. *Industrial Engineering Chemistry*, **54**(8), 30–37.
30. Rosato, D. V. (1959) *Asbestos*, Van Nostrand Reinhold.
31. Rosato, D. V. (1986) *Testing and statistical quality control*, University of Lowell Seminars.
32. Rosato, D. V. and Lawrence, J. R. (1973) *Plastics Industry Safety Handbook*, Cahners.
33. Rosato, D. V. (1967) 25 years of polyethylene. *Plastics World*, Jan.
34. Rosato, Capt. D. V. (1944) Theoretical potential for polyethylene. US–British correspondence, USAF Materials Laboratory, Wright-Patterson Air Force Base.
35. Rosato, D. V. (1970) Insulation news. *Wire and Wire Products*, Nov.
36. Rosato, D. V. (1970) Wire and wire products – insulation news, *Wire and Wire Products*, Dec.
37. Rosato, D. V. (1991) *Design features that influence performance: detractors/constraints*, SPE-ANTEC, May.
38. Rosato, D. V. (1990) Materials selection, in *Concise Encyclopedia of Polymer Science and Engineering* (ed. J. I. Kroschwitz), Wiley.
39. Rosato, D. V. (1997) Additives, plastics, in *Wiley Encyclopedia of Packaging Technology*, 2nd edn, Wiley.
40. Rosato, D. V. (1994) *Designing With Plastics – Materials, Processes, Products*, Open University Seminars, Geneva, Switzerland, May.
41. Rosato, D. V. (1989) *Injection Mold Design and Product Design*, China's Chemical Planning Institute Seminar, Beijing, China.
42. Rosato, D. V. (1985–1991) *Advanced Engineering Design, Etc.*, Open University Seminars, Northampton, UK.
43. Rosato, D. V. (1990) Reinforced plastics: thermosets, in *Concise Encyclopedia of Polymer Science and Engineering* (ed. J. I. Kroschwitz), Wiley.
44. Rosato, D. V. (1990) *Product Design – Plastics Selection Guide*, SPE-ANTEC, May.
45. Rosato, D. V. (1990) Current and future trends in the use of plastics for blow molding. *SME Tech. Paper MS90–198*, June.
46. Rosato, D. V. (1989) Plastics and solid waste. Rhode Island School of Design (presentation), Oct.

47. Rosato, D. V. (1988) *Injection Molding Technology: Economics and Markets*, SPE-ANTEC, May.
48. Rosato, D. V. (1987) Plastics replaced aorta permits living normal long life. Newton-Wellesley Hospital, Mar.
49. Rosato, D. V. (1987) Role of additives in plastics: function of processing aids. *SPE-IMD Newsletter*, Nov.
50. Rosato, D. V. (1987) *Blow Molding Expanding Technologywise and Marketingwise*, SPE-ANTEC, May.
51. Rosato, D. V. (1987) History-injection molding machine and reciprocating screw plasticizer. *SPE-IMD Newsletter*, No. 15.
52. Rosato, D. V. (1986) *Optimize Performance of Injection Molding Machine: Interrelate Machines/Mold/Material Performance*, SPE-ANTEC, May.
53. Rosato, D. V. (1974–1986) *Quality Control and Statistical Control*, University of Lowell Seminars.
54. Colby, P. N. (1992) Screw and barrel technology. *Spirex Bulletin*, Spirex Corp., Youngstown, OH 44512, USA.
55. DuBois, J. H. and Rosato, D. V. (1968) From laminates to composites. *Plastics World*, Apr.
56. Cook, F. (1996) Injection molding with Ericson. *World Plastics and Rubber Technology*, No. 8.
57. Busgen, A. (1995) Fiber-reinforced shapes for component structures. *Kunststoffe*, Oct.
58. Dorgham, M. A. and Rosato, D. V. (1986) *Designing with Plastics Composites*, Interscience Enterprises Ltd, World Trade Center Bldg, 110 Avenue Louis Casai, Case Postale 306, CH-1215, Geneva Airport, Switzerland.
59. Griffith, A. A. (1921) The phenomena of rupture and flow in solids. *Philosophical Transactions of the Royal Society of London, Series A*, **221**, 163–98. Griffith, A. A. (1924) The theory of rupture. In *Proceedings of the 1st International Congress on Applied Mechanics*, p. 55.
60. Schwartz, S. S. and Goodman, S. (1982) *Plastics Materials and Processes*, Chapman & Hall.
61. *Engineering Plastics*, Vol. 2, ASM International (1988).
62. Chanda, M. and Roy, S. K. (1987) *Plastics Technology Handbook*, Marcel Dekker.
63. Rauwendaal, C. (1986) *Polymer Extrusion*, Hanser.
64. Bakker, M. (1986) *Wiley Encyclopedia of Packaging Technology*, Wiley.
65. Rao, N. S. (1981) *Designing Machines and Dies for Polymer Processing with Computer*, Hanser.
66. **Authors** (1988) Melt temperature revealed. *Plastics Today*, Jul.
67. Berins, M. L. (1991) *SPI Plastics Engineering Handbook*, Chapman & Hall.
68. Kroschwitz, J. I. (1990) *Concise Encyclopedia of Polymer Science and Engineering*, Wiley.
69. Rubin, I. I. (1990) *Handbook of Plastic Materials and Technology*, Wiley.
70. Deming, E. W. (1986) *Out of Crisis*, MIT Center for Advanced Engineering Study.
71. Wright, R. E. (1995) *Injection/Transfer of Thermoset Plastics*, Hanser.
72. LNP processing tips for molding fiber glass reinforced TPs. *Bulletin 249-1-195* (1995).
73. Kirkland, C. (1995) Hinterspritzen: backmolding come back home again. *Injection Molder*, Sep.
74. Kennedy, P. (1993) *Flow Analysis*, Moldflow PTY Ltd.
75. Avenas, J. F. (1990) *Modeling Polymer Processing*, Hanser.

76. Lubin, G. and Rosato, D. V. (1964) Applications of reinforced plastics. In *Proceedings of the 4th International RP Conference*, British Plastics Federation, London, Nov. 25–27.
77. Szycher, M. and Robinson, W. J. (1980) *Synthetic Biomedical Polymers*, Technomic.
78. Brady, G. S. and Clauser, H. R. (1985) *Materials Handbook*, McGraw-Hill.
79. Boyer, H. E. (1987) *Selection of Materials for Service Environments*, ASM.
80. *International Selector*, Condura, San Diego CA, annually updated.
81. *Plastiserve-Plastics Information Systems*, Data Service Inc., annually updated.
82. *Plastics Technology Data Bank*, Bill Publications, annually updated.
83. Adolfsson, G. (1995) Why a tube? *Tube Topics*, **29**(4).
84. Throne, J. L. (1987) *Thermoforming*, Hanser.
85. *Structural Plastics Selection Manual*, No. 66, ASCE (1984).
86. Miles, D. C. *et al.* (1965) *Polymer Technology*, Chemical Publications.
87. Weismantel, R. (1994) IM part geometry, quality factors and cycle times affect equipment choices. *Modern Plastics*, Nov.
88. Pickel, H. *et al.* (1995) Electric and hydraulic = hybrid. *Kunststoffe*, Sep.
89. Canby, T. Y. and O'Rear, C. (1989) Reshaping our lives – advanced materials. *National Geographic*, Dec.
90. Hunkar, D. B. (1995) What you should know before buying an injection press, *Plastics World*, Oct.
91. Hulbert, W. (1994) Screw architecture and control systems influence extrusion product quality. *Modern Plastics*, Nov.
92. Frederick, C. D. (1996) The voice of rotational molding. *World Plastics & Rubber Technology*, No. 8.
93. Bevis, M. J. (1996) Looking to the future – processing. *World Plastics & Rubber Technology*, No. 8.

Index